IET ENERGY ENGINEERING 160

Microgrids for Rural Areas

Other volumes in this series:

Volume 1	**Power Circuit Breaker Theory and Design** C.H. Flurscheim (Editor)
Volume 4	**Industrial Microwave Heating** A.C. Metaxas and R.J. Meredith
Volume 7	**Insulators for High Voltages** J.S.T. Looms
Volume 8	**Variable Frequency AC Motor Drive Systems** D. Finney
Volume 10	**SF_6 Switchgear** H.M. Ryan and G.R. Jones
Volume 11	**Conduction and Induction Heating** E.J. Davies
Volume 13	**Statistical Techniques for High Voltage Engineering** W. Hauschild and W. Mosch
Volume 14	**Uninterruptible Power Supplies** J. Platts and J.D. St Aubyn (Editors)
Volume 15	**Digital Protection for Power Systems** A.T. Johns and S.K. Salman
Volume 16	**Electricity Economics and Planning** T.W. Berrie
Volume 18	**Vacuum Switchgear** A. Greenwood
Volume 19	**Electrical Safety: A guide to causes and prevention of hazards** J. Maxwell Adams
Volume 21	**Electricity Distribution Network Design, 2nd Edition** E. Lakervi and E.J. Holmes
Volume 22	**Artificial Intelligence Techniques in Power Systems** K. Warwick, A.O. Ekwue and R. Aggarwal (Editors)
Volume 24	**Power System Commissioning and Maintenance Practice** K. Harker
Volume 25	**Engineers' Handbook of Industrial Microwave Heating** R.J. Meredith
Volume 26	**Small Electric Motors** H. Moczala *et al.*
Volume 27	**AC-DC Power System Analysis** J. Arrillaga and B.C. Smith
Volume 29	**High Voltage Direct Current Transmission, 2nd Edition** J. Arrillaga
Volume 30	**Flexible AC Transmission Systems (FACTS)** Y.-H. Song (Editor)
Volume 31	**Embedded generation** N. Jenkins *et al.*
Volume 32	**High Voltage Engineering and Testing, 2nd Edition** H.M. Ryan (Editor)
Volume 33	**Overvoltage Protection of Low-Voltage Systems, Revised Edition** P. Hasse
Volume 36	**Voltage Quality in Electrical Power Systems** J. Schlabbach *et al.*
Volume 37	**Electrical Steels for Rotating Machines** P. Beckley
Volume 38	**The Electric Car: Development and future of battery, hybrid and fuel-cell cars** M. Westbrook
Volume 39	**Power Systems Electromagnetic Transients Simulation** J. Arrillaga and N. Watson
Volume 40	**Advances in High Voltage Engineering** M. Haddad and D. Warne
Volume 41	**Electrical Operation of Electrostatic Precipitators** K. Parker
Volume 43	**Thermal Power Plant Simulation and Control** D. Flynn
Volume 44	**Economic Evaluation of Projects in the Electricity Supply Industry** H. Khatib
Volume 45	**Propulsion Systems for Hybrid Vehicles** J. Miller
Volume 46	**Distribution Switchgear** S. Stewart
Volume 47	**Protection of Electricity Distribution Networks, 2nd Edition** J. Gers and E. Holmes
Volume 48	**Wood Pole Overhead Lines** B. Wareing
Volume 49	**Electric Fuses, 3rd Edition** A. Wright and G. Newbery
Volume 50	**Wind Power Integration: Connection and system operational aspects** B. Fox *et al.*
Volume 51	**Short Circuit Currents** J. Schlabbach
Volume 52	**Nuclear Power** J. Wood
Volume 53	**Condition Assessment of High Voltage Insulation in Power System Equipment** R.E. James and Q. Su
Volume 55	**Local Energy: Distributed generation of heat and power** J. Wood
Volume 56	**Condition Monitoring of Rotating Electrical Machines** P. Tavner, L. Ran, J. Penman and H. Sedding
Volume 57	**The Control Techniques Drives and Controls Handbook, 2nd Edition** B. Drury
Volume 58	**Lightning Protection** V. Cooray (Editor)
Volume 59	**Ultracapacitor Applications** J.M. Miller

Volume 62	**Lightning Electromagnetics** V. Cooray
Volume 63	**Energy Storage for Power Systems, 2nd Edition** A. Ter-Gazarian
Volume 65	**Protection of Electricity Distribution Networks, 3rd Edition** J. Gers
Volume 66	**High Voltage Engineering Testing, 3rd Edition** H. Ryan (Editor)
Volume 67	**Multicore Simulation of Power System Transients** F.M. Uriate
Volume 68	**Distribution System Analysis and Automation** J. Gers
Volume 69	**The Lightening Flash, 2nd Edition** V. Cooray (Editor)
Volume 70	**Economic Evaluation of Projects in the Electricity Supply Industry, 3rd Edition** H. Khatib
Volume 72	**Control Circuits in Power Electronics: Practical issues in design and implementation** M. Castilla (Editor)
Volume 73	**Wide Area Monitoring, Protection and Control Systems: The enabler for smarter grids** A. Vaccaro and A. Zobaa (Editors)
Volume 74	**Power Electronic Converters and Systems: Frontiers and applications** A.M. Trzynadlowski (Editor)
Volume 75	**Power Distribution Automation** B. Das (Editor)
Volume 76	**Power System Stability: Modelling, analysis and control** B. Om P. Malik
Volume 78	**Numerical Analysis of Power System Transients and Dynamics** A. Ametani (Editor)
Volume 79	**Vehicle-to-Grid: Linking electric vehicles to the smart grid** J. Lu and J. Hossain (Editors)
Volume 81	**Cyber-Physical-Social Systems and Constructs in Electric Power Engineering** S. Suryanarayanan, R. Roche and T.M. Hansen (Editors)
Volume 82	**Periodic Control of Power Electronic Converters** F. Blaabjerg, K. Zhou, D. Wang and Y. Yang
Volume 86	**Advances in Power System Modelling, Control and Stability Analysis** F. Milano (Editor)
Volume 87	**Cogeneration: Technologies, optimisation and implementation** C.A. Frangopoulos (Editor)
Volume 88	**Smarter Energy: From smart metering to the smart grid** H. Sun, N. Hatziargyriou, H.V. Poor, L. Carpanini and M.A. Sánchez Fornié (Editors)
Volume 89	**Hydrogen Production, Separation and Purification for Energy** A. Basile, F. Dalena, J. Tong and T.N. Veziroğlu (Editors)
Volume 90	**Clean Energy Microgrids** S. Obara and J. Morel (Editors)
Volume 91	**Fuzzy Logic Control in Energy Systems With Design Applications in MATLAB®/Simulink®** I.H. Altaş
Volume 92	**Power Quality in Future Electrical Power Systems** A.F. Zobaa and S.H.E.A. Aleem (Editors)
Volume 93	**Cogeneration and District Energy Systems: Modelling, analysis and optimization** M.A. Rosen and S. Koohi-Fayegh
Volume 94	**Introduction to the Smart Grid: Concepts, technologies and evolution** S.K. Salman
Volume 95	**Communication, Control and Security Challenges for the Smart Grid** S.M. Muyeen and S. Rahman (Editors)
Volume 96	**Industrial Power Systems with Distributed and Embedded Generation** R. Belu
Volume 97	**Synchronized Phasor Measurements for Smart Grids** M.J.B. Reddy and D.K. Mohanta (Editors)
Volume 98	**Large Scale Grid Integration of Renewable Energy Sources** A. Moreno-Munoz (Editor)
Volume 100	**Modeling and Dynamic Behaviour of Hydropower Plants** N. Kishor and J. Fraile-Ardanuy (Editors)
Volume 101	**Methane and Hydrogen for Energy Storage** R. Carriveau and D.S.-K. Ting
Volume 104	**Power Transformer Condition Monitoring and Diagnosis** A. Abu-Siada (Editor)
Volume 106	**Surface Passivation of Industrial Crystalline Silicon Solar Cells** J. John (Editor)
Volume 107	**Bifacial Photovoltaics: Technology, applications and economics** J. Libal and R. Kopecek (Editors)

Volume 108	**Fault Diagnosis of Induction Motors** J. Faiz, V. Ghorbanian and G. Joksimović
Volume 110	**High Voltage Power Network Construction** K. Harker
Volume 111	**Energy Storage at Different Voltage Levels: Technology, integration, and market aspects** A.F. Zobaa, P.F. Ribeiro, S.H.A. Aleem and S.N. Afifi (Editors)
Volume 112	**Wireless Power Transfer: Theory, technology and application** N. Shinohara
Volume 115	**DC Distribution Systems and Microgrids** T. Dragičević, F. Blaabjerg and P. Wheeler
Volume 117	**Structural Control and Fault Detection of Wind Turbine Systems** H.R. Karimi
Volume 119	**Thermal Power Plant Control and Instrumentation: The control of boilers and HRSGs, 2nd Edition** D. Lindsley, J. Grist and D. Parker
Volume 120	**Fault Diagnosis for Robust Inverter Power Drives** A. Ginart (Editor)
Volume 123	**Power Systems Electromagnetic Transients Simulation, 2nd Edition** N. Watson and J. Arrillaga
Volume 124	**Power Market Transformation** B. Murray
Volume 125	**Wind Energy Modeling and Simulation Volume 1: Atmosphere and plant** P. Veers (Editor)
Volume 126	**Diagnosis and Fault Tolerance of Electrical Machines, Power Electronics and Drives** A.J.M. Cardoso
Volume 128	**Characterization of Wide Bandgap Power Semiconductor Devices** F. Wang, Z. Zhang and E.A. Jones
Volume 129	**Renewable Energy from the Oceans: From wave, tidal and gradient systems to offshore wind and solar** D. Coiro and T. Sant (Editors)
Volume 130	**Wind and Solar Based Energy Systems for Communities** R. Carriveau and D.S.-K. Ting (Editors)
Volume 131	**Metaheuristic Optimization in Power Engineering** J. Radosavljević
Volume 132	**Power Line Communication Systems for Smart Grids** I.R.S. Casella and A. Anpalagan
Volume 139	**Variability, Scalability and Stability of Microgrids** S.M. Muyeen, S.M. Islam and F. Blaabjerg (Editors)
Volume 145	**Condition Monitoring of Rotating Electrical Machines** P. Tavner, L. Ran and C. Crabtree
Volume 146	**Energy Storage for Power Systems, 3rd Edition** A.G. Ter-Gazarian
Volume 147	**Distribution Systems Analysis and Automation, 2nd Edition** J. Gers
Volume 155	**Energy Generation and Efficiency Technologies for Green Residential Buildings** D. Ting and R. Carriveau (Editors)
Volume 157	**Electrical Steels, 2 Volumes** A. Moses, K. Jenkins, P. Anderson and H. Stanbury
Volume 172	**Lighting Interaction with Power Systems, 2 Volumes** A. Piantini (Editor)
Volume 905	**Power System Protection, 4 Volumes**

Microgrids for Rural Areas

Research and case studies

Edited by
Rajeev Kumar Chauhan, Kalpana Chauhan
and Sri Niwas Singh

The Institution of Engineering and Technology

Published by The Institution of Engineering and Technology, London, United Kingdom

The Institution of Engineering and Technology is registered as a Charity in England & Wales (no. 211014) and Scotland (no. SC038698).

First published 2020

The Institution of Engineering and Technology
Michael Faraday House
Six Hills Way, Stevenage
Herts, SG1 2AY, United Kingdom

www.theiet.org

British Library Cataloguing in Publication Data
A catalogue record for this product is available from the British Library

ISBN 978-1-78561-998-4 (Hardback)
ISBN 978-1-78561-999-1 (PDF)

Typeset in India by MPS Limited

Contents

About the editors

Rajeev Kumar Chauhan is working as Assistant Professor in the Department of Electrical Engineering, Dayalbagh Educational Institute, Agra, India. He was associated with University of Texas at Austin, United States, and University of Manchester, United Kingdom, as a visiting scientist and a research scientist, respectively. He is a senior member of IEEE and a member of IE (India). He has received some reputed national and international awards for his research contributions in the area of microgrid systems.

Kalpana Chauhan is Assistant Professor in the Department of Electrical Engineering, Central University of Haryana, Mahendragarh, India. She was associated with University of Texas at Austin as a visiting scientist. She is a senior member of IEEE, WIE, IAS and an associate member of IE (India). She has received some reputed national and international awards for her research contributions in electrical engineering.

Sri Niwas Singh is Vice-Chancellor at MMMUT Gorakhpur and on leave from Professor (HAG), Department of Electrical Engineering, IIT Kanpur, India. He is an IEEE Region 10 (Asia-Pacific) Vice-Chairman, Technical Activities 2019–2020. Prof. Singh is Fellows of IEEE, INAE, IET(UK), IE(I), IETE and life member of ISCEE. He has received various reputed national and international awards for his research contributions in electrical engineering and services.

Preface

The concept of microgrid (MG) is proposed for better penetration of renewable energy into the utility grid and consequently helping the energy management to simplify important grid issues, like savings for peak demand by taking advantage of dynamic pricing and reduced energy cost. In the case of DC distribution, the MGs are the best selection for designing a smart grid. MGs can be operated by multi-loads and multisources. The combination may be of renewable and traditional sources. MG can be utilized as a single user by operating it in off-grid and be utilized as grid-connected configuration. MGs are having the ability to control renewable sources to interact with smart grid for power balancing of utility grid by introducing some advanced energy management. The focus of this book is the case study–based research and solution for rural electrification. This book also deals with the low-voltage DC distribution system for various applications like charging of electric vehicle (EV).

Chapter 1 of this book is an overview of the rural electrification with DC microgrid (DCMG) and the introduction to EVs. Best option for rural electrification is a reliable and standalone system. DCMG is a such kind of solution, as it requires less maintenance that is difficult in the rural areas. The most significant development in DCMG increases the utilization of batteries and EV as energy storage (ES) devices in the MG. The utilization of such sources improves the system reliability and efficiency as these balance the system power. This requires a proper load or source scheduling. DCMG provides horizontal infrastructures to integrate distributed generation (DG) and loads. Chapter 2 is giving a general description of MGs, illustrating basic characteristics and structures of MGs with the objective of providing a general background information on MGs that will help one to better understand the best promising structure and architecture in designing such systems. In Chapter 3, different methods and tools for optimal MG investment and planning problem have been discussed. The specific aspects of the optimization formulation that addresses the challenges of rural MGs design have been discussed. This chapter ends with an application of the methodology to a real remote MG in Alaska. Chapter 4 assesses the impact of different EV charging models on the energy demand reduction (EDR) performance. The coordinated controls of voltage reduction technique are considered for both an urban and a rural MG, respectively. The Monte Carlo method is applied to analyze the impact of three different EV charging models (i.e., uncoordinated charging, V2G charging and mixed charging) on the grid EDR performance. Chapter 5 studies the different types of ES device technologies considering characteristics and operation constraints for MG use. The technology of energy storage systems (ESSs) has the important role on the application of ESSs for MG use. The battery ES has poor life cycle and influence on the MG's future development. The main characteristics of ESSs technologies are the important in the future development of MG. The ESSs are used as peak shaving, renewable energy systems integration, spinning reserve, load following, frequency regulation and black start in the MG application. In Chapter 6, authors present the control technique for integration of smart DCMG to the utility grid. A control technique of the bidirectional three-phase voltage source inverter into two rotating *d–q* synchronous reference frames, using the feed-back and feed-forward control loops, has been proposed. The performance analysis of the smart DCMG along with the developed control technique is also carried out under different operating scenarios in this chapter. Chapter 7 focuses on the energy management system (EMS) for an MG. The hierarchy of the various controllers utilized in the EMS is presented. Various programming methods can be used for optimizing the behavior of the EMS such that the MG can operate in a safe and reliable manner to match the demanded load with the available energy sources. The requirements for a communication system within the MG are briefly discussed. A test case with a particle swarm optimization technique is provided. Chapter 8 presents a concept of using augmented grey wolf optimization algorithm to reconfigure the distribution system aiming to reducing the power losses. The developed algorithm is successfully applied on IEEE 33-bus distribution system which proved its ability to find optimal solutions through lower computational time. In Chapter 9, the impacts of ESS, a novel multi-objective intelligent scheduling formulation, are proposed, which considers batteries life loss cost, operation cost and uncertainty costs associated with the variability of renewable sources. The test system used corresponds to an MG in an island with similar features of a real MG in Dongfushan Island in China. Chapter 10 aims to provide a tutorial of distributed droop control for accurate active and reactive power sharing in isolated AC MGs. Two recently developed distributed droop control mechanisms will be studied for improving the slow convergence rate of the conventional consensus-based droop control method. First, the alternating direction multipliers method-based consensus control is proposed. In Chapter 11, an extendable multiple output hybrid converter is presented for AC/DCMG applications. The proposed hybrid converter is capable of giving single DC and *n*-number of AC outputs simultaneously. This type of converter topologies is most suited for rural-based MG application where requirements are less maintenance, high power density and less cost. In Chapter 12, the topologies of various multi-input DC/DC converters have been compared from different aspects such as the battery life, the soft-switching, the source utilization, and the isolation between the input and the output. The chapter provides an information source in the selection guide on multi-input DC/DC converters of distributed energy resources in MGs for researchers, designers and application engineers. In Chapter 13, the steady-state and dynamic models of the buck converter connected to photovoltaic (PV) systems are derived. Finally, a control design algorithm is discussed to regulate the DC bus voltage in the presence of variations in PV voltage, DC busload and reference DC bus voltage. The simulation and experimental results are demonstrated in order to validate the controller performance. Chapter 14 aims to show the value of each MG system in providing energy efficiency, ancillary services, demand response and electricity bill reduction by using several case studies of MGs. In particular, the possibility offered by such systems to satisfy the energy load requirements of several classes of electric users, ranging from buildings and business facilities to rural communities, has been investigated. Subsequently, the techno-economic feasibility and the optimal sizing of components, as well as the economic performances of each MG system, have been compared to those achievable with traditional load feeding by the public distribution network (DN). Chapter 15 proposes modeling and model simulation for target design and evaluation—smart energy grid (SEG). MG-distributed production structures are assigned to a geographic information system that provided a bottleneck for the advancement and realization of real-world SEGs. In addition, there is no technology to support the evaluation of different web applications at different scales inside the grid, such as a hybrid EV. Chapter 16 discusses good practices and proposed some solutions to overcome challenges detected in some nanogrid projects developed in Colombia. Colombia is making an approach to energy MG projects through small projects that we will call nanogrids. These will show the behavior of users in the presence of energy for the solution of daily needs, the challenges in design and operation and the social changes that they have. The analysis of nanogrids might help one to predict behavior in larger networks in sectors with similar characteristics. In Chapter 17, the effects associated with the integration of DGs into the Libyan medium voltage-DN have been highlighted and discussed in detail. In particular, the insufficiency of the utilized protection practices are thoroughly investigated and explained. Generally applicable approaches are recommended, which can enhance the applied protection schemes in the interconnected DN.

Acknowledgements

We want to give huge thanks to our supporter during the editing of this book. Editing a book is harder than we thought and more rewarding than we could have ever imagined. This would not have been possible without the adjustment made by children. They have cooperated a lot and gave their continuous emotional support during this journey.

We are eternally grateful to our parents, who taught us discipline, love, manners, respect and so much more that have helped us succeed in life. They ever encouraged us to do hard work. We would like to thank all our friends and colleagues for their direct and indirect support.

"Thanks to everyone in our publishing team."

Part I

Fundamentals

Chapter 1

Overview of microgrids for rural areas and low-voltage applications

Rajeev Kumar Chauhan[1]*, Kalpana Chauhan*[2] *and Sri Niwas Singh*[3,4]

The chapter deals with an overview of the rural electrification with DC microgrid and the introduction to electric vehicles (EVs). The best option for rural electrification is the reliable and standalone system. DC microgrid requires less maintenance, which is advantageous in the rural areas. The most significant development in DC microgrid increases the utilization of batteries and EV as energy storage (ES) devices in the microgrid. The utilization of such sources improves the system reliability and efficiency as these balance the system power. This requires a proper load or source scheduling. DC microgrid provides the horizontal infrastructures to integrate distributed generation and loads.

1.1 Overview

The overall electrification rate in rural areas is 64.5%, and 35.5% of the population still lives without any access to the electricity. But the technical and economic developments during the last decades have established opportunities to create a new competitive distribution system based on modern power electronic technology. From a technological point of view, the DC distribution system is a new concept in electrical distribution, and it generates a new area of business to power electronic device manufacturers. Renewable energy sources (RES) such as photovoltaic (PV) system can be the best solution for helping in getting good electrification in rural areas with the implementation of the DC distribution system. The RES such as the PV panel can be directly utilized at the value of microgrid without converting to alternating current (AC) and feeding into the main grid. Generally, the output of RES is DC and this DC power is injected into an electric grid by using

[1]Department of Electrical Engineering, Dayalbagh Educational Institute, Agra, India
[2]Department of Electrical Engineering, Central University of Haryana, Mahendragarh, India
[3]Madan Mohan Malaviya University of Technology, Gorakhpur, India
[4]Department of Electrical Engineering, Indian Institute of Technology, Kanpur, India

suitable power converters for converting DC to AC. But most of the load in residential/commercial buildings is based on DC loads or AC loads that can be easily converted to DC loads. The rapid increment of DC-compatible appliances in the buildings helps the PV energy to become the fastest growing source of renewable energy, and it is expected to see continued strong growth in the immediate future. The PV power is, therefore, required to be provided with a certain reliability of supply and a certain level of stability. Motivated by the above issues, many grid operators have to develop DC microgrids, which treat PV power generation in a special manner.

The concept of microgrids is proposed for better penetration of renewable energy into the utility grid and consequently for helping the energy management to simplify important grid issues such as savings for peak demand by taking advantage of the dynamic pricing and reduced energy cost. In the case of DC distribution, the microgrids are the best selection for designing a smart grid. Microgrids can be operated by multi-loads and multi-sources. The combination may be of renewable and traditional sources. Microgrid can be utilized as a single user by operating it in off-grid and be utilized as grid-connected configuration. Microgrids are having the ability to control renewable sources to interact with smart grid for power balancing of utility grid by introducing some advanced energy management.

1.2 Importance of microgrids

A microgrid refers to a power distribution system integrated with the distributed energy resources (DERs) and controllable loads, which can either operate with the main grid (i.e., grid-connected mode) or use the DERs to supply the loads without the main grid (i.e., islanded mode) [1]. It can be viewed as a small-scale grid, and it is widely considered as one of the key components to integrate renewable DERs and save transmission losses for efficiency.

In a conventional power system, the power flow is single directional from bulk generation, e.g., a power plant, to load via a transmission and distribution system. As more DERs such as roof-top PVs, small wind turbines (WTs), and battery systems (e.g., Tesla Powerwall) are becoming available on the distribution side, bidirectional power flow is now possible and it makes a conventional power distribution system into a microgrid that can use only the DERs as the power supply without the main grid (i.e., islanded mode) or even sell back its surplus power to the main grid. The size of a microgrid may vary depending on the application. For example, the concept of microgrids can be implemented as a smart home, a smart building, or a smart campus. Figure 1.1 shows a microgrid example consisting of PVs, WTs, diesel generators, batteries, and loads. Two-way communications are available between the local controllers and the microgrid central controller. There are a few on-going projects focusing on building the microgrid infrastructure such as the smart power infrastructure demonstration for energy reliability and security from Sandia National Labs [2] and the UCSD microgrid project [3].

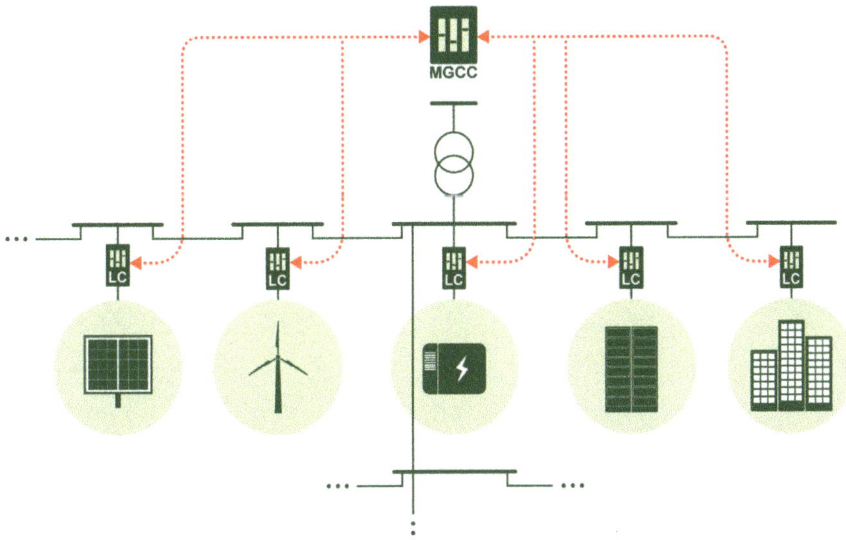

Figure 1.1 A microgrid

1.3 Classification of microgrids

1.3.1 AC microgrid system

A typical AC microgrid system interconnected with medium voltage (MV) system at the point of common coupling (PCC) is shown in Figure 1.2. The main system could be an AC or DC bulk system. The DG units and ESS are connected at some points within the distribution networks. Part of the network consisting of the DG units and load circuits can form a small isolated AC electric power system, i.e., an "AC microgrid." During normal operating conditions, the two networks are interconnected at the PCC, while the loads are supplied from the local sources (e.g., the RES-based DG units) and if necessary from the utility. If the load demand power is less than the power produced by the DG units, excess power can be exported to the utility.

1.3.2 Hybrid AC–DC microgrid system

The conceptual layout of the hybrid AC–DC microgrid is shown in Figure 1.3. After staying on AC technology in the area of electric power supply, the DC power joins it with increasing technology advancements in power conversion, generation, transmission, and consumption. However, there are many challenges in DC technologies, so they should be integrated into the system by applying many algorithms in some or every step of the process. The scope to explore the DC technologies with its specific advantages is the microgrid. The hybrid AC–DC microgrids facilitate benefits of both the technologies and are having the integration of AC technology with the DC technology. In leading hybrid AC–DC systems, the AC and DC buses

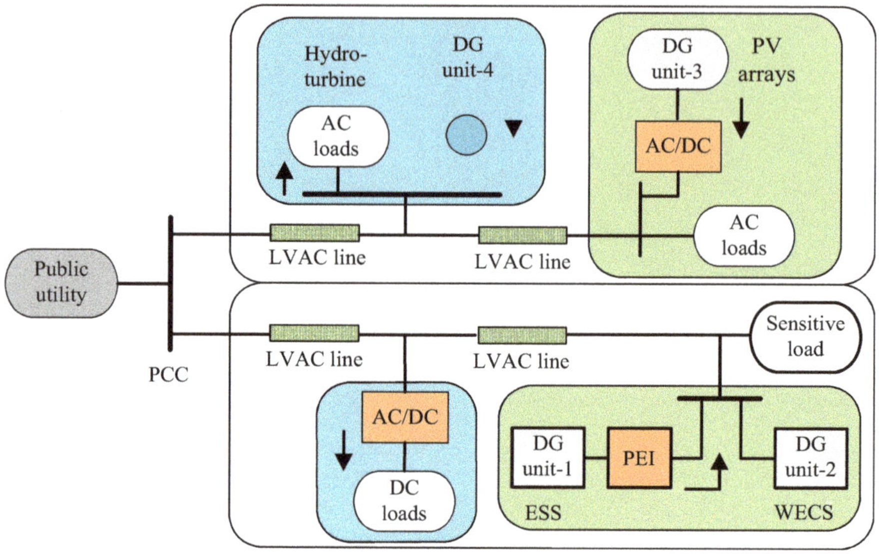

Figure 1.2 AC microgrid structure with the DG units and mixed types of loads

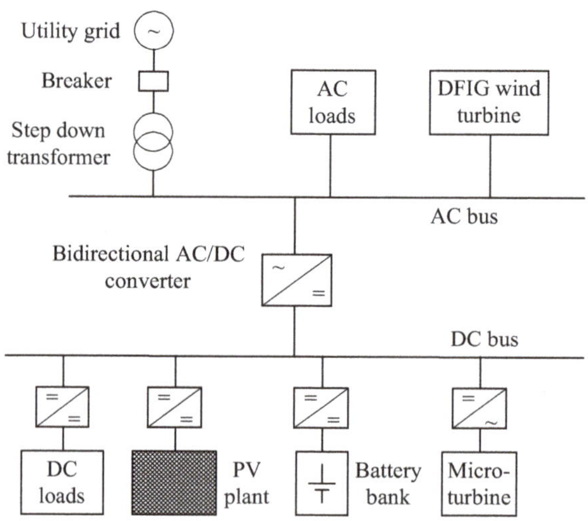

Figure 1.3 Conceptual layout of a hybrid AC–DC microgrid

are connected through interlinking bidirectional converters. However, this interlinking creates a stability issue and requires control algorithms to maintain the power quality. Microgrids, which are having different types of sources, and loads are the type of AC–DC systems.

1.3.3 DC microgrid system

Traditional electric power system was designed to move the central station AC power, via high-voltage AC transmission lines and lower voltage distribution lines to households and businesses that use the power in incandescent lights, AC motors, and other AC equipment. Meanwhile, the DC power system has been used in industrial power distribution system, telecommunication infrastructures, and point-to-point transmissions over long distances or via sea cables and for interconnecting AC grids with different frequencies. Today's consumer equipment and tomorrow's DG units are dominated by power electronics devices. These devices (such as computers, fluorescent lights, variable speed drives, households, businesses, industrial appliances, and equipment) need the DC power for their operation. However, all these DC devices require a conversion of the available AC power into DC for use, and the majority of these conversion stages typically use inefficient rectifiers. Moreover, the power from DC-based DG units must be converted to AC to tie with the existing AC electric network, only later to be converted to DC for many end users. These DC–AC–DC power conversion stages result in substantial energy losses. Using the positive experiences in the high-voltage DC (HVDC) operation and the advances in power electronics technology, interests in pursuit of effective solutions have increased. Figure 1.4 shows the typical DC MG systems interconnected with the main systems at PCC that can be MVAC network from the conventional power plants or an HVDC transmission line connecting an offshore wind farm.

1.4 Electric vehicles

The EVs are the low carbon emission vehicles that fully or partially are operated or energized with electricity [4,5].

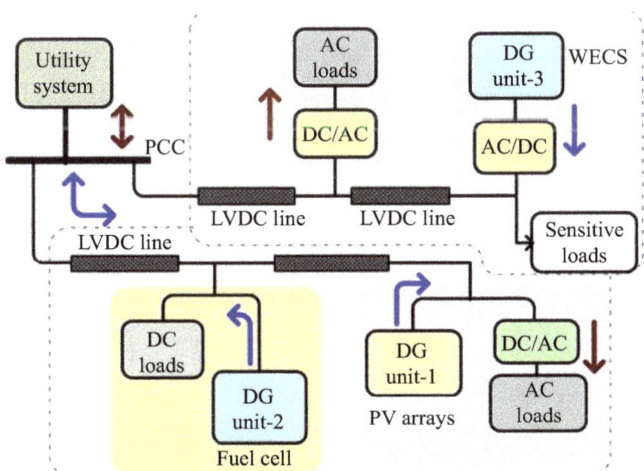

Figure 1.4 Concept of a DC microgrid system with the DG units and mixed types of loads

1.4.1 What are the electric vehicles?

EVs' concept has prevailed since 1837, known as electric car. After the invention of rechargeable batteries, these were used in EVs during 1859. EVs, partially or fully on electricity, are powered by electric motors for propulsion. The electric motor helps one to get energy from rechargeable batteries, solar panels, or other fuels. The EV may be the solution to the problem of pollution in urban and populated areas. It also solves the problems in crises of conventional fuels.

EVs can be classified as follows:

1. Hybrid EVs (HEVs): EVs that employ both electric and other fossil fuels. The rechargeable battery helps one to use the other fuels efficiently. During no availability or costly electric charging, the back fuel helps one to operate the vehicle.
2. Battery EVs (BEVs): EVs that are independently operated by the electricity with the inclusion of the other fuels. Charged batteries derive the vehicle independently [6,7].
3. Plug-EVs (PEVs): These types of EVs having rechargeable batteries which are recharged by the external supply while plugged-in.
4. Plug-in hybrids EVs (PHEVs): PEVs can operate in combination with other fuels then these called as PHEVs.

Figure 1.5 shows the broad classification of the EVs.

1.4.2 Hybrid electric vehicle control system

Mostly used HEVs require control strategies for reliable and efficient operations. The control strategies broadly may be classified as rule-based control and optimization

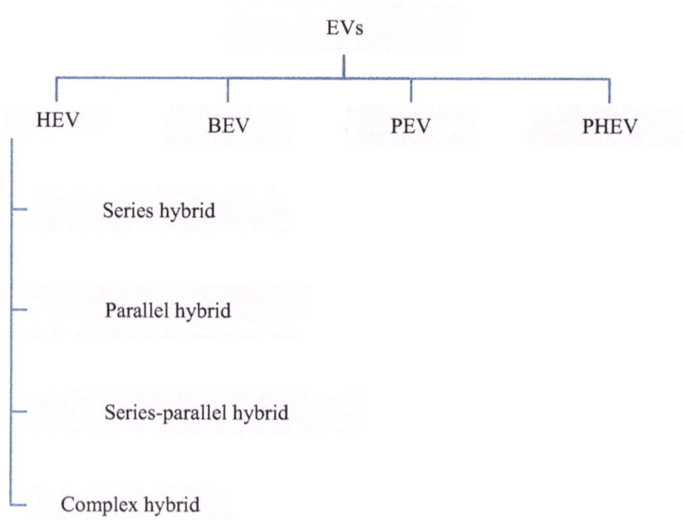

Figure 1.5 Classification of EVs

control. Rule-based control may be deterministic (thermostatic control and electrical assist) and fuzzy (conventional, adaptive, and predictive). On the other hand, optimization-based control may be real time (equivalent consumption minimization, neural network, particle swarm optimization, and model prediction) and global (linear program, dynamic program, and stochastic control system). On the other hand, all EVs run totally on electricity.

1.4.3 Advantages and disadvantages of electric vehicles over conventional vehicles

Advantages:
EVs have a number of benefits compared to conventional vehicles:

1. *Environment friendly:* EVs do not emit any tailpipe pollutants, although they may be emitted by the energy plant that produces electricity. No air pollutants are caused by electricity from nuclear, hydro, solar, or wind-powered crops.
2. *Energy efficient:* EVs convert approximately 59%–62% of the grid's electricity to wheel power. Conventional petrol cars only convert around 17%–21% of the electricity stored in petrol to wheel power.
3. *Performance benefits:* Electric motors provide quiet, smooth operation, and speed, requiring less maintenance than inner combustion engines.
4. *Health benefits:* Reduced emissions of damaging exhaust are excellent news for our health. Better air quality will lead to fewer air pollution-related health issues and expenses. EVs are also slower than cars for petrol/diesel, which implies less pollution from noise. Reducing damaging emissions of the exhaust is excellent news for our health. Better air quality will result in fewer health issues and air pollution expenses. Also, EVs are louder cars for petrol/diesel, which implies less noise.
5. *Safety improvements:* Recent results have shown that security can be improved by several characteristics. EVs tend to have a reduced gravity center that makes them less likely to roll over. They can also have a lower risk of significant fires or explosions, and the building and durability of EVs can make them safer in a crash.
6. *Energy security:* EVs can assist with Australia's energy security at domestic level. Australia is currently heavily dependent on other nations for imports of petroleum. EVs can be powered easily from local and renewable sources of energy, decreasing our reliance on foreign oil. There are also better job advantages for Australians by using local electricity.

Drawbacks comparison with conventional vehicles:

1. Initial cost is very high.
2. It needs charging in a shorter range within 100 or some more miles.
3. Charging takes more time.
4. Till now charging stations are limited.
5. Require battery replacement every 5–7 years depending on the brand.

1.5 Standards

The growing infrastructure of the DC distribution system requires some standards to choose the proper strategy for the installation and interconnection of the microgrids. There are many standards published by the Institute of Electrical and Electronics Engineers (IEEE) and the International Electrotechnical Commission (IEC), but we will elaborate here some selected ones that are mostly related to the research work. Some industry-related standards and their status are also mentioned here.

1.5.1 Standards for microgrid with PV and battery interconnection

In this, the battery (i.e., lead–acid batteries) is used in the interconnection of PV to get the uninterrupted power supply in the microgrid. The standards related to this interconnection are mentioned in Table 1.1. In the starting there are three standards, the first IEEE Std. 937-2007 is related to design, installation, storage, mounting, maintenance, and safety precautions. The second IEEE Std. 1013-2007 is related to energy capacity determination, and the third IEEE Std. 1361-2014 is related to battery test procedure of lead–acid secondary batteries connected to the PV power systems. However, the next four standards are of 1547 series as IEEE Std. 1547.2-2008, IEEE Std. 1547.3-2007, IEEE Std. 1547.6-2011, and IEEE Std. 1547.7-2013, giving standards for interconnecting the distributed resources with the electric power system.

Table 1.1 Battery interconnection with PV

Practice for installation and maintenance of lead–acid batteries for photovoltaic (PV) systems	IEEE Std. 937-2007
Recommended practice for sizing lead–acid batteries for standalone photovoltaic (PV) systems	IEEE Std. 1013-2007
Guide for selecting, charging, testing, and evaluating lead–acid batteries used in stand-alone photovoltaic (PV) systems	IEEE Std. 1361-2014
Application guide for IEEE Std. 1547(TM), IEEE standard for interconnecting distributed resources with electric power systems	IEEE Std. 1547.2-2008
Guide for monitoring, information exchange, and control of distributed resources interconnected with electric power systems	IEEE Std. 1547.3-2007
Recommended practice for interconnecting distributed resources with electric power systems distribution secondary networks	IEEE Std. 1547.6-2011
Guide for conducting distribution impact studies for distributed resource interconnection	IEEE Std. 1547.7-2013

Table 1.2 Electric vehicle supply equipment (EVSE)/installation

EV conductive charging system, AC-, DC-supply (IEC TC 69)	IEC 61851-1, -23
EV wireless power transfer system (IEC TC 69) (draft)	IEC 61980
In-cable protection device mode 2 supply equipment (IEC SC23E)	IEC 62752-1
Plug, socket-outlet, vehicle connector, and inlet (IEC SC23H)	IEC 62196
Electrical installations: supply of electric vehicle (IEC TC64)	IEC 60364-7-722

Table 1.3 Agency status/standards

ETSI EN 300 132-3-0	Power interface standard	Released
ETSI EN 301605	Earthing and bonding for 400-V DC systems	Released
ITU-(ITU-T 1.1200)	Adopted ETSI voltage levels	Released
IEC/IEEE (SG22H)	Working group in place—new DC UPS standard	Released 2015
ATIS-0600315.01.2015	Voltage levels standard	Released

1.5.2 Standards for electric vehicles when they are used as a source

Indeed, for comfortable and secure vehicle utilization, there are many aspects of EVs that need to be standardized as, for example, the plugs for charging, the chargers voltages, the communication between the vehicle and the chargers, fast charging systems and respective billing, security measures for vehicle safety and safety of persons against electric shock, on-board electric ES for propulsion, and vehicle energy performance and energy measurements. Some of the standards when EV is used as the source are mentioned in Table 1.2.

The IEC 61851 is related to the standards for on-board and off-board equipment and IEC 61980-1:2015 for the wireless transfer, respectively, to charge EVs up to 1,000-V AC and voltages up to 1,500-V DC. On the other hand, the IEC 62752:2016 is for cable control and protection for charging the EVs, and IEC 62196 is based on charging, connection and other requirements for both EV and its related equipment. IEC 60364-7-722 is related to supply the EVs.

1.5.3 Agency-related standards

Some standards are provided on the basis of companies' requirements or consumer side demand. These may be for power interfacing or to welcome other incoming sources. Some of these are mentioned in Table 1.3.

1.6 Conclusion

The electrical distribution infrastructure is not very well extended in rural areas. But the technical and economic developments during the last decades have established opportunities to create a new competitive distribution system based on modern power electronic technology. From a technological point of view, the low-voltage DC distribution system is a new concept in electrical distribution, and it generates a new area of business to power electronic device manufacturers. RES such as solar PV (SPV) system can be the best solution for helping in getting good electrification in rural areas with the implementation of the DC distribution system. The RES such as the SPV panel can be directly utilized at the value of microgrids without converting to AC and feeding into the main grid. Generally, output power of RES is DC, and this DC power is injected into an electric grid by using suitable power converters by converting DC to AC. But most of the load in the residential/commercial buildings is based on DC loads or AC loads that can be easily converted to DC loads. The rapid increment of DC-compatible appliances in the buildings helps the PV energy to become the fastest growing source of renewable energy, and it is expected to see continued strong growth in the immediate future. The PV power is, therefore, required to be provided with a certain reliability of supply and a certain level of stability. Motivated by the above issues, many grid operators have to develop DC microgrids, which treat PV power generation in a special manner.

References

[1] X. Fang, S. Misra, G. Xue, and D. Yang, "Smart grid – the new and improved power grid: a survey," *IEEE Communications Surveys and Tutorials*, vol. 14, no. 4, pp. 944–980, 2012.

[2] J. Stamp, "The SPIDERS project – smart power infrastructure demonstration for energy reliability and security at US military facilities," in *Proc. IEEE PES Innovative Smart Grid Technologies*, pp. 1, January 2012.

[3] B. Washom, J. Dilliot, D. Weil, *et al.*, "Ivory tower of power: microgrid implementation at the University of California, San Diego," *IEEE Power and Energy Magazine*, vol. 11, no. 4, pp. 28–32, 2013.

[4] F. Sun, "Current situation and trend of development of electric vehicle," *Science Chinese People*, vol. 8, pp. 44–47, 2006.

[5] C. Gao, and Z. Liang, "Overview of the impact of electric vehicle charging on the grid," *Grid Technology*, vol. 2, pp. 127–131, 2011.

[6] J. Zhao, F. Wen, and A. Yang, "Effect of electric vehicle on power system and its scheduling and control problem," *Power System Automation*, vol. 14, pp. 2–10, 2011.

[7] B. Cao, C. Zhang, and Z. Bai, "Development and development trend of electric vehicle technology," *Journal of Xi'an Jiao Tong University*, vol. 1, pp. 1–5, 2014.

Chapter 2
Microgrid architectures

Enrico Elio De Tuglie[1], Alessia Cagnano[1], Emilio Ghiani[2], Susanna Mocci[2] and Fabrizio Pilo[2]

This chapter presents a general description of microgrids (MGs), illustrating basic characteristics and structures of MGs with the objective of providing general background information on MGs that will help one to better understand the best promising structure and architecture in designing such systems. At present, AC MGs are still the main arrangement of MGs, even if they involve some drawbacks such as the need for synchronization of DG units, the circulation of reactive power and electromagnetic compatibility problems due to the intensive use of AC/direct current (DC) converters. Recently, DC MGs are emerging as a possible solution to avoid the aforementioned problems for a few isolated DC devices that must be connected into ex-novo networks. Although DC-powered components for residential and industrial applications are going to be more and more developed, the vast majority of existing devices currently in use in every area are fed in AC.

2.1 Introduction to microgrids

The persisting growth of variable renewable generation—predominantly on distribution level—is gradually altering the operating conditions of both electricity transmission and distribution networks.

The MG concept was introduced as a viable solution for integrating distributed energy resources (DERs) and, more properly, of renewable ones into distribution network. The basic characteristics and structures of MGs can be identified by means of an insightful literature review of the existing forms of these systems. For instance, the analysis of the literature [1–6] points out that there is no univocally recognized knowledge in designing such systems. In fact, depending on planning objectives and the way of power supply, MGs are built with different structures and control architectures.

An MG comprises a portion of an electric power distribution network that aggregates a variety of DERs along with local loads. Also, differently from

[1]Department of Electrical and Information Engineering, Polytechnic University of Bari, Bari, Italy
[2]Department of Electrical and Electronic Engineering, University of Cagliari, Cagliari, Italy

traditional distribution networks, it is normally supervised and controlled by a monitoring, supervision and control system (SCADA) that is able to coordinate and dispatch all MG components so as this system can act as a single controllable entity for the upstream distribution grid.

Commonly, an MG is built so as to have enough generating capacity to meet local load demand. Therefore, if properly managed, it can be operated as a self-sufficient energy system that can interact with the main distribution grid or disconnect from it to operate in island/autonomous mode.

In general, depending on the general structure, an MG can be subdivided into two main categories: AC and DC. At present, AC MGs are the main form of MGs, even if they involve some drawbacks such as the need for synchronization of DG units, the circulation of reactive power and electromagnetic compatibility problems due to the intensive use of AC/DC converters. Recently, DC MGs are emerging as a promising solution to avoid the aforementioned problems for a few isolated DC devices that must be connected to ex-novo networks.

In the following sections, details on both AC and DC MG architectures and their main applications are given.

2.2 AC microgrids architectures

2.2.1 Structures

The AC MG is the widely used architecture because it is easier to design and implement as it mostly exploits the existing AC network infrastructure. In this architecture, all DERs and loads are connected to a common AC main bus through power converters. Figure 2.1 illustrates the typical configuration of an AC MG and its basic elements.

The electric power is produced by small generating plants normally located near the load and then transmitted to consumers through a local distribution network. Typical generation technologies applicable for MGs can be classified into two main categories: programmable and nonprogrammable generating units. The first category includes induction, synchronous and diesel engine generators, small hydro, microturbines, combined heat and power units. The second category includes renewable energy sources (RESs) such as photovoltaic systems and wind turbine generators [1].

Besides the aforementioned generation sources, storage systems are also integrated into an MG for ensuring the load balance in the real time. The most common types of storage devices available for MG applications are batteries, flywheels, super and ultracapacitors, pumped hydroelectric, thermal energy and compressed air energy storage [6]. More recently, the development of vehicle-to-grid systems is promoting the adoption of electric vehicle batteries as storage systems. In fact, as outlined in [7] these devices, providing mobile energy storage points can be a successful solution for the load frequency control of isolated MGs.

Smart power meters are, in general, the basis in the new conception of the smart grid; in MGs they provide detailed information about loads energy

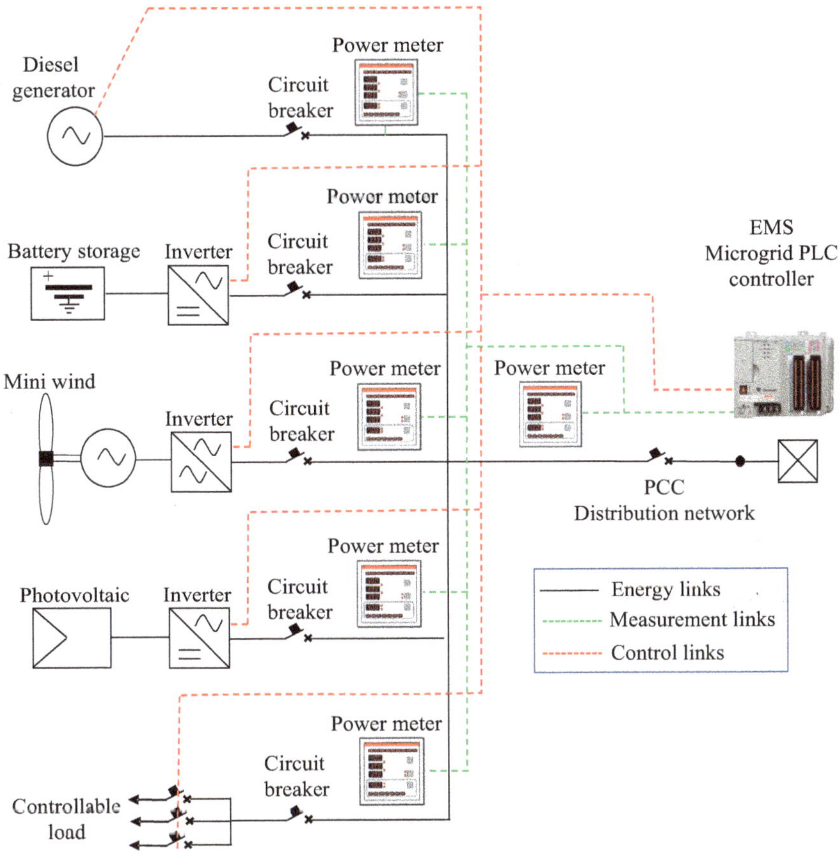

Figure 2.1 Single-line diagram of a typical AC microgrid

consumption and play a big role in the control of the MG since the recorded power and energy profiles can be integrated in energy management systems (EMS), used to coordinate the operation of generation sources and loads.

Moreover, an AC MG can supply different types of consumers that are normally classified into critical and noncritical loads. The first ones are those loads that are noninterruptible such as commercial and industrial loads. The second ones are not essential loads such as residential ones, and thus they can be disconnected at the time of crisis [8].

Besides the aforementioned components, a typical AC MG requires other infrastructures such as the interconnection point and the distribution network.

Several MGs have been built all over the world [1–5]. The analysis of these demonstration projects points out that AC MGs vary in size and structural components, even if they all have the same basic characteristics.

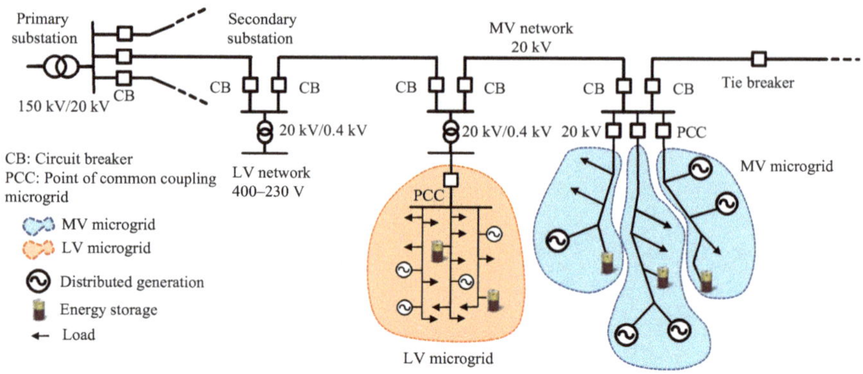

Figure 2.2 MV and LV AC MGs interconnected to upstream distribution network

2.2.2 MG point of common coupling

One of the crucial aspects in developing an MG is its interconnection point with the distribution network, usually denoted by the term point of common coupling (PCC). PCC is the point in the electrical system in which the cluster of single/multiple customers or multiple electrical loads may be connected to creating the MG aggregate. Depending on the position in the power system and the voltage level, this interconnection point originates medium-voltage (MV) or low-voltage (LV) MGs (Figure 2.2).

In LV MGs the PCC is downstream of the distribution substation through a circuit breaker equipped with a protection relay capable of autonomously disconnecting the MG following to grid outages.

This is, for example, the case of systems such as PrInCE Lab MG installed at the Polytechnic University of Bari [9] and the BC Hydro installed at the Boston Bar system in Canada [10]. Anyway, with the aim to speed up the disconnection of the MG from the main grid, new interconnection devices have been developed during the last years or are already at test stage. Among them, the most popular is the interface static switch (ISS) that is usually equipped with conventional protective functions to automatically detect islanding conditions occurrence. Moreover, in order to make this device capable of autonomously reconnecting the MG to the main grid when particular conditions occur, synchronization functions have been added. In order to test the ability of this device to realize a seamless disconnection and reconnection of an MG, it is currently installed in some popular MGs such as the CERTS MG testbed [11,12], Santa Rita Jail [13], UW MG [14], Laboratory scale MG in Hong Kong and MG at CSIRO Energy Centre [14]. In some other applications, such as the Sendai MG, a power converter has been added to the ISS with the ultimate goal to enhance its performances.

2.2.3 MG distribution network configuration

AC MGs basically consist of distribution feeders that connect different DERs along with loads at different places. Depending on the DER types implemented and on the voltage level, MGs can have different network configurations.

Most of the existing MGs are configured radially because this configuration guarantees a better coordination of the protection devices [15]. Moreover, by exploiting the existing distribution network infrastructures they are easier and less costly to implement because they do not require several additional hardware devices. Despite their simplicity, the radial systems are characterized by low service reliability. For this reason, different meshed MGs are realized around the world such as the DeMoTec test MG developed by ISET laboratories in Germany [16], the AM Steinweg microgrid [17], the CERTS testbed [11], the Bronsbergen Holiday Park [18], the smart polygeneration MG at the University of Genova [19] and the University Seville Spain testbed [20].

It is worth noting that, even if this MG topology is considered to be promising, in some circumstances and especially in residential areas, the ring configuration is preferred. In this configuration, the MG is designed in closed loop in which the start and the end of the feeder coincide. Therefore, if a system failure occurs by sectioning the feeder from one end, the supply is preserved from the other end, improving thus the system reliability.

A main three-phase feeder at which various three-phase and single-phase lateral branches are connected typically constitutes most of the aforementioned MGs. Industrial and larger commercial loads are inevitably served from three-phase feeders, while most of the residential and commercial loads are single phase.

2.2.4 MG control system

MG control system is normally designed to perform several tasks and meet the following fundamental requirements:

- MG should be able to disconnect itself from the main grid following grid outages.
- MG in islanded mode should be able to balance generation and demand (storage) in the real time; to meet with this requirement adequate "spinning reserve of active and reactive power" should be always guaranteed.
- Resynchronization of the MG to the main grid must be performed in order to provide uninterrupted power to MG loads.
- MG, in the grid-connected mode, should supply energy at minimum cost by taking care of the cost of electricity flows in and out of the MG and any main grid constraints.
- MG should be able to provide ancillary services to the grid.

In order to achieve these requirements, different MG control architectures have been realized. These architectures mainly depend upon the MG control modes that can be, respectively, master–slave, peer-to-peer and droop–control mode.

Among the most widely used architectures, the hierarchical one has proved to be the most adequate in managing all aspects of the MG operation in all its operating modes, and transitions. This is mainly structured on three control levels [21]:

1. local source controllers (LSC) and load controllers (LC)
2. MGC system central controller
3. central autonomous management controller.

Each MC follows requests from the central controller, when connected to the power grid, and performs local optimization of the active and reactive power production, and fast load tracking following an islanding situation. LCs installed at the controllable loads provide load control capabilities following orders from the MGC for load management.

The MGC is responsible for the optimization of the MG operation. It uses the market prices of electricity and gas and grid security concerns to determine the amount of power that the MG should draw from or inject into the distribution system, thus optimizing the local production capabilities.

The top control level is responsible for planning the MG operation in order to ensure the safe and economic operation of the distribution network. In order to determine the operation schedule of the MG at every hour interval, meeting the needs of the external power grid, the unit commitment problem is usually adopted. The resulting power exchange plan with the utility grid is passed to the second control level where they are processed to coordinate and dispatch programmable MG components so that the amount of power exchanged at the PCC is kept as close as possible to the planned one. The optimal control actions are centrally evaluated by solving economic dispatch problems aimed at maintaining the MG in a safe and economic operation. The evaluated control actions are then sent to the local controllers of DER—inverters where they are actuated to manage the associated element [22]. Therefore, the secondary controller will move power set points of dispatchable generators and storage devices in order to restore the fixed amount of the primary reserve of the master controller.

2.2.5 Control philosophy of AC microgrids

In order to design appropriate control systems, it is helpful to conceptually classify the MG operating conditions into all operating states and transitions as suggested in [9]. As outlined in this chapter, an MG normally operates in rid-connected mode or in islanded mode if an unpredicted event or a planned disconnection occurs. In both these operating modes, the system lies in a normal or a secure state in which economic objectives and operational issues are the main concerns. Anyway, following a particular dispatching policy, the MG can be moved to an unsafe operating point (alert state) where its safety margins are dangerously reduced. To restore its safety margins, preventive control actions need to be implemented. Anyway, if the actuated preventive actions are not succeeded, the MG will remain in the alert state. Therefore, if an unpredicted event occurs, the MG could enter in the emergency state where some limits are violated. In order to promptly recover the system security, corrective control actions need to be quickly activated. If the evaluated corrective actions are not successful, a general blackout will occur. The blackout can be recovered by actuating the on-grid or off-grid black start procedures.

It must be further noticed that the MG can directly move from one normal state to another. The transition from the on-grid to the off-grid mode can follow a system perturbation or it can be planned. The opposite transition (from off-grid to the on-grid) can be intentionally activated by the system operator through a synchronization procedure.

2.3 DC and hybrid AC–DC microgrid architectures

2.3.1 DC microgrids

DC MGs are usually connected through an AC/DC converter to the main distribution network and it can feed AC and DC loads [23]. The AC/DC converter can operate also as static switch for the connection and disconnection of the MG (Figure 2.3). The DC MG architecture presents some advantages over the AC MG: a reduced number of (and simpler) power converters (DC/DC and rectifiers), the possibility of adapting the DC bus voltage to the MG requirements and a very high quality of the DC bus voltage, so that some DC loads can be connected directly to the DC bus.

In order to make the DC MG capable of exporting the excess in the generated power, the power electronic interface must be bidirectional and the converter is composed of two main systems:

1. AFE (active front end) bidirectional AC/DC input converter connected to distribution network;
2. Bidirectional DC/DC output converter connected to the DC load/generation.

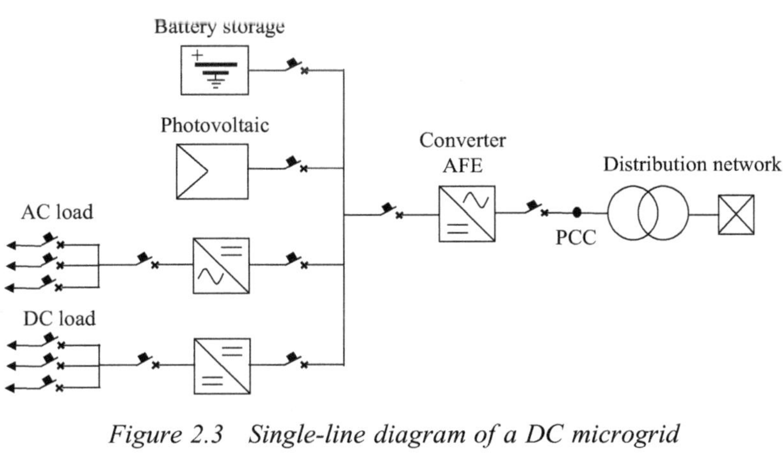

Figure 2.3 Single-line diagram of a DC microgrid

The first converter takes power from the distribution network, delivering it to an intermediate DC bus. The current demanded is sinusoidal, with very low harmonic distortion and with unity or variable power factor, capable of exchanging active and reactive power. The second converter takes the power of that bus and generates a clean and stable continuous voltage at its adjustable output, with an extremely low current ripple. The DC MG presents then a DC bus with a regulated voltage; mostly distributed generators need a DC/DC or AC/DC power electronic interfaces to be connected to the bus. The AC loads require a DC/AC converter to adapt the bus voltage to the required conditions. DC loads sometimes can be connected directly to the DC bus or may need a DC/DC converter, depending on the bus voltage.

Promising applications of DC MGs are in the sector of smart homes and buildings and in data centers [24]. In fact, in many homes and buildings the consumer loads are today, at least internally, DC supplied as well as electronic-based office and home appliances. Newer more efficient lighting technologies such as compact fluorescent fixtures and solid-state lighting involve a DC stage and hence it is more efficient to utilize them in a DC distribution system. DC power is used in variable speed drives for pumps, heating, ventilation and air conditioning systems, fans, elevators, mills and traction systems.

2.3.2 Hybrid AC–DC microgrids

The main drawback of the DC MG architecture is the series-connected bidirectional AC/DC converter handling the whole power flow from/to the distribution grid, since it reduces the reliability of the MG in the case of failures in the converter. Besides, it requires a specific installation, being impossible to use the existing AC cabling and devices. Another disadvantage of the DC architecture is that AC loads cannot be directly connected to the MG and the voltage of DC loads is not homogeneous, frequently the ratings of 110, 48, 24 and 12 V are used in the same context (e.g., data centers). Thus, additional power stages would be needed to eventually adapt the DC bus level or to generate AC voltages.

For this reason, the AC–DC hybrid MG architecture that consists of an AC MG with a DC subgrid, tied together by a bidirectional AC/DC converter, is preferred. In AC–DC hybrid MG architecture, DG can be connected to the AC or to the DC feeders. AC loads are connected to the AC feeder, whereas DC loads are connected to the DC feeder, using a power converter to adapt the voltage level if it is necessary. The DC subgrid can act as a generator or a load of the AC MG, depending on the power balance at the DC feeder (Figure 2.4).

This architecture combines the advantages of AC and DC MGs. There is a direct connection to the grid, thus presenting a high reliability; the AC feeder allows using the existing equipment; the DC feeder allows the use of a reduced number of simpler converters. Moreover, some DC loads can be connected directly to the DC feeder, without the need of any power converter. This MG architecture is suitable for installations with critical loads (at the DC feeder) in combination with more robust loads (at the AC feeder).

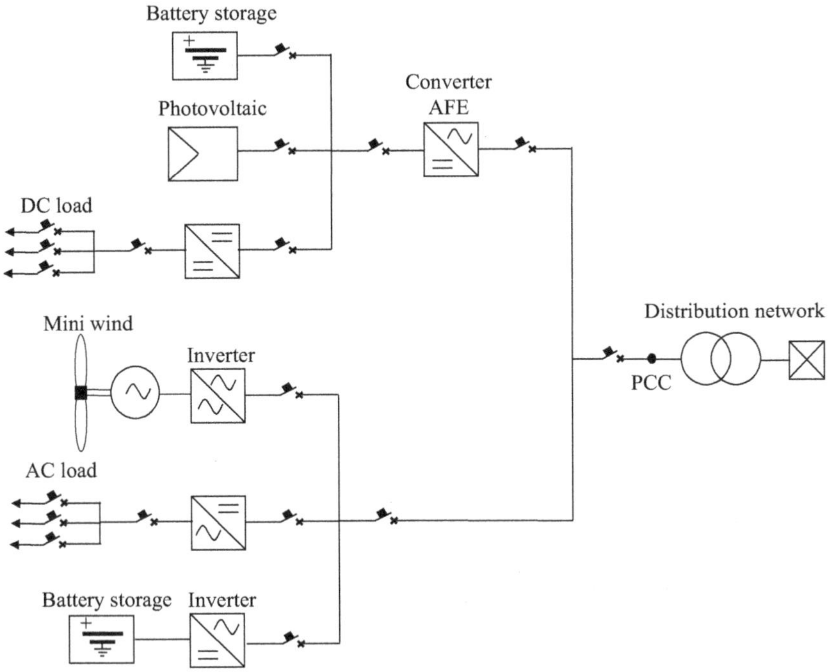

Figure 2.4 Single-line diagram of hybrid AC–DC microgrid architecture

Finally, given that DC distribution grid is a feasibility option studied by several researchers in the recent years [25,26]. Another option of hybrid MG is the AC MG derived from a DC distribution network. In Figure 2.5, such hybrid configuration is represented considering a smart house with computers, laptops, tablets, phones, printers, TVs, microwave ovens and lighting that consume electricity in AC form, although they have a DC converter inside them.

2.4 Multi-MGs in smart distribution networks

The EU long-term goal for the no-carbon society is promoting the development of local MG with suitable regulatory and governmental acts. Innovative network schemes and operation policies are proposed to help facing the new challenges mainly based on the exploitation of communication and information technologies to increase the observability and controllability of the distribution system. Recognizing the benefit that one can get by exploiting the MG concept, as an active part of the LV network comprising several distributed generation sources, controllable loads and storage devices, is a key issue towards the MG concept deployment [27,28]. Furthermore, the MG concept is extended into multi-microgrids (MMGs) concept, or networked MGs [29–31], identifying the benefits that can be obtained at MV level.

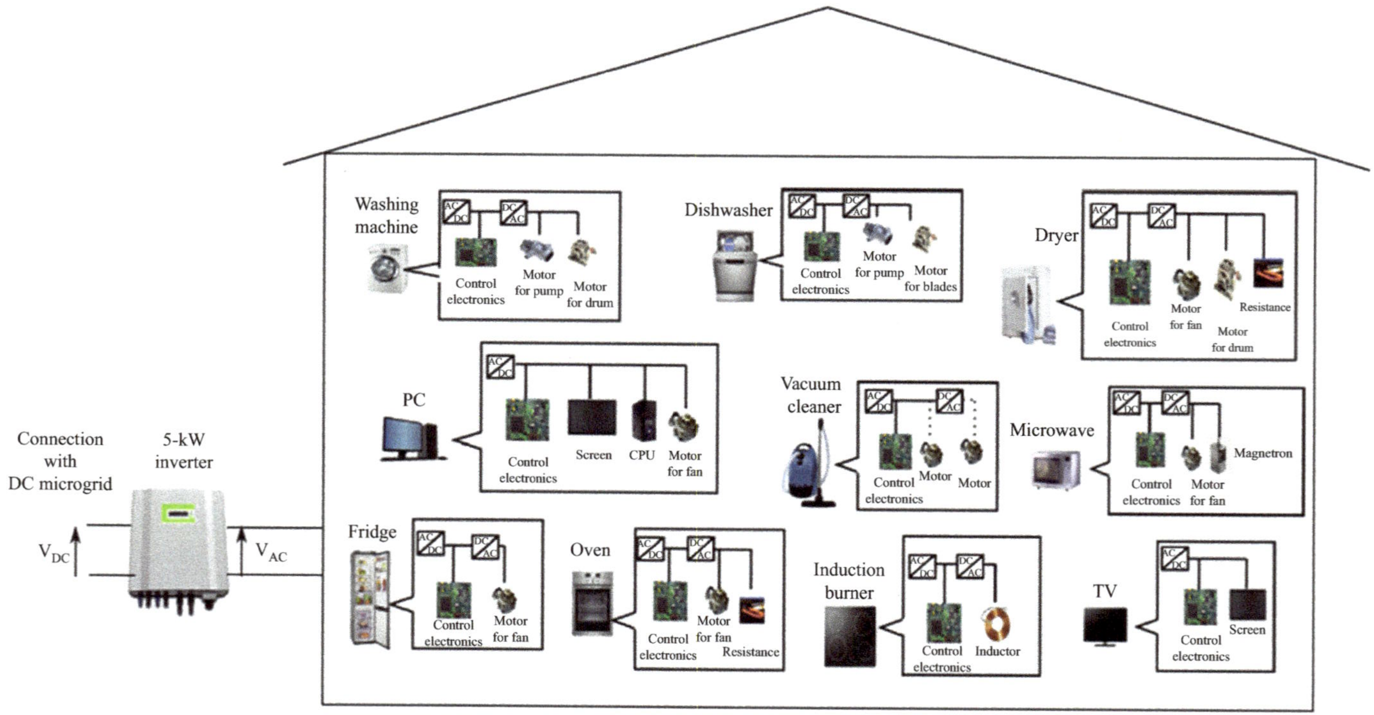

Figure 2.5 AC microgrid in a smart home fed by a DC network [32]

The reorganization of the distribution system into MMGs can determine significant benefits to the electric power consumers, with the opportunity of reducing the energy bill and increasing the perceived reliability, to the distribution system operator (DSO) that may postpone or avoid investments or activate new business and to the environment, thanks to the improved exploitation of RESs.

The advantages of the reconfiguration of distribution networks into interconnected MGs, autonomously operated, are summarized in the following points:

- *Deferment of investments*. The MGs take advantage of all benefits related to the energy sources properly placed in proximity of loads. In fact, in a distribution network, the production of energy close to the point of consumption can reduce power losses, relieve congestions and defer DSO capital expenditures (CAPEX).
- *Reliability and power quality*. The opportunity of using local generators and storage during network faults can improve the reliability. Intentional islanding, i.e., the use of DG to supply portions of distribution network during upstream line faults and scheduled interruptions, can be a valuable option to reduce the number of service interruptions. MGs can improve the power quality of the distribution system by exploiting local resources and demand flexibility.
- *Environmental impact*. The control of dispatchable DG and storage devices and the active participation of demand increase the penetration of RES. The local management makes the customers more aware of the importance of the smart use of energy. The result is an increase in self-consumption and an improvement in energy efficiency.

In particular, in rural areas the development of MMGs is very attractive, and specific planning methodologies that consider the opportunity to develop sustainable MMG have been proposed [21,27].

MMG allows resupplying MV and LV customers with a network reconfiguration that intentionally isolates some portions of the network with enough generation and distributed energy storage (DES). Generation and storage can be connected to the MV network or it can come from the private LV MGs already connected.

MG mostly operates interconnected to the main distribution grid but can operate also islanded in the case of external faults. Then, in a distribution scheme that allows the decomposition into small, autonomous portions of network like MGs, customers that are not directly involved in the fault can be supplied. In this case, the distribution system acts as a backbone that provides connectivity to several MGs.

In Figure 2.6 the architecture of the innovative smart distribution network with MMGs is represented. According to transformative architecture for the normal operation and self-healing of networked MGs, MGs can support and interchange electricity with each other in the proposed infrastructure. The networked MGs are connected to the main power distribution backbone with a designed communication network. In each MG, the MG energy management system (MEMS) schedules the MG operation and all MEMSs are globally optimized. In the normal operation mode, the objective is to schedule dispatchable DG, storage systems, and

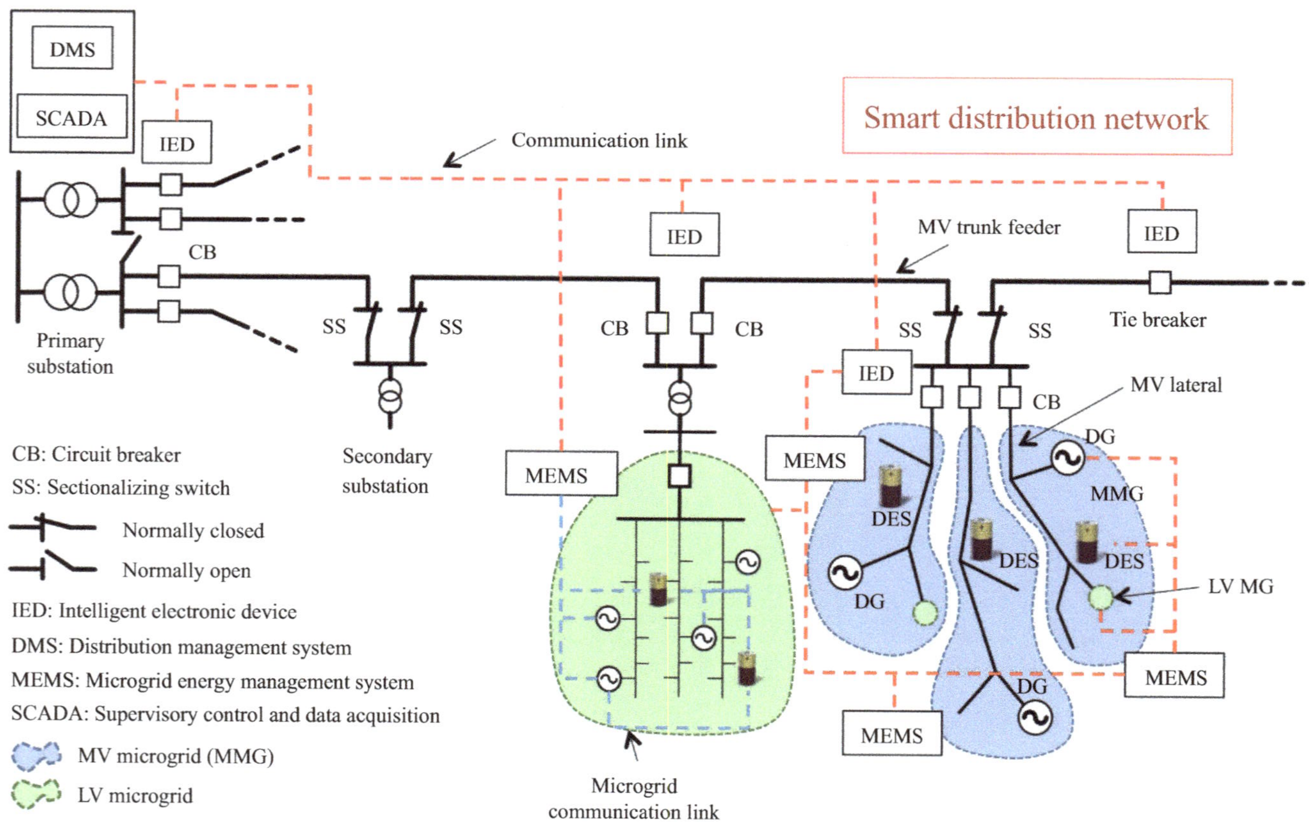

Figure 2.6 Future smart distribution network scenario with MMGs

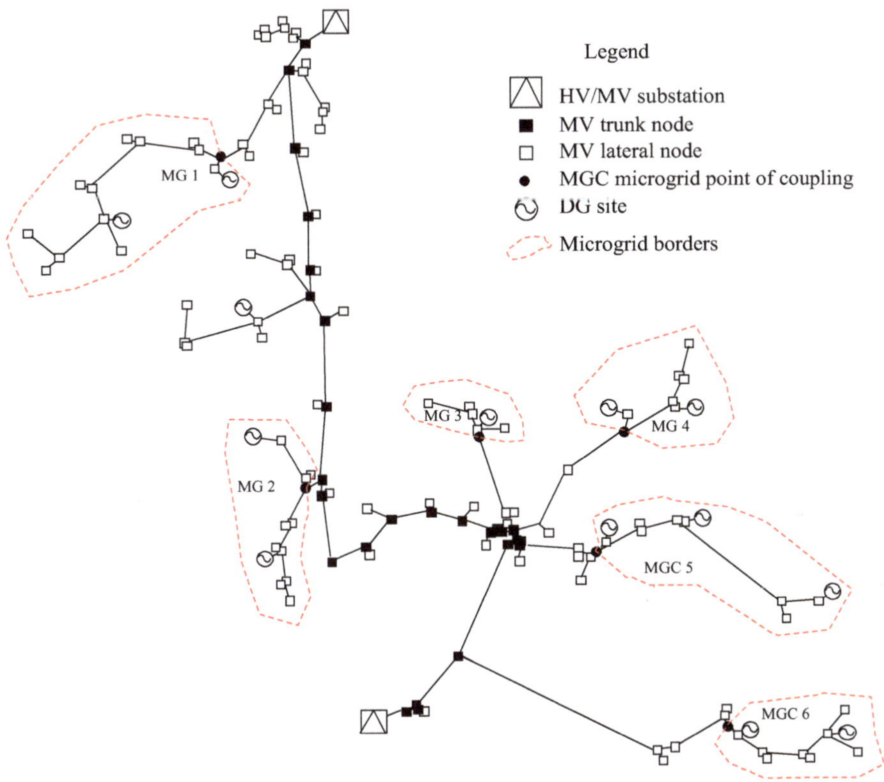

Figure 2.7 Reorganization of a power distribution system with MMGs

controllable loads to minimize the operation costs and maximize the supply adequacy of each MG. When a generation deficiency or fault happens in an MG, the architecture can switch to the self-healing mode and the local generation capacities of other MGs can be used to support the on-emergency portion of the system [28].

It can be expected that in the near future the distribution network will look like an MMG and mainly used as a backup infrastructure.

The need for rethinking the planning goals and methodologies is evident. For this reason, the development of methodologies capable of producing expansion plans for the MV distribution system with MMG can be very valuable. The planning methodology allows the distribution planner to find the optimal set of sustainable MGs that considers the availability of generators, lines and switching devices (Figure 2.7). Planning procedure may be useful to DSOs that strive to offer new high-level services to the customers supplied by their network. Indeed, DSOs can decide to favorite the MG to postpone network investments and to open new markets (the market of MG building and maintenance), without reducing the quality of service. The planning strategies may also be used by regulators to assess the savings achievable by customers operating in the free energy market through the MG aggregator and to define subsidies for distributors that facilitate the development of a new distribution system.

MMGs can offer significant benefits to the customers, thanks to the expected increase in the quality of service. Responsive loads and DESs play a significant role in MMGs, since flexible management of load and system operation lead to different advantages: the passing of the load from its passive state to an active one and increasing its capability to participate in demand side management programs, and the fast development of DESs for load and generation balancing.

Moreover, DSOs may also postpone or avoid investments. From the social and economic point of view, it contributes to the reduction of pollution, thanks to the improved integration of RES, and it pushes for the introduction of new business opportunities related to local market for ancillary services.

In order to make the MMG a viable planning alternative, costs and benefits have to be properly addressed by distinguishing those among the various players involved. In this way, when a development plan has to be prepared, different possible alternatives can be compared in the sake of optimal solutions.

The DSO is generally oriented to the maximum exploitation of existing assets and infrastructures by operating them close to their physical limits. The optimization process minimizes a multiobject function with CAPEX (e.g., for building new lines) and operating expenditures (OPEX) (e.g., cost of energy losses and cost of interruptions). The optimization is always limited by technical constraints (e.g., rated capacity of assets, maximum voltage drop and maximum allowable voltage) and topological constraints (e.g., radiality of networks in normal operation and redundancy of paths). Recently, probabilistic approaches have been proposed to deal with the growing riskiness of the distribution-planning problem [33].

In fact, with so much new DG being installed, the level of uncertainties that characterizes the planning environment increases particularly focused on the energy production from RES. The most advanced optimal planning techniques include the active operation of available resources in the list of actions when constraints are violated.

In order to define innovative expansion plans for the MV distribution system, novel network schemes are needed based on the coexistence of main feeders that interconnect sustainable MGs, using probabilistic approaches to take into account the variability of loads and DG (Figure 2.8). Generation and storage can be connected directly to the MV network or it can come from the private MGs already connected.

2.5 Review of MGs applications and research purposes

Several real-operating and laboratory MGs have been developed around the world for a variety of applications and research purposes. Most of these MGs can operate in the grid-connected mode only such as smart polygeneration MG built at the University of Genova, Italy [20], the KEPRI MG located in Korea [14,34] and the Kyoto eco-energy virtual MG [35]. These facilities provide useful testbeds for developing and testing optimal dispatching strategies.

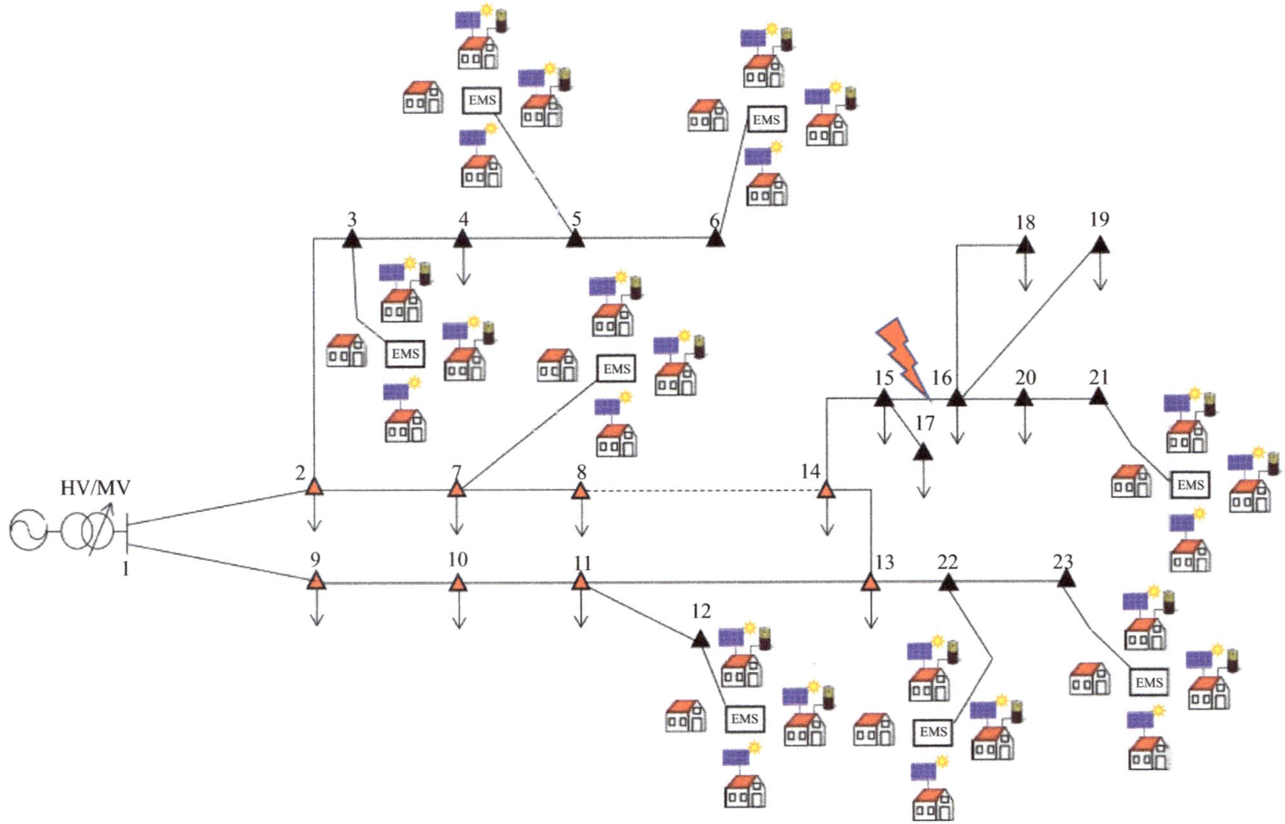

Figure 2.8 Planning of a power distribution system with MMGs

Other MGs [34,36,37] can operate only in the island mode and they have been mainly used to develop proper control functions able to ensure the reliable supply of electricity.

More specifically, the objective of Kythnos island MG [36,37] was to test centralized as well as decentralized control strategies for autonomous systems. Additionally, it also provides a good testbed facility to develop new protection schemes and investigate on voltage and frequency stability. The main goal of the Huatacondo MG [36] was to develop optimal dispatching strategies aimed at ensuring the economic, reliable and secure operation of an isolated MG. The Continuon's MV/LV [34] is instead used to develop control strategies able to regulate the frequency and the voltage of an isolated system.

The Bornholm MG built in Denmark [34,36] provides a powerful testbed for developing resynchronization control strategies. Moreover, it is also adopted to set up off-grid black start procedures.

Many other MGs can be operated in either grid-connected or islanded modes, with or without power interruption. For instance, the MGs developed at the University of Manchester [38] and at the Bronsbergen Holiday Park [19] focused mainly on the development of new control strategies that are able to stabilize the MG during the nonbumpless islanding transition.

The CERTS MG also provides a good testbed facility for developing and testing new control strategies to stabilize the MG during its transition from the on-grid to the off-grid operation and vice versa. Additionally, it is also focused on the development of new protection schemes for MGs.

The two MV MGs built in Canada: the Hydro Quebec and the BC Hydro MGs [39,40] provide useful testbeds for developing and testing new control strategies able to ensure a bumpless transition from the on-grid to the off-grid modes.

Also, the MG developed at the National Technical University of Athens (NTUA) in Greece focused on developing new bumpless islanding procedures to activate when a planned disconnection from the main grid is going to happen. Moreover, the NTUA MG also provides a useful testbed for developing optimal black start procedures.

The PrInCE Lab experimental MG built at the Polytechnic University of Bari in Italy provides a powerful testbed to develop and test new bumpless as well as non-bumpless islanding procedures [41]. It was also focused on developing new reserve management control strategies for ensuring MG stability when it is operating in isolated mode [42]. Additionally, the PrInCE Lab MG can be also adopted for developing new optimal dispatching strategies able to guarantee the economic, reliable and secure operation of the MG in both grid-connected and islanded modes [9].

2.6 Conclusion

The purpose of this chapter is to provide a general description of MGs, beginning with a historical overview of their evolution. Initially, the factors that have given an impetus to the development of such systems will be analyzed. Then, the basic characteristics and structures of MGs will be identified by means of an insightful literature review of the existing forms of these systems.

References

[1] N.W.A. Lidula, and A.D. Rajapakse, MGs research: A review of experimental MGs and test systems. Renewable and Sustainable Energy Reviews, Vol. 15, No. 1, pp. 186–202. 2011.

[2] T.S. Ustun, C. Ozansoy, and A. Zayegh, Recent developments in MGs and example cases around the world—A review. Renewable and Sustainable Energy Reviews, Vol. 15, No. 8, pp. 4030–4041. 2011.

[3] L. Mariam, M. Basu, and M.F. Conlon, A review of existing MG architectures. Journal of Engineering, Vol. 2013, pp. 1–8. 2013. doi:10.1155/2013/937614.

[4] M. Agrawal, and A. Mittal, Deployment of MGs in the developed countries: An appraisal. International Journal of Electrical and Electronics Engineering Research (IJEEER), Vol. 4, No. 3, pp. 23–34. 2014.

[5] L.I. Qing, X.U. Zhao, and Y. Li, Recent advancements on the development of MGs. Journal of Modern Power Systems and Clean Energy, Vol. 2, No. 3, pp. 206–211. 2014.

[6] E. Planas, J. Andreu, J.I. Gárate, I.M. de Alegría, and E. Ibarra, AC and DC technology in MGs: A review. Renewable and Sustainable Energy Reviews, Vol. 43, pp. 726–749. 2015.

[7] W. Eshetu, P. Sharma, and C. Sharma, ANFIS based load frequency control in an isolated micro grid. In 2018 IEEE International Conference on Industrial Technology (ICIT), pp. 1165–1170. IEEE. February 2018.

[8] I. Patrao, E. Figueres, G. Garcerá, and R. González-Medina, MG architectures for low voltage distributed generation. Renewable and Sustainable Energy Reviews, Vol. 43, pp. 415–424. 2015.

[9] A. Cagnano, E. De Tuglie, and L. Cicognani, Prince—Electrical energy systems lab: A pilot project for smart MGs. Electric Power Systems Research, Vol. 148, pp. 10–17. 2017.

[10] B. Kroposki, R. Lasseter, T. Ise, S. Morozumi, S. Papatlianassiou, and N. Hatziargyriou, Making MGs work. IEEE Power and Energy Magazine, Vol. 6, No. 3, pp. 54–65. 2008.

[11] J. Eto, R Lasseter, B. Schenkman, *et al.*, Overview of the CERTS microgrid laboratory test bed. In Integration of Wide-Scale Renewable Resources Into the Power Delivery System, 2009 CIGRE/IEEE PES Joint Symposium, pp. 1–7. IEEE, Calgary, AB. July 2009.

[12] R.H. Lasseter, J.H. Eto, B. Schenkman, *et al.*, CERTS Microgrid laboratory test bed. IEEE Transactions on Power Delivery, Vol. 26, No. 1, pp. 325–332. 2011.

[13] N. Hatziargyriou, (Ed.) Microgrids: Architectures and Control. Hoboken, NJ: Wiley/IEEE Press, 2014.

[14] E. Hossain, E. Kabalci, R. Bayindir, and R. Perez, MG testbeds around the world: State of art. Energy Conversion and Management, Vol. 86, pp. 132–153. 2014.

[15] S. Elsaiah, M. Benidris, and J. Mitra. A method for reliability improvement of microgrids. In Proceedings of 2016 Power Systems Computation Conference (PSCC), pp. 1–7. IEEE. June 2016.

[16] M. Barnes, A. Dimeas, A. Engler, *et al.*, Microgrid laboratory facilities. In IEEE 2005 International Conference on Future Power System, pp. 1–6. IEEE, Amsterdam. November 2005.

[17] T. Loix, and K.U. Leuven, The residential micro grid of Am Steinweg in Stutensee, Germany. Available online on February 2009. http://www.leonardo-energy.org/sites/leonardo-energy/files/root/pdf/2009/article3.pdf.

[18] T. Loix, and K.U. Leuven, The first micro grid in the Netherlands: Bronsbergen. Available online on February 2009. http://www.leonardo-energy.org/sites/leonardo-energy/files/root/pdf/2009/article2.pdf.

[19] S. Bracco, F. Delfino, F. Pampararo, M. Robba, and M. Rossi, The University of Genoa smart polygeneration MG test-bed facility: The overall system, the technologies and the research challenges. Renewable and Sustainable Energy Reviews, Vol. 18, pp. 442–459. 2013.

[20] M. Petrollese, L. Valverde, D. Cocco, G. Cau, and J. Guerra, Real-time integration of optimal generation scheduling with MPC for the energy management of a renewable hydrogen-based MG. Applied Energy, Vol. 166, pp. 96–106. 2016.

[21] J. Vasiljevska, J.A. Peças Lopes, and M.A. Matos, Evaluating the impacts of the multi-MG concept using multicriteria decision aid. Electric Power Systems Research, Vol. 91. pp. 44–51. 2012.

[22] F. Li, R. Li, and F. Zhou. Microgrid Technology and Engineering Application. Amsterdam: Elsevier, 2015.

[23] I Patrao, E. Figueres, G. Garcerá, and R. González-Medina. Microgrid architectures for low voltage distributed generation. Renewable and Sustainable Energy Reviews, Vol. 43. pp. 415–424. 2015.

[24] A.T. Elsayed, A.A. Mohamed, and O.A. Mohammed, DC microgrids and distribution systems: An overview. Electric Power Systems Research, Vol. 119, pp. 407–417. 2015.

[25] D.J. Hammerstrom, AC versus DC distribution systems: Did we get it right? In 2007 IEEE Power Engineering Society General Meeting, pp. 1–5. IEEE. 2007.

[26] M.R. Starke, L.M. Tolbert, and B. Ozpineci. AC vs. DC distribution: a loss comparison. 2008 IEEE/PES Transmission and Distribution Conference and Exposition (T&D), pp. 1–7. IEEE. 2008.

[27] E. Ghiani, S. Mocci, and F. Pilo, Optimal reconfiguration of distribution networks according to the MG paradigm. In 2005 IEEE International Conference on Future Power Systems. IEEE, Amsterdam. 16–18 November 2005.

[28] Z. Wang, B. Chen, J. Wang, and C. Chen, Networked MGs for self-healing power systems. IEEE Transactions on Smart Grid, Vol. 7, No. 1, pp. 310–319. 2016.

[29] G. Celli, E. Ghiani, S. Mocci, F. Pilo, G.G. Soma, and C. Vertuccio, Probabilistic planning of multi-MGs with optimal hybrid multi-generation sets. In 46th International Conference on Large High Voltage Electric Systems 2016, Paris. 21–26 August 2016.

[30] G. Celli, E. Ghiani, S. Mocci, and F. Pilo, Distributed generation and intentional islanding: effects on reliability in active networks. In Proc. of CIRED 2005-18th International Conference and Exhibition on Electricity Distribution, Turin, Italy, pp. 1–5. 6–9 June 2005.

[31] G. Celli, S. Mocci, F. Pilo, and G.G. Soma, Multi-MGs for innovative distribution networks in rural areas. In Proceedings of the Mediterranean Conference on Power Generation, Transmission, Distribution and Energy Conversion (MedPower 2016), Belgrade. 6–9 November 2016.

[32] E. Planas, J. Andreu, J. Ignacio Gárate, I. Martínez de Alegría, and E. Ibarra, AC and DC technology in microgrids: A review. Renewable and Sustainable Energy Reviews, Vol. 43, pp. 726–749. 2015.

[33] A.A. Khan, M. Naeem, M. Iqbal, S. Qaisar, and A. Anpalagan, A compendium of optimization objectives, constraints, tools and algorithms for energy management in MGs. Renewable and Sustainable Energy Reviews, Vol. 58. 2016.

[34] M. Qu, C. Marnay, and N. Zhou, MG policy review of selected major countries, regions, and organizations. Final report, Ernest Orlando Lawrence, Berkeley National Laboratory. pp. 1664–1683. 2014.

[35] S. Morozumi, H. Nakama, and N. Inoue, Demonstration projects for grid-connection issues in Japan. Elektrotechnik and Informationstechnik, Vol. 125, No. 12, pp. 426–431. 2008.

[36] R. Palma-Behnke, D. Ortiz, L. Reyes, G. Jimenez-Estevez, and N. Garrido, A social SCADA approach for a renewable based MG—The Huatacondo project. In 2011 IEEE Power and Energy Society General Meeting, Detroit, MI, USA, USA. pp. 1–7. IEEE. July 2011.

[37] C. Marnay, N. Zhou, M. Qu, and J. Romankiewicz, International Microgrid Assessment: Governance, INcentives and Experience (IMAGINE). Berkeley, CA, USA: China Energy Group, Environmental Energy Technologies Division, Lawrence Berkeley National Laboratory. 2014.

[38] G. Kyriakarakos, D.D. Piromalis, A.I. Dounis, K.G. Arvanitis, and G. Papadakis, Intelligent demand side energy management system for autonomous polygeneration MGs. Applied Energy, Vol. 103, pp. 39–51. 2013.

[39] F. Katiraei, C. Abbey, S. Tang, and M. Gauthier, Planned islanding on rural feeders—utility perspective. In Power and Energy Society General Meeting-Conversion and Delivery of Electrical Energy in the 21st Century, Pittsburgh, PA, pp. 1–6. IEEE. July 2008.

[40] R.H. Lasseter, Microgrids. In Conference Proceedings 2002 IEEE Power Engineering Society Winter Meeting, New York, NY, pp. 305–308. IEEE. 2002.

[41] A. Cagnano, E. De Tuglie, M. Trovato, L. Cicognani, and V. Vona, A simple circuit model for the islanding transition of microgrids. In 2016 IEEE 2nd International Forum on Research and Technologies for Society and Industry Leveraging a better tomorrow (RTSI), Bologna, pp. 1–6. IEEE. Sept. 2016.

[42] A. Cagnano, A. Caldarulo Bugliari, and E. De Tuglie, A cooperative control for the reserve management of isolated MGs. Applied Energy, Vol. 218, pp. 256–265. 2018.

Chapter 3

The microgrid investment and planning in rural locations

Miguel Heleno[1], Carman Bas Domenech[2], Gonçalo Cardoso[1] and Salman Mashayekh[3]

This chapter presents different methods and tools for microgrid optimal investment and planning problem, focusing on specific methodological aspects addressing the challenges of rural microgrids design. In particular, three aspects of rural microgrids planning are analyzed: (1) the multi-energy nature of rural microgrids, where electricity coexists with other energy vectors (such as heat distribution); (2) the occupation of large portions of the rural territory, which requires planning methods to consider the microgrid internal network constraints; (3) the remote (and sometimes off-the-grid) locations of rural microgrids, which require security criteria and multi-objective approaches to be considered in planning problem. These three methodological aspects are discussed using the example of a real microgrid in Alaska.

3.1 Introduction

A growing global interest in reducing environmental impacts, expanding access, and improving resiliency and reliability of energy systems is driving the integration of distributed energy resources (DER) in otherwise centralized energy networks. Microgrids in urban environments, defined as dense clusters of interconnected electricity loads and DER that present themselves to the larger electric utility as single, flexible, and controllable entities, have emerged as an advanced solution for DER integration and are drawing increasing interest from the power industry.

In contrast, rural microgrids have emerged a long time ago from the practical need to increase energy access and reduce energy costs in small-scale rural communities, where the main utility grid was unable to provide secure and reliable access to power. Rudimental forms of what we call today microgrids have been

[1]Energy Storage and Distributed Resources Division, Lawrence Berkeley National Laboratory, Berkeley, California, USA
[2]Department of Electrical and Electronic Engineering, University of Melbourne, Parkville, Australia
[3]Globality, Inc., Menlo Park, California, USA

installed during the last decades in rural areas across the globe to power isolated or weakly connected systems, typically supplying significant amount of load distributed throughout large portions of the territory. In these remote and semiremote rural environments, microgrids are intrinsically multi-energy, linking electricity generation and consumption with other energy vectors, such as heat and gas.

Due to their inherent decentralized characteristics based on local energy resources, an ability to island from the grid, rural microgrids can contribute to more resilient, reliable, secure, cost-effective, and environment-friendly access to energy and can be a valuable resource to the hosting infrastructure by providing utility grid supportive services precisely in the geographical areas where typically the distribution systems face more challenges. However, taking full advantage of the microgrid value chain requires a diversified energy infrastructure and coordination between multiple energy vectors. Multi-energy microgrids are complex energy systems that tie together electrical, heating and cooling energy flows, and this complexity introduces significant challenges to the investment and planning of rural microgrids.

In this chapter, we presented different methods and tools for optimal microgrid investment and planning problem and focus our attention on the specific aspects of the optimization formulation that addresses the challenges of rural microgrids design. This chapter ends with an application of the methodology to a real remote microgrid in Alaska.

3.2 Microgrid planning problem

3.2.1 Microgrid investment planning tools and methods

The main challenge of microgrids investment problem is to determine the optimal size and location of DER within a microgrid in order to supply the microgrid loads in a cost-effective and reliable manner.

The key inputs of microgrid planning problems include electric, heating, and cooling end-use customer loads; techno-economic data of DER technologies (including capital costs, operation, and maintenance costs; electric efficiency; heat-to-power ratio; sprint capacity; maximum operating hours among others); weather and renewable sources data (such as ambient temperature, solar radiation, and wind speed); investment constraints (capital investment limits, minimum internal rate of return, maximum payback period, etc.); in interconnected microgrids, another important component of operational costs is related to the utility tariffs, including electric and natural gas prices, peak charges, and interconnected costs.

The main outputs include the optimal DER investment portfolio, the optimal sizing of each DER, the optimal dispatch of all DER present in the solution, including any load management decisions such as load-shifting, peak-shaving, or load prioritized curtailments in the event of outages. Additional outputs may include extensive information on site-wide costs, energy consumption, and greenhouse gas emissions.

Thus, the main goal of several microgrid investment models is to find the optimal combination of technology portfolio, sizing (sometimes placement), and operation to supply all energy services required by the site under consideration, while optimizing the electric and heat energy flows to minimize overall investment and operational costs and/or other alternative objectives, such as minimizing carbon dioxide (CO_2) emissions.

Significant efforts have been carried out to address the problem of optimal sizing in multi-energy microgrids. Among the predominant tools and methods, simulation-based [1–5] and optimization-based approaches [6–12] emerge as the most common. Discussions and review on different methods used in optimal microgrid sizing have been presented in [12–15]. Key takeaways are that simulation-based methods provide fast solutions and straightforward implementations that potentially capture nonlinear behaviors in detail but do not guarantee optimality and rely heavily on user input and prior knowledge of candidate solutions. Furthermore, simulation methods can typically require separate algorithms to address different sizing objectives, and modeling topologies with multiple nodes may not be possible. *HOMER* [2,3] and *RETScreen* [4] are common examples of publicly available microgrid sizing tools that rely on simulation techniques.

Conversely, optimization algorithms aim to guarantee optimality and require low-user intervention to define the space of feasible solutions. However, optimization models can become high dimensional requiring heavy computational power and memory allocation or become intractable. Moreover, their classification and characteristics are dictated by the nature of the formulation used, with a fundamental distinction being made with regards to linearity. Specifically, the two most common optimization approaches for DER and microgrid sizing problems are Mixed Integer Linear Programming (MILP) [6,7,9,10] and Mixed Integer Nonlinear Programming models [11,12]. Finally, optimization approaches by heuristic and metaheuristics methods are becoming popular, especially for the problem of sitting and sizing of DER into microgrids. Metaheuristics, in particular, have increased significantly in the past few years, as they offer methods to address nonlinear formulations in a reasonable computational time [16]. Particle swarm optimization has become very popular metaheuristic method to solve a different kind of planning problems [17].

Linear and mixed integer linear models guarantee optimality and tend to solve quickly but require approximations to describe nonlinear effects. On the other hand, nonlinear, mixed integer nonlinear models, and metaheuristic approaches may capture additional detail in nonlinear phenomena at the cost of not guaranteeing optimality. Moreover, these nonlinear methods may require intensive computational resources and significant time to solve, especially when a large number of investment technologies are considered in the microgrid investment and planning problem. Therefore, MILP methods tend to be more flexible, being adequate for different types of microgrid investment problems and capable of dealing with diverse sizes of investment portfolio within a reasonable time, while guaranteeing optimality. For instance, a MILP formulation is behind two of the most complete microgrid design and investment planning tools currently available to the public: *Distributed Energy Resources Customer Adoption Model (DER-CAM)* and the *Microgrid Design Toolkit* (which shares some of the DER-CAM formulation).

Thus, the rest of this chapter will be dedicated to the application of MILP formulations in the context of investment and planning of rural microgrids.

3.3 Basic formulation

The simplest approaches to the microgrid investment planning problem are to represent the microgrid infrastructure in a single node, where all the generation and loads are connected. This approach minimizes the investment and operation costs by selecting an optimal technology portfolio DERs and choosing the most adequate sizes at an aggregated level, without considering DER placement within the microgrid network. Under this representation, the formulation assumes a single bus with constant voltage in which all the loads of the microgrid are connected. This bus is also the point of common coupling with the utility grid, in scenarios of interconnected microgrids, where the power exchange with a main grid is measured.

The DER capacities to be installed from the different types of technologies considered in the investment problem can be modeled using either continuous or discrete variables. If a technology is available in small enough modules and capital costs can be represented by a linear cost function, it is generally assumed that this technology can be modeled as a continuous variable, which significantly decreases the computational time. That is the case, for example, of photovoltaic (PV) or electricity storages that usually have linear capacity costs (per kW or kW h) and are sold in relatively small modules, allowing to be represented as continuous variables without affecting the error of the microgrid investment solution.

The opposite case is referred to as discrete technologies that are modeled using discrete variables. This is the case of all diesel and gas fired generators and combined heat and power (CHP) as well as wind or hydroturbines that are generally sold in modules of significant kW capacity that do not allow treating them as continuous.

Equation (3.1) presents an objective function that minimizes the overall investment and operation costs of a microgrid, considering units of a discrete technology (d) with a maximum operating power and an investment cost. Also, it takes into account the investment of continuous technologies (k)—requiring a fixed cost for the installation (C_k^{Fx})—associated with a binary variable that represents the decision of investing (or not) in a certain technology. The amount of capacity to be installed is defined by the continuous variable and multiplied by the variable price per kW or kWh (C_k^{Vr}) of the technology. The operation costs considered include generation costs for each technology j, the curtailment of load costs for each kW of load curtailed, and the costs of buying electricity from the utility grid, in the case of interconnected microgrids:

$$C^{\text{Cost}} = \sum_d \text{ni}_d \cdot \overline{P}_d \cdot \text{InvCost}_d + \sum_k \left(C_k^{\text{Fx}} \cdot \text{pur}_k + C_k^{\text{Vr}} \cdot \text{Cap}_k\right) + \sum_{j,t} \text{Pg}_{j,t} \cdot \text{GC}_j + \sum_t \text{Pl}_t^{\text{Cur}} \cdot \text{CC} + \sum_t \text{Pex}_t \cdot \text{CTariff}_t \tag{3.1}$$

The optimization is subjected to a set of constraints, representing the different technical and operational limits. Equation (3.2) ensures the power balance of the microgrid, considering aggregated load in each time step, the generation of the different technologies, curtailment of loads, and the discharging and charging powers of the storage installed:

$$\sum_t \text{Pl}_t = \sum_{j,t} \text{Pg}_{j,t} + \sum_t \text{Pl}_t^{\text{Cur}} + S_t^{\text{Out}} \eta_s^D - \frac{S\text{In}_t}{\eta_s^C} \tag{3.2}$$

Equation (3.3) limits the generation capability of the renewable generation technologies based on the available primary resource. Equation (3.4) imposes technical limits to generation of discrete technologies, and (3.5) represents the option of the microgrid owner to limit the number of discrete technologies to be installed. Equations (3.6) and (3.7) constraint the generation limits of the continuous technologies, making them within the capacity:

$$\text{Pg}_{c,t} \leq \text{Cap}_r \cdot \text{AvRes}_{r,t} \tag{3.3}$$

$$\text{no}_{d,t} \cdot \underline{P}_d \leq \text{Pg}_{d,t} \leq \text{no}_{d,t} \cdot \overline{P}_d \tag{3.4}$$

$$\text{no}_{d,t} \leq \text{ni}_{d,t} \tag{3.5}$$

$$\text{Cap}_k \leq \text{pur}_k \cdot M \tag{3.6}$$

$$\text{Pg}_{r,t} \leq \text{Cap}_{n,r} \tag{3.7}$$

Equations (3.8)–(3.10) model the electrical storage technologies, using the well-known state of charge (SOC) reservoir model for the storage operations, taking into account self-discharge, the battery capacity line limits, and the discharging and charging rate limits. Regarding these storage equations, it is important to stress two aspects: first, as the efficiencies when charging and discharging are considered less than 100 percent (nonideal processes), there is no need to generate a binary variable to model the fact that charge and discharge cannot take place simultaneously: the optimization will inherently pick one "operation mode," decreasing the computing time; the second is the fact that the storage equations used are generic for the storage of other kinds of energy, for instance cooling or heating storage, which will be useful in further formulations of the microgrid design problem through this chapter:

$$\text{SOC}_{s,t} = (1 - \phi_s) \cdot \text{SOC}_{s,t-1} + S_{s,t}^{\text{In}} \cdot \Delta t - S_{s,t}^{\text{Out}} \cdot \Delta t \tag{3.8}$$

$$\underline{\text{SOC}}_s \leq \text{SOC}_{n,s,t} \leq \overline{\text{SOC}}_s \tag{3.9}$$

$$S_{s,t}^{\text{In}} \leq \text{Cap}_s \cdot \overline{\text{SRt}}_s^{\text{ch}}, S_{s,t}^{\text{Out}} \leq \text{Cap}_s \cdot \overline{\text{SRt}}_s^{\text{dch}} \tag{3.10}$$

3.4 Addressing the specific characteristics of rural microgrids

When planning a rural microgrid, key attributes need to be accounted to generate useful and adequate microgrid design. Specifically, this chapter will review three characteristics of rural microgrids that will require an expansion to the formulation presented above:

- Rural microgrids are intrinsically multi-energy, linking electricity generation, and consumption with other energy vectors. Unlike urban environments, where most of the thermal loads tend to be supplied by electricity (e.g., heating, ventilating, and air conditioning systems), in rural environments these loads are typically provided by small-scale gas, diesel, or biomass units. Thus, modeling the interplay between electrical and heat sources can capture efficiency gains that cannot be neglected in the microgrid investment and planning.
- Rural microgrids tend to have large portions of load dispersed throughout the territory. Thus, the microgrid optimal investment and planning problem should consider the impact of the dispatch on the voltage drops and the line limits of the internal microgrid network. Therefore, a multi-node formulation of the microgrid planning problem should be adopted.
- A significant share of rural microgrids are supplying remote off-grid locations. In this case, the microgrid investment problem should ensure that the selected portfolio of DER is capable of supplying the load in the case of an outage of a generation unit. Moreover, the operation of an off-grid microgrid goes beyond the economic and reliability criteria. In fact, in isolated rural locations the microgrid operation responds to a variety of objectives related to specific social and environmental concerns of the community. Therefore, the microgrid planning in rural remote communities is intrinsically a multi-objective problem.

3.4.1 Multi-energy characteristic

The first characteristic of rural microgrids to be added to the formulation is the need to serve heating and cooling loads (CL) as well as electric, conforming to a multi-energy microgrid. Furthermore, accurately modeling a microgrid using a multi-energy approach can lead to better overall system efficiency as the excess heat can be considered. The heating balance assures the supply to heating loads (HT) and the heat needed for the absorption chiller (AC), with a coefficient of performance of CP^a. To do so, the system will use heating generation (HG) such as gas-fired boilers, useful heat recovered from CHP technologies of electric generation (EG), modeled using α_j, and heat storage (HS). On the other hand, the cooling balance assures supply to CL using cooling generation, e.g., ACs or electrical chillers (EC) and cooling storage (CS).

$$\begin{aligned}&\textstyle\sum_{j\in \mathrm{HG}}\mathrm{Pg}_{j,t}-\sum_{j\in \mathrm{EG}}\alpha_j\cdot \mathrm{Pg}_{j,t}-\mathrm{Pl}_{\mathrm{HT},t}-\frac{1}{\mathrm{CP}^a}\cdot \mathrm{Pg}_{j=\mathrm{AC},t}\\&-\frac{1}{\eta^{\mathrm{ch}}_{s=\mathrm{HS}}}\cdot S^{\mathrm{In}}_{s=\mathrm{HS},t}+\eta^{\mathrm{dch}}_{s=\mathrm{HS}}\cdot S^{\mathrm{Out}}_{s=\mathrm{HS},t}=0\end{aligned} \tag{3.11}$$

$$\sum\nolimits_{c\in CG}\mathrm{Pg}_{c,t}-\mathrm{Pl}_{\mathrm{CL},t}-\frac{1}{\eta^{\mathrm{ch}}_{s=\mathrm{CS}}}\cdot S^{\mathrm{In}}_{s=\mathrm{CS},t}+\eta^{\mathrm{dch}}_{s=\mathrm{HS}}\cdot S^{\mathrm{Out}}_{s=\mathrm{CS},t}=0 \tag{3.12}$$

3.4.2 *Multi-node characteristic*

As mentioned above, a single-node representation of a microgrid in the investment and planning problem might not be adequate to model atypical rural infrastructure with large distances between the nodes. In these particular applications, the topology of the existing energy network (heat and electricity) can be taken into account to allow a representation of the microgrid internal voltage drops and line security limits in the investment and planning model. Furthermore, modeling the internal microgrid network allows one to study the problem of optimally placing the DER, ensuring technical feasibility during operation (e.g., meeting voltage limits or cable capacities), which cannot be assured in a single-node formulation.

3.4.2.1 Heat network

When taking into account a multi-node network in a multi-energy microgrid, a set of equations is needed to model the internal thermal network and the energy flows between n and n' nodes. To account for an arbitrary piping network, (3.11) is replaced by (3.13) that models the heat transfer between nodes n and n' taking into account heat losses, using the coefficient $\gamma_{n,n'}$. On the other hand, (3.14) imposes heat pipes capacities $\overline{H_{n,n'}}$

$$\begin{aligned}0=&\textstyle\sum_{j\in \mathrm{HG}}\mathrm{Pg}_{n,j,t}+\sum_{j\in \mathrm{EG}}\alpha_j\cdot \mathrm{Pg}_{n,j,t}-\mathrm{Pl}_{n,\mathrm{HT},t}-\frac{1}{\mathrm{CP}^a}\cdot \mathrm{Pg}_{n,j=\mathrm{AC},t}\\&-\frac{1}{\eta^{\mathrm{ch}}_{s=\mathrm{HS}}}\cdot S^{\mathrm{In}}_{n,s=\mathrm{HS},t}+\eta^{\mathrm{dch}}_{s=\mathrm{HS}}\cdot S^{\mathrm{Out}}_{n,s=\mathrm{HS},t}\\&-\textstyle\sum_{n'}H_{n,n',t}+\sum_{n'}\left(1-\gamma_{n,n'}\right)H_{n,n',t}\end{aligned} \tag{3.13}$$

$$0\le H_{n,n',t}\le \overline{H_{n,n'}} \tag{3.14}$$

3.4.2.2 Electricity network

A MILP formulation requires linear equations, whereas the network model or power flow constraints entail a nonlinear, nonconvex model. To solve this problem, a linearization model of the power flow equations, called *LinDistFlow*, is commonly used. This set of equations approximates the power flow constraints and allows modeling radial networks with voltage magnitude deviations between the nodes and considers line resistances and reactances as well as active and reactive power flows in the network.

Equation (3.15) takes the difference between square of voltage magnitude VSq on n and n' nodes, connected through a line with an impedance made of $r_{n,n'}$ and $x_{n,n'}$ with the following power flow: $P_{n,n',t}$, $Q_{n,n',t}$. Equation (3.16) sets a fixed voltage for the slack bus. Equations (3.17) and (3.18) ensure that the active $(\mathrm{Pi}_{n,t})$ and reactive $(\mathrm{Qi}_{n,t})$ power injected in a node n equals the power flow between that node and the rest of the buses in the microgrid. To transition from the electric balance for a single-node to a multi-node network, (3.12) needs to be replaced by (3.19) that takes into account the injection of a node and specifies electric chillers consumption. In (3.20) the voltage of each bus is limited, and (3.21) imposes a line capacity constraint, through a linear approximation. Instead of the real constraint that is modeled by a circle that relates apparent, reactive, and active power flow, in this approximation an octagon is used, as it can be seen in Figure 3.1.

$$\mathrm{VSq}_{n,t} - \mathrm{VSq}_{n',t} = 2\left(r_{n,n'} \cdot P_{n,n',t} + x_{n,n'} \cdot Q_{n,n',t}\right) \tag{3.15}$$

$$\mathrm{VSq}_{n=1,t} = V_0^2 \tag{3.16}$$

$$\mathrm{Pi}_{n,t} = \sum\nolimits_{n'} P_{n,n',t} \tag{3.17}$$

$$\mathrm{Qi}_{n,t} = \sum\nolimits_{n'} Q_{n,n',t} \tag{3.18}$$

$$\mathrm{Pi}_{n,t} = \sum\nolimits_{n,j,t} \mathrm{Pg}_{n,j,t} - \mathrm{Pl}_j - \frac{1}{\mathrm{CP}^e}\mathrm{Pg}_{n,\mathrm{EC},t} + S_{n,t}^{\mathrm{Out}}\eta_s^D - \frac{S_{n,t}^{\mathrm{In}}}{\eta_s^C} \tag{3.19}$$

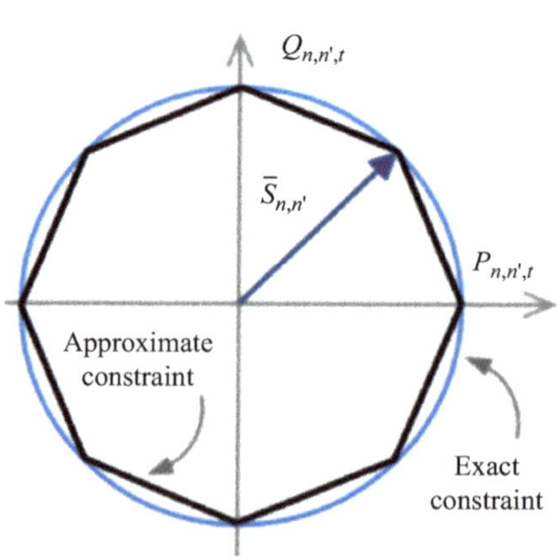

Figure 3.1 Approximation of line capacity constraints

$$\underline{V^2} \leq \mathrm{VS}_{n,t} \leq \overline{V^2} \tag{3.20}$$

$$\pm Q_{n,n',t} \leq \operatorname{cotan}\left(\left(\frac{1}{2} - \mathrm{e}\right)\frac{\pi}{4}\right) \cdot \left(P_{n,n',t} - \cos\left(e\frac{\pi}{4}\right) \cdot \overline{S}_{n,n'}\right) + \sin\left(e\frac{\pi}{4}\right) \cdot \overline{S}_{n,n'} \quad e \in \{1, \ldots, 4\} \tag{3.21}$$

3.4.2.3 Illustrative case study

The following example aims to demonstrate how the investment options in a microgrid design and planning problem can differ between single-node and multi-node modeling in a multi-energy microgrid. In order to do so, the case study is composed by a 12-kV microgrid representing an interconnected rural community in California with a total energy peak of 10 MW. The main loads are considered to be the community services (primary school, restaurant, and hospital) as well as the community warehouse. Their annual electrical, heating, and CL are listed in Table 3.1. The details of the microgrid can be found in Figure 3.2. For the electrical network, a cable with an impedance of $(64 + i1.4) \times 10^{-6}$ pu/m and ampacity of 0.4 pu was arbitrarily considered. For the heating network, pipes with thermal loss coefficient of $\gamma = 4 \times 10^{-5}$ (in %/m) and capacity of 3,000 $\mathrm{kW_{th}}$ were considered. Investments in PV, battery, CHP-enabled internal combustion engine (ICE), AC, gas-fired boiler, and electric chiller were allowed (characteristics in Tables 3.2 and 3.3).

To generate the single-node case study, building loads were aggregated and electrical and thermal networks were not considered. The DER portfolio and sizes that result from the optimization are at an aggregated level. The results of both carrying out a single-node analysis and a multi-node analysis of the same microgrid are gathered in Table 3.4.

To compare results, one has to consider multi-node aggregated values and the single-node values. The most noticeable differences can be found in the heating and cooling power installed. As (3.13) shows, the heating network modeling considers losses, whereas in a single-node approach those losses are neglected. Consequently,

Table 3.1 Building annual electrical, cooling, and heating loads

Node	Annual electrical load		Annual cooling load		Annual heating load	
	Energy (MWh)	**Max power (kW)**	**Energy (MWh)**	**Energy (MWh)**	**Energy (MWh)**	**Max power (kW)**
1	1,999.0	645.4	2,004.1	103.5	2,001.0	2,563.1
2	1,814.2	521.6	190.1	546.1	2,017.3	5,732.3
3	1,447.3	268.2	2,553.9	322.9	3,999.7	991.5
4	1,852.1	271.3	40.3	314.3	1,998.5	1,529.7
Aggregated	7,112.6	1,706.5	4,788.4	1,286.8	10,016.5	10,816.6

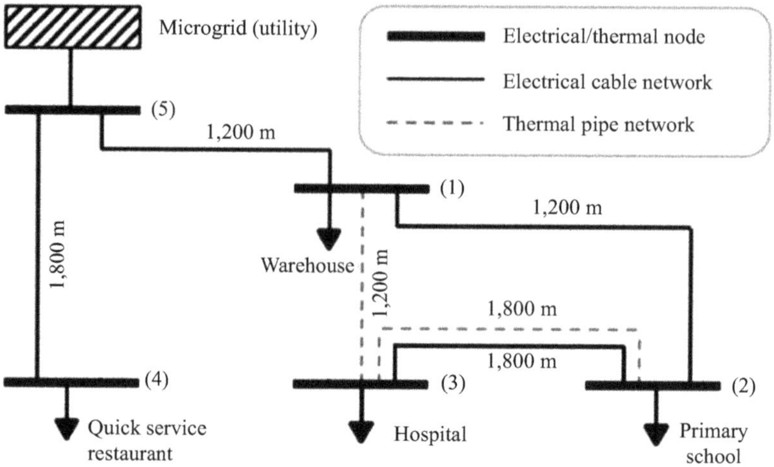

Figure 3.2 Electrical and thermal networks for the example five-node microgrid

Table 3.2 Discrete technology option characteristics

	Capacity (kW)	Lifetime (years)	Capital cost ($/kW)	Efficiency (%)	Heat recovery (kW/kW)
ICE-1	1,000	20	4,969	0.368	1.019
ICE-2	2,500	20	4,223	0.404	0.786
ICE-3	5,000	20	3,074	0.416	0.797

Table 3.3 Continuous technology option characteristics

Technology	Fixed cost ($)	Variable cost ($/kW or $/kWh)	Lifetime (years)
Battery	500	500	5
PV	2,500	2,500	30
Gas boiler	6,000	45	10
Electric chiller	2,300	230	10
Absorption chiller	250	250	20

the power installed needed to supply the same load decreases, which can lead to underestimating the microgrid design.

On the other hand, as the equations from *LinDistFlow* (3.15)–(3.21) show, the voltage and capacity line limitations are represented but the losses are not considered. This justifies the similar results between a single-node example and a

Table 3.4 Technology investment results for single node and multi-node

Node	Gas boiler (kW)	Abs chiller (kW)	PV (kW)	ICE–CHP (kW)	Electric chiller (kW)	Yearly utility purchase (kWh)
1	0	104	0	0	0	
2	7,213	415	0	1,000	131	
3	0	304	1,619		309	
4	1,538	221	0		93	
Multi-node—aggregated	8,751	1,044	1,619	1,000	533	82,801
Single node	8,515	929	1,588	1,000	374	80,518

Table 3.5 Investment required in single-node and multi-node case

Case	Investment cost (k$)
Single node	4,687
Multi-node	4,841 (+3.27%)

multi-node example for microgrid technologies. Recently a new method to approximate losses in the network while keeping a linearized approach was presented in [18]. Finally, regardless the loss representation, a multi-node network solves the problem of placement of technologies, whereas in a single-node approximation this problem, which is not straightforward, is not solved.

In Table 3.5, the differences between the investment costs required in both cases are shown, as expected, the multi-node microgrid results in higher investment costs, as it has more installed power.

3.4.3 Remote microgrids

Although the concept of microgrids is relatively new for well-connected utility customers, microgrids have been the only solution for rural and remote communities, which have comprised off-grid power systems for a long time, simply because such communities do not have access to the main grid. These off-grid communities, however, rely mostly on diesel generation to supply their loads, despite much higher fuel prices in the remote areas. Renewable-based microgrids are suitable replacements for diesel-based communities that are not environmentally friendly due to their carbon emissions. However, the design of renewable-based isolated microgrids in remote locations adds a layer of complexity compared to just rural microgrids. In this application, one of the main challenges is the need of the microgrid operator (whose functions are comparable to a utility) for a secure and reliable microgrid that can tolerate $N-1$ generator contingencies to guarantee that the microgrids are self-sufficient at all times when an outage occurs.

3.4.3.1 $N-1$ contingencies

The following formulation, extending the previous equations (3.1)–(3.19), presents a comprehensive microgrid investment and planning optimization that addresses power generation mix selection, resource sizing and allocation, and operation scheduling, simultaneously considering electricity, cooling, and heating loops in the microgrid to take full advantage of excess heat. In the context of remote microgrids, the contribution relies on considering $N-1$ security constraints in the design and operation of the microgrid as a MILP optimization that ensures that the system has enough online generation reserve to make up for the loss of any single unit, considering generator ramping constraints.

Figure 3.3 shows assumptions about timing of a contingency. When an outage happens, the remaining generators (and storage devices) are given some time (Δt^{ctgrmp}) to ramp up. After this period, the system generation must be enough to meet the loads. This temporary post-contingency dispatch must be sustainable for a period of (Δt^{ctg}). This period (Δt^{ctg}) gives the system operator enough time to determine the optimal dispatch for the post-contingency state of the system.

As has been mentioned when presenting the single-node formulation, generation technologies are modeled using continuous or discrete variables. Thus, the treatment of a contingency will be modeled differently depending on which technology causes the outage.

When considering a continuous technology outage at a node, its entire generation $Pg_{n,c,t}$ at the node will be lost. That is because it is assumed that the entire capacity $Cap_{n,c}$ is installed in one unit. In contrast, several units from a discrete technology can be installed at a node. Hence, only a portion (generation of a single unit) of the entire generation is lost in an outage. Only the total generation of a technology at a node is relevant for power flow modeling, but knowledge about generation of individual units is needed for contingency analysis. This can lead to a disaggregation of the generation from a technology $Pg_{n,g,t}$ into the maximum number of units that can be installed from a technology at a bus, which will impose

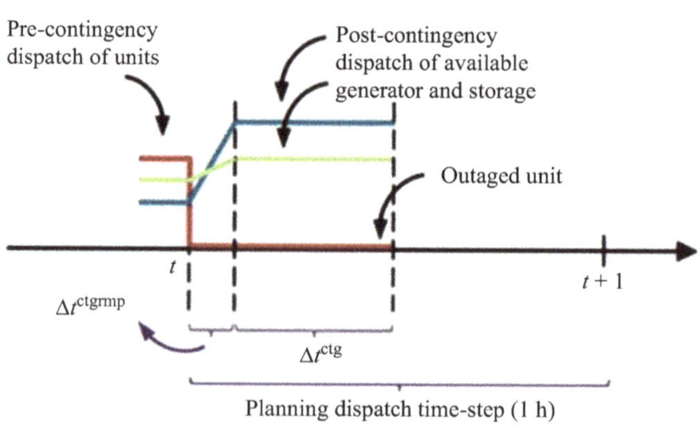

Figure 3.3 Timing assumptions for the contingency analysis

a significant computational burden on the solver. To address this challenge, the units from each discrete technology type (at a node) are classified into three categories and the aggregate generation of each category is tracked (as opposed to the generation of each unit).

The three categories are the units operating at the minimum load, the units operating at the maximum load and the units operating at partial load. This classification captures the trade-off between running the units at maximum load vs. minimum load. On the one hand, a generator providing maximum load has a higher efficiency. On the other hand, consideration of generator outages motivates the optimization to run more units at their minimum load, to increase the reserve in the system.

The following equations show how we disaggregate $\mathrm{Pg}_{n,g,t}$ and $\mathrm{no}_{n,g,t}$ into the number of units operating at the min load ($\mathrm{nm}_{n,g,t}$), the number of units operating at the max load ($\mathrm{nx}_{n,g,t}$), and generation power of the unit operating at a partial load ($\mathrm{PgPrt}_{n,g,t}$). Binary variables $\mathrm{bp}_{n,g,t}$ determine whether a partial load unit exists. To simplify the contingency constraints, it is assumed that a partial unit always exists ($\mathrm{bp}_{n,g,t} = 1$), if $\mathrm{no}_{n,g,t} > 0$.

$$\mathrm{no}_{n,g,t} = \mathrm{nm}_{n,g,t} + \mathrm{nx}_{n,g,t} + \mathrm{bp}_{n,g,t} \tag{3.22}$$

$$\mathrm{Pg}_{n,g,t} = \mathrm{PgPrt}_{n,g,t} + \mathrm{nm}_{n,g,t} \cdot \underline{P}_g + \mathrm{nx}_{n,g,t} \cdot \overline{P}_g \tag{3.23}$$

$$\mathrm{bp}_{n,g,t} \cdot \underline{P}_g \leq \mathrm{PgPrt}_{n,g,t} \leq \mathrm{bp}_{n,g,t} \cdot \overline{P}_g \tag{3.24}$$

$$\mathrm{bp}_{n,g,t} \cdot M \geq \mathrm{no}_{n,g,t} \tag{3.25}$$

Next, we determine the contribution of each DER to the post-contingency state of the system. The maximum output change for minimum load units after a contingency happens, $\Delta\mathrm{PgMin}_{n,g,t}$, is limited by the unused capacity of the units and their maximum rate of change in generation $\overline{P_g R_g}$. Similarly, the maximum contribution to the reserve of a partial load unit, $\Delta\mathrm{PgPrt}_{n,g,t}$, is limited by its unused capacity in (3.28) and the maximum rate of change in (3.29). Maximum load units cannot increase their generations if a contingency happens. The output change for a battery unit is limited by its maximum ramp rate, $\overline{\mathrm{SRt}}^{\mathrm{dch}}_{s=\mathrm{ES}}$, according to (3.30), and it must be able to sustain its post-contingency output for Δt^{ctg} as shown in (3.31).

$$\Delta\mathrm{PgMin}_{n,g,t} \leq \left(\overline{P}_g - \underline{P}_g\right) \cdot \mathrm{nm}_{n,g,t} \tag{3.26}$$

$$\Delta\mathrm{PgMin}_{n,g,t} \leq \overline{P_g R_g} \cdot \overline{P}_g \cdot \Delta t^{\mathrm{ctgrmp}} \cdot \mathrm{nm}_{n,g,t} \tag{3.27}$$

$$\Delta\mathrm{PgPrt}_{n,g,t} \leq \overline{P}_g - \mathrm{PgPrt}_{n,g,t} \tag{3.28}$$

$$\Delta\mathrm{PgPrt}_{n,g,t} \leq \overline{P_g R_g} \cdot \overline{P}_g \cdot \Delta t^{\mathrm{ctgrmp}} \cdot \mathrm{bp}_{n,g,t} \tag{3.29}$$

$$\Delta S^{\mathrm{Out}}_{n,s=ES,t} \leq \overline{\mathrm{SRt}}^{\mathrm{dch}}_{s=\mathrm{ES}} \cdot \mathrm{Cap}_{n,s=\mathrm{ES}} - S^{\mathrm{Out}}_{n,s=\mathrm{ES},t} \tag{3.30}$$

$$\Delta t^{\mathrm{ctg}} \cdot \left(\Delta S^{\mathrm{Out}}_{n,s=\mathrm{ES},t} + S^{\mathrm{Out}}_{n,s=\mathrm{ES},t} \right) \leq \mathrm{SOC}_{n,s=\mathrm{ES},t} \tag{3.31}$$

The dynamic allocation of system reserve can offer a less conservative and more economical solution than traditional approaches. To this end, the following set of equations dynamically allocates reserve in the system depending on the generation of different technologies, storage system SOC, technology ramping constraints, etc. For each outage, the pre-contingency generation power of the outaged unit must be less than the available reserve in the system. The available reserve is composed of the generation increase in discrete technologies (minimum load units and partial load unit), the increase in storage output, and finally the post-contingency load curtailment, including electric chiller load and storage charging load (since they will not be met during the contingency period).

The reserve capacity in the system must be larger than the generation of each continuous unit at each bus n at any given time t, where all of the system discrete generators and batteries can contribute to the system reserve. Post-contingency electrical load curtailments $\mathrm{PlCur}_{n',t}$ as well as pre-contingency storage charging and shedding of electric chiller loads can increase the system reserve. Equation (3.33) imposes a similar constraint considering the outage of the storage system at any bus n. Note that all of the system batteries, except for the battery whose outage is being constrained (hence, the negative term in the reserve calculation), can contribute to the reserve:

$$\begin{aligned} \mathrm{Pg}_{n,c=\mathrm{PV},t} \leq {} & \sum\nolimits_{n',g,t} \left(\Delta\mathrm{PgMin}_{n',g,t} + \Delta\mathrm{PgPrt}_{n',g,t} \right) + \sum\nolimits_{n'} \Delta S^{\mathrm{Out}}_{n',s=\mathrm{ES},t} \\ & + \sum\nolimits_{n'} \mathrm{PlCur}_{n',t} + \sum\nolimits_{n'} S^{\mathrm{In}}_{n',s=\mathrm{ES},t} + \sum\nolimits_{n'} \frac{1}{\mathrm{CP}^{e}} \cdot \mathrm{Pg}_{n',c=\mathrm{EC},t} \end{aligned} \tag{3.32}$$

$$\begin{aligned} S^{\mathrm{Out}}_{n,s=\mathrm{ES},t} \leq {} & \sum\nolimits_{n',g,t} \left(\Delta\mathrm{PgMin}_{n',g,t} + \Delta\mathrm{PgPrt}_{n',g,t} \right) \\ & + \sum\nolimits_{n'} \Delta S^{\mathrm{Out}}_{n',s=\mathrm{ES},t} - \Delta S^{\mathrm{Out}}_{n,s=\mathrm{ES},t} \\ & + \sum\nolimits_{n'} \mathrm{PlCur}_{n',t} + \sum\nolimits_{n'} S^{\mathrm{In}}_{n',s=\mathrm{ES},t} + \sum\nolimits_{n'} \frac{1}{\mathrm{CP}^{e}} \cdot \mathrm{Pg}_{n',c=\mathrm{EC},t} \end{aligned} \tag{3.33}$$

$$\begin{aligned} \mathrm{PgLst}_{n,g,t} \leq {} & \sum\nolimits_{n',g',t} \left(\mathrm{PgMin}_{n',g',t} + \Delta\mathrm{PgPrt}_{n',g',t} \right) \\ & - \Delta\mathrm{PgPrtLst}_{n,g,t} + \sum\nolimits_{n'} \Delta S^{\mathrm{Out}}_{n',s=\mathrm{ES},t} \\ & + \sum\nolimits_{n'} \mathrm{PlCur}_{n',t} + \sum\nolimits_{n'} S^{\mathrm{In}}_{n',s=\mathrm{ES},t} + \sum\nolimits_{n'} \frac{1}{\mathrm{CP}^{e}} \cdot \mathrm{Pg}_{n',c=\mathrm{EC},t} \end{aligned} \tag{3.34}$$

In (3.34), it is imposed that the security constraint for discrete generator outages $\mathrm{PgLst}_{n,g,t}$ is the largest generation power among all of the units from technology g connected to bus n. Hence, with (3.35)–(3.37) $\mathrm{PgLst}_{n,g,t}$ is equal to the maximum rated power if there is a unit operating at maximum load, otherwise, is the power of the partial load $\mathrm{PgPrt}_{n,g,t}$. The variable $\Delta\mathrm{PgPrtLst}_{n,g,t}$ is used in (3.38) to negate the contribution of the outaged unit from the total contribution:

$$\mathrm{PgLst}_{n,g,t} \geq \mathrm{bx}_{n,g,t} \cdot \overline{P}_g \tag{3.35}$$

$$\mathrm{bx}_{n,g,t} \leq \mathrm{nx}_{n,g,t} \leq \mathrm{bx}_{n,g,t} M \tag{3.36}$$

$$\mathrm{PgLst}_{n,g,t} \geq \mathrm{PgPrt}_{n,g,t} \tag{3.37}$$

$$\Delta\mathrm{PgPrtLst}_{n,g,t} \geq \Delta\mathrm{PgPrt}_{n,g,t} - \mathrm{bx}_{n,g,t} \cdot \overline{P}_g \tag{3.38}$$

3.4.4 Multi-objective

A common objective for the microgrid design is to minimize the overall investment and operation cost, as shown in (3.1). However, multi-objective optimization, in which environmental aspects are taken into account, can be of interest too. In isolated, remote, rural areas in which reliability is an issue, renewable generation that is non-dispatchable might not be the most cost-effective solution. However, policy and regulation are pushing toward the decarbonization of the grid, thus, including a multi-objective optimization in which the overall microgrid CO_2 emissions are minimized as well, can be of interest. Therefore, the following multi-objective formulation can replace (3.1) when treating with rural and remote microgrid design. The cost formulation is different from the one proposed in (3.1) as no exchange between main grid and microgrid can take place due to remoteness. Then, the CO_2 emissions by the grid are accounted taking into the account the rate of CO_2 production of each technology. Depending on the microgrid planning problem to be solved, the objective function can be only for cost, emissions, or a combination of both as a multi-objective problem:

$$\begin{aligned} C^{\mathrm{Cost}} = & \textstyle\sum_{n,g} \mathrm{ni}_{n,g} \cdot \overline{P}_g \cdot \mathrm{InvCost}_g \\ & + \textstyle\sum_{n,k} \left(C_k^{\mathrm{Fx}} \cdot \mathrm{pur}_{n,k} + C_k^{\mathrm{Vr}} \cdot \mathrm{Cap}_{n,k} \right) \\ & + \textstyle\sum_{n,j,t} \mathrm{Pg}_{n,j,t} \cdot \mathrm{GC}_j + \sum_{n,t} \mathrm{Pl}_{n,t}^{\mathrm{Cur}} \cdot \mathrm{CC}_n \end{aligned} \tag{3.39}$$

$$C^{\mathrm{CO_2}} = \sum\nolimits_{n,j,u,t} \mathrm{Pg}_{n,j,u,t} \cdot \mathrm{CO_2Rt}_j \tag{3.40}$$

3.5 Remote multi-energy microgrid case study

The end of this chapter aims at illustrating the planning of a rural and remote microgrid with the formulation that has been detailed throughout the chapter. To simulate the proposed methodology for microgrid design, taking into account security and reliability constraints in a multi-energy microgrid, a remote system representative of the city of Nome in Alaska isolated will be used. The network model was reduced to 19 nodes, as Figure 3.4 shows.

To analyze the effects on different characteristics of remote and rural microgrids, three study cases are considered. In Case I, the objective is to minimize overall costs, without considering reliability issues, which means that $N - 1$

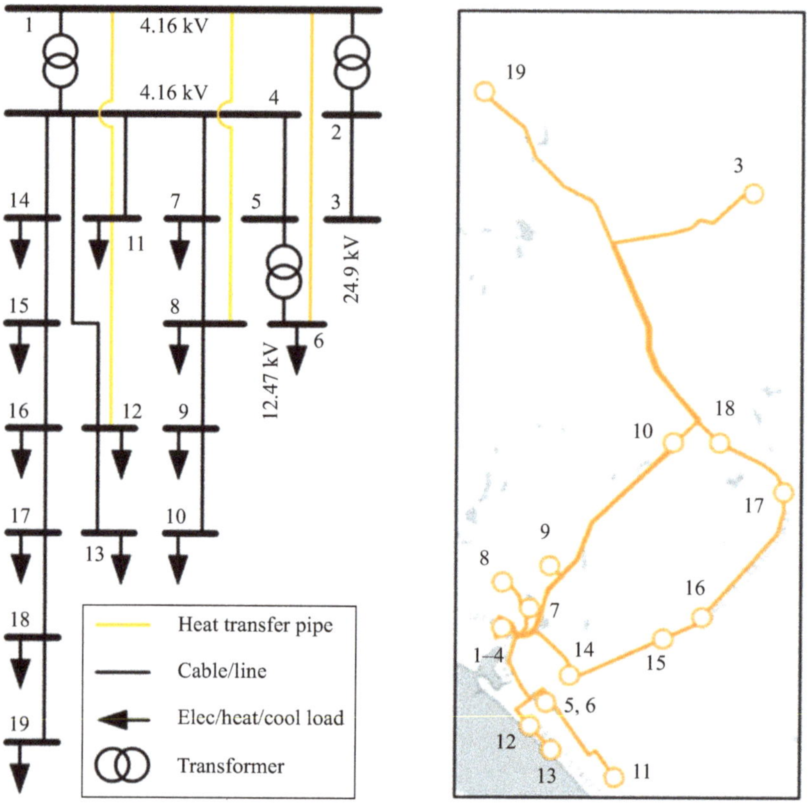

Figure 3.4 Single-line diagram and view of the analyzed microgrid

contingency constraints are not included. In Case II, we minimize the overall costs while including the contingency constraints. In Case III, we include the multi-objective merging a 50 percent weight for cost objective and 50 percent weight for CO_2 emission objective. For each case, the optimization model determines the optimal mix and size of technologies that can be installed at nodes 1, 8, and 18, which are the microgrid's central power plant, a hospital, and a residential neighborhood toward the end of a long feeder, respectively.

The results of the case studies, including DER capacities at each node, annualized investment and annual operation costs, post-contingency load curtailment costs, and annual CO_2 emissions are summarized in Table 3.6.

The study shows that a cost-minimization design, when taking into account security constraints in a remote and rural microgrid, might not lead to the adoption of distributed generation based on renewable energies. This can change when taking into account emissions of CO_2 in the objective function as well.

A more thorough analysis regarding the optimal electrical dispatch from Case III for a typical September weekday will be carried out, and results are shown in Figure 3.5 for nodes 1 and 8. The generation at node 1 is entirely exported to other

Table 3.6 Case study results

		Nodes	Case I	Case II	Case III
DER capacities	Photovoltaic (kW)	n1			4,274
		n8			
		n18			
	Battery (kWh)	n1		2,883	8,220
		n8			
		n18			
	CHP-enabled diesel engine (units × kW)	n1	1 × 5,000	3 × 1,000	2 × 1,000 + 1 × 5,000
		n8			
		n18			
Costs	Operation (k$)		8,133	8,662	7,486
	Investment (k$)		962	1,240	4,291
	Curtailment (k$)		–	31	39
	Total (k$)		9,096	9,932	11,816
	CO_2 (million tons)		16,104	16,650	14,329

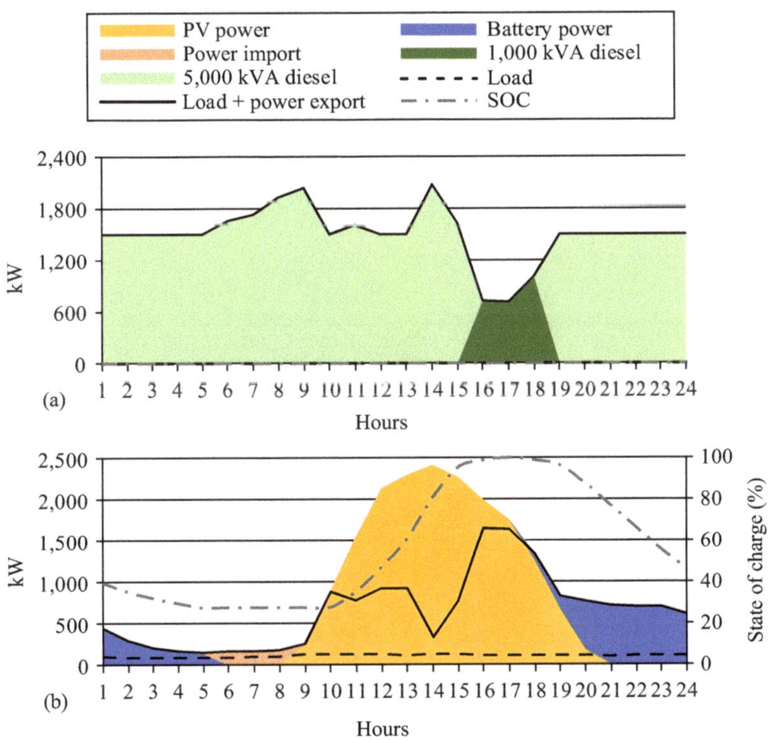

Figure 3.5 Electricity dispatch during a typical September weekday—Case III: (a) node 1 and (b) node 8

nodes, since the node does not have any loads of its own. Node 8 is equipped with a PV and a battery system to supply its own load and to export the extra power to other nodes. The battery is charged during peak PV hours and discharged at morning and afternoon. The total electric load is the summation of electrical and CL, since the latter can only be met by EC in this example.

The optimal heating dispatch from Case III for a typical July weekday is shown in Figure 3.6 for nodes 1, 11, and 17. The heat from diesel units at node 1 is recovered and exported to other nodes connected to node 1 through pipes. The HT at node 11 are met by the heat imported to the node. This node also exports the extra heat (from import) to other microgrid nodes. Node 17 has a gas-boiler heater installed locally, which is used to meet its HT.

The second analysis regarding the solution of Case III focuses in the contingency treatment. Figure 3.7 shows the generation outage power vs. system reserve for various generation contingencies during a peak day at September in

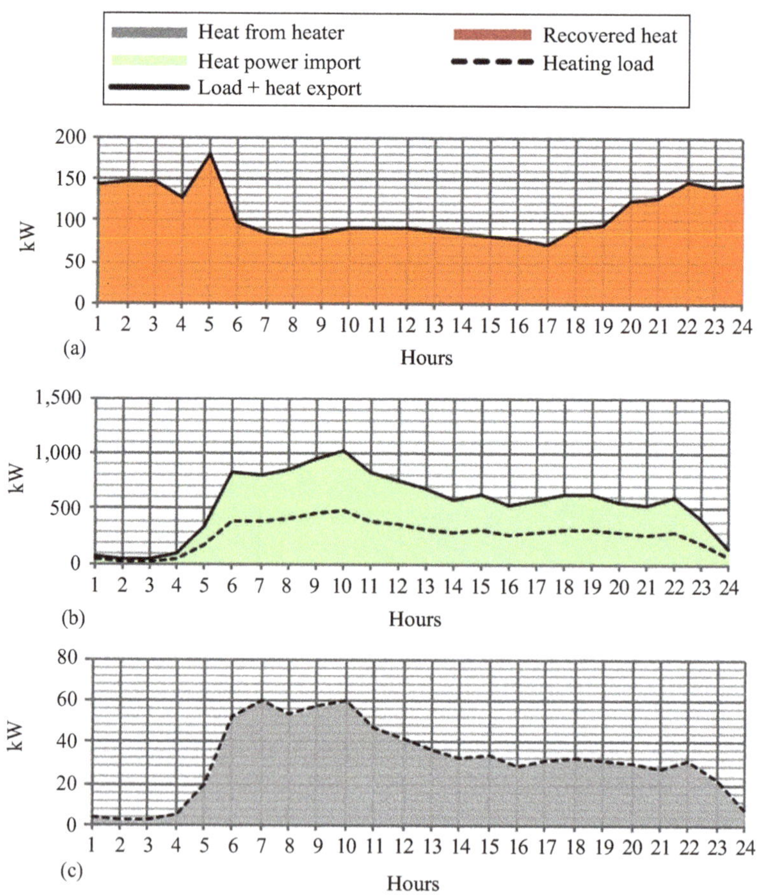

Figure 3.6 Heating dispatch during a typical July weekday—Case III: (a) node 1, (b) node 11, and (c) node 17

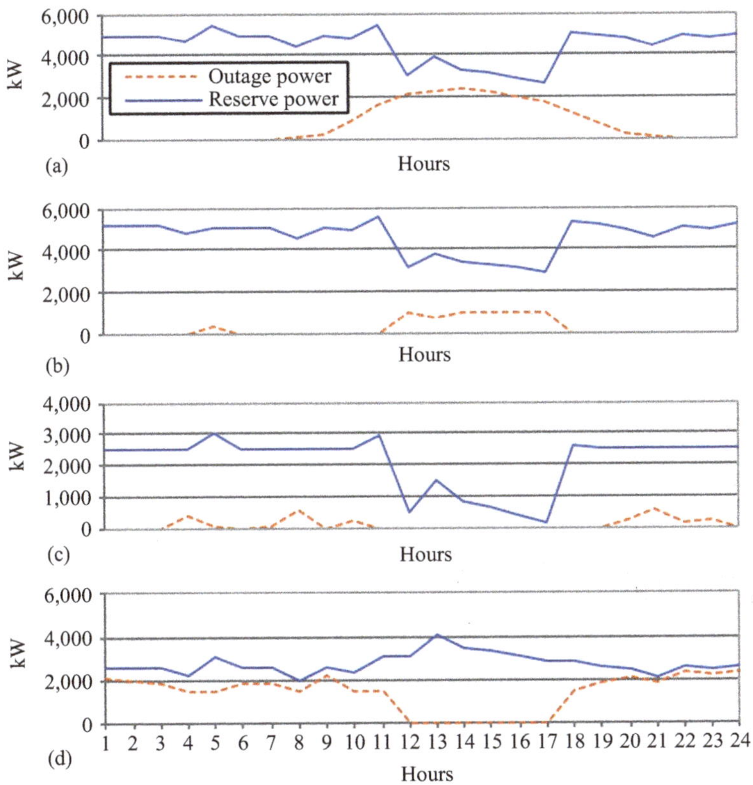

Figure 3.7 Outage power vs. system reserve for various system contingencies during a September peak day in Case III: (a) outage of PV at node 8, (b) outage of battery at node 8, (c) outage of one of the two 1,000 kVA diesel units at node 1, and (d) outage of the 5,000 kVA diesel unit at node 1

Case III. The integration of security constraints in the system forces the system reserve to be more than the outage power at any given time, as shown in Figure 3.6. In this example, outage of the 5,000 kVA diesel unit at node 1 is the most severe contingency, since it is much larger than all other dispatchable units in the system. Figure 3.7 confirms this intuition and shows that although the system maintains enough reserve against this contingency, the difference between the outage power and the reserve in the system is much smaller compared to the other three contingencies.

Finally, as was discussed when the linearized power flow equations were showed, it is interesting to carry out a post-optimization analysis to study the implications of linearizing the equations, a main issue in MILP formulations for microgrid design. Therefore, the voltage variations are studied (during the year) of each microgrid node for Case III, comparing the approximate power flow solution

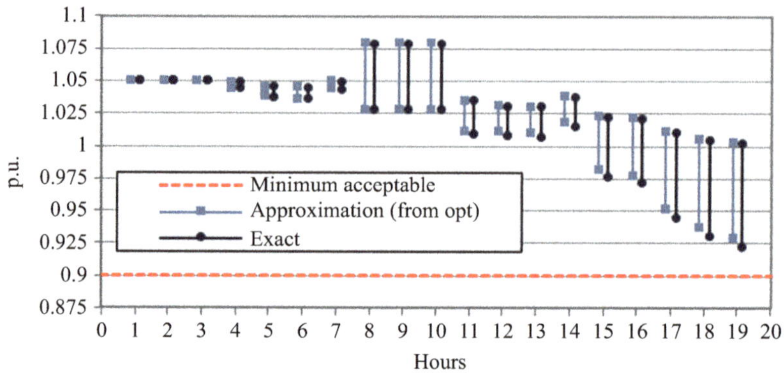

Figure 3.8 Comparison of voltage variations at each bus between approximate and exact power flow solutions—Case III

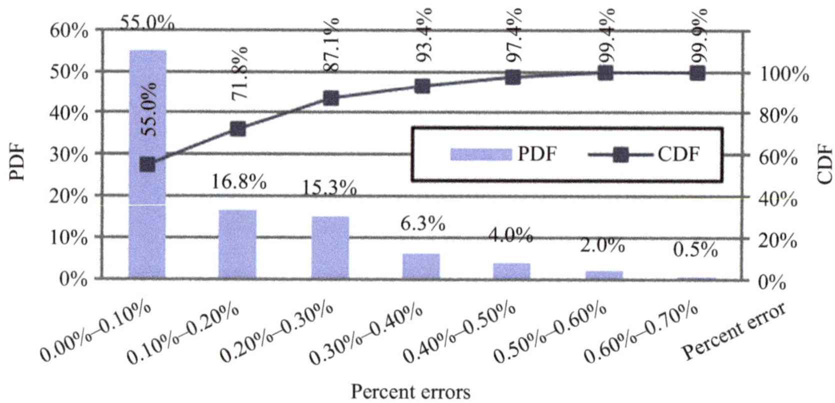

Figure 3.9 PDF and CDF of voltage magnitude error—Case III

(from the optimization) and the exact power flow solution (using Newton–Raphson algorithm). To assess the accuracy of the approximate power flow solution, the probability distribution function (PDF) and cumulative distribution function (CDF) for voltage magnitude error in Case III were calculated, showing that more than 97 percent of data points have an error less than 0.5 percent. Thus, the linearized model used is very accurate (Figures 3.8 and 3.9).

3.6 Conclusions

This chapter provided an overview of the standard methodologies used to optimally solve microgrid planning and design problems. In particular, three aspects of rural microgrids planning were analyzed: (1) the multi-energy nature of microgrids, where a significant portion of the HT is not supplied by electricity and depends on other vectors, such as heat distribution; (2) the significant sizes of rural microgrids and the consequent occupation of large portions of the territory, which requires planning methods that consider internal network voltage drops and line flow limits; and (3) the remote location and off-grid organization of rural microgrids (with minimal or no connection to a main grid), which needs reliability constraints and multi-objective approaches to be included in the planning problem. Changes in typical microgrid planning formulation addressing these specific challenges of rural microgrids were introduced in this chapter. The case studies show that these changes have a significant impact on the microgrid planning results and are able to better represent the reality of rural microgrids.

References

[1] J. Neubauer and M. Simpson, "Deployment of Behind-The-Meter Energy Storage for Demand Charge Reduction," *NREL/TP-5400-63162*. 2015, p. 30.

[2] O. Hafez and K. Bhattacharya, "Optimal planning and design of a renewable energy based supply system for microgrids," *Renewable Energy*. 2012, vol. 45, pp. 7–15.

[3] Hybrid Optimization of Multiple Energy Resources (HOMER), Software Tool, available: https://www.homerenergy.com/

[4] RETScreen, Energy Management Software, available: www.retscreen.net

[5] E. Hittinger, T. Wiley, J. Kluza, and J. Whitacre, "Evaluating the value of batteries in microgrid electricity systems using an improved energy systems model," *Energy Conversion and Management*. 2015, vol. 89, pp. 458–472.

[6] M. Stadler, M. Kloess, M. Groissböck, *et al.*, "Electric storage in California's commercial buildings," *Applied Energy*. 2013, vol. 104, pp. 711–722.

[7] C. Marnay, G. Venkataramanan, M. Stadler, A. S. Siddiqui, and R. Firestone, "Optimal technology selection and operation of commercial-building microgrids," *IEEE Transaction on Power Systems*. 2008, vol. 23, no. 3, pp. 975–982.

[8] G. Cardoso, M. Stadler, M. C. Bozchalui, *et al.*, "Optimal investment and scheduling of distributed energy resources with uncertainty in electric vehicle driving schedules," *Energy*. 2014, vol. 64, pp. 17–30.

[9] S. X. Chen, H. B. Gooi, and M. Q. Wang, "Sizing of energy storage for microgrids," *IEEE Transaction on Smart Grid*. 2012, vol. 3, no. 1, pp. 142–151.

[10] S. Bahramirad, W. Reder, and A. Khodaei, "Reliability-constrained optimal sizing of energy storage system in a microgrid," *IEEE Transaction on Smart Grid*. 2012, vol. 3, no. 4, pp. 2056–2062.

[11] C. Chen, S. Duan, T. Cai, B. Liu, and G. Hu, "Optimal allocation and economic analysis of energy storage system in microgrids," *IEEE Transaction on Power Electronics*. 2011, vol. 26, no. 10, pp. 2762–2773.

[12] T. Logenthiran, D. Srinivasan, A. M. Khambadkone, and T. Sundar Raj, "Optimal sizing of Distributed Energy Resources for integrated microgrids

using Evolutionary Strategy," *2012 IEEE Congress on Evolutionary Computation*, Brisbane, Australia, 2012. pp. 1–8.

[13] G. Mendes, C. Ioakimidis, and P. Ferrão, "On the planning and analysis of integrated community energy systems: a review and survey of available tools," *Renewable and Sustainable Energy Reviews*. 2011, vol. 15, no. 9. pp. 4836–4854.

[14] C. Gamarra and J. M. Guerrero, "Computational optimization techniques applied to microgrids planning: a review," *Renewable Sustainable Energy Review*. 2015, vol. 48, pp. 413–424.

[15] H. Fathima and K. Palanisamy, "Optimization in microgrids with hybrid energy systems – a review," *Renewable Sustainable Energy Review*. 2015, vol. 45, pp. 431–446.

[16] P. S. Georgilakis, S. Member, and N. D. Hatziargyriou, "Power distribution networks: models, methods, and future research," *IEEE Transactions on Power Systems*. 2013, vol. 28, no. 3, pp. 3420–3428.

[17] A. El-Zonkoly, "Optimal placement of multi-distributed generation units including different load models using particle swarm optimisation," *IET Generation, Transmission and Distribution*. 2011, vol. 5, pp. 760–771.

[18] S. Bolognani, S. Zampieri, and O. C. Feb, "On the existence and linear approximation of the power flow solution in power distribution networks," *IEEE Transactions on Power Systems*. 2016, vol. 31, no. 1, pp. 1–10.

Part II

Optimization, control and storage

Chapter 4

Assessment of energy saving for integrated CVR control with large-scale electric vehicle connections in microgrids

Haiyu Li[1], *Yuken Shen*[1] *and Rajeev Kumar Chauhan*[2]

The coordinated controls of voltage reduction (CCVR) technique as one of the functions of demand side management has been rolled out in many power utilities in the United Kingdom to provide fast energy demand reduction (EDR) for a microgrid when the distributed generation is low. With the advance of new low-carbon technologies (LCTs), different load types, such as heat pumps and electric vehicles (EVs), are emerging. This paper assesses the impact of different EV charging models on the EDR performance. The CCVR technique is considered for both an urban and a rural microgrid, respectively. The Monte Carlo (MC) method is applied to analyze the impact of three different EV charging models (i.e., uncoordinated charging (UC), vehicle-to-grid (V2G) charging and mixed charging (MIX)) on the grid EDR performance. In comparison with different EV charging strategies, the impact of V2G strategy on the grid EDR capability is the least, while the MIX strategy can help the grid to increase EV penetration capability more effectively.

4.1 Introduction

With the rapid increase in new electricity loads and generations from the LCTs, various challenges of voltage management on the distributed microgrids have been created. It has been found that most of the electricity utilities supply in the upper half of the acceptable voltage range in order to keep the lowest customer utilization voltage consistent with the legal voltage standards. This not only causes energy wastes and reduces the efficiency of most electrical appliances and the distributed microgrid but also leads to negative environmental impacts [1,2]. One of the most effective ways to deal with this situation is by applying the flexible reduced voltage control, a technique that provides the energy reduction by lowering the supply

[1]School of Electrical and Electronic Engineering, The University of Manchester, Manchester, United Kingdom
[2]Department of Electrical Engineering, Dayalbagh Educational Institute, Agra, India

voltage without violating the voltage limits [3]. However, in the modern distributed microgrids, the increasing integration of large-scale EVs has interfered the control scheme and performance of conservation voltage reduction (CVR) in different ways. First, the EV charging process brings more loading pressure to the distributed microgrids, which limits the potential headroom of applying CVR. Second, the different charging and discharging behaviors of EV make the distributed microgrids more vulnerable to the voltage variation. Third, the EV penetration changes the overall load composition of the distributed microgrid, which directly impacts the CVR effectiveness. Therefore, there is a strong requirement of implementing CVR with different EV models; analyzing the CVR performance on different types of low-voltage (LV)-distributed microgrids (i.e., for urban areas or rural area) and generalizing the impact of EVs on CVR effectiveness.

The CVR technique was first tested by American Electric Power System in 1978 [4] and then being widely implemented by many utilities around the world. According to various simulation-based studies [5–9] and demonstration projects [10,11] taken by various worldwide utilities over the last 5 years, CVR has provided an average energy reduction of 0.73% for each 1% voltage reduction. An in-depth review of implementing and assessing CVR has been carried out by Wang and Wang, and CVR researches have been summarized in [12]. With the substantial increase in using EVs, integrating EV into the distributed microgrids has become an emerging research topic. In the recent years, several researches have turned attention to analyze the impact of EVs on CVR and voltage control of the distributed microgrids. Singh [13] has pointed out that the discharging process of EV could reverse the direction of power flow, thus create huge voltage variations on the distributed microgrids and cause a large impact on the CVR. Higgins *et al.* [7] have applied an off-load CVR control model to the UK-distributed microgrid in 2030 with high-penetration EVs. They have observed a larger voltage drop along the feeder with the increase of EV penetration which, in consequence, limits the potential headroom of implementing CVR. In their research, however, only off-load voltage control is considered.

In the recent years, the on-load tap changer (OLTC) transformer became available at LV substations and also the smart monitoring devices were installed. This allows the CVR to be implemented on the LV microgrids and makes the CVR control as close as possible to the customers. In this case, an integrated CVR model, also known as the coordinated voltage reduction control (CVRC) model, has been proposed and implemented on a typical UK-distributed microgrid with considering different EV penetrations [14]. It has been found that the higher the EV penetrations, the less the energy reduction that CVRC can achieve due to the constant-power characteristic of simple EV charging method.

This paper presents the probability analysis of the impact of different EV charging models on CVRC performance. High-granularity realistic load models (in 1-min resolution) are generated by using the CREST tool [15] and then modified to time-varying ZIP models. The probability approach with the consideration of various uncertain parameters is applied to generate three different EV models: the UC model, the V2G model and the MIX model. The CVRC algorithm is proposed that applies and coordinates two voltage control methods over each predetermined control interval based on the voltage constraints and dynamic voltage profile of nominal power-flow analysis. A typical urban and a typical rural LV-distributed microgrid are modeled to implement the proposed CVRC and EV models. The simulation results are analyzed by using the MC method and the impacts of various EV penetrations and different charging strategies on CVRC performance are compared and analyzed.

4.2 Coordinated voltage reduction control model

To implement the CVRC algorithm with different EV models to the LV microgrid, an MC-based CVRC model is considered that consists of four main blocks as described in Figure 4.1.

First, the LV microgrid models are built in the Open DSS; the load profiles and EV profiles are randomly assigned to all customers. Second, the time series is introduced and driven by MATLAB®, since the Open DSS is only able to solve the power flow at one particular point of time. The nominal power flow is carried out and the voltage profile is analyzed. Then, the CVRC algorithm is applied in which optimal operations of control devices are calculated and stored in the control matrix. After that, the CVRC power flow is carried out and all control devices operate based on the control matrix. With all results from the nominal and CVRC power flows being stored in the MATLAB, the impact metrics are calculated and the simulation continues to the next MC iteration. The load and EV profiles are randomly reassigned at the beginning of each MC iteration and the total of i_t MC iterations are carried out. Finally, the mean values of each impact metric are

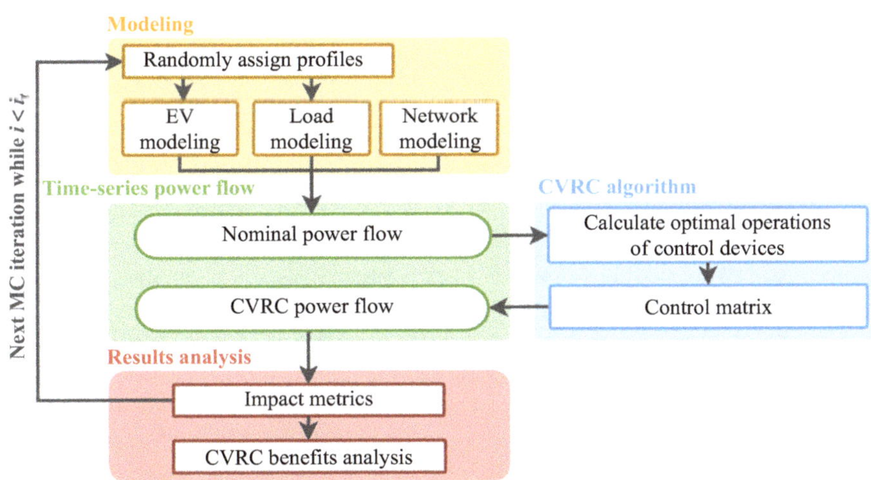

Figure 4.1 Flowchart of the coordinated voltage reduction control model

calculated. This helps one to minimize the bias toward the simulation results and provide a generalized CVRC analysis.

4.2.1 Coordinated voltage reduction control algorithm

The CVRC algorithm is described in detail in the flowchart shown in Figure 4.2. It applies two voltage control methods and coordinates them over each predetermined control interval based on the voltage constraints and voltage profile of nominal power-flow analysis. After the coordination in each control interval, the energy consumption is compared with all other power flows to ensure that the optimal energy consumption is achieved. With all control intervals analyzed, a control matrix is generated to record the optimal operations of control devices in all intervals. The control matrix is finally applied to drive the CVRC power-flow analysis. In this study, the OLTC and circuit capacitors are applied as two voltage control methods.

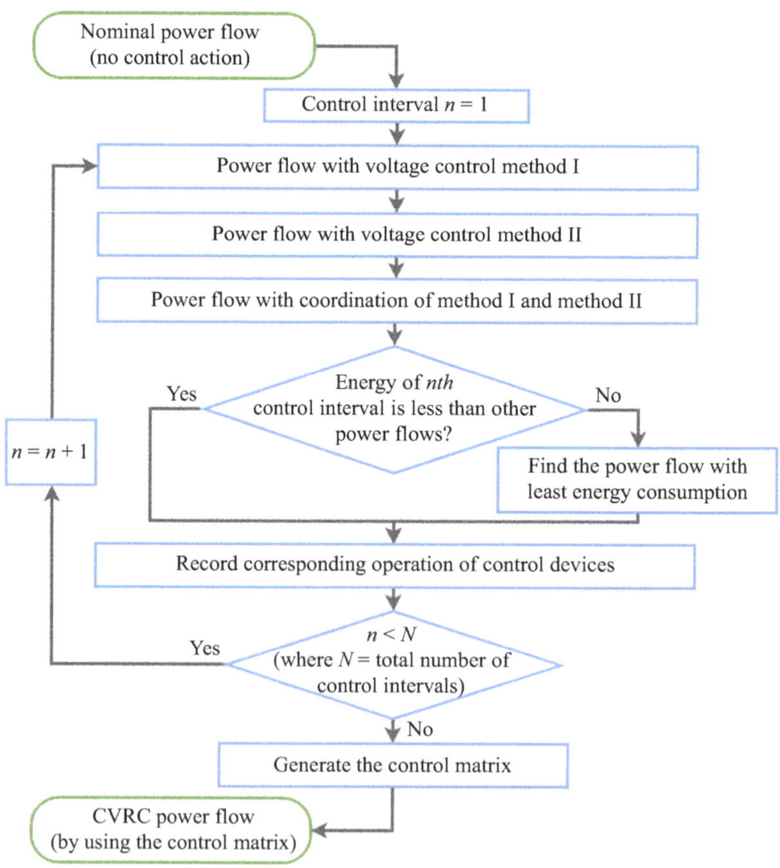

Figure 4.2 Flowchart of CVRC algorithm

4.2.2 OLTC control

The purpose of controlling OLTC is to select the optimal tap position that provides the lowest possible feeder voltage without violating the voltage limits. In each specific control interval, the tap ratio of the optimal tap position is determined by considering the lowest feeder voltage and the potential headroom of voltage reduction. The lowest feeder voltage is always a continuous variable, while the tap ratio is selected from a range of discrete values. In this case, the tap ratio is first treated as continuous variables during the calculations and it is rounded off to the nearest discrete value (i.e., the nearest tap position) in the last step.

The OLTC is installed on the primary side of the transformer, which adjusts the primary voltage from −8% to +8% in nine steps of 2% [16]. The tap positions and the corresponding tap ratios are shown in Table 4.1.

In the nominal power flow, the tap changer is set to its nominal position (i.e., tap position 5). When CVRC is applied, the tap position of OLTC is adjusted over each predetermined control interval. In this case, the interval between two control actions, Δt (in min), is defined and the total number of control intervals N is calculated as $N = 1{,}440/\Delta t$. Considered the voltage limits, the potential range of tap ratio in the nth control interval $TR(n)$ is calculated as

$$TR(n) \leq \frac{V_s(n)}{V_s(n) - [V_{min}(n) - V_{limit}]} \tag{4.1}$$

where n ($n = 1, 2, \ldots, N$) is an integer representing the index number of control intervals and N is the total number of control intervals; $V_s(n)$ is the voltage on the secondary side of the transformer in the nth control interval of nominal power flow; $V_{min}(n)$ is the lowest feeder voltage in the nth control interval of nominal power flow and V_{limit} is the lower limit of feeder voltage (i.e., 0.94 pu). For all tap ratios that satisfy (4.1), the largest one is selected for the nth control interval and the corresponding tap position is stored in the control matrix.

4.2.3 Capacitor control

In large radial distributed microgrids, especially in those heavily loaded microgrids, high reactive power flowing through the feeders may cause a large voltage drop. This limits the effect of CVRC benefits, especially during the peak time. In this case, the switchable circuit capacitors are implemented to compensate local reactive power, reduce the voltage drop and flatten the feeder voltage. This may provide more headroom for voltage and energy reductions and, as a result, more significant CVRC benefits are expected. Two types of circuit capacitors are considered in this study: the 100 kVAr capacitor and the 150 kVAr capacitor. Each

Table 4.1 Tap positions and tap ratios of OLTC

Tap position	1	2	3	4	5	6	7	8	9
Tap ratio	0.92	0.94	0.96	0.98	1.00	1.02	1.04	1.06	1.08

capacitor is connected in three phases and can be switched in stages as shown in Tables 4.2 and 4.3.

In the nominal power flow, all capacitors are switched off (i.e., switch stage 1). When CVRC is applied, the switch stage of each capacitor is adjusted over each predetermined control interval. Under the traditional control methods, the capacitor banks are set to turn on when the current in a feeder (on which they are installed) exceeds a given threshold. This is typical of how the capacitor banks are operated by many utilities [17]. In the CVRC model used in this paper, however, the purpose of adding capacitors is to flatten the voltage profile and reduce the voltage drop. It is more straightforward to operate the capacitors on the basis of the variation of voltage drop. Therefore, the capacitors are switched on when the voltage drop along the feeder (on which they are installed) exceeds a threshold:

$$Q(n,f) = \begin{cases} Q_{cap}(n,f), & VD(n,f) > VD_{th} \\ 0, & VD(n,f) \le VD_{th} \end{cases} \tag{4.2}$$

where $f\,(f = 1, 2, \ldots, F)$ is an integer representing the index number of feeders and F is the total number of feeders; $Q(n,f)$ is the reactive power compensated to the fth feeder in the nth control interval; $Q_{cap}(n,f)$ is the reactive power provided by the capacitor installed on the fth feeder (i.e., the selected stage size); $VD(n,f)$ is the voltage drop along the fth feeder in the nth control interval and VD_{th} is the threshold voltage drop.

In each control interval, the peak reactive power on which the capacitor is installed, $Q_{peak}(n,f)$ can be calculated from the nominal power flow. The selected stage size of each capacitor $Q_{cap}(n,f)$ should be smaller than and roughly equal to the peak reactive loading of the feeder:

$$Q_{cap}(n,f) \le Q_{peak}(n,f) \tag{4.3}$$

For all stage sizes that satisfy (4.3), the one that provides reactive power closest to the peak reactive loading is selected as $Q_{cap}(n,f)$ and the corresponding switch stage is stored in the control matrix.

Table 4.2 Switch stages and stage sizes of 100 kVAr capacitor

Switch stage	1	2	3
Stage size (kVAr)	0	50	100

Table 4.3 Switch stages and stage sizes of 150 kVAr capacitor

Switch stage	1	2	3	4
Stage size (kVAr)	0	50	100	150

4.2.4 Coordination

The coordination between OLTC and capacitors is applied on the basis of the following rules:

1. Apply the OLTC control prior to the capacitor control.
2. The threshold voltage drop (VD_{th}) that is used to switch on the capacitors is calculated on the basis of the nominal voltage profile and the maximum tap ratio of OLTC:

$$VD_{th} = \frac{V_s(n)}{TR_{max}} - V_{limit} \tag{4.4}$$

   If the voltage drop is so large that the tap ratio calculated from (4.1) cannot reach its maximum tap ratio TR_{max}, this means the microgrid cannot operate at the lowest possible voltage without violating the voltage limits and therefore extra reactive power should be locally compensated.
3. After determining the switch stage of all capacitors, recalculate the optimal tap position and update the control matrix because adding capacitors may provide more headroom for voltage reduction.

4.2.5 Impact metrics

In the proposed CVRC model, impact metrics are calculated to quantify and generalize the CVRC benefits. These metrics are calculated for each 24-h simulation and the mean values of each impact metric are calculated after the total of i_t MC iterations finishes for each specified EV charging strategy.

4.2.5.1 Voltage reduction, *dV* (%)

This metric represents the percentage of voltage reduction at the secondary side of the transformer:

$$dV = \frac{1}{i_t} \times \sum_{i=1}^{i=i_t} \frac{V_n^i - V_{CVRC}^i}{V_n^i} \times 100\% \tag{4.5}$$

where V_n^i and V_{CVRC}^i are the daily average substation voltage (i.e., second side) of the ith MC iteration in the nominal power flow and CVRC power flows, respectively.

4.2.5.2 Energy reduction, *dE* (%)

This metric indicates the percentage reduction of total active energy consumption in the microgrid:

$$dE = \frac{1}{i_t} \times \sum_{i=1}^{i=i_t} \frac{E_n^i - E_{CVRC}^i}{E_n^i} \times 100\% \tag{4.6}$$

where E_n^i and E_{CVRC}^i are the total energy consumptions of the ith MC iteration in the nominal and CVRC power flows, respectively.

4.2.5.3 CVRC factor, $CVRC_f$

The CVRC factor measures how effectively the voltage reduction can be converted to energy reduction. This metric registers the average CVRC factor over i_t MC iterations:

$$CVRC_f = \frac{1}{i_t} \times \sum_{i=1}^{i=i_t} \frac{V_n^i \cdot \left(E_n^i - E_{CVRC}^i\right)}{E_n^i \cdot \left(V_n^i - V_{CVRC}^i\right)} \tag{4.7}$$

4.2.5.4 Transformer utilization factor, *UF* (%)

This metric represents the utilization factor of the microgrid. It is calculated at the primary side of the substation as

$$UF = \frac{1}{i_t} \times \sum_{i=1}^{i=i_t} \frac{S_{peak}^i}{S_{rated}} \times 100\% \tag{4.8}$$

where S_{peak}^i is the maximum hourly average power flowing through the transformer in the *i*th MC iteration and S_{rated} is the rated power of transformer.

4.2.5.5 Voltage problems, *VP* (%)

The standard BS EN50160 [18] is adopted as the constraints of voltage. The customers whose voltages are outside the limits imposed by BS EN50160 are considered as having voltage problems. This metric shows the percentage of customers with voltage problems:

$$VP = \frac{1}{i_t} \times \sum_{i=1}^{i=i_t} \frac{N_{VP}^i}{N_t} \times 100\% \tag{4.9}$$

where N_{VP}^i is the number of customers with voltage problems of the *i*th MC iteration and N_t is the total number of customers.

4.3 Modeling

4.3.1 Microgrid modeling

In this study, two typical LV-distributed microgrids in the United Kingdom are modeled. One microgrid is located in the urban area, while the other is located in the rural area. The rural area is a lightly loaded small radial LV microgrid, while the urban area is a heavily loaded large radial LV microgrid. Compared to the rural area, it can be found that the urban area contains less, but longer feeders, feeding more customers and street lights.

The microgrid details are summarized in Table 4.4. The microgrid sketches are shown in Figures 4.3 and 4.4. On both the microgrids, the transformers provide an approximately 8% voltage upgrade at the secondary side; thus all feeders operate at 433 V instead of 400 V. The OLTC is installed on the primary side of each

Table 4.4 Details of microgrid

Microgrid	Typical rural microgrid	Typical urban microgrid
Transformer power	200 kVA	500 kVA
Transformer voltage	11 kV/433 V	6.6 kV/433 V
Number of feeders	6	4
Feeder voltage	433 V	433 V
Feeder length	263 m (on average)	435 m (on average)
Control devices	1 OLTC 3 capacitors (100 kVAr)	1 OLTC 1 capacitor (100 kVAr) 1 capacitor (150 kVAr)
Loads	148 domestic loads 5 nondomestic loads 34 street lights	282 domestic loads 12 nondomestic loads 59 street lights

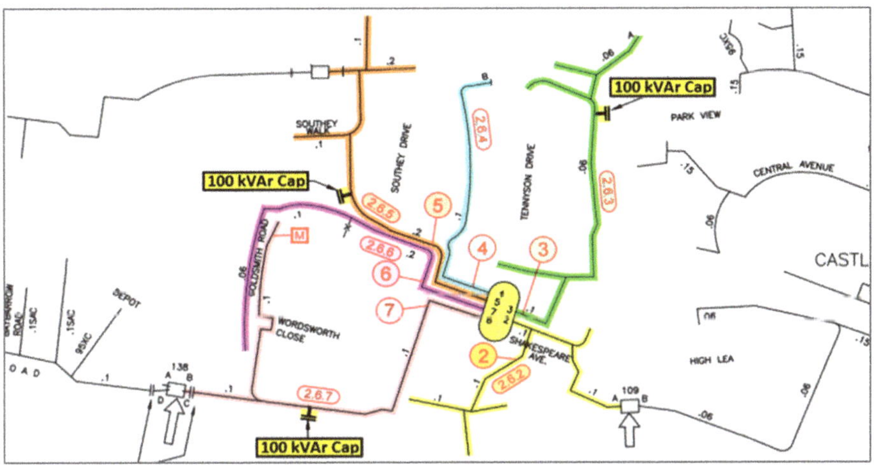

Figure 4.3 A sketch of the typical rural distributed microgrid

transformer and the circuit capacitors are installed as shown in the sketches. The microgrid models are built on the basis of the original microgrid data. A translation tool is generated in MATLAB that reads the original data in extensible markup language format; categorizes the data by types; writes the categorized data into different comma-separated values (CSV) files; identifies valuable CSV files and writes the corresponding data to text files in a readable Open DSS format. Once the microgrid model is completed, a three-step validation introduced in [14] is carried out before performing the time series power-flow analysis. It aims to ensure that all elements are connected in the right position and no isolated zone exists.

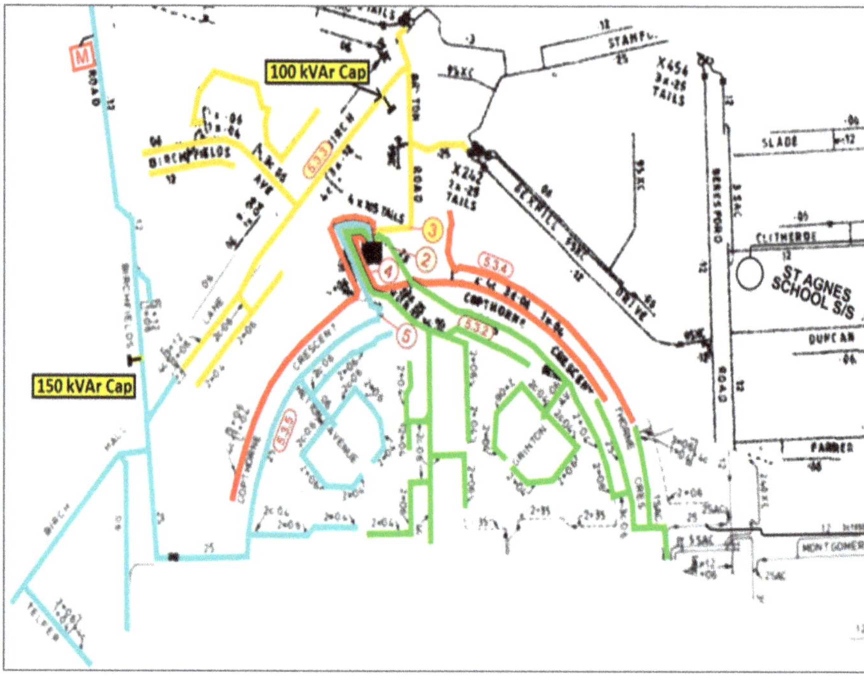

Figure 4.4 A sketch of the typical urban distributed microgrid

4.3.2 Load modeling

Three types of load are modeled that include the domestic loads, the nondomestic loads and the street lights. The domestic loads that represent the residential loads are modeled as dynamic ZIP loads with time-varying ZIP parameters. The non-domestic loads that represent the commercial and industrial loads are modeled as constant current for real power and constant impedance for reactive power. The street lights are assumed to be new compact fluorescent lamps “Philips FGS225 Residium” and are modeled as traditional ZIP loads with fixed ZIP parameters.

According to the classification made by ELEXON [19], the distribution load profiles are classified into eight profile classes (PC). PC1 and PC2 represent the domestic load profiles and PC3–8 represents the nondomestic load profiles. For domestic load profiles, the CREST tool is applied to generate high-granularity realistic ZIP load profiles for each appliance. The profiles of all appliances under use in a dwelling at each particular minute are aggregated by using the method introduced in [20]. Thus, the equivalent dynamic ZIP load profiles with time-varying ZIP parameters are obtained. The total of 1,000 individual dwelling profiles in 1-min resolution is generated for each PC. Based on the load composition, these profiles are randomly assigned to domestic loads at the beginning of each MC iteration in the CVRC model. For nondomestic load profiles, the diversified ELEXON profiles are applied. The original ELEXON profiles are converted from 30 to 1-min resolution through the interpolation method. In the CVRC model, it assumes that all nondomestic loads of each PC share the same profile. For street light profiles, the profiles are generated by assuming that all street lights are switched on and off according to the local times of sunset and sunrise [21].

4.3.3 Electric vehicle modeling

Three charging/discharging strategies are considered in this paper: the UC strategy, the coordinated V2G strategy and the MIX strategy. The EV model of each charging strategy is developed on the basis of the probability density functions (PDF) of four uncertain parameters: the battery capacity [22–24], the power charging/discharging level [25], the end-use time of EVs [22] and the state of charge (SOC). Based on the probability studies of those four uncertain parameters carried out in [26], a pool of 1,000 profiles is generated for each charging strategy through the MC method. All profiles are first generated at 1-h resolution and then converted to 1-min resolution through the interpolation method. The average profiles of each strategy are shown in Figures 4.5–4.7.

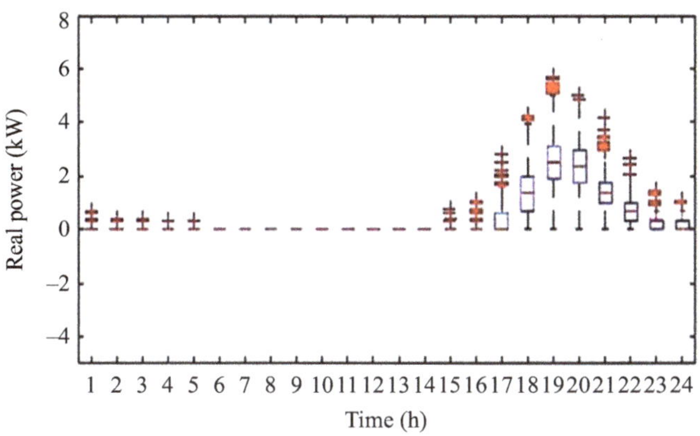

Figure 4.5 Average uncoordinated charging profile of electric vehicle

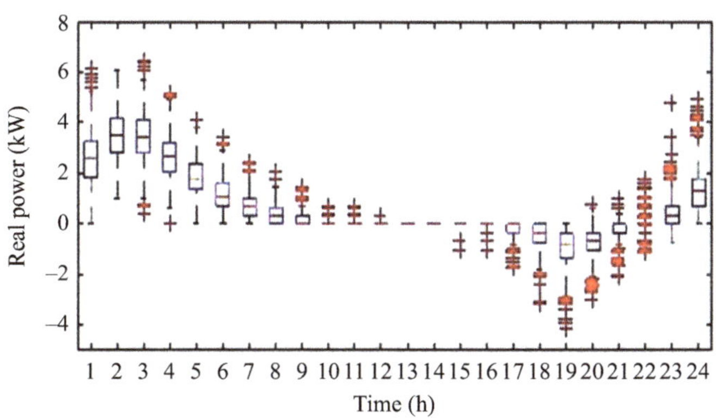

Figure 4.6 Average coordinated V2G profile of electric vehicle

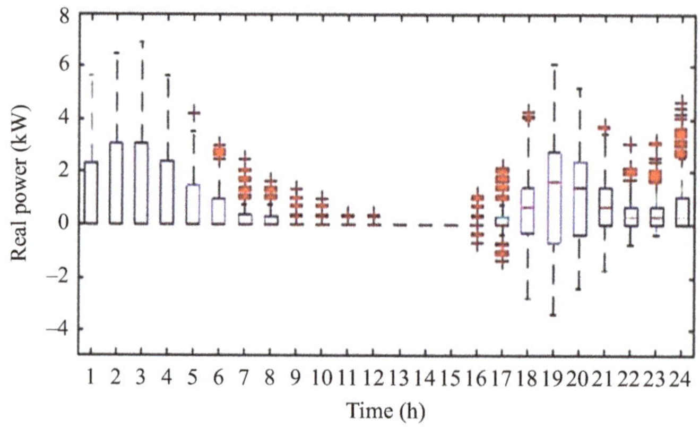

Figure 4.7 Average mixed profile of EV

For the UC strategy, all customers are supposed to charge their EVs once they arrive home after work and stop charging at 100% power (i.e., SOC = 1). For the V2G strategy, it assumes that all customers discharge their EVs after arriving home till 25% power left (i.e., SOC = 0.25); then the EVs are recharged at the midnight from 25% power to full power. The MIX strategy assumes that the customers use either UC or V2G strategy. According to the comparison of different proportion arrangements in [26], the most stable case is found when 60% customers use the UC and 40% customers use the V2G strategy. Thus, this proportion arrangement is applied to the MIX strategy in this study.

4.4 Result analysis

To analyze the impact of different EV charging strategies on CVRC performance, the CVRC model with EV connected is applied on both urban and rural microgrids. On each microgrid, three EV charging strategies introduced in Section 4.3 are applied and the EV penetration is increased from 0% to 100% in the steps of 10%. The MC loop is applied for each specific charging strategy and penetration, which randomly assigns EV profile and load profile to the customers and runs the CVRC model in each MC iteration till 100 iterations are completed. The mean value and the standard deviation of each scenario are calculated and the CVRC benefits are analyzed as follows.

4.4.1 Analysis of energy reduction

Figure 4.8 shows the voltage reduced by the CVRC model. In the case of baseload (i.e., without EV connected), the CVRC model reduces the average supply voltage by 7.4% on the rural microgrid and 3.8% on the urban microgrid. This provides 5.8% and 2.3% daily energy reduction on the rural and the urban microgrid, respectively, as shown in Figure 4.9.

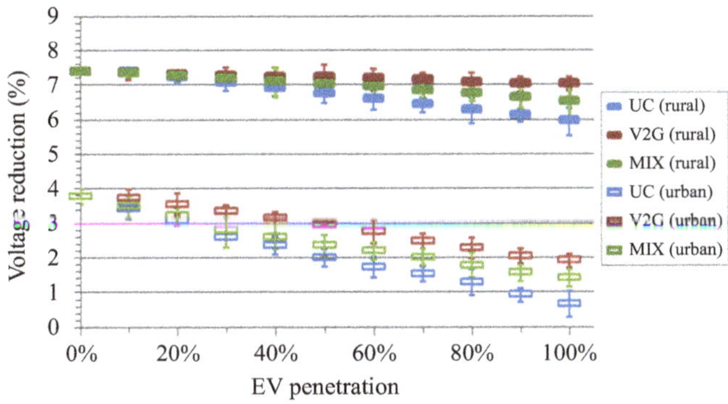

Figure 4.8 Voltage reduction with different EV charging strategies and penetrations on the rural and urban microgrids

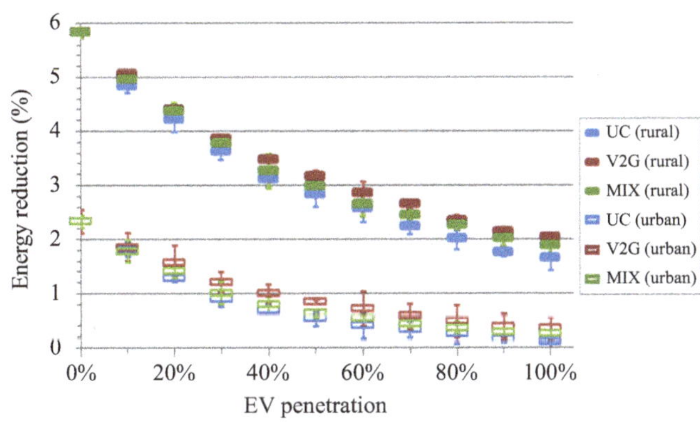

Figure 4.9 Energy reduction with different EV charging strategies and penetrations on the rural and urban microgrids

With the growth of EV penetration, the increasing demand during EV charging periods leads to a larger voltage drop along the feeders. This limits the potential headroom of voltage reduction in the CVRC model. As a result, the energy reduction decreases to about 2% in the rural microgrid and to 0.3% in the urban microgrid with 100% EV integrated. Among the three charging strategies, the V2G strategy always provides the most voltage reduction, while the UC always provides the least one. This is because, with the V2G and MIX strategies applied, the discharging process of EVs can locally compensate power to the microgrid. It reduces the voltage drop along the feeders and the loading pressure of the microgrid during the discharging period. As a result, among the three charging strategies, charging the EVs by V2G strategy allows the CVRC to provide deeper voltage reduction and achieve more energy reduction than the other two strategies.

Figure 4.10 shows the CVRC factor of both microgrids with different EV charging strategies and penetrations. The CVRC factor is directly affected by the load type and load composition of the microgrid. The constant impedance loads and constant current loads generally have larger CVRC factors, while the constant energy loads and constant power loads have relatively smaller CVRC factors. Due to the constant-power characteristic of EV, integrating more EVs changes the overall load composition of the microgrid and results in a smaller CVRC factor. Therefore, the CVRC factor is reduced from 0.8 (with baseload) to 0.3 (with 100% EV) in the rural microgrid and is reduced from 0.6 (with baseload) to 0.2 (with 100% EV) in the urban microgrid. Therefore, the CVRC is less effective with the increase of EV penetration on both microgrids. On the other hand, since the charging strategy does not change the load type of EVs, there is no significant difference among the CVRC factors obtained with three charging strategies.

In summary, the percentage of energy reduction and the daily kWh of energy reduction released by CVRC have been quantified as shown in Table 4.5. The implementation of CVRC provides energy reduction for all scenarios. The energy reduction capability gradually decreases with the increase of EV penetration. In comparison with different EV charging strategies, the impact of V2G strategy on CVRC energy reduction capability is the least.

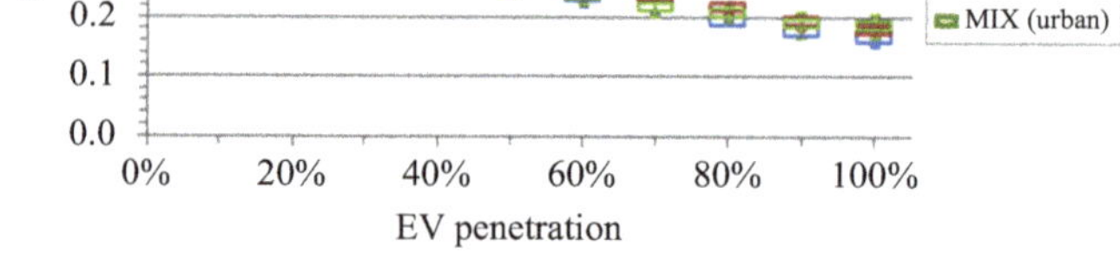

Figure 4.10 CVRC factor with different EV charging strategies and penetrations on the rural and urban microgrids

Table 4.5 Daily energy reduction by CVRC

	Daily energy reduction in rural microgrid		**Daily energy reduction in urban microgrid**	
	0% EV	**100% EV**	**0% EV**	**100% EV**
UC	5.81% (113 kWh)	1.65% (56 kWh)	2.33% (86 kWh)	0.11% (7 kWh)
V2G		2.02% (69 kWh)		0.34% (23 kWh)
MIX		1.87% (64 kWh)		0.26% (18 kWh)

4.5 Analysis of EV penetration capability

With the increase of EV penetration, there will be various problems occurring in the distributed microgrid, such as the overload of transformer and the overvoltage at the customer sides. This section analyzes these potential problems with the increasing EV penetration and different charging strategies on both the microgrids. The EV penetration capability of each microgrid is calculated, which represents the highest EV penetration that can be connected to a microgrid without having any overload or overvoltage problem.

Figure 4.11 shows the transformer utilization factor of both the microgrids with different EV charging strategies and penetrations. Overall, the implementation of CVRC can effectively reduce the transformer utilization factor in all scenarios. The utilization factor is reduced by approximately 2.5% in the rural microgrid (in equivalent of 5 kVA) and 1.5% in the urban microgrid (in equivalent of 7.5 kVA) with small-scale EV connected. As the EV penetration increases, the reduction in the utilization factor becomes less obvious. From this perspective, the CVRC can help to reduce the loading pressure of the microgrid and release more headroom for EV integration, but the effect is limited when EV penetration goes higher. On the other hand, the utilization factor can be effectively reduced by using the MIX strategy instead of the other two strategies. According to the EV profiles shown in Figures 4.3–4.5, the charging demand of the MIX strategy is separated into two periods rather than concentrated in one period. This effectively reduces the peak load of the microgrid, especially when operates with high EV penetrations. As a result, applying the MIX strategy allows up to 40% and 50% EVs to be integrated into the rural and the urban microgrid, respectively, without overloading the transformer.

Figure 4.12 shows the percentage of customers having voltage problems with different EV charging strategies and penetrations. With low-penetration EVs

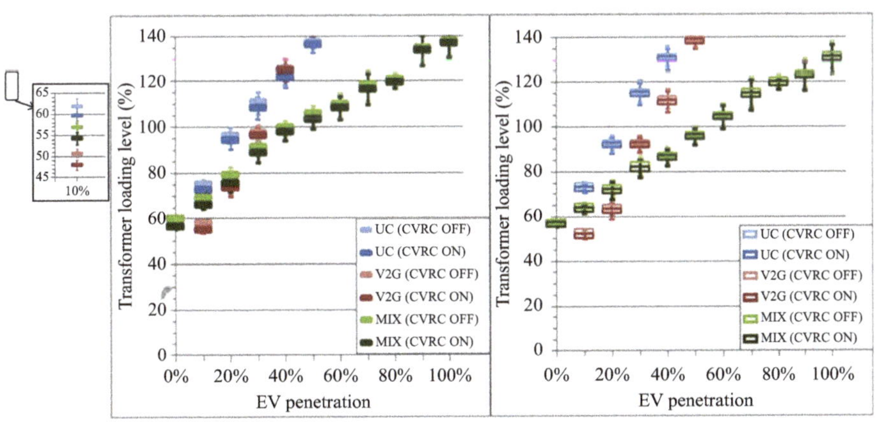

Figure 4.11 Transformer utilization factor with different EV charging strategies and penetrations on the rural (left) and urban (right) microgrids

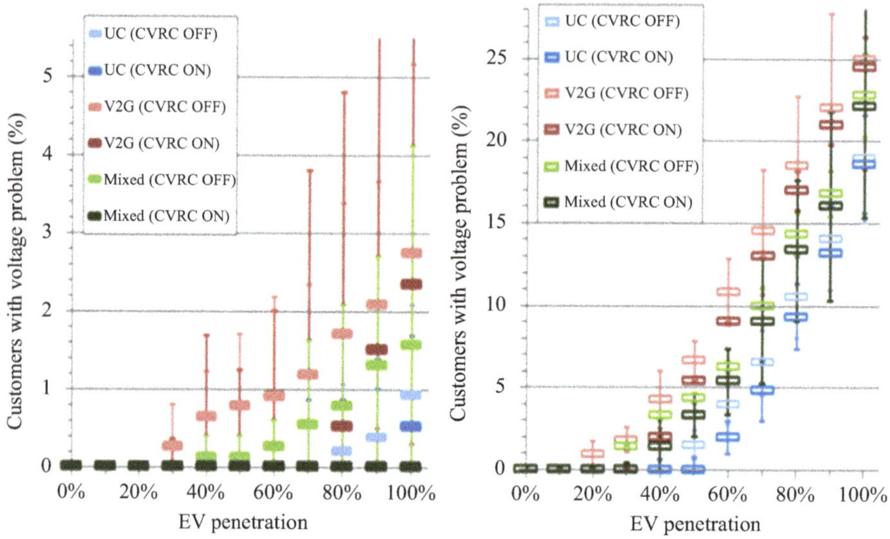

Figure 4.12 Customers having voltage problems with different EV charging strategies and penetrations on rural (left) and urban (right) microgrids

connected, the voltage problem mainly happens in the V2G and MIX strategies. The discharging process increases the feeder voltage that makes the customer voltage higher than the upper limit. In this case, applying CVRC can effectively eliminate the over-high voltage problem, especially in the rural microgrid (i.e., no voltage problem occurs with up to 70% EVs connected). As the EV penetration increases, more voltage problems start to occur in all three charging strategies. At this time, the voltage problems are mainly caused by over-LV due to the heavy charging demand. With the EV penetration increased to 100%, there are up to 2.5% customers in the rural microgrid and up to 25% customers in the urban microgrid having voltage problems even with the CVRC applied. Therefore, CVRC can only alleviate but cannot completely eliminate the voltage problems when high-penetration EVs are connected.

The EV penetration capability of a microgrid refers to the highest EV penetration that can be connected to a microgrid without overloading the transformer or causing any voltage problem to the customers. By analyzing Figures 4.11 and 4.12, the EV penetration capability of three charging strategies on both rural and urban microgrids is summarized in Table 4.6. Clearly, applying the CVRC model can effectively improve the EV penetration capability on both microgrids. In addition, using the MIX strategy allows up to 40% and 30% EVs to be integrated into the rural and the urban microgrid, respectively. Therefore, in comparison with different EV charging strategies, the MIX strategy can most effectively help CVRC to increase EV penetration capability in the distributed microgrids.

Table 4.6 Electric vehicle penetration capability improved by CVRC

	EV penetration capability in rural microgrid		EV penetration capability in urban microgrid	
	CVRC OFF (%)	CVRC ON (%)	CVRC OFF (%)	CVRC ON (%)
UC	20	20	20	20
V2G	20	30	10	30
MIX	30	40	20	30

4.6 Conclusion

This paper has proposed the MC-analysis-based CVRC model; implemented it on the typical LV-distributed microgrids and analyzed the impact of increasing EV penetration and various charging strategies on CVRC performance. The control algorithm proposed in the CVRC model applies two voltage control methods and coordinates them over each predetermined control interval based on the voltage constraints and voltage profile of nominal power-flow analysis. To analyze the impact of increasing EV penetrations and various charging strategies on CVRC, high-granularity EV models with three charging/discharging strategies have been built based on the PDF of various uncertain parameters.

The result shows that the implementation of CVRC can effectively improve the energy reduction in all scenarios. However, the energy reduction capability gradually decreases with the increase of EV penetration. In comparison with different EV charging strategies, the impact of V2G strategy on CVRC energy reduction capability is the least. In addition, the implementation of CVRC also helps to improve the EV penetration capability on both microgrids. Considered the limitation of voltage problems and transformer overloading, applying CVRC with the MIX strategy allows the largest number of EVs to be integrated into the microgrids. The analysis in this paper could guide the utilities to decide how to reasonably apply the CVRC in the future distributed microgrids with the substantial growth of EVs. Further work will consider the demand growth and more LCTs (i.e., heat pump and photovoltaic) to investigate the potential CVRC benefits on the future distributed microgrids.

References

[1] R.W. Uluski, "Volt/VAR Control and Optimization Concepts and Issues", Electric Power Research Institute (EPRI), 2011.

[2] M. Tufflex, "Voltage Optimisation: Reducing Energy, Saving Money", 2015. [Online]. Available at: http://www.marshalltufflexenergy.com/dyn/_assets/_pdfs/Voltageoptimisationbrochure-EY138.pdf. [Accessed 13 Dec 2017].

[3] W. Ellens, A. Berry, and S. West, "A quantification of the energy savings by conservation voltage reduction," in *Proc. 2012 IEEE International Conference on Power System Technology (POWERCON)*, Auckland, 2012, pp. 1–6.
[4] R. Preiss and V. Warnock, "Impact of voltage reduction on energy and demand," *IEEE Transactions on Power Apparatus and Systems*, vol. 97, no. 5, pp. 1665–1671, 1978.
[5] E. Diskin, T. Fallon, G. O'Mahony, and C. Power, "Conservation voltage reduction and voltage optimisation on Irish distribution networks," in *CIRED 2012 Workshop: Integration of Renewables into the Distribution Grid*, 2012.
[6] T.A. Short and R.W. Mee, "Voltage reduction field trials on distributions circuits," in *Proc. IEEE Transmission and Distribution Conference and Exposition (T&D)*, Orlando, FL, 2012, pp. 1–6.
[7] C. Higgins, B. Patel, and S. Ingram, "The application of conservation voltage reduction to distribution networks with high uptake of heat pumps and electric vehicles," in *Proc. CIRED Workshop*, Rome, 2014, Paper 396.
[8] L. Gutierrez-Lagos and L. Ochoa, "CVR assessment in UK residential low voltage networks considering customer types," in *Proc. IEEE Innovative Smart Grid Technologies – Asia (ISGT-Asia)*, Melbourne, VIC, Australia, 2016.
[9] J. Wang, A. Raza, T. Hong, A.C. Sullberg, F. de León and Q. Huang, "Analysis of energy savings of CVR including refrigeration loads in distribution systems," *IEEE Transactions on Power Delivery*, vol. 33, no. 1, pp. 158–168, 2018.
[10] P.K. Sen and K.H. Lee, "Conservation voltage reduction technique: an application guideline for smarter grid," in *Proc. IEEE Rural Electric Power Conference (REPC)*, Fort Worth, TX, 2014, pp. B1-1–B1-8.
[11] E. Vega-Fuentes, S. Leon-del Rosario, J. Cerezo-Sanchez, and A. Vega-Martinez, "Combined comparison-regression method for assessment of CVR effects," in *Proc. 55th International Scientific Conference on Power and Electrical Engineering of Riga Technical University (RTUCON)*, 2014.
[12] Z. Wang and J. Wang, "Review on implementation and assessment of conservation voltage reduction," *IEEE Transactions on Power Systems*, vol. 29, no. 3, pp. 1306–1315, 2014.
[13] R. Singh, F. Tuffner, J. Fuller, and K. Schneider, "Effects of distributed energy resources on conservation voltage reduction (CVR)," in *Proc. 2011 IEEE Power and Energy Society General Meeting*, Detroit, MI, USA, Oct 2011, pp. 1–7.
[14] Y. Shen, H. Li, K. Hoban, and B. Ingham, "Analysis of integrated CVR control performance with high EV penetrations," in *Proc. IEEE Powe and Energy Society General Meeting*, Chicago, Jul 2017, pp. 1–5.
[15] I. Richardson and M. Thomson, "Integrated domestic electricity demand and PV micro-generation model," Loughborough University Institutional Repository, Dspace.lboro.ac.uk, 2016. [Online]. Available at: https://dspace.lboro.ac.uk/dspace-jspui/handle/2134/7773. [Accessed 13 Dec 2017].

[16] "Smart Street", Enwl.co.uk, 2018. [Online]. Available: https://www.enwl.co.uk/innovation/smart-street/. [Accessed: 30 Jul 2018].

[17] H. Johal, W. Ren, Y. Pan, and M. Krok, "An integrated approach for controlling and optimizing the operation of a power distribution system," in *2010 IEEE PES Innovative Smart Grid Technologies Conference Europe (ISGT Europe)*, Gothenberg, 2010, pp. 1–7.

[18] BSI, "BS EN 50160:2007 – Voltage characteristics of electricity supplied by public distribution networks," BSI, 2007. [Online]. Available at: https://shop.bsigroup.com/ProductDetail/?pid=000000000030139621.

[19] ELEXON Limited, "Technical Operation Profiling," 2013. [Online]. Available at: https://www.elexon.co.uk/reference/technical-operations/profiling/.

[20] F. Lamberti, C. Dong, V. Calderaro, and L. Ochoa, "Estimating the load response to voltage changes at UK primary substations," in *Proc. 4th IEEE/PES Innovative Smart Grid Technologies Europe (ISGT EUROPE)*, Lyngby, Oct 2013, pp. 1–5.

[21] UK Hydrographic Office, "Astronomical data," [Online]. Available at: http://astro.ukho.gov.uk/psp/index_beta.html. [Accessed 8 Dec 2017].

[22] H. Wang, F. Wen, J. Huang, L. Zhang, and Q. Song, "Load characteristics of electric vehicles in charging and discharging states and impacts on distribution systems," in *Proc. International Conference on Sustainable Power Generation and Supply (SUPERGEN 2012)*, Hangzhou, Sep 2012, pp. 1–7.

[23] T. Au and M. Ortega-Vazquez, "Assessment of plug-in EV charging on distribution networks," in *Proc. IEEE PES General Meeting*, Vancouver, Jul 2013, pp. 1–5.

[24] PluginCars, "Compare Electric Cars and Plug-in Hybrids by Features, Price, Range," [Online]. Available at: http://www.plugincars.com/cars?field_isphev_value_many_to_one=pure+electric. [Accessed 13 Dec 2017].

[25] A. Anastasiadis, E. Voreadi, and N. Hatziargyriou, "Probabilistic load flow methods with high integration of renewable energy sources and electric vehicles-case study of Greece," in *Proc. IEEE Trondheim PowerTech*, Jun 2011, pp. 1–8.

[26] Y. Su, "Probability approach of profiling electric vehicles characteristics and impact on distribution systems," MSc dissertation, Department of Electronic and Electrical Engineering, University of Manchester, United Kingdom, 2016.

Chapter 5
Application of energy storage technologies on energy management of microgrids

Farshad Kalavani[1], *Behnam Mohammadi-Ivatloo*[1,2] *and Farshid Kalavani*[3]

A microgrid (MG) is a localized group of electricity loads and distributed energy resources (DERs). The DERs consist of the renewable energy resources, in which the wind farm and solar technologies have gained increasing interest and growth in the recent years. The renewable energy sources (RESs) such as wind and solar ones have the variable output power because of their stochastic nature. So, the energy storage systems (ESSs) have been investigated as achievable solution to solve the output power fluctuation of RESs. This chapter studies the different types of energy storage device technologies considering characteristics and operation constraints for MG use. The technologies of ESSs have the important role on the application of ESSs for MG use. The battery energy storage has a poor life cycle and influences the MG future development. The main characteristics of ESS technologies are the important things in the future development of MG. The ESSs are used as peak shaving, renewable energy systems integration, spinning reserve, load following, frequency regulation and black start in the MG application.

5.1 Introduction

An MG is a localized small scale of power network, which consists of controllable loads, DERs and storage devices [1]. The MG can be made of two operation modes: stand alone and grid connected. In the stand-alone mode of operation, the frequency control of MG is critical and the MG is as off-grid system. However, in the grid-connected mode, the frequency control of grid is improved by exchanging the power between MG and main grid [2]. The power quality of energy production can significantly be improved and the environmental emissions can mainly be reduced through the MG [3,4]. Different alternatives have been suggested instead of fossil fuels

[1]Faculty of Electrical and Computer Engineering, University of Tabriz, Tabriz, Iran
[2]Department of Energy Technology, Aalborg University, Aalborg, Denmark
[3]Faculty of Electrical Engineering, University of Shahed, Tehran, Iran

to gain sustainable energy resources [5,6]. The DER consists of MG-based renewable energy resources, in which the wind farms and solar technologies have gained increasing interest and growth in the recent years [7,8]. So, the periodical shortage of RESs output power develops problems in serving the load demands by electric generation on MG [9]. So, the power system in the MG occasionally loses to serve the load demand and leads to the grid frequency oscillation [10,11]. Therefore, the ESSs flatten the RES output power variation and cause high-power supply quality [12,13]. So, the integration of ESSs to RESs improves the reliability of serving the load demand on MG [14]. The integration of ESSs and RESs significantly reduces the abnormality of power system network [15]. The aggregated benefits and applications of ESSs technologies for stand-alone and grid-connected MGs have been studied in [16–18]. The proper management and selection of ESSs and RESs improve the energy serving of MG. This chapter discusses different energy storage technologies, which include the power generation and operation constraints of ESSs for efficient usage in the MG. Finally, this study consists of great information for implementing ESSs in MG energy management.

The configuration of ESS devices for MG applications is categorized as distributed and aggregated ESSs, in which, the distributed ESSs are directly connected to the DERs and control the power flow of DERs [19]. The distributed ESSs are commonly small-scale storage devices. Increasingly, the PV systems are being integrated with distributed ESSs for energy shifting [20]. In the aggregated ESSs, the storage systems are directly connected to the point of common coupling of MG and smooth the fluctuation of grid power flow [21].

The applications of ESSs in the power system consist of peak shaving [22], energy balance [23], spinning reserve [24], frequency control [10], distribution network black start [25], power system reliability [26], power quality improvement [27], RESs energy shifting [28], smoothing the fluctuations [29], voltage support in distribution systems [30] and off-grid services [31].

The ESSs are categorized due to the form of energy usage and the material of storage. So, ESSs are classified as a thermal, electrical, electrochemical, chemical, mechanical and hybrid energy storage. The compressed air energy storage (CAES) [32], cryogenic energy storage (CES) [26,33], pumped hydroelectric energy storage (PHES) [34,35], batteries [36], thermal energy storage (TES) [37], flywheel energy storage (FES) [38], supercapacitor energy storage [39], hydrogen storage [40], superconducting magnetic energy storage (SMES) [41] and hybrid energy storage [42] are the most popular ESS technologies, which are used in the MG applications.

5.2 Energy storage technologies

The different types of ESS technologies are studied in this section on the purpose of energy management of MG.

5.2.1 Compressed air energy storage

The CAES system is used in different industries since the nineteenth century. In the CAES technology, the cleaned air is compressed by compressors into the underground reservoirs (rock structures, abandon mines and salt cavern) or the aboveground reservoirs. The simplified diagram of CAES system is depicted in Figure 5.1. As shown in Figure 5.1, the compressors run to inject the air into the CAES reservoir during the off-peak periods. During peak periods, the stored air is transferred to be heated by heat sources. Then, the pressurized air is mixed with the fuel to burn in the combustion chamber. So, the compressors have been emitted from gas turbine process and the compressed air of the CAES plant is utilized in the combustion chamber.

The first unit of CAES system in the world was built in the Germany, Huntorf power plant, in 1978. The plant reservoirs consisted of two salt caverns with a capacity of 300,000 m^3. The plant runs daily with an operation cycle of 8 h to charge the reservoirs by compressed air and recovers electricity for 2 h with a power rating of 290 MW. The second CAES plant was built in the USA, McIntosh power plant, in 1991. The plant reservoirs are consisting of one solution-mined salt cavern with a capacity of 285,000 m^3. The operation cycle of plant is 26 h and the rating power is 110 MW.

The CAES system is developed for small-to-large scale of capacities. The CAES power plant can be available for use at the moderate response time and desirable performance. The utility for large-scale usage of CAES plants consists of grid applications for peak shaving, frequency and voltage control and load shifting [43]. The CAES plant can be integrated with renewable energy resources, mostly in wind power, to smooth the variation of output power [44]. The CAES system's major limitations are the round-trip efficiency and geographical location. The

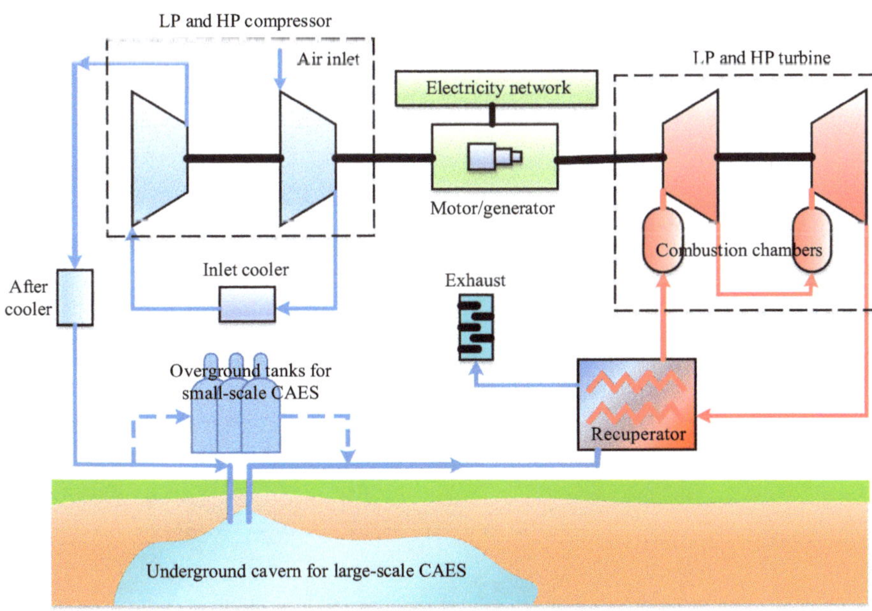

Figure 5.1 The simplified diagram of a CAES system [43]

CAES plant's overall efficiency is lower than battery and PHS energy storages. For a large-scale CAES plant, the investment cost of overground reservoirs is high. So, the geographical location limits the construction of large-scale CAES plants [45].

5.2.2 Pumped hydroelectric energy storage

In the PHES, the energy is stored in the form of gravitational potential energy of water. The water is pumped from a lower elevation reservoir to a higher level reservoir. The PHES consumes the energy in the off-peak periods to pump the water from the lower level reservoir to the higher level reservoir. In the peak periods or when the energy demand is high, the water flows down to a turbine, drives the generator and generates power. There are two types of pumped hydro storage: open loop and closed loop. In the closed loop PHES, the water of the reservoir is not connected to an outside water source. In the open loop PHES, the water of the reservoir is connected to a nature body of water such as river or lake as shown in Figure 5.2 [46].

The application of PHES system in isolated island grids decreases the electricity production cost and power system reliability [47,48]. The PHES system has the efficiency between 65% and 80% [49]. In the worldwide, there are 300 PHES plants with a total capacity of 127 GW [50].

5.2.3 Cryogenic energy storage

The CES stores the energy in the form of liquid. The charging process of CES consists of liquefying air and storing the liquid air in reservoirs. In peak periods when power is needed, the stored liquid air is pumped to be pressurized. After that, the high-pressure liquid air is expanded by ambient heat or waste heat of industries.

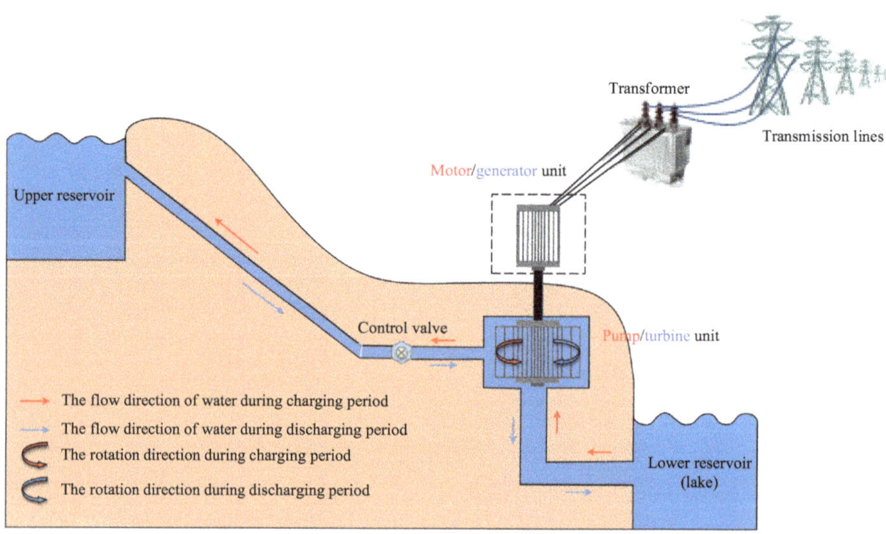

Figure 5.2 The open loop and closed loop types of PHES

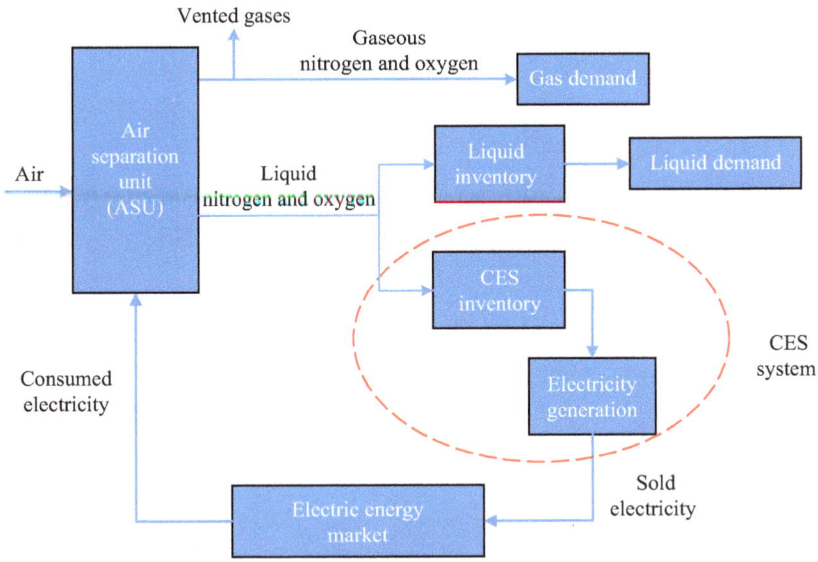

Figure 5.3 The model of integrated ASU-CES and electricity market [33]

Finally, the expanded high-pressure air drives the turbine and recovers the electricity. The round-trip efficiency in a CES plant is about 70% [33].

The Highview company in the United Kingdom is the pioneer in the world, which contributes to the construction of CES plants. The first CES plant (350 kW/ 2.5 MWh) was developed by the researchers at the University of Leeds and Highview company in 2015. The new CES plant (5 MW/15 MWh) is under construction in the United Kingdom [51].

The air liquefaction plant is required for the CES plant, in which the air separation plant is used in industry for years to serve O_2 and N_2 inquiry. Therefore, the integration of a CES plant to the existing air separation units (ASU) is so economical. The integration of air separation and a CES plant is described in Figure 5.3. As shown in Figure 5.3, the surplus-produced liquid products of ASU are stored in the CES reservoirs and recovered electricity is used to sell the electricity market in the high electricity price periods. The integrated ASU-CES system improves the reliability of serving energy of consumers [26]. The optimal CES sizes using stochastic method are determined for the MG application as described in [52]. The CES system is utilized for load shifting of nuclear power plants as described in [53].

5.2.4 Battery energy storage systems

The battery energy storage systems (BESS) are used to balance the fluctuations of frequency in the power system with a high share of renewable energy resources. The comprehensive review of BESS technologies is studied in [54,55]. In most cases, the BESS is used to control the frequency of the MG. A framework for selecting suitable BESS technology for demand response program of insolated MG is presented in [56]. The contribution of BESS to the power balance of MG to maintain the frequency within reasonable amount is depicted in Figure 5.4, in which the power profile in positive amount denotes discharging power to the MG and the power profile in negative amount denotes charging power to the BESS [57].

The BESS capacity is defined as the total quantity energy that the BESS can discharge within once discharge of the battery. The state of charge of the BESS is formulated as (5.1), which depicts the proportion of remaining capacity to the nominal capacity.

$$\text{State of charge} \triangleq \frac{\text{Remaining capacity}}{\text{Nominal capacity}} \tag{5.1}$$

In the most studies, the parallel operation of BESS can be possible by quasi *Z*-source inverters in the MG application [58]. The different technologies of BESS are described in the subsections.

5.2.4.1 The nickel–cadmium (Ni–Cd) battery

The Ni–Cd battery is a type of rechargeable battery, in which the electrodes are made of metallic cadmium and nickel oxide hydroxide.

The Ni–Cd batteries were invented in 1899. The terminal voltage of an Ni–Cd battery is about 1.2 V, which diminishes little when fully discharged. The Ni–Cd batteries are manufactured in a wide range of capacities and sizes.

The Ni–Cd batteries have been used in a wide diversity of power system applications. The largest Ni–Cd BESS with a power of 26 MW rated for 15 min was applied in the transmission project of Alaska for providing the spinning reserve capability. The Ni–Cd batteries can be applied as utility applications such as smoothing the wind farm output power fluctuation [59].

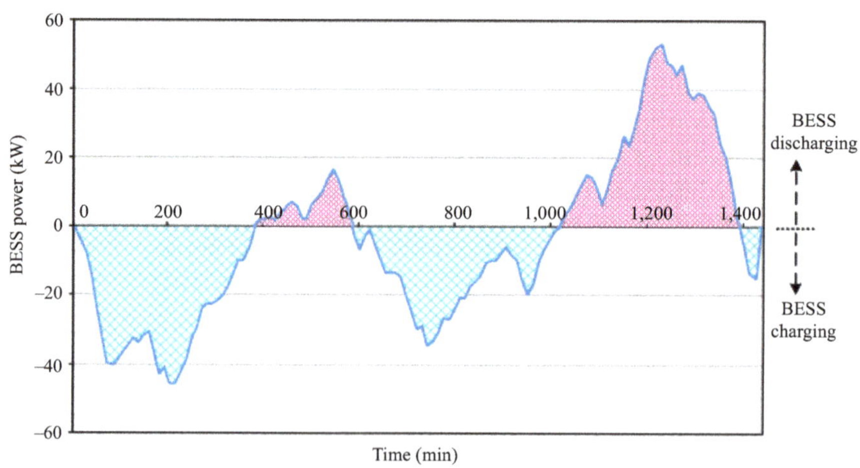

Figure 5.4 The power profile of BESS in the MG within a day time horizon [57]

5.2.4.2 Lithium-ion (Li-ion) battery

In the recent years, the Li-ion technology of battery is a fast growing technology in the world. The Li-ion battery energy storages are large-scale storage devices, which have maintenance and high energy density. The charging and discharging operation and the schematic view of a Li-ion battery is presented in Figure 5.5. The Li-ion battery anode and cathode materials are mostly manufactured from graphite carbon cell and lithium metal oxide. In charging process, lithium-ion moves from cathode to anode. The process is inversed in the discharging period. The drawbacks of Li-ion batteries are their expensive manufacturing cost and transportation requirement, as for larger quantity these need regulatory control for shipment.

The MGs operate independently from the main network and are the small-scale power system. The Li-ion batteries are the most appropriate battery technology for application in the isolated MG [60]. Recently, the Tesla company has implemented the largest plant of Li-ion battery for the Hornsdale wind farm with a total capacity of 100 MW.

5.2.4.3 Lead–acid (PbA) battery

The PbA batteries were designed in 1859 and are the most widely used type of rechargeable batteries. The anode and cathode are manufactured using Pb and PbO_2, respectively. Sulfuric acid is used as the electrolyte in the PbA batteries. The cost of PbA batteries is less expensive in comparison with other battery technologies. The PbA batteries have high efficiency (70%–80%). Therefore, the lead–acid batteries are suitable for MG application as large-scale energy storage. The PbA batteries have the ability of supplying a high amount of current, which means a high weight-to-power ratio. In the fully charged situation, the cathode plate consists of lead dioxide and the anode plate is of lead. The schematic views of the current

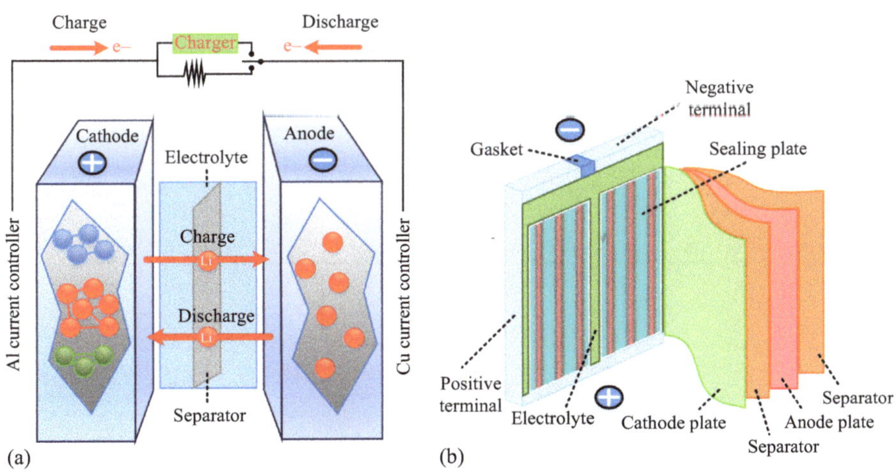

Figure 5.5 (a) The Li-ion battery operation in the charging and discharging modes and (b) the Li-ion battery structure [60]

potential of anode and cathode electrodes of a PbA battery are described in Figure 5.6.

The improvement in the material of PbA batteries makes it suitable to integrate as energy storage with the renewable resources in the MG application. The application of a PbA battery in an islanded MG is investigated in [61].

5.2.4.4 Sodium–sulfur (NaS) battery

The NaS batteries are a kind of molten-salt batteries. The electrodes comprise liquid sulfur (S) and sodium (Na). The electrolyte of NaS battery is beta alumina solid (BAS). The negative electrode is sodium, whereas the positive electrode is sulfur. The charging and discharging schematic of an NaS battery is depicted in Figure 5.7. In the discharging process, the sodium is oxidized through the passing of the BAS electrolyte to form the sodium ion. The manufactured sodium ion is incorporated with sulfur to produce sodium polysulfide. The mechanism is reversed during the charging periods [63].

This NaS battery technology has the advantages of long life cycle, high efficiency, high-energy density and is constructed of the inexpensive materials. The NaS batteries are suitable for smoothing the fluctuation of renewable energy power generation. However, the operation temperature of this type of energy storage is high (350 °C/623 K), which is needed to maintain the sulfur and sodium in the phase of liquid. This limitation leads to make obstacle in the utilizing of the NaS battery in different applications. The largest plant of NaS battery energy storage in the world with an output power of 34 MW and a storage capacity of 244.8 MWh was constructed in Japan, project of Rokkasho wind farm, for stabilizing the fluctuations of wind farm output power [64].

The characteristic comparison of different types of battery energy storage devices is presented in Table 5.1. The Ni–Cd and Li-ion are the most efficient batteries in the discharging process.

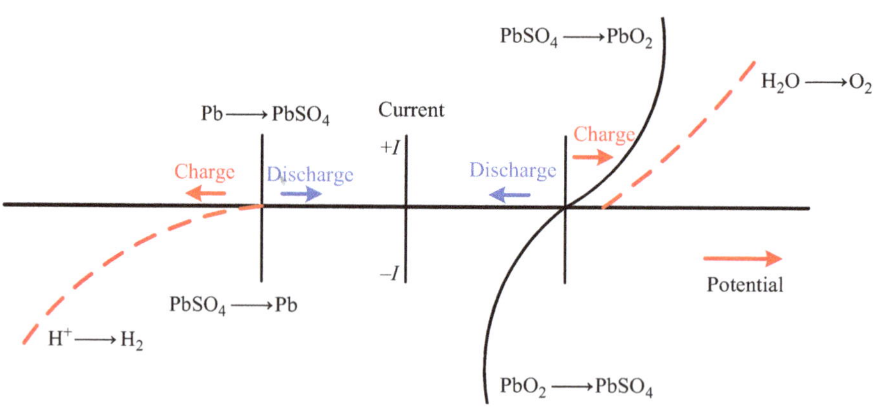

Figure 5.6 The current potential of anode and cathode electrodes of a PbA battery [62]

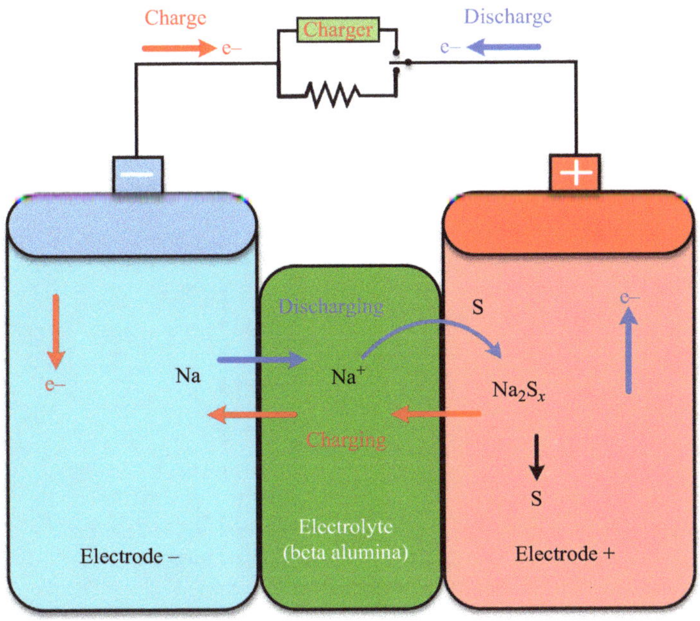

Figure 5.7 The charging and discharging schematic of an NaS battery [63]

Table 5.1 The characteristics of different technologies of battery energy storage devices

Technology name	Output power (MW)	Storage capacity (MW h)	Efficiency (%)	Lifetime (years)	Discharge time (h)	Response time
Ni–Cd	≤26	≤100	≤90	≤20	≤6	ms
Li-ion	0.25–25	≤100	≤90	≤15	≤1	ms
Lead–acid	0.25–50	≤100	≤85	≤20	≤4	ms
NaS	≤300	≤50	≤80	≤15	≤6	ms
Vanadium redox	≤250	≤50	≤80	≤10	≤8	≤10 min

5.2.5 Thermal energy storage

The TES technologies allow us to provisionally store energy in the form of ice or heat for using at different time. In the recent years, the usage of TES system is grown by 11% in the residential, industrial, power generation and load-shifting applications.

The most used type of TES system is ice storage system. The ice storage system can be integrated in to air conditioner system to provide cooling requirement in the

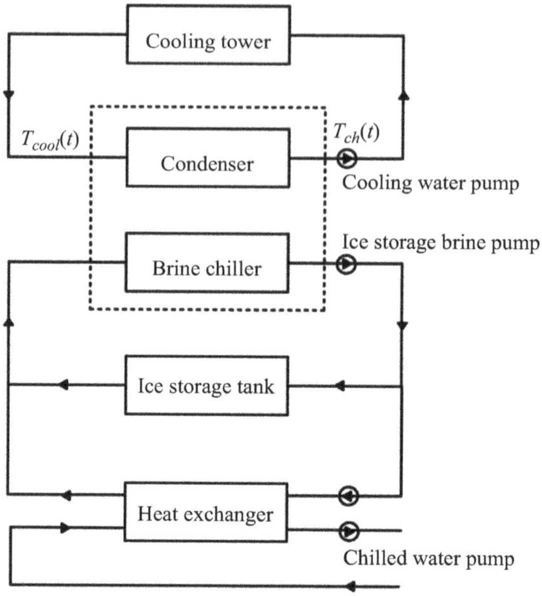

Figure 5.8 The mechanism of ice storage air conditioner system [65]

summer and lessen the energy consumption of big commercial buildings. The ice storage is an innovational technology, which makes the customers able to shift the power consumption of the air conditioner systems to the off-peak periods. The technology of ice storage stores energy in the form of ice because of large heat of fusion for water and large heat of solidification for ice. The schematic view of ice storage air conditioner system is depicted in Figure 5.8. In the melting point of ice, a relatively large amount of energy can be stored or released. Therefore, much energy will be released when water freezes and much energy can be captured when ice melts. In which, one metric ton of water can lose 334 MJ (equivalent to 93 kWh) of energy to change into ice [65]. The ice storage air conditioner system has been applied in several states of America as the national master plan [66].

5.2.6 Flywheel energy storage

FES stores energy in the form of rotational energy by accelerating a massive rotating cylinder (flywheel) to a very high speed. The FES can be categorized into types of low-speed and high-speed ones [67]. The low-speed flywheels have the speed below 10,000 rpm, which are mostly utilized in industries. The high amount of energy can be stored in the high-speed FES systems, which have a rotational speed of 100,000 rpm. The schematic view of FES system and the flywheel rotational cylinder are depicted in Figure 5.9. In the FES system, the electrical energy is used to drive the flywheel and the energy is stored in the form of kinetic energy at high speed. In the discharge mode of FES system, the flywheel speed is

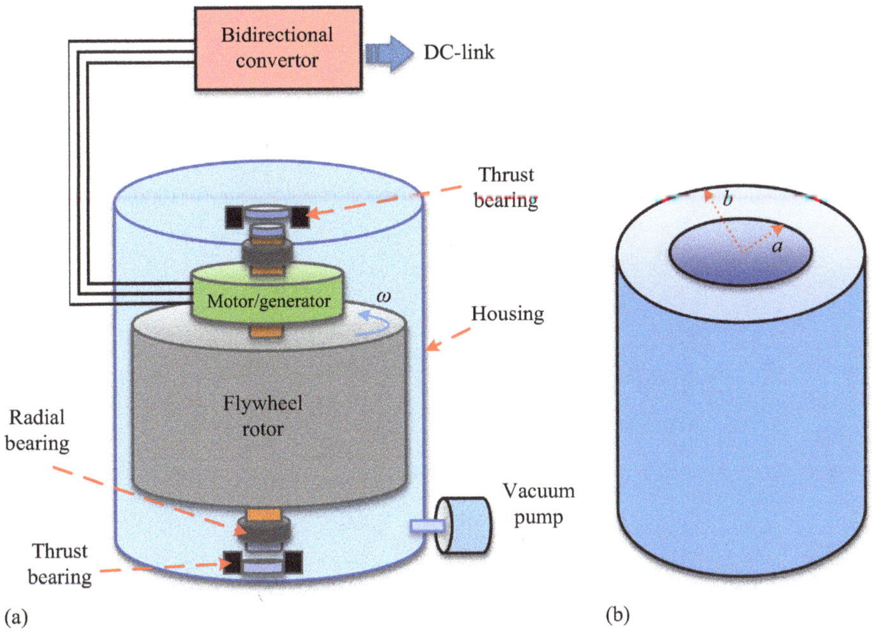

Figure 5.9 (a) The schematic view of FES system and (b) the flywheel rotational cylinder [72]

decreased and the kinetic energy is transmitted to the generator. Finally, the electrical energy is generated by decelerating rotor torque [68].

The largest losses of energy in the rotational rotors are related to the mechanical bearings. In the FES mechanism, much of this energy loss is resulted from the flywheel changing orientation due to the rotation of the earth. So, aligning the flywheel axis of rotation in parallel with the earth rotational axis decreases the energy losses in the mechanical bearings. The overall round-trip efficiency of FES system is in the range of 85%–95% and the rated output power of FES systems is in the range of 0–50 MW [69]. The FES system improves the wind power penetration in the MG by controlling the voltage, frequency and power stability [70]. In [71], the FES system studied is to be utilized as uninterruptible power supply for supplying the critical loads in the MG.

5.2.7 Hydrogen energy storage

Hydrogen energy storage (HES) system is a type of chemical energy storage system, which is suitable for storing energy for a longtime period. Hydrogen gas has the great energy quantity between fuels, which is suitable for storing and dispatching energy. Hydrogen gas has the ability to generate 120 MJ/kg, which makes possible to store large amount of energy in the small amount of hydrogen. The other important privilege of hydrogen gas is the stable chemistry, which causes to store energy for longer time period [73].

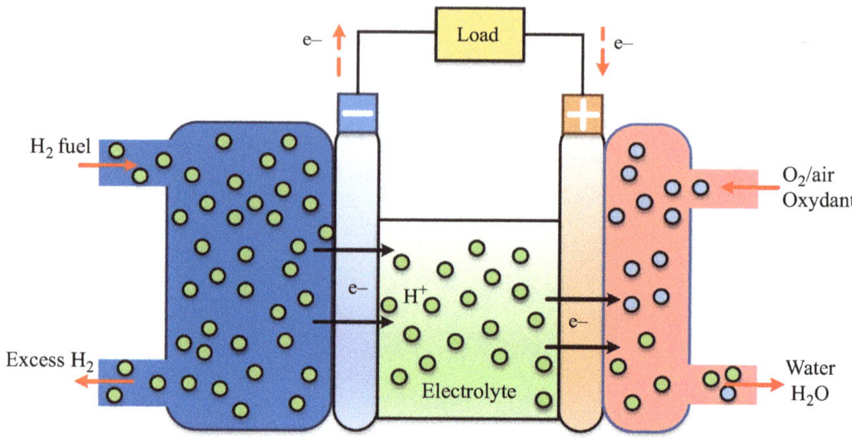

Figure 5.10 The power recovery mechanism of hydrogen fuel cell [75]

Hydrogen gas can be produced by the electrolyte, photolytic or thermal processes from a large number of different raw materials such as water, natural gas, coal, hydrogen sulfide, biomass and others. Water electrolysis technology consumes electricity to produce the hydrogen gas and stores electrical energy in the form of chemical energy.

In the power recovery process, the hydrogen gas is burned and the water is released to the environment. Therefore, the HES system is an emission-free power generation. The HES has low-energy density by volume and high-energy density by weight. The hydrogen gas burns faster and produces a large amount of energy per mass in comparison with other types of fuels [74].

The most applicable type of HES system is hydrogen fuel cell (HFC). In the HFC process, there are two electrodes, in which hydrogen ions move from one electrode through the electrolyte to reach the oxygen electrode. Finally, the electrons, oxygen and hydrogen ions are incorporated to produce water. Figure 5.10 shows the operation mechanism of HFC system in the power recovery process.

The HFC system in the presence of renewable energy resources is utilized for energy management in a low-voltage MG [76]. The integrated HES system and demand response program simultaneously reduce the daily operation cost and wind curtailment [77]. Reference [78] presents a new technique for HFC islanding detection in the MG. The hydrogen storage system is used for the performance improvement of MG by using model predictive control in [79].

5.2.8 Superconducting magnetic energy storage

In the SMES system, the electrical energy is stored in the magnetic field. In the charging mode, an AC to DC converter provides the current of superconducting coil. However, the stored energy in the magnetic field is discharged to the power grid or the connected loads using the DC to AC converter. The superconducting coil operates at the cryogenic temperature. The schematic diagram of the SMES system is shown in Figure 5.11.

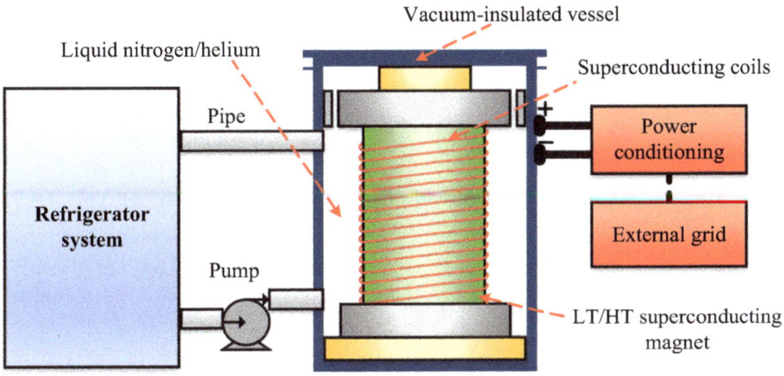

Figure 5.11 The schematic diagram of the SMES system [80]

The SMES systems are categorized based on the operating temperature to two types as follows: low-temperature and high-temperature SMES, which are working at around 7 K and 70 K, respectively.

The SMES system has long lifetime around 30 years, high efficiency (95%–98%) and high-energy density (4 kW/l). Equation (5.2) expresses the stored energy in the SMES devices, which is as follows:

$$W_{LS} = \frac{1}{2} L \times I^2 \tag{5.2}$$

where W_{LS} is the stored energy amount in the superconducting coil, I is the current amount through the superconducting coil and L is the self-inductance amount of superconducting coil.

The SMES systems are utilized with a rated output power of 0.1–10 MW. The SMES system improves the dynamic performance of the MG by controlling the power flow of renewable energy resources [81]. In [82], the hybrid energy storage comprising SMES and battery energy storage is designed to improve the lifetime of battery energy storage and to compensate the power fluctuation in the MG. The impact of SMES system to improve the frequency stability from renewable energy resources, uncertain output power and irregular demand loads in the MG is studied in [83].

5.3 Comparison of energy storage technologies

The ESS technologies named CAES and PHES are large-scale energy storages, whereas CES, BESS, TES, FES, HES and SMES are efficient for small-scale energy management in the MG. By controlling the SOC and minimizing overall loss of ESS, the reserve is provided for future demand and the efficiency is optimized in the energy management of MG, respectively. The characteristic comparison of different technologies of energy storage devices is provided in Table 5.2.

Table 5.2 The characteristic comparison of different technologies of energy storage devices

ESS technology name	Storage scale	Efficiency (%)	Advantage	Disadvantage
CAES	Large scale	70–80	• Low-cost storage reservoir • Longer lifetime compared with BESS • Backup power capability	• Geographical location limitation • Low round-trip efficiency compared with BESS
PHES	Large scale	65–80	• Low-cost storage reservoir • Longer lifetime compared with BESS • Backup power capability	• Geographical location limitation • Low round-trip efficiency compared with BESS
CES	Small-to-large scale	70	• Nondependence at geographical location limitation • Low-cost increasing the reservoir capacity • Lower dissipation rate	• Low round-trip efficiency • High energy dissipation rate
BESS	Small-to-large scale	80–90	• Energy arbitrage • Backup power capability • Peak demand reduction • Ancillary service	• The disadvantages of different technologies of BESS are considered in Table 5.1
TES	Small-to-large scale	60–95	• Rapid charge and discharge times • Negligible environmental impacts • Increasing the grid resiliency • Load shifting of air conditioner systems • Green building certification	• High energy dissipation rate • Difficult to store and transfer

(Continues)

Table 5.2 (Continued)

ESS technology name	Storage scale	Efficiency (%)	Advantage	Disadvantage
FES	Small-to-large scale	85–90	• Low maintenance cost • Negligible environmental impacts • Short recharge time • High energy density • High-power output • Low-energy dissipation	• Mechanical stress • Short discharge time • High capital cost
HES	Small-to-large scale	30–50	• No environmental emissions • Store energy for long lifetime period • High energy density by volume	• Low round-trip efficiency • High capital cost • Tricky to transfer • Low energy density by weight
SMES	Small-to-large scale	>90	• Low-power loss compared to other energy storage technologies • Short-time delay during charge and discharge periods • Improving the power quality for critical loads	• High capital cost • High-dissipation rate for long time-periods (10%–15% per day)

5.4 Conclusion

ESSs technologies are used for solving the uncertain output power of renewable energies in the MG application. The ESSs are utilized in energy management in MG by storing energy during off-peak periods in which the demand and cost of energy is low. In the peak periods, the stored energy in the ESSs is recovered and injected to the MG. However, the selection of an efficient ESS for the application of energy management in the MG is a challenging issue. The current chapter presents the study of different ESSs technologies, characteristics and considers the advantages and disadvantages of the types of ESSs for energy management in the MG. In this chapter, the implementation of ESS technologies, such as CAES, PHES, CES, BESS, TES, FES, HES and SMES, is studied in the MG application.

In addition, the output power rate, storage capacity, efficiency, response time, ESS scale, capital cost, life cycle and the operation mechanism of ESS technologies have been explained in this chapter.

References

[1] M. Marzband, A. Sumper, A. Ruiz-Álvarez, J. L. Domínguez-García, and B. Tomoiagă, "Experimental evaluation of a real time energy management system for stand-alone microgrids in day-ahead markets," *Applied Energy*, vol. 106, pp. 365–376, 2013.

[2] M. Schifani, E. Waffenschmidt, and R. Iravani, "Supervisory control of microgrids in grid-connected and islanding mode—Investigations using a real-time digital simulation platform," in *Energy and Sustainability Conference (IESC), 2017 International*, Farmingdale, NY, 2017, pp. 1–9: IEEE.

[3] V. S. Tabar, M. A. Jirdehi, and R. Hemmati, "Sustainable planning of hybrid microgrid towards minimizing environmental pollution, operational cost and frequency fluctuations," *Journal of Cleaner Production*, Vol. 203, pp. 1187–1200, 2018.

[4] A. Rabiee, M. Sadeghi, and J. Aghaei, "Modified imperialist competitive algorithm for environmental constrained energy management of microgrids," *Journal of Cleaner Production*, Vol. 202, pp. 273–292, 2018.

[5] A. Aktas, K. Erhan, S. Ozdemir, and E. Ozdemir, "Experimental investigation of a new smart energy management algorithm for a hybrid energy storage system in smart grid applications," *Electric Power Systems Research*, vol. 144, pp. 185–196, 2017.

[6] A. K. Arani, H. Karami, G. Gharehpetian, and M. Hejazi, "Review of flywheel energy storage systems structures and applications in power systems and microgrids," *Renewable and Sustainable Energy Reviews*, vol. 69, pp. 9–18, 2017.

[7] Y. Ren, P. Suganthan, and N. Srikanth, "Ensemble methods for wind and solar power forecasting—A state-of-the-art review," *Renewable and Sustainable Energy Reviews*, vol. 50, pp. 82–91, 2015.

[8] D. Laslett, C. Carter, C. Creagh, and P. Jennings, "A large-scale renewable electricity supply system by 2030: Solar, wind, energy efficiency, storage and inertia for the South West Interconnected System (SWIS) in Western Australia," *Renewable Energy*, vol. 113, pp. 713–731, 2017.

[9] A. Castillo and D. F. Gayme, "Grid-scale energy storage applications in renewable energy integration: A survey," *Energy Conversion and Management*, vol. 87, pp. 885–894, 2014.

[10] J. Li, R. Xiong, Q. Yang, F. Liang, M. Zhang, and W. Yuan, "Design/test of a hybrid energy storage system for primary frequency control using a dynamic droop method in an isolated microgrid power system," *Applied Energy*, vol. 201, pp. 257–269, 2017.

[11] M. R. Aghamohammadi and H. Abdolahinia, “A new approach for optimal sizing of battery energy storage system for primary frequency control of islanded microgrid,” *International Journal of Electrical Power & Energy Systems*, vol. 54, pp. 325–333, 2014.

[12] F. Diaz-Gonzalez, F. D. Bianchi, A. Sumper, and O. Gomis-Bellmunt, “Control of a flywheel energy storage system for power smoothing in wind power plants,” *IEEE Transactions on Energy Conversion*, vol. 29, no. 1, pp. 204–214, 2014.

[13] J. M. Guerrero, M. Chandorkar, T.-L. Lee, and P. C. Loh, “Advanced control architectures for intelligent microgrids-Part I: Decentralized and hierarchical control,” *IEEE Transactions on Industrial Electronics*, vol. 60, no. 4, pp. 1254–1262, 2013.

[14] J. Han, S. K. Solanki, and J. Solanki, “Coordinated predictive control of a wind/battery microgrid system,” *IEEE Journal of Emerging and Selected Topics in Power Electronics*, vol. 1, no. 4, pp. 296–305, 2013.

[15] M. Y. Suberu, M. W. Mustafa, and N. Bashir, “Energy storage systems for renewable energy power sector integration and mitigation of intermittency,” *Renewable and Sustainable Energy Reviews*, vol. 35, pp. 499–514, 2014.

[16] M. Katsanevakis, R. A. Stewart, and J. Lu, “Aggregated applications and benefits of energy storage systems with application-specific control methods: A review,” *Renewable and Sustainable Energy Reviews*, vol. 75, pp. 719–741, 2017.

[17] S. Kouro, J. I. Leon, D. Vinnikov, and L. G. Franquelo, “Grid-connected photovoltaic systems: An overview of recent research and emerging PV converter technology,” *IEEE Industrial Electronics Magazine*, vol. 9, no. 1, pp. 47–61, 2015.

[18] O. Ellabban, H. Abu-Rub, and F. Blaabjerg, “Renewable energy resources: Current status, future prospects and their enabling technology,” *Renewable and Sustainable Energy Reviews*, vol. 39, pp. 748–764, 2014.

[19] T. Morstyn, B. Hredzak, and V. G. Agelidis, “Control strategies for microgrids with distributed energy storage systems: An overview,” *IEEE Transactions on Smart Grid*, vol. 9, no. 4, pp. 3652–3666, 2018.

[20] C. Restrepo, A. Salazar, H. Schweizer, and A. Ginart, “Residential battery storage: Is the timing right?,” *IEEE Electrification Magazine*, vol. 3, no. 3, pp. 14–21, 2015.

[21] W. Li and G. Joos, “Comparison of Energy Storage System Technologies and Configurations in a Wind Farm,” in *2007 IEEE Power Electronics Specialists Conference*, Orlando, FL, 2007, pp. 1280–1285: IEEE.

[22] A. Oudalov, R. Cherkaoui, and A. Beguin, “Sizing and optimal operation of battery energy storage system for peak shaving application,” in *Power Tech, 2007 IEEE Lausanne*, Lausanne, 2007, pp. 621–625: IEEE.

[23] B. V. Mathiesen, H. Lund, D. Connolly, *et al.*, “Smart energy systems for coherent 100% renewable energy and transport solutions,” *Applied Energy*, vol. 145, pp. 139–154, 2015.

[24] E. Mashhour and S. M. Moghaddas-Tafreshi, "Bidding strategy of virtual power plant for participating in energy and spinning reserve markets—Part I: Problem formulation," *IEEE Transactions on Power Systems*, vol. 26, no. 2, pp. 949–956, 2011.

[25] A. Di Giorgio, A. Giuseppi, F. Liberati, and A. Pietrabissa, "Controlled electricity distribution network black start with energy storage system support," in *Control and Automation (MED), 2017 25th Mediterranean Conference on*, Valletta, 2017, pp. 781–786: IEEE.

[26] M. Zamani-Gargari, F. Kalavani, M. Abapour, and B. Mohammadi-Ivatloo, "Reliability assessment of generating systems containing wind power and air separation unit with cryogenic energy storage," *Journal of Energy Storage*, vol. 16, pp. 116–124, 2018.

[27] Q. Tabart, I. Vechiu, A. Etxeberria, and S. Bacha, "Hybrid energy storage system microgrids integration for power quality improvement using four-leg three-level NPC inverter and second-order sliding mode control," *IEEE Transactions on Industrial Electronics*, vol. 65, no. 1, pp. 424–435, 2018.

[28] A. S. Subburaj, B. N. Pushpakaran, and S. B. Bayne, "Overview of grid connected renewable energy based battery projects in USA," *Renewable and Sustainable Energy Reviews*, vol. 45, pp. 219–234, 2015.

[29] Q. Jiang and H. Hong, "Wavelet-based capacity configuration and coordinated control of hybrid energy storage system for smoothing out wind power fluctuations," *IEEE Transactions on Power Systems*, vol. 28, no. 2, pp. 1363–1372, 2013.

[30] M. Kashem and G. Ledwich, "Energy requirement for distributed energy resources with battery energy storage for voltage support in three-phase distribution lines," *Electric Power Systems Research*, vol. 77, no. 1, pp. 10–23, 2007.

[31] A. Saez-de-Ibarra, A. Milo, H. Gaztañaga, *et al.*, "Analysis and comparison of battery energy storage technologies for grid applications," in *PowerTech (POWERTECH), 2013 IEEE Grenoble*, Grenoble, 2013, pp. 1–6: IEEE.

[32] M. Abbaspour, M. Satkin, B. Mohammadi-Ivatloo, F. H. Lotfi, and Y. Noorollahi, "Optimal operation scheduling of wind power integrated with compressed air energy storage (CAES)," *Renewable Energy*, vol. 51, pp. 53–59, 2013.

[33] F. Kalavani, B. Mohammadi-Ivatloo, and K. Zare, "Optimal stochastic scheduling of cryogenic energy storage with wind power in the presence of a demand response program," *Renewable Energy*, vol. 130, pp. 268–280, 2019.

[34] A. Foley, P. Leahy, K. Li, E. McKeogh, and A. Morrison, "A long-term analysis of pumped hydro storage to firm wind power," *Applied Energy*, vol. 137, pp. 638–648, 2015.

[35] B. Steffen and C. Weber, "Optimal operation of pumped-hydro storage plants with continuous time-varying power prices," *European Journal of Operational Research*, vol. 252, no. 1, pp. 308–321, 2016.

[36] A. Bernardes, D. C. R. Espinosa, and J. S. Tenório, "Recycling of batteries: A review of current processes and technologies," *Journal of Power Sources*, vol. 130, no. 1–2, pp. 291–298, 2004.

[37] F. Kalavani, M. Nazari-Heris, and B. Mohammadi-Ivatloo, "Evaluation of peak shifting and saving energy of ice storage air conditioning system in Iran," *Journal of Operation and Automation in Power Engineering*, vol. 5, no. 2, pp. 163–170, 2017.

[38] J. R. Hull, T. M. Mulcahp, K. L. Uherka, R. A. Erck, and R. G. Abboud, "Flywheel energy storage using superconducting magnetic bearings," *Applied Superconductivity*, vol. 2, no. 7–8, pp. 449–455, 1994.

[39] W. Jing, C. H. Lai, W. S. Wong, and M. D. Wong, "Dynamic power allocation of battery-supercapacitor hybrid energy storage for standalone PV microgrid applications," *Sustainable Energy Technologies and Assessments*, vol. 22, pp. 55–64, 2017.

[40] M. Aguado, E. Ayerbe, C. Azcárate, *et al.*, "Economical assessment of a wind–hydrogen energy system using WindHyGen® software," *International Journal of Hydrogen Energy*, vol. 34, no. 7, pp. 2845–2854, 2009.

[41] T.-T. Nguyen, H.-J. Yoo, and H.-M. Kim, "Applying model predictive control to SMES system in microgrids for eddy current losses reduction," *IEEE Transactions on Applied Superconductivity*, vol. 26, no. 4, pp. 1–5, 2016.

[42] G. Oriti, A. L. Julian, N. Anglani, and G. D. Hernandez, "Novel hybrid energy storage control for a single phase energy management system in a remote islanded microgrid," in *Energy Conversion Congress and Exposition (ECCE), 2017 IEEE*, Cincinnati, OH, 2017, pp. 1552–1559: IEEE.

[43] X. Luo, J. Wang, M. Dooner, and J. Clarke, "Overview of current development in electrical energy storage technologies and the application potential in power system operation," *Applied Energy*, vol. 137, pp. 511–536, 2015.

[44] H. Guo, Y. Xu, H. Chen, and X. Zhou, "Thermodynamic characteristics of a novel supercritical compressed air energy storage system," *Energy Conversion and Management*, vol. 115, pp. 167–177, 2016.

[45] H. Chen, T. N. Cong, W. Yang, C. Tan, Y. Li, and Y. Ding, "Progress in electrical energy storage system: A critical review," *Progress in Natural Science*, vol. 19, no. 3, pp. 291–312, 2009.

[46] S. Rehman, L. M. Al-Hadhrami, and M. M. Alam, "Pumped hydro energy storage system: A technological review," *Renewable and Sustainable Energy Reviews*, vol. 44, pp. 586–598, 2015.

[47] J. Kaldellis, M. Kapsali, and K. Kavadias, "Energy balance analysis of wind-based pumped hydro storage systems in remote island electrical networks," *Applied Energy*, vol. 87, no. 8, pp. 2427–2437, 2010.

[48] G. Caralis, D. Papantonis, and A. Zervos, "The role of pumped storage systems towards the large scale wind integration in the Greek power supply system," *Renewable and Sustainable Energy Reviews*, vol. 16, no. 5, pp. 2558–2565, 2012.

[49] J. G. Levine, "Pumped hydroelectric energy storage and spatial diversity of wind resources as methods of improving utilization of renewable energy sources," University of Colorado, Boulder, 2007.

[50] D. A. Katsaprakakis, D. G. Christakis, A. Zervos, D. Papantonis, and S. Voutsinas, "Pumped storage systems introduction in isolated power production systems," *Renewable Energy*, vol. 33, no. 3, pp. 467–490, 2008.

[51] "GE storage agreement with Highview. Modern power systems," available online at: http://viewer.zmags.com/publication/f997bf2b#/f997bf2b/4, p. 4, April 2014.

[52] F. Kalavani, B. Mohammadi-Ivatloo, A. Karimi, and F. Kalavani, "Stochastic optimal sizing of integrated cryogenic energy storage and air liquefaction unit in microgrid," *Renewable Energy*, vol. 136, pp. 15–22, 2019.

[53] Y. Li, H. Cao, S. Wang, *et al.*, "Load shifting of nuclear power plants using cryogenic energy storage technology," *Applied Energy*, vol. 113, pp. 1710–1716, 2014.

[54] J. Cho, S. Jeong, and Y. Kim, "Commercial and research battery technologies for electrical energy storage applications," *Progress in Energy and Combustion Science*, vol. 48, pp. 84–101, 2015.

[55] C. S. Lai, Y. Jia, L. L. Lai, Z. Xu, M. D. McCulloch, and K. P. Wong, "A comprehensive review on large-scale photovoltaic system with applications of electrical energy storage," *Renewable and Sustainable Energy Reviews*, vol. 78, pp. 439–451, 2017.

[56] T. Khalili, A. Jafari, M. Abapour, and B. Mohammdi-Ivatloo, "Optimal battery technology selection and incentive-based demand response program utilization for reliability improvement of an insular microgrid," *Energy*, Vol. 169, pp. 92–104, 2018.

[57] H. Jia, Y. Mu, and Y. Qi, "A statistical model to determine the capacity of battery–supercapacitor hybrid energy storage system in autonomous microgrid," *International Journal of Electrical Power & Energy Systems*, vol. 54, pp. 516–524, 2014.

[58] J. Khajesalehi, M. Hamzeh, K. Sheshyekani, and E. Afjei, "Modeling and control of quasi Z-source inverters for parallel operation of battery energy storage systems: Application to microgrids," *Electric Power Systems Research*, vol. 125, pp. 164–173, 2015.

[59] X. Tan, Q. Li, and H. Wang, "Advances and trends of energy storage technology in Microgrid," *International Journal of Electrical Power & Energy Systems*, vol. 44, no. 1, pp. 179–191, 2013.

[60] H. Qian, J. Zhang, J.-S. Lai, and W. Yu, "A high-efficiency grid-tie battery energy storage system," *IEEE Transactions on Power Electronics*, vol. 26, no. 3, pp. 886–896, 2011.

[61] E. Radu, P. Dorin, P. Toma, and L. Eniko, "An islanded renewable energy microgrid emulator for geothermal, biogas, photovoltaic and lead acid battery storage," in *Industrial Electronics (ISIE), 2017 IEEE 26th International Symposium on*, Edinburgh, 2017, pp. 2109–2114: IEEE.

[62] D. Rand and P. Moseley, "Lead–acid battery fundamentals," in *Lead-Acid Batteries for Future Automobiles*. Amsterdam: Elsevier, 2017, pp. 97–132.

[63] N. Kawakami, Y. Lijima, M. Fukuhara, *et al.*, "Development and field experiences of stabilization system using 34MW NAS batteries for a 51MW wind farm," in *Industrial Electronics (ISIE), 2010 IEEE International Symposium on*, Bari, 2010, pp. 2371–2376: IEEE.

[64] B. Yang, Y. Markarov, J. Desteese, *et al.*, "On the use of energy storage technologies for regulation services in electric power systems with significant penetration of wind energy," in *European Electricity Market (EEM), 2008 5th International Conference on*, Lisboa, 2008, pp. 1–6: IEEE.

[65] M. Nazari-Heris and F. Kalavani, "Evaluation of peak shifting and energy saving potential of ice storage based air conditioning systems in Iran," *Journal of Operation and Automation in Power Engineering*, vol. 5, no. 2, pp. 163–170, 2017.

[66] F. Sehar, S. Rahman, and M. Pipattanasomporn, "Impacts of ice storage on electrical energy consumptions in office buildings," *Energy and Buildings*, vol. 51, pp. 255–262, 2012.

[67] I. Hadjipaschalis, A. Poullikkas, and V. Efthimiou, "Overview of current and future energy storage technologies for electric power applications," *Renewable and Sustainable Energy Reviews*, vol. 13, no. 6–7, pp. 1513–1522, 2009.

[68] M. G. Molina, "Distributed energy storage systems for applications in future smart grids," in *Transmission and Distribution: Latin America Conference and Exposition (T&D-LA), 2012 Sixth IEEE/PES*, Montevideo, 2012, pp. 1–7: IEEE.

[69] Y. Yuan, Y. Sun, and Y. Huang, "Design and analysis of bearingless flywheel motor specially for flywheel energy storage," *Electronics Letters*, vol. 52, no. 1, pp. 66–68, 2015.

[70] G. O. Suvire, M. G. Molina, and P. E. Mercado, "Improving the integration of wind power generation into AC microgrids using flywheel energy storage," *IEEE Transactions on Smart Grid*, vol. 3, no. 4, pp. 1945–1954, 2012.

[71] R. Arghandeh, M. Pipattanasomporn, and S. Rahman, "Flywheel energy storage systems for ride-through applications in a facility microgrid," *IEEE Transactions on Smart Grid*, vol. 3, no. 4, pp. 1955–1962, 2012.

[72] M. Amiryar and K. Pullen, "A review of flywheel energy storage system technologies and their applications," *Applied Sciences*, vol. 7, no. 3, p. 286, 2017.

[73] S. Niaz, T. Manzoor, and A. H. Pandith, "Hydrogen storage: Materials, methods and perspectives," *Renewable and Sustainable Energy Reviews*, vol. 50, pp. 457–469, 2015.

[74] R. Bhandari, C. A. Trudewind, and P. Zapp, "Life cycle assessment of hydrogen production via electrolysis—A review," *Journal of Cleaner Production*, vol. 85, pp. 151–163, 2014.

[75] M. Hannan, M. Hoque, A. Mohamed, and A. Ayob, "Review of energy storage systems for electric vehicle applications: Issues and challenges," *Renewable and Sustainable Energy Reviews*, vol. 69, pp. 771–789, 2017.

[76] S. A. Konstantinopoulos, A. G. Anastasiadis, G. A. Vokas, G. P. Kondylis, and A. Polyzakis, "Optimal management of hydrogen storage in stochastic smart microgrid operation," *International Journal of Hydrogen Energy*, vol. 43, no. 1, pp. 490–499, 2018.

[77] M. A. Mirzaei, A. S. Yazdankhah, and B. Mohammadi-Ivatloo, "Integration of demand response and hydrogen storage system in security constrained unit commitment with high penetration of wind energy," in *Electrical Engineering (ICEE), Iranian Conference on*, Mashhad, 2018, pp. 1203–1208: IEEE.

[78] Z. Liang, Y. Qin, F. Wu, *et al.*, "Standalone condition diagnosing of fuel cell in microgrid composed of wind turbine/fuel cell/combined heat & power using variational mode decomposition analysis model," *International Journal of Hydrogen Energy*, vol. 43, no. 39, pp. 18452–18462, 2018.

[79] D. Morin, Y. Stevenin, C. Grolleau, and P. Brault, "Evaluation of performance improvement by model predictive control in a renewable energy system with hydrogen storage," *International Journal of Hydrogen Energy*, vol. 43, no. 45, pp. 21017–21029, 2018.

[80] R. Amirante, E. Cassone, E. Distaso, and P. Tamburrano, "Overview on recent developments in energy storage: Mechanical, electrochemical and hydrogen technologies," *Energy Conversion and Management*, vol. 132, pp. 372–387, 2017.

[81] M. G. Molina and P. E. Mercado, "Power flow stabilization and control of microgrid with wind generation by superconducting magnetic energy storage," *IEEE Transactions on Power Electronics*, vol. 26, no. 3, p. 910, 2011.

[82] J. Li, R. Xiong, H. Mu, *et al.*, "Design and real-time test of a hybrid energy storage system in the microgrid with the benefit of improving the battery lifetime," *Applied Energy*, vol. 218, pp. 470–478, 2018.

[83] S. Singh, R. K. Verma, A. K. Shakya, and S. P. Singh, "Frequency regulation of micro-grid connected hybrid power system with SMES," *Technology and Economics of Smart Grids and Sustainable Energy*, vol. 2, no. 1, p. 13, 2017.

Chapter 6

Control technique for integration of smart dc microgrid to the utility grid

Mahesh Kumar[1] *and Sri Niwas Singh*[2,3]

In this chapter, authors present the control technique for integration of smart dc microgrid (DCMG) to the utility grid. A ring-type architecture of "smart DCMG" for integrating various renewable and nonrenewable distributed generators (DGs) and different dc and ac loads, using power electronic converters, is presented. A bidirectional three-phase voltage source inverter (VSI) has been used to integrate the DCMG to the utility grid as well as the three-phase loads. In this chapter, a control technique of the bidirectional three-phase VSI into two rotating *d–q* synchronous reference frames (SRFs), using the feedback and feed-forward control loops, has been proposed. In the proposed control technique, the dual controllers have been suggested for controlling the positive-sequence components as well as the negative-sequence components of both three-phase ac voltage and current, independently, in its own SRF. The performance analysis of the smart DCMG along with the developed control technique is also carried under different operating scenarios, in this chapter.

6.1 Introduction

Nowadays, green energy sources such as wind, solar, fuel cells, geothermal, biomass, and tidal are being used for the power generation. The direct connection of various renewable and nonrenewable DGs to the utility grid causes the various problems such as voltage and frequency variations, network contingency, more stress on electrical network, reduced power quality of the utility grid, stability, and protection issues. Therefore, the concept of the microgrid (ac or dc) is a most promising solution, which offers certain advantages such as accelerated use of the renewable energy sources, reduced transmission losses resulting in higher efficiency, higher flexibility, and better controllability [1–3]. The microgrids are

[1]Electrical and Electronics Engineering Department, Amity School of Engineering and Technology, Amity University Haryana, Gurugram, India
[2]Madan Mohan Malaviya University of Technology, Gorakhpur, India
[3]Department of Electrical Engineering, Indian Institute of Technology Kanpur, Kanpur, India

conceptualized as a low/medium voltage distribution network for the integration of various DGs, energy storage systems (ESSs), and loads, using the several power electronic converters. A microgrid can be operated in both islanded and grid-connected modes. A DCMG is being preferred over an ac microgrid due to its several benefits such as (1) better current carrying; (2) no need of controlling frequency, reactive power, and phase angle; (3) reduced total loss due to the absence of corona and reactive power, i.e., higher efficiency; and (4) no need of frequency synchronization for interconnections of multiple DGs [1,2,4]. Therefore, the smart DCMG is a promising option for facilitating the connections of several DGs, ESSs, and loads. A ring-type architecture of the smart DCMG with autonomous controls, based on dc power pool, is presented and discussed in this chapter.

The conversion of dc power supply to ac power at desired output voltage and frequency requires an inverter, which is broadly classified into two types: (1) current source inverters and (2) VSIs. The pulse width modulation (PWM) based voltage source converters (VSCs) are being widely used for interfacing various DGs, ESSs, loads, and utility grid to the microgrid, because of their several advantages such as effective parallel operation, no dc filter required due to the presence of dc link capacitor, reduced ac filter requirement for higher order harmonics elimination, no need of separate reactive power source as VSC can operate in any of the four quadrants, no need of external circuit for commutation purpose, no risk of commutation failure, fast response due to high switching frequency, controlling of active and reactive power independently, ability to supply the dead load, and require only unidirectional voltage blocking [1,5]. The VSCs allow effective voltage and power flow control, better quality of power conversion, power factor correction, system balancing, fault protection, and maximum power point tracking of the DG sources [1,2,6].

Three-phase VSIs are being widely used in adjustable-speed drives, uninterruptible power supply, industrial ac drive systems, high-voltage DC systems, flexible AC transmission system controllers, and for integrating the three-phase loads and utility grid to the DCMG (or dc bus) system. A bidirectional three-phase VSI has been used to integrate the DCMG to the utility grid as well as the three-phase loads in this chapter. The various control schemes of the three-phase VSI for integrating the three-phase loads and/or the utility grid to the DCMG (or dc bus) system are proposed in the literature [1,7–24] such as hysteresis band current control, dead-beat current control, proportional-resonant (PR) current control, predictive current control, linear proportional–integral (PI) controllers, and the control strategies into single rotating *d–q* SRF as well as into two *d–q* SRFs. These control techniques have several drawbacks as follows:

- The problems in "hysteresis band current control" schemes are as the issues of interphase distortion, poor steady-state performance with errors up to twice the hysteresis band. These also require variable switching frequency for variable load conditions resulting into poor harmonic performance and more stress on the power switching devices, and need of high switching frequency for fast dynamic response resulting in more switching losses [7,8].

- The drawbacks of the "dead-beat current control" are more sensitivity to the variations of the load parameters and require an estimation of the load parameters, band-width limitations due to inherent plant delay, and sensitivity to plant uncertainties, which cause the robustness and stability issues in the system [9,10].
- The PR control strategy suffers from the sensitivity to frequency variations, exponentially decaying transients during step changes, and being pushed toward instability margins when a small phase shift occurs [11,12].
- The main issues with the "predictive current control" are complexity in implementation, poor transient performance, not providing implicit current limit, delay in the computation of the optimal switching state, need of high sampling rate for good performance, and large algorithm processing time [13–15].
- The "linear PI controllers" cause the problem of optimal dynamic response due to relatively slow transient response [10,16].
- The main drawbacks of the control schemes based on the single rotating *d–q* SRF include limited bandwidth, the presence of dc component of the reactive power, and poor performance of the current controller in the case of unbalanced load due to the presence of the negative-sequence components [17–20].
- To address the above problems, it is necessary to control the negative-sequence component along with the positive-sequence component, separately, which requires dual current controllers. Thus, the control strategies of the three-phase VSI in the stand-alone mode or in grid-connected mode under unbalanced load or nonideal source voltage, with dual current controllers, are proposed and implemented into two rotating *d–q* SRFs, namely, positive SRF and negative SRF for controlling the positive and the negative-sequence current components, independently, in its own SRF, in [21–24]. The problem with all the existing control schemes using the *d–q* transformation is that the VSC/VSI model contains two cross-coupling terms linking the *d* and *q*-axes dynamics, which has been resolved by using the inner current feedback control loop. All these existing control schemes into two rotating *d–q* SRFs eliminate dc link voltage harmonics only, while the negative-sequence current components on ac side terminal of the three-phase VSI are ignored. All the existing control techniques of the VSCs/VSIs based on the rotating *d–q* SRF have used the feedback control loops, which behave as constant power loads (CPLs) with the negative-incremental resistance (or impedance) characteristics at the input terminals. This causes the problems of the stability degradation of the VSCs/VSIs, i.e., reduced stability margins, reduction in the power quality, and difficulty in dc voltage stabilization [25,26].

To overcome these problems, a control technique of the three-phase VSI into two rotating *d–q* SRFs has been proposed in this chapter, for integration of the DCMG to the utility grid and the three-phase loads. The proposed control strategy utilizes dual PI controllers for controlling the positive-sequence components as well as the negative-sequence components of both three-phase ac voltage and current, independently, in its own SRF. The combination of the feed-forward and feedback control loops has been used in the proposed control technique. The proposed control technique is aimed at maintaining the rated sinusoidal three-phase ac voltage with low total harmonic distortion (THD), bidirectional power flow between the DCMG and utility grid, and dc voltage stabilization of the microgrid. The performance analyses of the proposed smart DCMG with the developed control technique are also carried out under various operating conditions in this chapter.

6.2 Proposed architecture of smart dc microgrid

The proposed architecture of a ring-type "smart DCMG" with autonomous controls, based on dc energy pool, is shown in Figure 6.1. Various renewable and nonrenewable DGs, such as wind turbine (WT), solar photovoltaic (SPV), solid oxide fuel cells (SOFCs), micro-turbine (MT), battery energy storage system (BESS), and dc as well as ac loads, have been integrated to the DCMG, and their interconnection to the utility grid for bidirectional power flow between the DCMG and the utility grid to balance the power in the DCMG, as shown in Figure 6.1. The smart DCMG is aimed at providing reliable and high-quality electric power supply. The autonomous coordinated control or decentralized control has been achieved by using only the dc voltage of the DCMG as common reference signal for all DG units. Each DG unit, connected to the DCMG, has been controlled autonomously without communicating with each other.

The variable speed WT with doubly fed induction generator (DFIG) is integrated to the DCMG through a bidirectional three-phase VSC. The DFIG is being widely used in the WT DG because of its several advantages such as high efficiency, better controllability, improved power quality, reduced mechanical stresses, independent control of active and reactive power, and low cost of the back-to-back DFIG converter due to its lower rating typically between 20% and 30% of rated power of the DFIG. The SPV system is connected to the DCMG through a dc–dc boost converter. The SOFC as a controllable DG, with peak power capacity of 1.8 times of its rated power capacity, has also been connected to the DCMG through a dc–dc boost converter. The MT DG is integrated to the DCMG through a voltage source rectifier. In the islanded mode, a BESS is connected to the DCMG through a bidirectional dc–dc converter to maintain the system power balance. The smart DCMG is connected to the utility grid as well as three-phase loads through a VSI. In the grid-connected mode, the power balance in the DCMG is managed by the utility grid. The power balance in the DCMG at any time is represented by the stable dc voltage of the microgrid.

6.3 Modeling of smart dc microgrid

The modeling of each components of the DCMG is very important part for the designing of the smart DCMG. The mechanical power extracted from the available

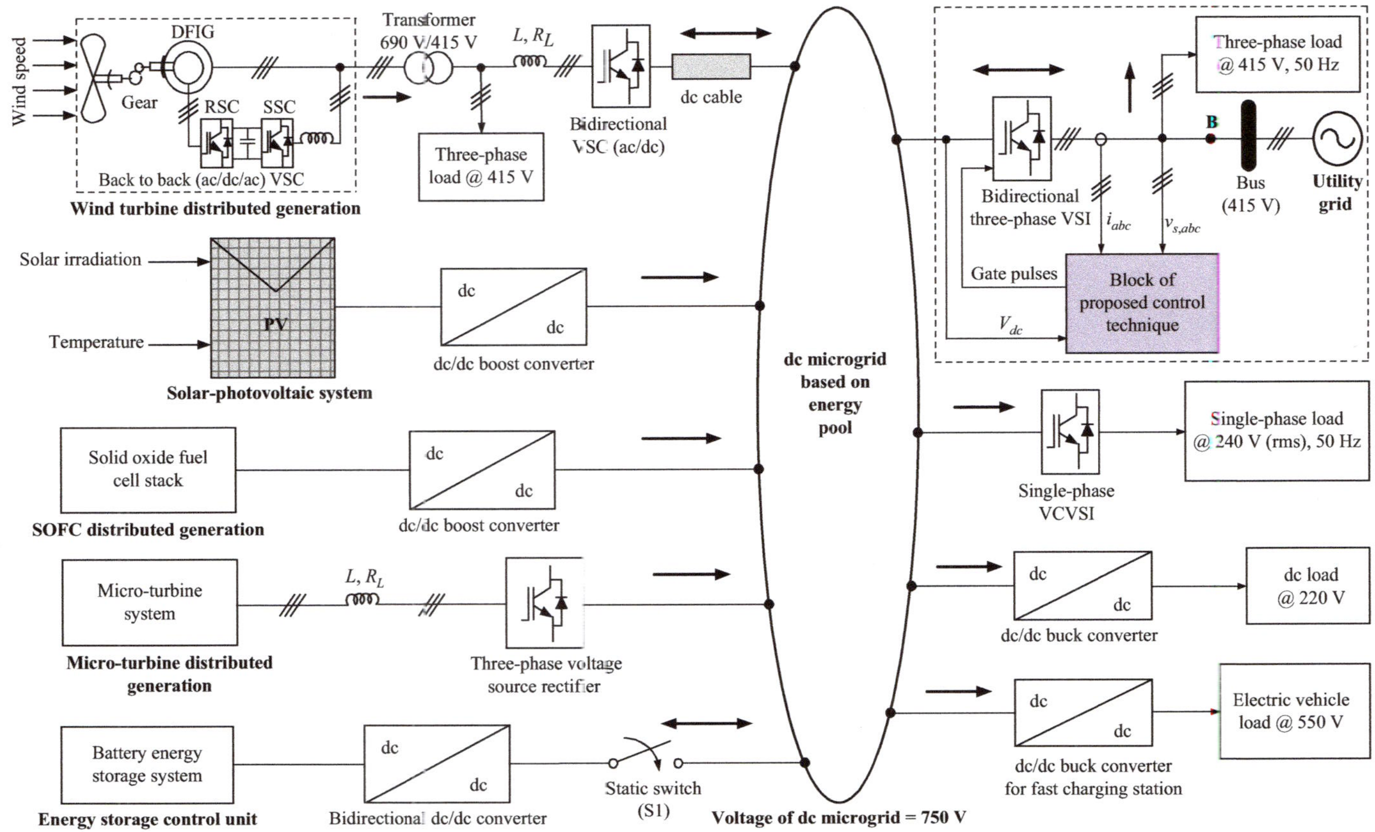

Figure 6.1 Proposed architecture of the smart dc microgrid for integration of various DGs and utility grid

wind in a WT DG is given as [1,2]

$$P_{mech} = \frac{\partial(KE_{wind})}{\partial t} C_P(\lambda, \theta) = \frac{1}{2} C_P(\lambda, \theta) \rho_{air} A_b v_{wind}^3 \tag{6.1}$$

where P_{mech} is the mechanical power generated by the WT, KE_{wind} is the kinetic energy of the wind, v_{wind} is the available wind speed, ρ_{air} is the air density, A_b is the swept area of the rotor blades, and $C_p(\lambda, \theta)$ is the power performance coefficient of the turbine that depends on the blade pitch angle θ and tip-speed ratio λ. The power coefficient, tip-speed ratio (λ), and its parameter (λ_i) are defined by a generic equation as

$$\left.\begin{aligned} &C_P(\lambda, \theta) = C_1 \left(\frac{C_2}{\lambda_i} - C_3\theta - C_4 \right) e^{-C_5/\lambda_i} + C_6\lambda \\ &\text{where, } \lambda = \frac{\omega_t R_b}{v_{wind}} \quad \text{and} \quad \frac{1}{\lambda_i} = \frac{1}{\lambda + 0.08\theta} - \frac{0.035}{\theta^3 + 1} \end{aligned}\right\} \tag{6.2}$$

where C_1 to C_6 are the coefficients taken as $C_1 = 0.5176$, $C_2 = 116$, $C_3 = 0.4$, $C_4 = 5$, $C_5 = 21$, and $C_6 = 0.0068$, ω_t is the electrical speed. The maximum power coefficient for a typical WT varies between 0.48 and 0.50, and theoretical maximum value of C_p is equal to 0.59.

The current–voltage characteristic of the SPV DG system is expressed mathematically by the following equation:

$$\left.\begin{aligned} I_{PV} = {} & n_p I_{ph} - n_p I_{sat} \left[\exp\left\{ q\left(V_{PV} + I_{PV}\left(n_{se} R_{se}/n_p \right) \right) / n_{se} K T_{cell} A \right\} - 1 \right] \\ & - \left[\left\{ \left(n_p V_{PV}/n_{se} \right) + I_{PV} R_{se} \right\} / R_{sh} \right] \end{aligned}\right\} \tag{6.3}$$

where I_{PV} is the output current of the SPV DG, I_{ph} is the photo-current, V_{PV} is the output voltage across the terminal of the SPV DG, I_{sat} is the diode saturation current, q is the charge of an electron, K is the Boltzmann constant, T_{cell} is the cell operating temperature, A is the diode ideality factor, R_{se} and R_{sh} are the internal series and parallel resistances of the SPV DG system, respectively; n_{se} is the number of series SPV cells in an SPV stack system, n_P is the number of stacks in parallel in the SPV DG system.

The SOFC stack output voltage is being determined by Nernst's equation as

$$V_{FC,stack} = N_{FC} \left(V_{0,FC} + \frac{RT}{2F} \left[\ln \left(\frac{p_{H_2} p_{O_2}^{0.5}}{p_{H_2O}} \right) \right] - r_{FC} I_{FC} \right) \tag{6.4}$$

where I_{FC} is the rated current of the SOFC, N_{FC} is the number of fuel cells connected in series in the SOFC stack, $V_{FC,stack}$ is the output voltage of the SOFC stack, $V_{0,FC}$ (=1.18 V) is the ideal standard potential of the SOFC, and r_{FC} is the ohmic loss of a single SOFC. p_{H_2}, p_{H_2O}, and p_{O_2} are the partial pressures of H_2,

H_2O, and O_2, respectively, and are described as

$$\left\{ p_{H_2} = \frac{\left(q_{H_2}^{in} - 2K_r I_{FC}\right)}{[K_{H_2}(1 + \tau_{H_2} s)]}, \quad p_{H_2O} = \frac{2K_r I_{FC}}{[K_{H_2O}(1 + \tau_{H_2O} s)]}, \quad p_{O_2} = \frac{\left(q_{O_2}^{in} - K_r I_{FC}\right)}{[K_{O_2}(1 + \tau_{O_2} s)]} \right\} \tag{6.5}$$

where $q_{H_2}^{r}$ is the hydrogen flow that reacts, $q_{H_2}^{in}$ is the input hydrogen flow, K_r is a constant dependent on Faraday's constant, K_{H_2}, K_{H_2O}, and K_{O_2} are the valve molar constants for hydrogen, water, and oxygen, respectively, and τ_{H_2}, τ_{H_2O}, and τ_{O_2} are the response time constants for flow of hydrogen, water, and oxygen, respectively.

The modeling of the split-shaft type MT DG with simplified synchronous generator has been considered as shown in Figure 6.2, where T_1 and T_2 are the time constants for fuel systems; T_3 is the load limit time constant; K_T is the temperature control loop gain; $D_{turbine}$ is the damping of turbine; V_{max} and V_{min} are the maximum and minimum valve positions, respectively; P_{ref} is the reference active power; ΔP is the difference between the power generation (P_G) and load demand (P_{Load}); ω is the rotor speed; ω_{ref} is the reference angular speed; $Load_{max}$ is the maximum loading capability, and P_{in} is the active power-controlled signal to the input of the MT DG.

The loads are generally modeled as constant resistance (CR), constant current (CC), or constant power (CP) type, depending on their characteristics. The steady-state characteristic of the loads as a function of the voltage is expressed as follows:

$$P_L(V_L) = K_{CR} V_L^2 + K_{CC} V_L + K_{CP} \tag{6.6}$$

where K_{CR} is the coefficient of the CR load, K_{CC} is the coefficient of the CC load, K_{CP} is the coefficient of the CPL, V_L is the load voltage, and P_L is the load power. The various loads connected to the DCMG, as shown in Figure 6.1, are mostly considered as CPLs. A theoretical CPL is modeled as $P_L = V_L \cdot I_L =$ constant, where I_L is the load current.

The lithium-ion battery is being widely used for the electric vehicle (EV) load. The EV batteries are modeled as CPLs.

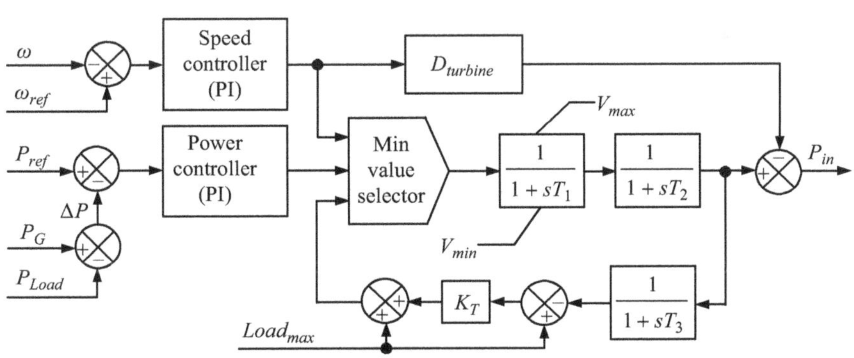

Figure 6.2 Block diagram of a split-shaft micro-turbine model

6.4 Modeling of three-phase VSI with LCL filter

A two-level, three-phase VSI with an Inductance – Capacitance – Inductance (LCL) output filter has been considered, which is shown in Figure 6.3. The voltage and current equations of the three-phase VSI with an LCL filter into three-phase stationary reference frame, two-phase (α–β) stationary reference frame, and rotating d–q SRF have been expressed by (6.7), (6.8), and (6.9), respectively.

$$\left.\begin{aligned} v_{conv,abc} &= R_{L_{f1}} i_{1,abc} + L_{f1}\frac{di_{1,abc}}{dt} + v_{C_f,abc} \\ v_{C_f,abc} &= L_{f2}\frac{di_{2,abc}}{dt} + R_{L_{f2}} i_{2,abc} + v_{s,abc} \end{aligned}\right\} \quad \text{and} \quad C_f\frac{dv_{C_f,abc}}{dt} = i_{1,abc} - i_{2,abc} \tag{6.7}$$

$$\left.\begin{aligned} v_{conv,\alpha\beta} &= R_{L_{f1}} i_{1,\alpha\beta} + L_{f1}\frac{di_{1,\alpha\beta}}{dt} + v_{C_f,\alpha\beta} \\ v_{C_f,\alpha\beta} &= L_{f2}\frac{di_{2,\alpha\beta}}{dt} + R_{L_{f2}} i_{2,\alpha\beta} + v_{s,\alpha\beta} \end{aligned}\right\} \quad \text{and} \quad C_f\frac{dv_{C_f,\alpha\beta}}{dt} = i_{1,\alpha\beta} - i_{2,\alpha\beta} \tag{6.8}$$

$$\begin{pmatrix} v_{conv,d} = R_{L_{f1}} i_{1,d} + L_{f1}\dfrac{di_{1,d}}{dt} + v_{C_f,d} - \omega L_{f1} i_{1,q} \\ v_{conv,q} = R_{L_{f1}} i_{1,q} + L_{f1}\dfrac{di_{1,q}}{dt} + v_{C_f,q} + \omega L_{f1} i_{1,d} \\ v_{C_f,d} = R_{L_{f2}} i_{2,d} + L_{f2}\dfrac{di_{2,d}}{dt} + v_{s,d} - \omega L_{f2} i_{2,q} \\ v_{C_f,q} = R_{L_{f2}} i_{2,q} + L_{f2}\dfrac{di_{2,q}}{dt} + v_{s,q} + \omega L_{f2} i_{2,d} \end{pmatrix}, \text{ and}$$
$$\begin{pmatrix} C_f\dfrac{dv_{C_f,d}}{dt} = i_{1,d} - i_{2,d} + \omega C_f v_{C_f,q} \\ C_f\dfrac{dv_{C_f,q}}{dt} = i_{1,q} - i_{2,q} - \omega C_f v_{C_f,d} \end{pmatrix} \tag{6.9}$$

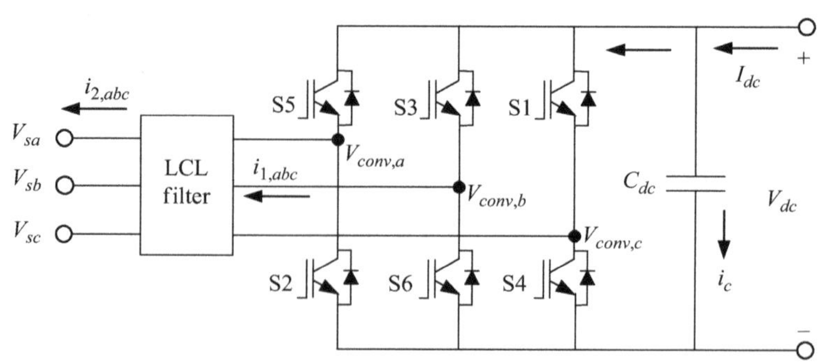

Figure 6.3 Three-phase, two-level, voltage source inverter with an LCL filter

where L_{f1} and L_{f2} are the inductors of the LCL filter, $R_{L_{f1}}$ and $R_{L_{f2}}$ are the resistances of the inductors (L_{f1}) and (L_{f2}), respectively, $i_{1,abc}$ and $i_{2,abc}$ are the output currents of three-phase VSI through the inductors L_{f1} and L_{f2} of the LCL filter, respectively, and $v_{C_f,abc}$ is the three-phase voltage across the filter capacitor (C_f), $v_{C_f,\alpha}$ and $v_{C_f,\beta}$ are the α-axis and β-axis voltage components of the filter capacitor voltage, respectively, $v_{C_f,d}$ and $v_{C_f,q}$ are the d-axis and q-axis voltage components of the filter capacitor voltage, respectively.

Under unbalanced load conditions, using a delayed signal cancellation (DSC) method, the measured output ac voltage/current of the three-phase VSI has been decomposed into two components, i.e., positive sequence and negative-sequence components, using two-phase α–β stationary reference frame. The positive and the negative-sequence components of the voltage/current are transformed from two-phase (α–β) stationary reference frame into two-phase (d–q) rotating SRFs. The expressions for the voltage drop across the LCL filter inductor (L_{f1}) and current through the filter capacitor, as shown in Figure 6.4, for the positive-sequence and the negative-sequence components into rotating d–q SRF can be given by (6.10)–(6.11) and (6.12)–(6.13), respectively:

$$\left.\begin{aligned}\frac{di^p_{1,d}}{dt} &= -\frac{R_{L_{f1}}}{L_{f1}}i^p_{1,d} + \frac{1}{L_{f1}}v^p_{conv,d} - \frac{1}{L_{f1}}v^p_{C_f,d} + \omega i^p_{1,q}\\ \frac{di^p_{1,q}}{dt} &= -\frac{R_{L_{f1}}}{L_{f1}}i^p_{1,q} + \frac{1}{L_{f1}}v^p_{conv,q} - \frac{1}{L_{f1}}v^p_{C_f,q} - \omega i^p_{1,d}\end{aligned}\right\} \tag{6.10}$$

$$\frac{dv^p_{C_f,d}}{dt} = \frac{i^p_{1,d}}{C_f} - \frac{i^p_{2,d}}{C_f} + \omega v^p_{C_f,q}, \quad \text{and} \quad \left.\frac{dv^p_{C_f,q}}{dt} = \frac{i^p_{1,q}}{C_f} - \frac{i^p_{2,q}}{C_f} - \omega v^p_{C_f,d}\right\} \tag{6.11}$$

$$\left.\begin{aligned}\frac{di^n_{1,d}}{dt} &= -\frac{R_{L_{f1}}}{L_{f1}}i^n_{1,d} + \frac{1}{L_{f1}}v^n_{conv,d} - \frac{1}{L_{f1}}v^n_{C_f,d} - \omega i^n_{1,q}\\ \frac{di^n_{1,q}}{dt} &= -\frac{R_{L_{f1}}}{L_{f1}}i^n_{1,q} + \frac{1}{L_{f1}}v^n_{conv,q} - \frac{1}{L_{f1}}v^n_{C_f,q} + \omega i^n_{1,d}\end{aligned}\right\} \tag{6.12}$$

$$\frac{dv^n_{C_f,d}}{dt} = \frac{i^n_{1,d}}{C_f} - \frac{i^n_{2,d}}{C_f} - \omega v^n_{C_f,q}, \quad \text{and} \quad \left.\frac{dv^n_{C_f,q}}{dt} = \frac{i^n_{1,q}}{C_f} - \frac{i^n_{2,q}}{C_f} + \omega v^n_{C_f,d}\right\} \tag{6.13}$$

The expressions for the voltage drop across the LCL filter inductor (L_{f2}), as shown in Figure 6.4, for the positive-sequence and the negative-sequence components into rotating d–q SRF can be given by (6.14) and (6.15), respectively:

$$\left.\begin{aligned}\frac{di^p_{2,d}}{dt} &= -\frac{R_{L_{f2}}}{L_{f2}}i^p_{2,d} + \frac{1}{L_{f2}}v^p_{C_f,d} - \frac{1}{L_{f2}}v^p_{s,d} + \omega i^p_{2,q}\\ \frac{di^p_{2,q}}{dt} &= -\frac{R_{L_{f2}}}{L_{f2}}i^p_{2,q} + \frac{1}{L_{f2}}v^p_{C_f,q} - \frac{1}{L_{f2}}v^p_{s,q} - \omega i^p_{2,d}\end{aligned}\right\} \tag{6.14}$$

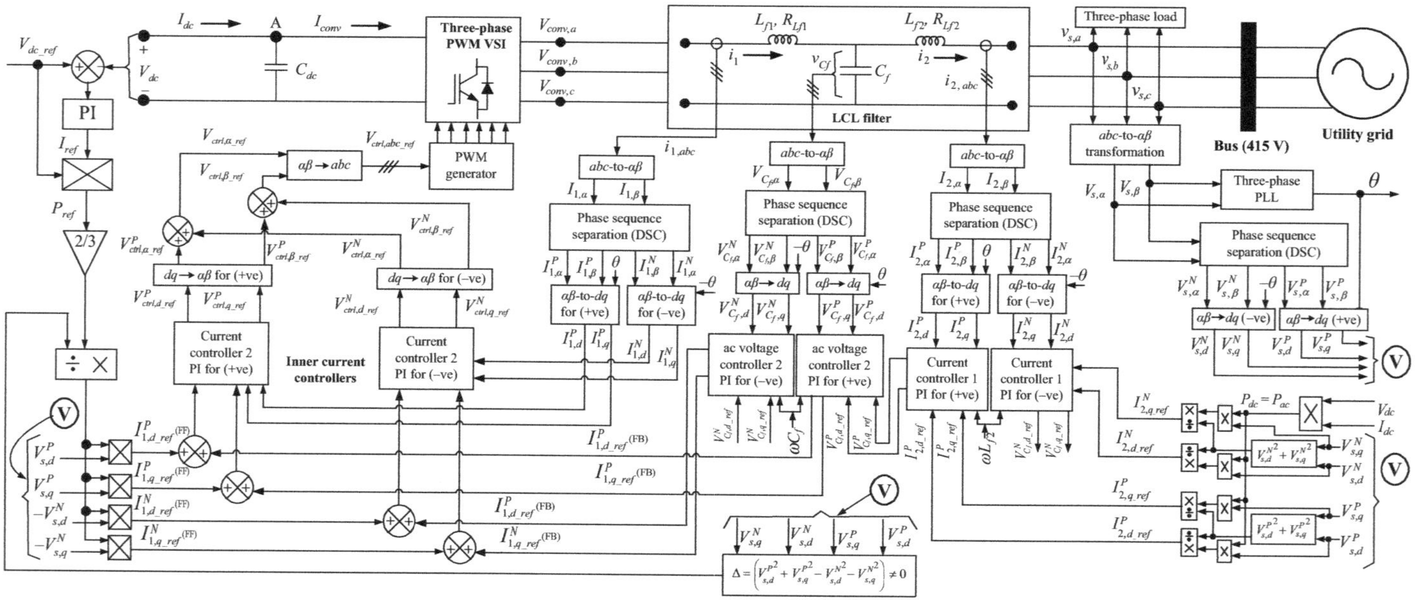

Figure 6.4 Proposed control strategy of three-phase VSI with LCL filter for integration of DCMG to the utility grid

$$\left.\begin{aligned}\frac{di^n_{2,d}}{dt} &= -\frac{R_{L_{f2}}}{L_{f2}}i^n_{2,d} + \frac{1}{L_{f2}}v^n_{C_f,d} - \frac{1}{L_{f2}}v^n_{s,d} - \omega i^n_{2,q}\\ \frac{di^n_{2,q}}{dt} &= -\frac{R_{L_{f2}}}{L_{f2}}i^n_{2,q} + \frac{1}{L_{f2}}v^p_{C_f,q} - \frac{1}{L_{f2}}v^n_{s,q} + \omega i^n_{2,d}\end{aligned}\right\} \tag{6.15}$$

where $i^p_{1,d}$ and $i^p_{1,q}$ are the d- and q-axis positive-sequence components of the output current (i_1) of three-phase VSI through the filter inductor (L_{f1}), respectively; $i^n_{1,d}$ and $i^n_{1,q}$ are the d- and q-axis negative-sequence components of the output current (i_1), respectively; $i^p_{2,d}$ and $i^p_{2,q}$ are the d- and q-axis positive-sequence components of the output current (i_2) through the filter inductor (L_{f2}), respectively; $i^n_{2,d}$ and $i^n_{2,q}$ are the d- and q-axis negative-sequence components of the output current (i_2), respectively; $v^p_{conv,d}$ and $v^p_{conv,q}$ are the d- and q-axis positive-sequence components of the three-phase VSI ac voltage, respectively; $v^n_{conv,d}$ and $v^n_{conv,q}$ are the d- and q-axis negative-sequence components of the VSI ac voltage, respectively; $v^p_{C_f,d}$ and $v^p_{C_f,q}$ are the d- and q-axis positive-sequence components of the filter capacitor voltage, respectively; $v^n_{C_f,d}$ and $v^n_{C_f,q}$ are the d- and q-axis negative-sequence components of the filter capacitor voltage, respectively; $v^p_{s,d}$ and $v^p_{s,q}$ are the d- and q-axis positive-sequence components of three-phase ac grid voltage, respectively; $v^n_{s,d}$ and $v^n_{s,q}$ are the d- and q-axis negative-sequence components of three-phase ac grid voltage, respectively.

The state space model of the three-phase VSI for the positive and negative-sequence d-axis components can be derived from (6.10)–(6.15) as follows:

$$\left.\begin{aligned}\frac{d}{dt}\begin{bmatrix} i^{p/n}_{1,d}\\ i^{p/n}_{2,d}\\ v^{p/n}_{C_f,d}\end{bmatrix} &= \begin{bmatrix} -R_{L_{f1}}/L_{f1} & 0 & -1/L_{f1}\\ 0 & -R_{L_{f2}}/L_{f2} & 1/L_{f2}\\ 1/C_f & -1/C_f & 0\end{bmatrix}\begin{bmatrix} i^{p/n}_{1,d}\\ i^{p/n}_{2,d}\\ v^{p/n}_{C_f,d}\end{bmatrix} + \begin{bmatrix} 1/L_{f1}\\ 0\\ 0\end{bmatrix} v^{p/n}_{conv,d}\\ &\pm \begin{bmatrix}\omega\\ 0\\ 0\end{bmatrix} i^{p/n}_{1,q} + \begin{bmatrix} 0\\ -1/L_{f2}\\ 0\end{bmatrix} v^{p/n}_{s,d} \pm \begin{bmatrix} 0\\ \omega\\ 0\end{bmatrix} i^{p/n}_{2,q} \pm \begin{bmatrix} 0\\ 0\\ \omega\end{bmatrix} v^{p/n}_{C_f,q}\end{aligned}\right\} \tag{6.16}$$

$$\left.\begin{aligned}\text{outputs, } \boldsymbol{Y}_1 &= \begin{bmatrix} i^p_{1,d}\\ i^p_{2,d}\\ v^p_{C_f,d}\end{bmatrix} = \begin{bmatrix} 1 & 0 & 0\\ 0 & 1 & 0\\ 0 & 0 & 1\end{bmatrix}\begin{bmatrix} i^p_{1,d}\\ i^p_{2,d}\\ v^p_{C_f,d}\end{bmatrix}, \text{ and}\\ \boldsymbol{Y}_3 &= \begin{bmatrix} i^n_{1,d}\\ i^n_{2,d}\\ v^n_{C_f,d}\end{bmatrix} = \begin{bmatrix} 1 & 0 & 0\\ 0 & 1 & 0\\ 0 & 0 & 1\end{bmatrix}\begin{bmatrix} i^n_{1,d}\\ i^n_{2,d}\\ v^n_{C_f,d}\end{bmatrix}\end{aligned}\right\} \tag{6.17}$$

The state space model of the three-phase VSI, the positive and negative-sequence q-axis components, can be derived from (6.10)–(6.15) as follows:

$$\left.\begin{aligned}\frac{d}{dt}\begin{bmatrix} i^{p/n}_{1,q} \\ i^{p/n}_{2,q} \\ v^{p/n}_{C_f,q}\end{bmatrix} &= \begin{bmatrix} -R_{L_{f1}}/L_{f1} & 0 & -1/L_{f1} \\ 0 & -R_{L_{f2}}/L_{f2} & 1/L_{f2} \\ 1/C_f & -1/C_f & 0\end{bmatrix}\begin{bmatrix} i^{p/n}_{1,q} \\ i^{p/n}_{2,q} \\ v^{p/n}_{C_f,q}\end{bmatrix} + \begin{bmatrix} 1/L_{f1} \\ 0 \\ 0\end{bmatrix} v^{p/n}_{conv,q} \\ &\mp \begin{bmatrix} \omega \\ 0 \\ 0\end{bmatrix} i^{p/n}_{1,d} + \begin{bmatrix} 0 \\ -1/L_{f2} \\ 0\end{bmatrix} v^{p/n}_{s,q} \mp \begin{bmatrix} 0 \\ \omega \\ 0\end{bmatrix} i^{p/n}_{2,d} \mp \begin{bmatrix} 0 \\ 0 \\ -\omega\end{bmatrix} v^{p/n}_{C_f,d}\end{aligned}\right\} \tag{6.18}$$

$$\left.\begin{aligned}\text{outputs, } \boldsymbol{Y}_2 &= \begin{bmatrix} i^{p}_{1,q} \\ i^{p}_{2,q} \\ v^{p}_{C_f,q}\end{bmatrix} = \begin{bmatrix} 1 & 0 & 0 \\ 0 & 1 & 0 \\ 0 & 0 & 1\end{bmatrix}\begin{bmatrix} i^{p}_{1,q} \\ i^{p}_{2,q} \\ v^{p}_{C_f,q}\end{bmatrix}, \text{ and} \\ \boldsymbol{Y}_4 &= \begin{bmatrix} i^{n}_{1,q} \\ i^{n}_{2,q} \\ v^{n}_{C_f,q}\end{bmatrix} = \begin{bmatrix} 1 & 0 & 0 \\ 0 & 1 & 0 \\ 0 & 0 & 1\end{bmatrix}\begin{bmatrix} i^{n}_{1,q} \\ i^{n}_{2,q} \\ v^{n}_{C_f,q}\end{bmatrix}\end{aligned}\right\} \tag{6.19}$$

Consider small perturbations in the various quantities as $i^p_{1,d} = I^p_{1,d} + \tilde{i}^p_{1,d}$, $i^p_{1,q} = I^p_{1,q} + \tilde{i}^p_{1,q}$, $i^n_{1,d} = I^n_{1,d} + \tilde{i}^n_{1,d}$, $i^n_{1,q} = I^n_{1,q} + \tilde{i}^n_{1,q}$, $i^p_{2,d} = I^p_{2,d} + \tilde{i}^p_{2,d}$, $i^p_{2,q} = I^p_{2,q} + \tilde{i}^p_{2,q}$, $i^n_{2,d} = I^n_{2,d} + \tilde{i}^n_{2,d}$, $i^n_{2,q} = I^n_{2,q} + \tilde{i}^n_{2,q}$, $v^p_{C_f,d} = V^p_{C_f,d} + \tilde{v}^p_{C_f,d}$, $v^p_{C_f,q} = V^p_{C_f,q} + \tilde{v}^p_{C_f,q}$, $v^n_{C_f,d} = V^n_{C_f,d} + \tilde{v}^n_{C_f,d}$, $v^n_{C_f,q} = V^n_{C_f,q} + \tilde{v}^n_{C_f,q}$, $v^p_{conv,d} = V^p_{conv,d} + \tilde{v}^p_{conv,d}$, $v^p_{conv,q} = V^p_{conv,q} + \tilde{v}^p_{conv,q}$, $v^n_{conv,d} = V^n_{conv,d} + \tilde{v}^n_{conv,d}$, $v^n_{conv,q} = V^n_{conv,q} + \tilde{v}^n_{conv,q}$, $v^p_{s,d} = V^p_{s,d} + \tilde{v}^p_{s,d}$, $v^p_{s,q} = V^p_{s,q} + \tilde{v}^p_{s,q}$, $v^n_{s,d} = V^n_{s,d} + \tilde{v}^n_{s,d}$, and $v^n_{s,q} = V^n_{s,q} + \tilde{v}^n_{s,q}$. The quantities in small letters, $i^{p,n}_{1,dq}$, $i^{p,n}_{2,dq}$, $v^{p,n}_{C_f,dq}$, $v^{p,n}_{conv,dq}$, and $v^{p,n}_{s,dq}$, are the instantaneous values, and the quantities in capital letters, $I^{p,n}_{1,dq}$, $I^{p,n}_{2,dq}$, $V^{p,n}_{C_f,dq}$, $V^{p,n}_{conv,dq}$, and $V^{p,n}_{s,dq}$ are the average values. The quantities with tilde, $\tilde{i}^{p,n}_{1,dq}$, $\tilde{i}^{p,n}_{2,dq}$, $\tilde{v}^{p,n}_{C_f,dq}$, $\tilde{v}^{p,n}_{conv,dq}$, and $\tilde{v}^{p,n}_{s,dq}$ are the perturbations in the various quantities.

By putting all these values in (6.16)–(6.19) and after solving these equations, the small signal state space models of three-phase VSI for the positive and negative-sequence d-axis and q-axis components can be derived as follows:

$$\left.\begin{aligned}\frac{d}{dt}\begin{bmatrix} \tilde{i}^{p/n}_{1,d} \\ \tilde{i}^{p/n}_{2,d} \\ \tilde{v}^{p/n}_{C_f,d}\end{bmatrix} &= \begin{bmatrix} -R_{L_{f1}}/L_{f1} & 0 & -1/L_{f1} \\ 0 & -R_{L_{f2}}/L_{f2} & 1/L_{f2} \\ 1/C_f & -1/C_f & 0\end{bmatrix}\begin{bmatrix} \tilde{i}^{p/n}_{1,d} \\ \tilde{i}^{p/n}_{2,d} \\ \tilde{v}^{p/n}_{C_f,d}\end{bmatrix} + \begin{bmatrix} 1/L_{f1} \\ 0 \\ 0\end{bmatrix} \tilde{v}^{p/n}_{conv,d} \\ &\pm \begin{bmatrix} \omega \\ 0 \\ 0\end{bmatrix} \tilde{i}^{p/n}_{1,q} + \begin{bmatrix} 0 \\ -1/L_{f2} \\ 0\end{bmatrix} \tilde{v}^{p/n}_{s,d} \pm \begin{bmatrix} 0 \\ \omega \\ 0\end{bmatrix} \tilde{i}^{p/n}_{2,q} \pm \begin{bmatrix} 0 \\ 0 \\ \omega\end{bmatrix} \tilde{v}^{p/n}_{C_f,q}\end{aligned}\right\} \tag{6.20}$$

$$\text{outputs,}\ \boldsymbol{Y}_1 = \begin{bmatrix} \tilde{i}^p_{1,d} \\ \tilde{i}^p_{2,d} \\ \tilde{v}^p_{C_f,d} \end{bmatrix} = \begin{bmatrix} 1 & 0 & 0 \\ 0 & 1 & 0 \\ 0 & 0 & 1 \end{bmatrix} \begin{bmatrix} \tilde{i}^p_{1,d} \\ \tilde{i}^p_{2,d} \\ \tilde{v}^p_{C_f,d} \end{bmatrix},\ \text{and}$$

$$\left.\boldsymbol{Y}_3 = \begin{bmatrix} \tilde{i}^n_{1,d} \\ \tilde{i}^n_{2,d} \\ \tilde{v}^n_{C_f,d} \end{bmatrix} = \begin{bmatrix} 1 & 0 & 0 \\ 0 & 1 & 0 \\ 0 & 0 & 1 \end{bmatrix} \begin{bmatrix} \tilde{i}^n_{1,d} \\ \tilde{i}^n_{2,d} \\ \tilde{v}^n_{C_f,d} \end{bmatrix}\right\} \tag{6.21}$$

$$\frac{d}{dt}\begin{bmatrix} \tilde{i}^{p/n}_{1,q} \\ \tilde{i}^{p/n}_{2,q} \\ \tilde{v}^{p/n}_{C_f,q} \end{bmatrix} = \begin{bmatrix} -R_{L_{f1}}/L_{f1} & 0 & -1/L_{f1} \\ 0 & -R_{L_{f2}}/L_{f2} & 1/L_{f2} \\ 1/C_f & -1/C_f & 0 \end{bmatrix} \begin{bmatrix} \tilde{i}^{p/n}_{1,q} \\ \tilde{i}^{p/n}_{2,q} \\ \tilde{v}^{p/n}_{C_f,q} \end{bmatrix} + \begin{bmatrix} 1/L_{f1} \\ 0 \\ 0 \end{bmatrix} \tilde{v}^{p/n}_{conv,q}$$

$$\left.\pm \begin{bmatrix} \omega \\ 0 \\ 0 \end{bmatrix} \tilde{i}^{p/n}_{1,d} + \begin{bmatrix} 0 \\ -1/L_{f2} \\ 0 \end{bmatrix} \tilde{v}^{p/n}_{s,q} \pm \begin{bmatrix} 0 \\ \omega \\ 0 \end{bmatrix} \tilde{i}^{p/n}_{2,d} \pm \begin{bmatrix} 0 \\ 0 \\ \omega \end{bmatrix} \tilde{v}^{p/n}_{C_f,d}\right\} \tag{6.22}$$

$$\text{outputs,}\ \boldsymbol{Y}_2 = \begin{bmatrix} \tilde{i}^p_{1,q} \\ \tilde{i}^p_{2,q} \\ \tilde{v}^p_{C_f,q} \end{bmatrix} = \begin{bmatrix} 1 & 0 & 0 \\ 0 & 1 & 0 \\ 0 & 0 & 1 \end{bmatrix} \begin{bmatrix} \tilde{i}^p_{1,q} \\ \tilde{i}^p_{2,q} \\ \tilde{v}^p_{C_f,q} \end{bmatrix},\ \text{and}$$

$$\left.\boldsymbol{Y}_4 = \begin{bmatrix} \tilde{i}^n_{1,q} \\ \tilde{i}^n_{2,q} \\ \tilde{v}^n_{C_f,q} \end{bmatrix} = \begin{bmatrix} 1 & 0 & 0 \\ 0 & 1 & 0 \\ 0 & 0 & 1 \end{bmatrix} \begin{bmatrix} \tilde{i}^n_{1,q} \\ \tilde{i}^n_{2,q} \\ \tilde{v}^n_{C_f,q} \end{bmatrix}\right\} \tag{6.23}$$

The plus sign in (6.20) and minus sign in (6.22) have been used for the positive-sequence components. The minus sign in (6.20) and plus sign in (6.22) have been used for the negative-sequence components.

6.5 Determination of transfer functions of three-phase VSI

In general, the state space equation in standard form is given as

$$\dot{\mathbf{X}} = \mathbf{AX} + \mathbf{B_1}U_1 + \mathbf{B_2}U_2 + \mathbf{B_3}U_3 + \mathbf{B_4}U_4 + \mathbf{B_5}U_5 \quad \text{and output,}\ \mathbf{Y} = \mathbf{CX} \tag{6.24}$$

The transfer function of the standard state space equation can be derived as

$$\mathbf{Y}(s) = \mathbf{C}[s\mathbf{I}\text{-}\mathbf{A}]^{-1}[\mathbf{B_1}U_1(s) + \mathbf{B_2}U_2(s) + \mathbf{B_3}U_3(s) + \mathbf{B_4}U_4(s) + \mathbf{B_5}U_5(s)] \tag{6.25}$$

Comparing (6.20) and (6.21) with the standard state space (6.24), one gets

$$\left.\begin{array}{l} \boldsymbol{X}=\begin{bmatrix}\tilde{i}^{p}_{1,d}\\ \tilde{i}^{p}_{2,d}\\ \tilde{v}^{p}_{C_f,d}\end{bmatrix},\boldsymbol{X}'=\begin{bmatrix}\tilde{i}^{n}_{1,d}\\ \tilde{i}^{n}_{2,d}\\ \tilde{v}^{n}_{C_f,d}\end{bmatrix},\boldsymbol{A}=\begin{bmatrix}-R_{L_{f1}}/L_{f1} & 0 & -1/L_{f1}\\ 0 & -R_{L_{f2}}/L_{f2} & 1/L_{f2}\\ 1/C_f & -1/C_f & 0\end{bmatrix},\boldsymbol{C}=\begin{bmatrix}1&0&0\\0&1&0\\0&0&1\end{bmatrix}\\ \boldsymbol{B_1}=\begin{bmatrix}1/L_{f1}\\0\\0\end{bmatrix},\ \boldsymbol{B_2}=\pm\begin{bmatrix}\omega\\0\\0\end{bmatrix},\ \boldsymbol{B_3}=\begin{bmatrix}0\\-1/L_{f2}\\0\end{bmatrix},\ \boldsymbol{B_4}=\pm\begin{bmatrix}0\\\omega\\0\end{bmatrix},\ \boldsymbol{B_5}=\pm\begin{bmatrix}0\\0\\\omega\end{bmatrix}\\ U_1=\tilde{v}^{p}_{conv,d},\ U_2=\tilde{i}^{p}_{1,q},\ U_3=\tilde{v}^{p}_{s,d},\ U_4=\tilde{i}^{p}_{2,q},\ U_5=\tilde{v}^{p}_{C_f,q}\\ U'_1=\tilde{v}^{n}_{conv,d},\ U'_2=\tilde{i}^{n}_{1,q},\ U'_3=\tilde{v}^{n}_{s,d},\ U'_4=\tilde{i}^{n}_{2,q},\ U'_5=\tilde{v}^{n}_{C_f,q} \end{array}\right\} \tag{6.26}$$

Comparing (6.22) and (6.23) with the standard state space, one gets

$$\left.\begin{array}{l} \boldsymbol{X}_1=\begin{bmatrix}\tilde{i}^{p}_{1,q}\\ \tilde{i}^{p}_{2,q}\\ \tilde{v}^{p}_{C_f,q}\end{bmatrix},\boldsymbol{X}'_1=\begin{bmatrix}\tilde{i}^{n}_{1,q}\\ \tilde{i}^{n}_{2,q}\\ \tilde{v}^{n}_{C_f,q}\end{bmatrix},\boldsymbol{A}=\begin{bmatrix}-R_{L_{f1}}/L_{f1} & 0 & -1/L_{f1}\\ 0 & -R_{L_{f2}}/L_{f2} & 1/L_{f2}\\ 1/C_f & -1/C_f & 0\end{bmatrix},\boldsymbol{C}=\begin{bmatrix}1&0&0\\0&1&0\\0&0&1\end{bmatrix}\\ \boldsymbol{B'_1}=\begin{bmatrix}1/L_{f1}\\0\\0\end{bmatrix},\ \boldsymbol{B'_2}=\begin{bmatrix}\mp\omega\\0\\0\end{bmatrix},\ \boldsymbol{B'_3}=\begin{bmatrix}0\\-1/L_{f2}\\0\end{bmatrix},\ \boldsymbol{B'_4}=\begin{bmatrix}0\\\mp\omega\\0\end{bmatrix},\ \boldsymbol{B'_5}=\begin{bmatrix}0\\0\\\mp\omega\end{bmatrix}\\ W_1=\tilde{v}^{p}_{conv,q},\ W_2=\tilde{i}^{p}_{1,d},\ W_3=\tilde{v}^{p}_{s,q},\ W_4=\tilde{i}^{p}_{2,d},\ W_5=\tilde{v}^{p}_{C_f,d}\\ W'_1=\tilde{v}^{n}_{conv,q},\ W'_2=\tilde{i}^{n}_{1,d},\ W'_3=\tilde{v}^{n}_{s,q},\ W'_4=\tilde{i}^{n}_{2,d},\ W'_5=\tilde{v}^{n}_{C_f,d} \end{array}\right\} \tag{6.27}$$

The inverse of the matrix **[sI-A]** can be determined as follows:

$$\left.\begin{array}{l} [\mathbf{sI\text{-}A}]=\begin{bmatrix}\left(S+\dfrac{R_{L_{f1}}}{L_{f1}}\right) & 0 & \dfrac{1}{L_{f1}}\\ 0 & \left(S+\dfrac{R_{L_{f2}}}{L_{f2}}\right) & -\dfrac{1}{L_{f2}}\\ -\dfrac{1}{C_f} & \dfrac{1}{C_f} & s\end{bmatrix}\\ [\mathbf{sI\text{-}A}]^{-1}=\dfrac{1}{\Delta_1}\begin{bmatrix}\left(s^2+\dfrac{R_{L_{f2}}}{L_{f2}}s+\dfrac{1}{C_fL_{f2}}\right) & \dfrac{1}{C_fL_{f1}} & -\dfrac{1}{L_{f1}}\left(s+\dfrac{R_{L_{f2}}}{L_{f2}}\right)\\ \dfrac{1}{C_fL_{f2}} & \left(s^2+\dfrac{R_{L_{f1}}}{L_{f1}}s+\dfrac{1}{C_fL_{f1}}\right) & \dfrac{1}{L_{f2}}\left(s+\dfrac{R_{L_{f1}}}{L_{f1}}\right)\\ \dfrac{1}{C_f}\left(s+\dfrac{R_{L_{f2}}}{L_{f2}}\right) & -\dfrac{1}{C_f}\left(s+\dfrac{R_{L_{f1}}}{L_{f1}}\right) & \left(s+\dfrac{R_{L_{f1}}}{L_{f1}}\right)\cdot\left(s+\dfrac{R_{L_{f2}}}{L_{f2}}\right)\end{bmatrix} \end{array}\right\} \tag{6.28}$$

where $\Delta_1 = s^3 + \left(\frac{R_{L_{f1}}}{L_{f1}} + \frac{R_{L_{f2}}}{L_{f2}}\right)s^2 + \left(\frac{1}{C_f L_{f1}} + \frac{1}{C_f L_{f2}} + \frac{R_{L_{f1}} R_{L_{f2}}}{L_{f1} L_{f2}}\right)s + \frac{R_{L_{f1}} + R_{L_{f2}}}{C_f L_{f1} L_{f2}}$

The controllable transfer functions of the positive and negative-sequence d-axis components can be derived from (6.20), (6.21), (6.25), (6.26), and (6.28) as follows:

$$G^p_{C,d}(s) = \left.\frac{\mathbf{Y}_1(s)}{U_1(s)}\right|_{\substack{\tilde{i}^p_{1,q}=0,\tilde{v}^p_{s,d}=0 \\ \tilde{i}^p_{2,q}=0,\tilde{v}^p_{C_f,q}=0}} = \mathbf{C}[\mathbf{sI\text{-}A}]^{-1}\mathbf{B}_1 = \left.\frac{\mathbf{Y}_3(s)}{U'_1(s)}\right|_{\substack{\tilde{i}^n_{1,q}=0,\tilde{v}^n_{s,d}=0 \\ \tilde{i}^n_{2,q}=0,\tilde{v}^n_{C_f,q}=0}} = G^n_{C,d}(s)$$

$$\left.\begin{array}{l} \text{where, } \mathbf{Y}_1(s) = \begin{bmatrix} \tilde{i}^p_{1,d}(s) \\ \tilde{i}^p_{2,d}(s) \\ \tilde{v}^p_{C_f,d}(s) \end{bmatrix}, \text{ and } \mathbf{Y}_3(s) = \begin{bmatrix} \tilde{i}^n_{1,d}(s) \\ \tilde{i}^n_{2,d}(s) \\ \tilde{v}^n_{C_f,d}(s) \end{bmatrix} \end{array}\right\} \quad (6.29)$$

Thus, by solving (6.26), (6.28), and (6.29), the controllable transfer functions of the positive-sequence and the negative-sequence d-axis components are given by (6.30) as follows:

$$\left.\begin{array}{l}
G_{C_1,d}(s) = \left.\frac{\tilde{i}^p_{1,d}(s)}{\tilde{v}^p_{conv,d}(s)}\right|_{\substack{\tilde{i}^p_{1,q}=\tilde{v}^p_{s,d}=0 \\ \tilde{i}^p_{2,q}=\tilde{v}^p_{C_f,q}=0}} = \left(\frac{1}{L_{f1}}\left(s^2 + \frac{R_{L_{f2}}}{L_{f2}}s + \frac{1}{C_f L_{f2}}\right)\Big/\Delta_1\right) = \left.\frac{\tilde{i}^n_{1,d}(s)}{\tilde{v}^n_{conv,d}(s)}\right|_{\substack{\tilde{i}^n_{1,q}=\tilde{v}^n_{s,d}=0 \\ \tilde{i}^n_{2,q}=\tilde{v}^n_{C_f,q}=0}} \\
G_{C_2,d}(s) = \left.\frac{\tilde{i}^p_{2,d}(s)}{\tilde{v}^p_{conv,d}(s)}\right|_{\substack{\tilde{i}^p_{1,q}=\tilde{v}^p_{s,d}=0 \\ \tilde{i}^p_{2,q}=\tilde{v}^p_{C_f,q}=0}} = \left(\frac{1}{C_f L_{f1} L_{f2} \Delta_1}\right) = \left.\frac{\tilde{i}^n_{2,d}(s)}{\tilde{v}^n_{conv,d}(s)}\right|_{\substack{\tilde{i}^n_{1,q}=\tilde{v}^n_{s,d}=0 \\ \tilde{i}^n_{2,q}=\tilde{v}^n_{C_f,q}=0}} \\
G_{C_3,d}(s) = \left.\frac{\tilde{v}^p_{C_f,d}(s)}{\tilde{v}^p_{conv,d}(s)}\right|_{\substack{\tilde{i}^p_{1,q}=\tilde{v}^p_{s,d}=0 \\ \tilde{i}^p_{2,q}=\tilde{v}^p_{C_f,q}=0}} = \left(\frac{1}{C_f L_{f1}}\left(s + \frac{R_{L_{f2}}}{L_{f2}}\right)\Big/\Delta_1\right) = \left.\frac{\tilde{v}^n_{C_f,d}(s)}{\tilde{v}^n_{conv,d}(s)}\right|_{\substack{\tilde{i}^n_{1,q}=\tilde{v}^n_{s,d}=0 \\ \tilde{i}^n_{2,q}=\tilde{v}^n_{C_f,q}=0}}
\end{array}\right\} \quad (6.30)$$

Since $V_{conv} = \delta V_{dc}$, then (6.30) can be expressed as

$$
\left.
\begin{aligned}
G_{C_1,d}(s) &= \left.\frac{\tilde{i}^p_{1,d}(s)}{\tilde{\delta}(s)}\right|_{\substack{\tilde{i}^p_{1,q}=\tilde{v}^p_{s,d}=0\\ \tilde{i}^p_{2,q}=\tilde{v}^p_{C_f,q}=0}} = \left(\frac{V_{dc}}{L_{f1}}\left(s^2+\frac{R_{L_{f2}}}{L_{f2}}s+\frac{1}{C_fL_{f2}}\right)\Big/\Delta_1\right) = \left.\frac{\tilde{i}^n_{1,d}(s)}{\tilde{\delta}(s)}\right|_{\substack{\tilde{i}^n_{1,q}=\tilde{v}^n_{s,d}=0\\ \tilde{i}^n_{2,q}=\tilde{v}^n_{C_f,q}=0}}\\
G_{C_2,d}(s) &= \left.\frac{\tilde{i}^p_{2,d}(s)}{\tilde{\delta}(s)}\right|_{\substack{\tilde{i}^p_{1,q}=\tilde{v}^p_{s,d}=0\\ \tilde{i}^p_{2,q}=\tilde{v}^p_{C_f,q}=0}} = \left(\frac{V_{dc}}{C_fL_{f1}L_{f2}\Delta_1}\right) = \left.\frac{\tilde{i}^n_{2,d}(s)}{\tilde{\delta}(s)}\right|_{\substack{\tilde{i}^n_{1,q}=\tilde{v}^n_{s,d}=0\\ \tilde{i}^n_{2,q}=\tilde{v}^n_{C_f,q}=0}}\\
G_{C_3,d}(s) &= \left.\frac{\tilde{v}^p_{C_f,d}(s)}{\tilde{\delta}(s)}\right|_{\substack{\tilde{i}^p_{1,q}=\tilde{v}^p_{s,d}=0\\ \tilde{i}^p_{2,q}=\tilde{v}^p_{C_f,q}=0}} = \left(\frac{V_{dc}}{C_fL_{f1}}\left(s+\frac{R_{L_{f2}}}{L_{f2}}\right)\Big/\Delta_1\right) = \left.\frac{\tilde{v}^n_{C_f,d}(s)}{\tilde{\delta}(s)}\right|_{\substack{\tilde{i}^n_{1,q}=\tilde{v}^n_{s,d}=0\\ \tilde{i}^n_{2,q}=\tilde{v}^n_{C_f,q}=0}}
\end{aligned}
\right\}
\tag{6.31}
$$

Similarly, the controllable transfer functions of the positive-sequence and negative-sequence q-axis components can be derived from (6.22), (6.23), (6.25), (6.27), and (6.28) as follows:

$$
\left.
\begin{aligned}
G^p_{C,q}(s) &= \left.\frac{\mathbf{Y}_2(s)}{W_1(s)}\right|_{\substack{\tilde{i}^p_{1,d}=0,\tilde{v}^p_{s,q}=0\\ \tilde{i}^p_{2,d}=0,\tilde{v}^p_{C_f,d}=0}} = \mathbf{C}[\mathbf{sI}\text{-}\mathbf{A}]^{-1}\mathbf{B}_1 = \left.\frac{\mathbf{Y}_4(s)}{W'_1(s)}\right|_{\substack{\tilde{i}^n_{1,d}=0,\tilde{v}^n_{s,q}=0\\ \tilde{i}^n_{2,d}=0,\tilde{v}^n_{C_f,d}=0}} = G^n_{C,q}(s)\\
G_{C1,q}(s) &= \left.\frac{\tilde{i}^p_{1,q}(s)}{\tilde{v}^p_{conv,q}(s)}\right|_{\substack{\tilde{i}^p_{1,d}=0,\tilde{v}^p_{s,q}=0\\ \tilde{i}^p_{2,d}=0,\tilde{v}^p_{C_f,d}=0}} = \left(\frac{1}{L_{f1}}\left(s^2+\frac{R_{L_{f2}}}{L_{f2}}s+\frac{1}{C_fL_{f2}}\right)\Big/\Delta_1\right) = \left.\frac{\tilde{i}^n_{1,q}(s)}{\tilde{v}^n_{conv,q}(s)}\right|_{\substack{\tilde{i}^n_{1,d}=0,\tilde{v}^n_{s,q}=0\\ \tilde{i}^n_{2,d}=0,\tilde{v}^n_{C_f,d}=0}}\\
G_{C2,q}(s) &= \left.\frac{\tilde{i}^p_{2,q}(s)}{\tilde{v}^p_{conv,q}(s)}\right|_{\substack{\tilde{i}^p_{1,d}=0,\tilde{v}^p_{s,q}=0\\ \tilde{i}^p_{2,d}=0,\tilde{v}^p_{C_f,d}=0}} = \left(\frac{1}{C_fL_{f1}L_{f2}\Delta_1}\right) = \left.\frac{\tilde{i}^n_{2,q}(s)}{\tilde{v}^n_{conv,q}(s)}\right|_{\substack{\tilde{i}^n_{1,d}=0,\tilde{v}^n_{s,q}=0\\ \tilde{i}^n_{2,d}=0,\tilde{v}^n_{C_f,d}=0}}\\
G_{C2,q}(s) &= \left.\frac{\tilde{v}^p_{C_f,q}(s)}{\tilde{v}^p_{conv,q}(s)}\right|_{\substack{\tilde{i}^p_{1,d}=0,\tilde{v}^p_{s,q}=0\\ \tilde{i}^p_{2,d}=0,\tilde{v}^p_{C_f,d}=0}} = \left(\frac{1}{C_fL_{f1}}\left(s+\frac{R_{L_{f2}}}{L_{f2}}\right)\Big/\Delta_1\right) = \left.\frac{\tilde{v}^n_{C_f,q}(s)}{\tilde{v}^n_{conv,q}(s)}\right|_{\substack{\tilde{i}^n_{1,d}=0,\tilde{v}^n_{s,q}=0\\ \tilde{i}^n_{2,d}=0,\tilde{v}^n_{C_f,d}=0}}
\end{aligned}
\right\}
\tag{6.32}
$$

6.6 Proposed control technique of three-phase VSI for integrating dc microgrid to utility grid

A control technique of a three-phase VSI with an LCL output filter, for the integration of the smart DCMG to the utility grid, has been proposed, as shown in Figure 6.4. The unbalanced power supply and ac faults on the utility grid may cause the problem of instability of the VSI and dc voltage stabilization of the microgrid. Due to the stability issue in the DCMG voltage, the output of the DGs connected to the DCMG would be affected, since each DG unit is controlled autonomously by using only DCMG voltage as common reference signal for all DGs. The purpose of the proposed control technique is to overcome the problem of the stability of the VSI and to maintain the sinusoidal three-phase ac voltage at its rated voltage with low THD, bidirectional power flow between the DCMG and utility grid, dc voltage stabilization of the microgrid, and power quality.

The proposed control strategy, with dual PI controllers for both ac voltage regulation and the inner current control, has been suggested and implemented into two rotating *d–q* SRFs, namely, positive SRF and negative SRF. The main contributions in the proposed control scheme are as follows:

1. Generally, only a single feedback current control loop with dual controllers is being used in the control strategies in the literature. However, the proposed control technique uses two feedback control loops (both current and voltage feedback control loops) along with dual PI controllers for controlling both the positive and negative-sequence components of the three-phase ac voltage/current, separately, in its own SRF.
2. The feed-forward control loop as a dc voltage controller has been used in the proposed control technique to provide the controlled feed-forward reference currents (both positive and negative-sequence *d–q* components) for inner current controllers.

As shown in Figure 6.4, in the proposed control technique, the measured output voltage and current of the three-phase VSI are transformed from three-phase (*abc*) stationary reference frame into two-phase (α–β) stationary reference frame, by using the Clarke transformation. The DSC method has been used for the phase sequence decomposition, referred to as positive-sequence components and negative-sequence components of the measured three-phase ac voltage/current into α–β stationary reference frame. The α–β positive and negative-sequence components of the voltage/current are transformed into two-phase (*d–q*) rotating SRF, by using the Park transformation. The gate pulse generation system is also shown in Figure 6.4.

In grid-connected DCMG system, the positive and negative-sequenced-*q* components of the utility grid ac voltage and the dc power supply (P_{dc}) provide the positive and negative-sequence reference currents ($I^{p,n}_{2,dq\_ref}$). These reference currents are used as feedback control signals for inner current (I_2) controllers of both the positive and the negative-sequence *d–q* current components, as shown in Figure 6.4.

The positive-sequence *d–q* components and negative-sequence *d–q* components of the current (I_2) have been controlled by the PI current controllers with cross coupling of filter inductor (L_{f2}), independently, in its own SRF, as shown in Figures 6.4 and 6.5(a) and (b). These current (I_2) controllers provide the positive and the negative-sequence reference voltages ($V^{p,n}_{C_f,dq\_ref}$) for ac voltage (V_{C_f}) controllers as feedback controllers. As shown in Figures 6.4 and 6.6(a) and (b), the ac

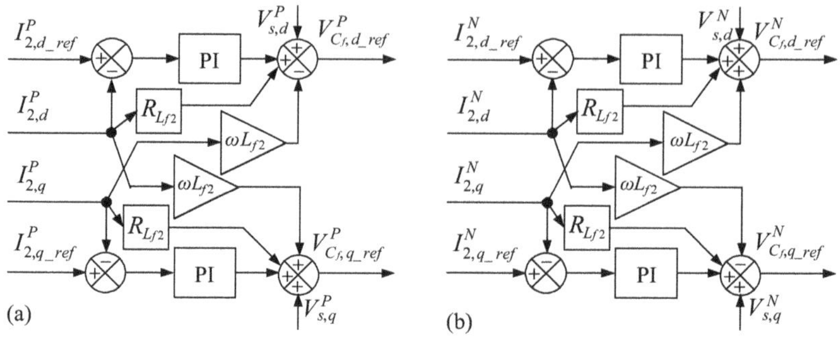

Figure 6.5 Inner current (I_2) controllers for (a) positive and (b) negative-sequence components

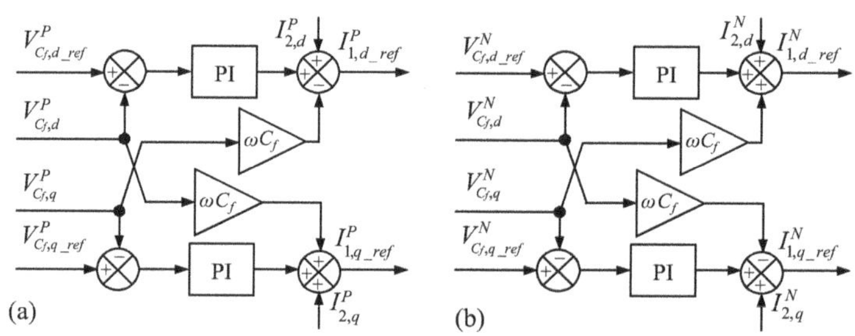

Figure 6.6 AC voltage (V_{C_f}) controllers for (a) positive and (b) negative-sequence components

voltage regulators, with cross-coupling of filter capacitor, have been used for controlling the positive-sequence d–q components and negative-sequence d–q components of voltage (V_{C_f}), independently, in its own SRF. These ac voltage controllers provide the positive and the negative-sequence reference currents ($I^{p,n}_{1,dq\_ref(\mathrm{FB})}$) as feedback control signals for inner current (I_1) controllers.

The dc voltage controller (as feed-forward control loop) provides a controlled reference current (I_{ref}) or reference power (P_{ref}) signal, as shown in Figure 6.4. The feed-forward positive and negative-sequence reference currents ($I^{p,n}_{1,dq\_ref(\mathrm{FF})}$), for inner current ($I_1$) controllers (both positive and negative sequences), have been generated by using the controlled reference power (P_{ref}) and the positive sequence as well as the negative-sequence d–q components of the utility grid voltage, as shown in Figure 6.4.

As shown in Figure 6.4, the combination of the feedback reference currents ($I^{p,n}_{1,dq\_ref(\mathrm{FB})}$) and feed-forward reference currents ($I^{p,n}_{1,dq\_ref(\mathrm{FF})}$) control signals has been used to generate the positive and the negative-sequence reference currents ($I^{p,n}_{1,dq\_ref}$) for the inner current (I_1) controllers. The inner current controllers, with cross-coupling of inductor (L_{f1}), have been used to control the positive sequence as well as the negative-sequence d–q components of current (I_1), independently, in its own SRF. The outputs of these current controllers provide the positive and the negative sequence controlled reference *d–q* voltages ($V^{p,n}_{ctrl,dq\_ref}$), which are transformed into two-phase (α–β) stationary reference frame ($V^{p,n}_{ctrl,\alpha\beta\_ref}$), by using the inverse-Park transformation, and, then, transformed into three-phase (*abc*) stationary reference frame to obtain controlled reference voltage (V_{ctrl,abc_ref}), by using inverse-Clarke transformation. This reference voltage (V_{ctrl,abc_ref}) is sent to the PWM generator to provide the controlled gate pulses for bidirectional three-phase VSI, as shown in Figure 6.4.

The feed-forward control loop allows one to improve the performance of three-phase VSI and stabilizes dc voltage. The feedback control loops eliminate the negative-sequence current in the VSI output to maintain good output voltage regulation. The inner current control loops allow faster transient/dynamic response under dc voltage fluctuations due to change in the power generation as well as the loads.

In balanced conditions, only the positive-sequence components of the voltage/current are present in the analysis, and the negative-sequence components (both *d* and *q*) of the voltage/current are not present (i.e., assumed to be zero). However, in the case of unbalanced or single-phase fault conditions, both the positive as well as negative-sequence *d–q* components of the voltage/current are present in the analysis. Separate controllers have been used for both the positive and negative-sequence *d–q* components of the three-phase ac voltage/current, as shown in Figures 6.4–6.7.

The controllable transfer functions have been used to determine the PI gains of the various controllers. The parameters of the PI controllers for stable closed loop system of the bidirectional three-phase VSI with 415 V, 50 Hz output are determined by using Bode plot based technique and are given in Table 6.1. The calculated parameters of an LCL filter for three-phase VSI are $L_{f1} = L_{f2} = 0.8472$ mH, $R_{Lf1} = R_{Lf2} = 2.6613$ mΩ, and $C_f = 30$ µF.

The voltage equations of current controllers for the current (I_1 or I_2) for both the positive and the negative-sequence components in *d–q* SRF are expressed by

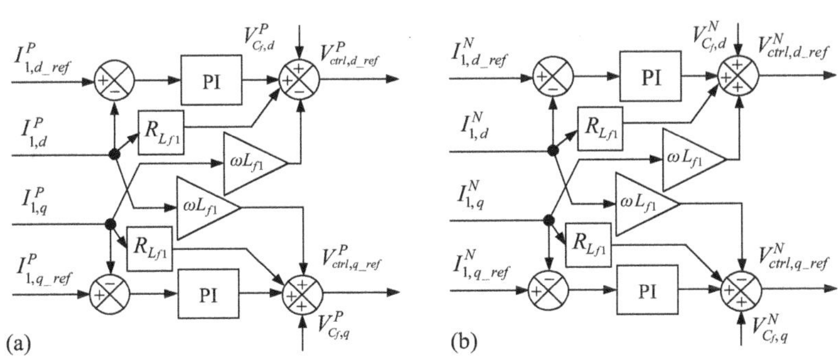

Figure 6.7 Inner current (I_1) controllers for (a) positive and (b) negative-sequence components

Table 6.1 Parameters of PI controllers for bidirectional three-phase VSI

PI controllers	Proportional gain (K_P)	Integral gain (K_I)
ac voltage controllers	$K^{p,n}_{P\_V_{ac}} = 0.6$	$K^{p,n}_{I\_V_{ac}} = 500\ \text{s}^{-1}$
Inner current controllers (I_1)	$K^{p}_{P_{C1}} = 3.1,\ K^{n}_{P_{C1}} = 1.4$	$K^{p}_{I_{C1}} = 9.74\ \text{s}^{-1},\ K^{n}_{I_{C1}} = 2.0\ \text{s}^{-1}$
Inner current controllers (I_2)	$K^{p}_{P_{C2}} = 1.6,\ K^{n}_{P_{C2}} = 1.0$	$K^{p}_{I_{C2}} = 20\ \text{s}^{-1},\ K^{n}_{I_{C2}} = 5.0\ \text{s}^{-1}$
dc voltage controllers	$K_{P\_V_{dc}} = 0.49$	$K_{I\_V_{dc}} = 27.42\ \text{s}^{-1}$

(6.33). The current equations of ac voltage regulators for both the positive and the negative-sequence components in d–q SRF are expressed by (6.34).

$$\left.\begin{aligned} V^{p,n}_{d\_ref} &= V^{p,n}_{d} + R_{L_f} I^{p,n}_{d} + \left(I^{p,n}_{d\_ref} - I^{p,n}_{d}\right)\left(K^{p,n}_{P_C} + \frac{K^{p,n}_{I_C}}{s}\right) \mp \omega L_f I^{p,n}_{q} \\ V^{p,n}_{q\_ref} &= V^{p,n}_{q} + R_{L_f} I^{p,n}_{d} + \left(I^{p,n}_{q\_ref} - I^{p,n}_{q}\right)\left(K^{p,n}_{P_C} + \frac{K^{p,n}_{I_C}}{s}\right) \pm \omega L_f I^{p,n}_{d} \end{aligned}\right\} \tag{6.33}$$

$$\left.\begin{aligned} I^{p,n}_{d\_ref} &= \left(V^{p,n}_{d\_ref} - V^{p,n}_{d}\right)\left(K^{p,n}_{P\_V_{ac}} + \frac{K^{p,n}_{I\_V_{ac}}}{s}\right) \mp \omega C_f V^{p,n}_{q} \\ I^{p,n}_{q\_ref} &= \left(V^{p,n}_{q\_ref} - V^{p,n}_{q}\right)\left(K^{p,n}_{P\_V_{ac}} + \frac{K^{p,n}_{I\_V_{ac}}}{s}\right) \pm \omega C_f V^{p,n}_{d} \end{aligned}\right\} \tag{6.34}$$

6.7 Operational analysis with simulation results

The performance analysis of the proposed DCMG with the proposed technique of the three-phase VSI has been carried out under various operating scenarios. Various case studies have been conducted to show the effectiveness of the proposed control technique and the smart DCMG. The parameters used for the WT, SPV, SOFC, and MT DGs, connected to the DCMG, are 600, 100, 100, and 150 kW, respectively.

The rated loads connected to the DCMG are (1) 100 kW three-phase load operating at 415 V, 50 Hz connected at ac bus of utility grid, (2) 200 kW single-phase load operating at 240 V, 50 Hz through a single-phase voltage controlled VSI, (3) 100 kW dc load operating at 220 V dc through a dc–dc buck converter, (4) fast dc charging station having 500 V dc through a dc–dc buck converter for 100 kW EV load, and (5) 100 kW three-phase load in the local area of the WT DG. The total rated load connected to the DCMG has been considered equal to the WT DG rated capacity, as given in, i.e., 600 kW in all the cases.

6.7.1 Case 1: variable generations and variable loads

In this case, the power generation by the WT DG varies at wind speeds 11.19, 2.4, 14, 10.1, 4.48, 16.27, 13.1, 5.7, 10.53, 10.04, 16.6, and 14 m/s, during 24 h/day. The variable power generation by the SPV system following the solar irradiation and temperature is shown in Figure 6.8. The nature of variation of the total load has

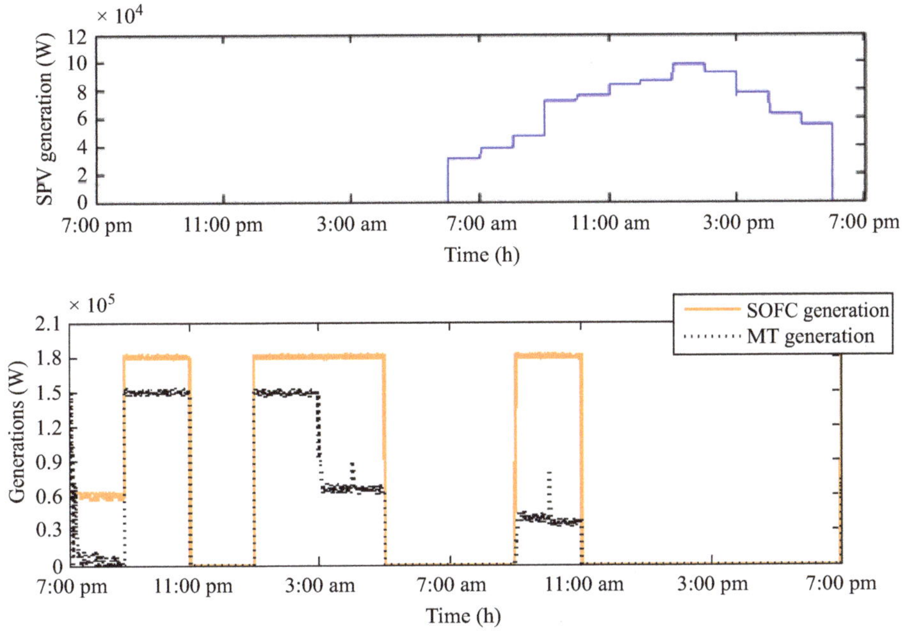

Figure 6.8 Power generations by SPV, SOFC, and MT DGs (Case 1)

been assumed the same as the residential load curve for 24 h, given in [27], and assumed to be constant for the next 2 h. The total power generation by all the DGs, total load demand, and grid output power to meet the power mismatch are shown in Figure 6.9. The power mismatch between total generation and total load has been fulfilled by transferring the power between the DCMG and the utility grid, as shown in Figure 6.9. The power rating of the three-phase VSI for the grid-connected mode is considered as 500 kV A. The proposed control strategy of three-phase VSI maintains sinusoidal load voltage at its rated value with low THD (2.6%–5%) and ensures required power flow between the DCMG and the utility grid, as shown in Figure 6.10. During the whole operation, the DCMG voltage is maintained almost constant as shown in Figure 6.11. The spikes in the DCMG voltage are due to the transients under changing of the power generation and/or the load.

6.7.2 *Case 2: fault on the utility grid*

In this case, the values and variations of all the loads connected to the DCMG are as given in Table 6.2. The power generations by the WT and SPV DGs are as shown in Figure 6.12. The variable power generations by the SOFC and MT DGs to meet the load demand are shown in Figure 6.13. The total generation, total load, grid output, and BESS output, to meet the mismatch power, are shown in Figure 6.14.

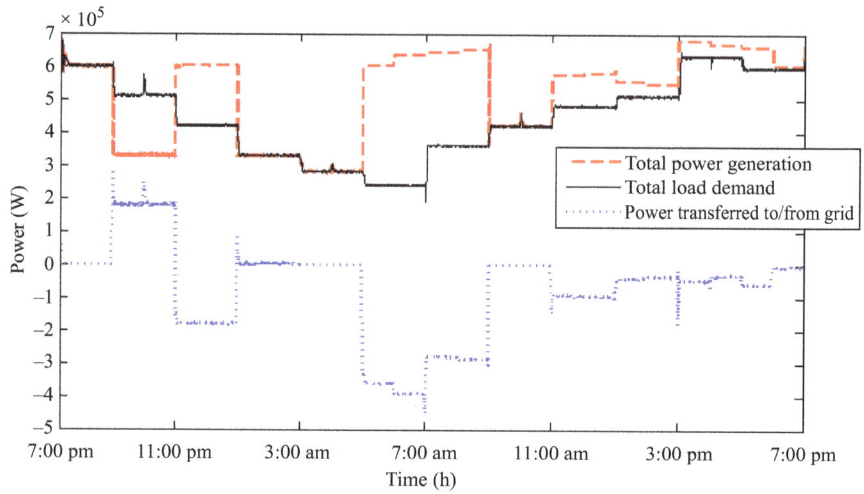

Figure 6.9 Total power generation, total load demand, and grid power (Case 1)

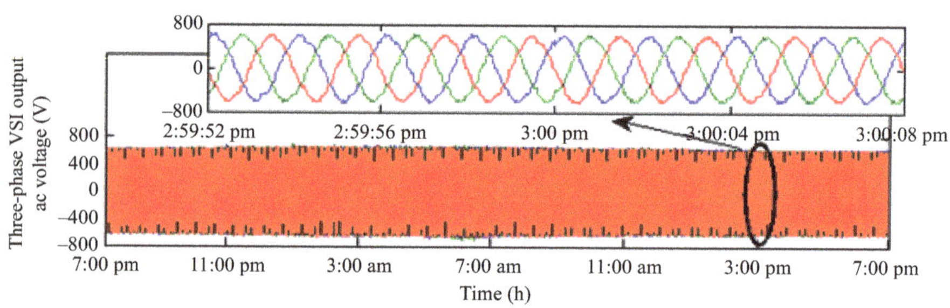

Figure 6.10 Three-phase load across VSI and three-phase VSI output ac voltage (Case 1)

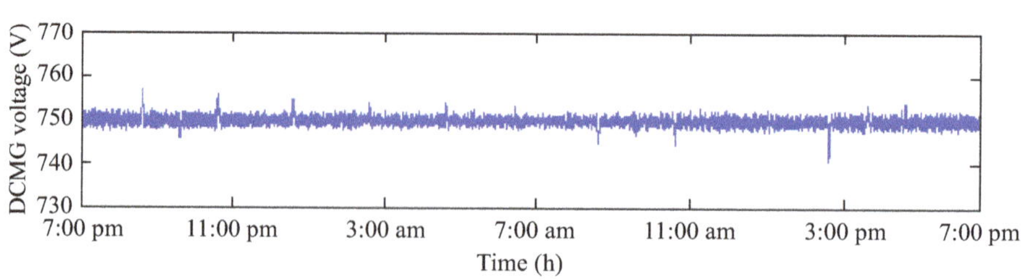

Figure 6.11 dc voltage of the microgrid (Case 1)

Table 6.2 Variations of the loads connected to the DCMG (kW)

Time (s)	Three-phase load in WTG area	Three-phase load connected at ac bus	Single-phase load	dc load	EV load	Total load
0–1	110	100	250	100	40	600
1–2	100	120	200	96	44	560
2–3	32	110	150	60	48	400
3–4	138	80	300	120	52	690
4–5	150	100	180	140	60	630

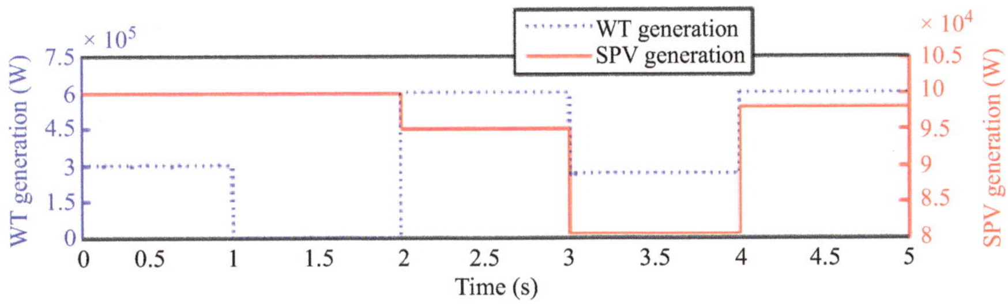

Figure 6.12 Power generations by WT and SPV DGs (Case 2)

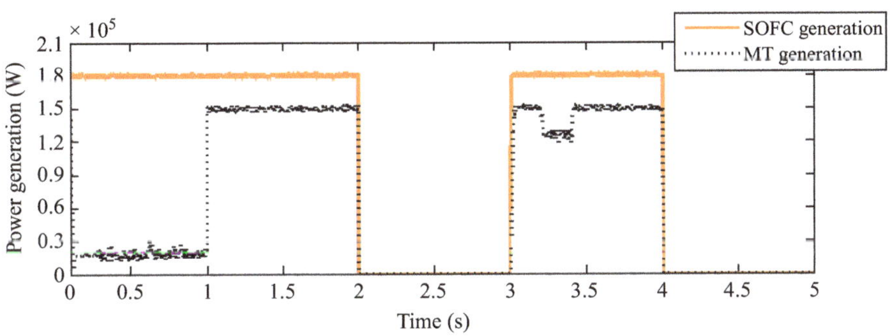

Figure 6.13 Power generations by SOFC and MT DGs (Case 2)

During t = 1–2 s, the total generation is less than the total load demand, and the deficit power is supplied by the grid, as shown in Figure 6.14. A symmetrical three-phase fault, with fault resistance ($R_f = 0.1$ mΩ), is simulated at $t = 1.2$ s on the grid side (at point B in Figure 6.1). The output ac voltage of three-phase VSI reduces from 1.0 to 0.056 pu, and three-phase load current reduces to zero, i.e., power consumed by three-phase load connected at ac bus is reduced from 120 kW to zero, as shown in Figure 6.15. During this fault, the total load demand has been reduced from 560 to 440 kW, while the total generation remains the same (430 kW

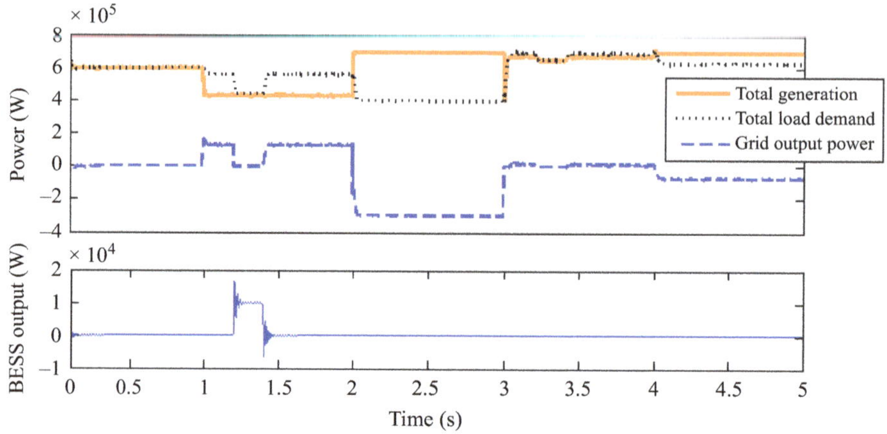

Figure 6.14 Total power generation, total load, grid output, and BESS output (Case 2)

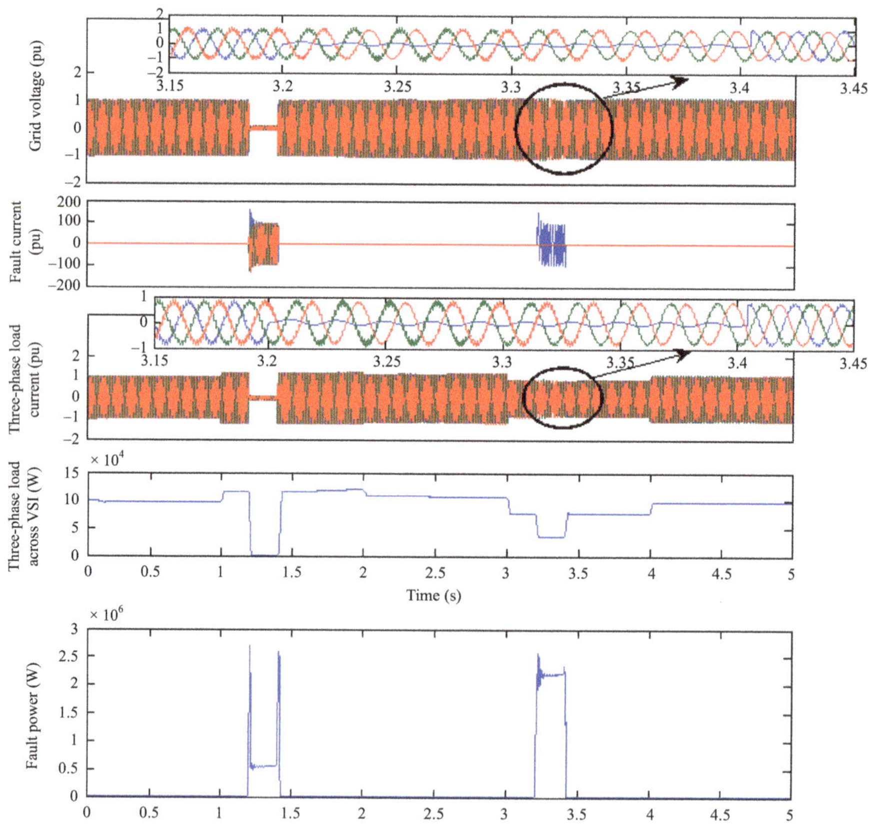

Figure 6.15 Grid voltage, fault current, three-phase load current, three-phase load across VSI, and fault power during fault, prefault, and postfault (Case 2)

as the prefault value), as shown in Figure 6.14, and the utility grid is incapable of supplying the power to the DCMG. Therefore, the BESS is connected immediately (as fault occurs) to the DCMG to meet the deficit power (10 kW), as shown in Figure 6.14. As the fault is cleared at $t = 1.4$ s, the BESS is disconnected from the DCMG, because utility grid is now capable of supplying the power to the DCMG (as in Figure 6.14), and the system regains the prefault conditions and operates under normal condition, as shown in Figures 6.13–6.15.

A single line-ground unsymmetrical fault at $t = 3.2$ s on phase-A of the grid (at point B in Figure 6.1), with fault resistance ($R_f = 0.1$ mΩ), is also simulated. The VSI output ac voltage of phase-A reduces from 1.0 to 0.11 pu, and current in phase-A reduces from 0.8 to 0.1 pu, i.e., total power consumed by three-phase load connected at ac bus reduces from 80 to 36 kW, as shown in Figure 1.15. During this fault, the total load and the total generation reduce from 690 to 646 kW and 673.13 to 646 kW, respectively, as shown in Figure 6.14. The generation from MT DG reduces from 150 to 122.87 kW, as given in Figure 6.13. As the fault is cleared at $t = 3.4$ s, the system regains the prefault conditions and operates under normal conditions, as shown in Figures 6.13–6.16. During the faults, the current flowing through the fault resistance and the power dissipated through the fault resistance are shown in Figure 6.15. As shown in Figure 6.16, at the time of the fault occurrence and fault clearing, the DCMG voltage undergoes transients. The transients die down in about 0.05 s, and the DCMG voltage attains almost constant value (750 V as prefault value) during the fault and the postfault periods, as shown in Figure 6.16. The transient surges in the voltages, as in Figures 6.16 and 6.17, are due to the rapid change in the power flow direction (which can be seen from Figure 6.14) between the utility grid and DCMG.

It can be seen that these sudden surges occur at the same instant when the grid power and individual loads connected to the DCMG suddenly change. Thus, the system, except the fault transient periods, remains under normal operation even during the fault, as shown in Figures 6.13–6.16.

6.7.3 Case 3: dc (L-G) fault on the dc microgrid

In this case, the power generations by the wind turbine generator (WTG) (202.5 kW) and the SPV system (100 kW) are constant, and outputs from the SOFC and MT DGs are variables, as shown in Figure 6.18. A dc (L-G) fault, with fault resistance ($R_f = 250$ mΩ), is simulated at $t = 0.6$ s on the DCMG bus. The DCMG voltage

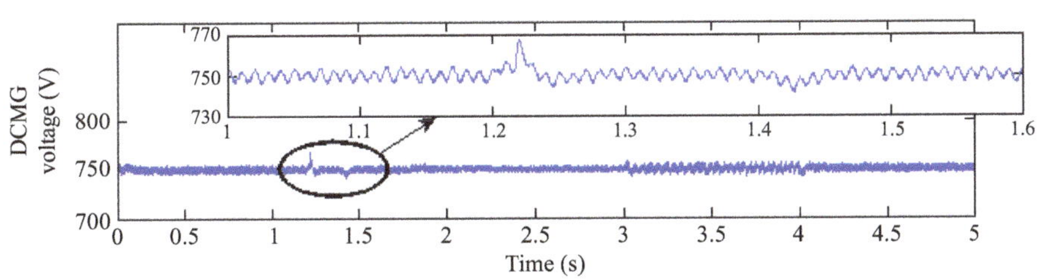

Figure 6.16 DCMG voltage during fault, prefault, and postfault (Case 2)

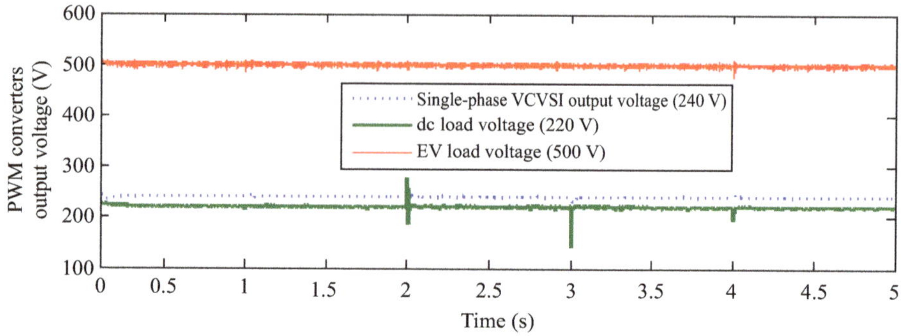

Figure 6.17 Output voltages of single-phase VCVSI, dc load buck converter, and EV load buck converter during fault, prefault, and postfault (Case 2)

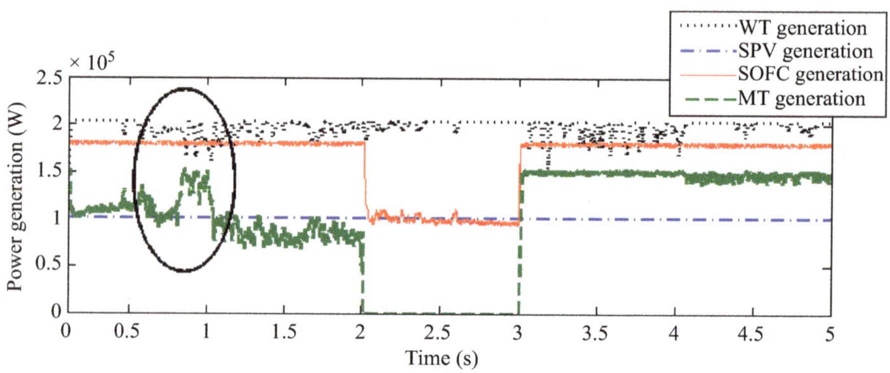

Figure 6.18 Power generations by WT, SPV, SOFC, and MT DGs (Case 3)

reduces from 750 to 620 V, as shown in Figure 6.19. Due to reduction in the DCMG voltage, the WTG output is also reduced, the load demand is fulfilled by changing the generation of the MT DG and power taking from utility grid, as shown in Figures 6.18 and 6.20. During the fault period, the proposed control technique does not allow any adverse impact on the power consumed by three-phase load connected at the ac bus, and power transfer between the DCMG and utility grid remains almost the same with slight transients, as shown in Figures 6.20 and 6.21. This also maintains the rated sinusoidal three-phase ac voltage during the fault and postfault, as shown in Figure 6.21. However, the power consumed by the three-phase load in WTG area, dc load, and EV load has reduced by 8, 17, and 2 kW, respectively, as shown in Figure 6.22. Thus, both the total load demand and total generation are reduced from 600 to 573 kW, as shown in Figure 6.20. The 70 kW

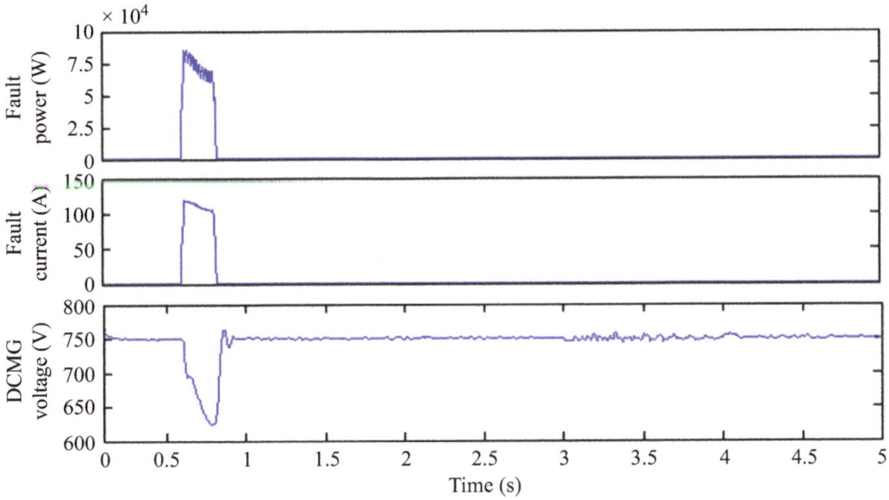

Figure 6.19 Fault power, fault current, and DCMG voltage during prefault, fault, and postfault (Case 3)

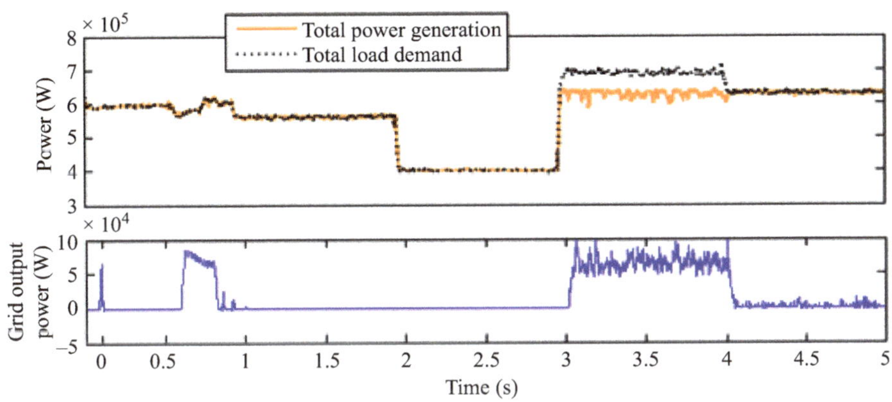

Figure 6.20 Total power generation, total load demand, and utility grid output (Case 3)

power is dissipated through the fault resistance (as in Figure 6.19), which is supplied by the utility grid, as shown in Figure 6.20. As the fault is cleared at $t = 0.8$ s, the system regains the prefault conditions. It operates under normal condition even during the fault, except the fault transient periods, as shown in Figures 6.18–6.22. After clearing the fault, during t = 3–4 s, the load demand is more than the generation and is met by the power transferring from the utility grid to the DCMG, as shown in Figure 6.20.

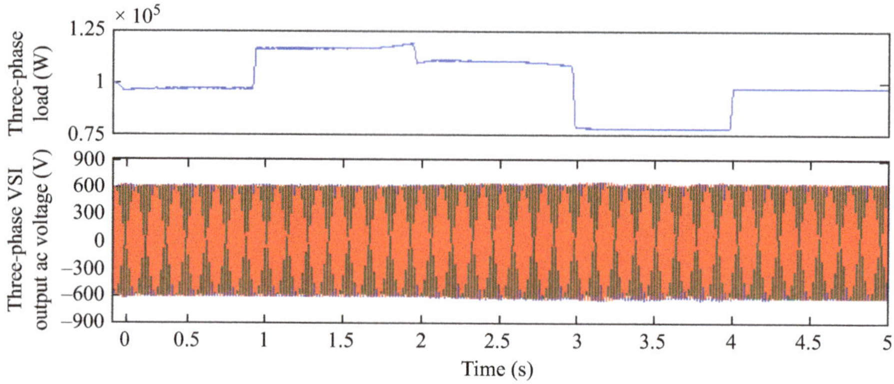

Figure 6.21 Three-phase load and three-phase VSI output ac voltage (Case 3)

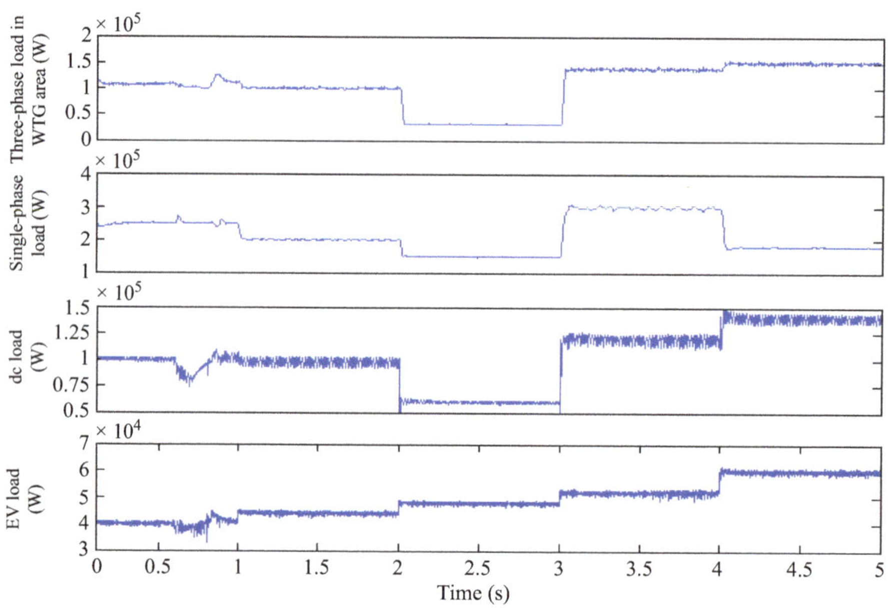

Figure 6.22 Power supplied to various loads connected to the DCMG during prefault, fault, and postfault (Case 3)

6.8 Conclusions

This chapter describes the modeling of the three-phase VSI with an output LCL filter using two rotating *d–q* SRFs and also a dynamic modeling of the smart DCMG. For integration of the utility grid to the DCMG, a control technique of the bidirectional three-phase VSI using two rotating *d–q* SRFs has been proposed. The proposed control technique uses dual PI controllers for controlling the positive sequence as well as the negative-sequence components of both the three-phase ac voltage and current. A combination of the feed-forward control and the feedback control loops has been used to generate the positive and negative-sequence *d–q* axis reference currents for the inner current controllers. The positive and the negative-sequence *d–q* components of the voltage as well as the current are controlled, independently, in its own SRF. The proposed control technique of the three-phase VSI has certain advantages such as

1. Use of the two feedback control loops for the output ac voltage and the current along with dual PI controllers eliminate the dc link voltage harmonics and the negative-sequence current from the output of the VSI. This maintains more effective output ac voltage regulation.
2. A dc voltage controller, as the feed-forward control loop, allows better dc voltage stabilization, improved performance, and faster transient recovery after the disturbances.

Various simulation results have been carried out to show the effectiveness and the robustness of the proposed DCMG and the proposed control technique of the three-phase VSI under different operating conditions, including the fault scenario. It has been observed that the proposed scheme of three-phase VSI maintains the rated sinusoidal load voltage with low THD and ensures the required power flow between the DCMG and the utility grid. During various ac faults on the utility grid, the proposed control technique does not allow any adverse impacts on the generations, loads in the DCMG, and on the DCMG voltage, except the load in faulted area.

During the dc fault on the DCMG bus, the proposed scheme allows one to maintain the rated sinusoidal three-phase ac voltage and ensures the required power flow between the DCMG and the utility grid. As the faults are cleared, the DCMG system regains prefault conditions with faster response and operates under normal mode even during the fault and postfault periods, using the proposed control technique. In addition, IEEE-1547 standard has also been complied by the proposed control technique, under different operating scenarios. The DCMG voltage has also been maintained almost constant at its desired value, under various operating conditions.

References

[1] Kumar M., 'Development of control schemes for power management and operation of dc microgrids', Ph.D. Thesis, Electrical Engineering Department, Indian Institute of Technology Kanpur, India, Jan. 2015.

[2] Kumar M., Singh S.N., and Srivastava S.C., 'Design and control of smart DC microgrid for integration of renewable energy sources', in *Proc. IEEE Power Energy Soc. Gen. Meeting*, IEEE, San Diego, CA, USA, 2012, pp. 1–7.

[3] Vandoorn T.L., Meersman B., Degroote L., Renders B., and Vandevelde L., 'A control strategy for islanded microgrids with dc-link voltage control', *IEEE Trans. Power Delivery*, 2011, vol. 26 (2), pp. 703–713.

[4] Khorsandi A., Ashourloo M., and Mokhtari H., 'A decentralized control method for a low-voltage dc microgrid', *IEEE Trans. Energy Convers.*, 2014, vol. 29 (4), pp. 793–801.

[5] Lindberg A. and Larsson T., 'PWM and control of three level voltage source converters in an HVDC back-to-back station', in *Proc. IET 6th International Conference on AC and DC Power Transmission*, IET, London, 23 Apr.–3 May 1996, pp. 297–302. Conference publication no. 423.

[6] Milasi R.M., Lynch A.F., and Li Y.W., 'Adaptive control of a voltage source converter for power factor correction', *IEEE Trans. Power Electron.*, 2013, vol. 28 (10), pp. 4767–4779.

[7] Yao Q. and Holmes D.G., 'A simple, novel method for variable-hysteresis-band current control of a three phase inverter with constant switching frequency', in *Proc. IEEE Conference on Industry Applications Society Annual Meeting*, vol. 2, IEEE, Toronto, ON, 2–8 Oct. 1993, pp. 1122–1129.

[8] Malesani L., Mattavelli P., and Tomasin P., 'Improved constant-frequency hysteresis current control of VSI inverters with simple feedforward bandwidth prediction', *IEEE Trans. Ind. Appl.*, 1997, vol. 33 (5), pp. 1194–1202.

[9] Buso S., Fasolo S., Malesani L., and Mattavelli P., 'A dead-beat adaptive hysteresis current control', *IEEE Trans. Ind. Appl.*, 2000, vol. 36 (4), pp. 1174–1180.

[10] Mohamed Y.A.R.I. and El-Saadany E.F., 'An improved deadbeat current control scheme with a novel adaptive self-tuning load model for a three-phase PWM voltage-source inverter', *IEEE Trans. Ind. Electron.*, 2007, vol. 54 (2), pp. 747–759.

[11] Teodorescu R., Blaabjerg F., Liserre M., and Loh P.C., 'Proportional-resonant controllers and filters for grid-connected voltage-source converters', *IEE Proc.—Electr. Power Appl.*, 2006, vol. 153 (5), pp. 750–762.

[12] Vidal A., Freijedo F.D., Yepes A.G., *et al.*, 'Assessment and optimization of the transient response of proportional-resonant current controllers for distributed power generation systems', *IEEE Trans. Ind. Electron.*, 2013, vol. 60 (4), pp. 1367–1383.

[13] Rodríguez J., Pontt J., Silva C.A., *et al.*, 'Predictive current control of a voltage source inverter', *IEEE Trans. Ind. Electron.*, 2007, vol. 54 (1), pp. 495–503.

[14] Sanchez P.M., Machado O., Peña E.J.B., Rodríguez F.J., and Meca F.J., 'FPGA-based implementation of a predictive current controller for power converters', *IEEE Trans. Ind. Inf.*, 2013, vol. 9 (3), pp. 1312–1321.

[15] Ramırez R.O., Espinoza J.R., Melın P.E., *et al.*, 'Predictive controller for a three-phase/single-phase voltage source converter cell', *IEEE Trans. Ind. Inf.*, 2014, vol. 10 (3), pp. 1878–1889.

[16] Bahrani B., Kenzelmann S., and Rufer A., 'Multivariable-PI-based *dq* current control of voltage source converters with superior axis decoupling capability', *IEEE Trans. Ind. Electron.*, 2011, vol. 58 (7), pp. 3016–3026.

[17] Serpa L.A., Ponnaluri S., Barbosa P.M., and Kolar J.W., 'A modified direct power control strategy allowing the connection of three-phase inverters to the grid through LCL filters', *IEEE Trans. Ind. Appl.*, 2007, vol. 43 (5), pp. 1388–1400.

[18] Rese L., Costa A.S., and Silva A.S., 'Enhanced modeling and control of VSIs in microgrids operating in grid-connected mode', in *Proc. IEEE PES International Conference on Innovative Smart Grid Technologies*, IEEE, Washington, DC, 16–20 Jan. 2012, pp. 1–8.

[19] Divshali P.H., Alimardani A., Hosseinian S.H., and Abedi M., 'Decentralized cooperative control strategy of microsources for stabilizing autonomous VSC-based microgrids', *IEEE Trans. Power Syst.*, 2012, vol. 27 (4), pp. 1949–1959.

[20] Reznik A., Simões M.G., Al-Durra A., and Muyeen S.M., 'LCL filter design and performance analysis for grid interconnected systems', *IEEE Trans. Ind. Appl.*, 2014, vol. 50 (2), pp. 1225–1232.

[21] Song H. and Nam K., 'Dual current control scheme for PWM converter under unbalanced input voltage conditions', *IEEE Trans. Ind. Electron.*, 1999, vol. 46 (5), pp. 953–959.

[22] Yazdani A. and Iravani R., 'A unified dynamic model and control for the voltage-sourced converter under unbalanced grid conditions', *IEEE Trans. Power Delivery*, 2006, vol. 21 (3), pp. 1620–1629.

[23] Hu J. and He Y., 'Modeling and control of grid-connected voltage-sourced converters under generalized unbalanced operation conditions', *IEEE Trans. Energy Convers.*, 2008, vol. 23 (3), pp. 903–913.

[24] Wang F., Duarte J.L., and Hendrix M.A.M., 'Design and analysis of active power control strategies for distributed generation inverters under unbalanced grid faults', *IET Gener. Transm. Distrib.*, 2010, vol. 4 (8), pp. 905–916.

[25] Radwan A.A.A. and Mohamed Y.A.R.I., 'Linear active stabilization of converter-dominated dc microgrids', *IEEE Trans. Smart Grid*, 2012, vol. 3 (1), pp. 203–216.

[26] Bottrell N., Prodanovic M., and Green T.C., 'Dynamic stability of a microgrid with an active load', *IEEE Trans. Power Electron.*, 2013, vol. 28 (11), pp. 5107–5119.

[27] Short T.A., *Electric Power Distribution Handbook*, CRC Press, Boca Raton, FL, USA, 2004, pp. 47.

Chapter 7

Energy management system of a microgrid using particle swarm optimization (PSO) and communication system

Vijay K. Sood[1]*, Mohammad Y. Ali*[1] *and Faizan Khan*[1]

This chapter focuses on the energy management system (EMS) for a microgrid (MG). The hierarchy of the various controllers utilized in the EMS is presented. Various programming methods can be used for optimizing the behavior of the EMS such that the MG can operate in a safe and reliable manner to match the demanded load with the available energy sources. The requirements for a communication system within the MG are briefly discussed. A test case with a particle swarm optimization (PSO) technique is provided.

Nomenclature

CHP	combined heat and power
DGs	distribution generations
DERs	distributed energy resources
DE	differential evolution
DSO	distribution system operator
EMS	energy management system
ESSs	energy storage systems
EVs	electric vehicles
GA	genetic algorithm
HMI	human machine interfaces
IEC	International Electrotechnical Commission
MAS	multi-agent system
MGs	microgrids
MGMS	microgrid management system
MILP	mixed-integer linear programing
NN	neural network
O&M	operation and maintenance

[1]Department of Electrical and Computer Engineering, Ontario Tech University (OTU), Oshawa, Canada

P2P	peer-to-peer
PCC	point of common coupling
PSO	particle swarm optimization
PV	photovoltaic
RESs	renewable energy sources
SOC	state of charge
SCADA	supervisory control and data acquisition
UC	unit commitment
VPP	virtual power plant
VSIs	voltage source inverters

7.1 Introduction

MGs are the new paradigm for the evolution of the distribution system. MGs are composed of distributed energy resources (DERs), such as generators, renewable energy sources (RESs), energy storage systems (ESSs) and a cluster of critical and noncritical loads. They provide stability to the main grid and offer optimal integration of these subsystems into the distribution system. MGs operate in one of two modes: grid-connected or islanded mode [1–5]. In grid-connected mode, an MG draws/supplies power from/to the main grid, depending on the generation and load requirements, and respecting certain suitable market policies to maximize the efficiency/cost, etc. Likewise, it can separate itself from the main grid whenever a fault occurs in the main grid and continues to supply power to connected critical loads. Furthermore, to ensure that the MG operates in an economical and reliable manner, it is equipped with a suitable monitoring and control system. The control system is responsible for scheduling and controlling of all DERs to warrant the stability, reliability and economical operation of the MG.

In this chapter, the MG control system is discussed. The MG control system functions at three levels: primary, secondary and tertiary. Section 7.2 describes the two different types of EMSs used in an MG (i.e., centralized and decentralized) and gives the advantages and disadvantages of both. Section 7.3 reviews various methods to assist the MG EMS to achieve optimal operation while minimizing operational costs. Section 7.4 deals with the different methods, i.e., using either linear, nonlinear, dynamic, rule-based, meta-heuristic, artificial intelligent or multi-agent methods, for establishing the EMS. Section 7.5 describes the role of communication systems in the area of MGs. In Section 7.6, a test case of a PSO technique for an MG model is presented. Section 7.7 provides the conclusions for this chapter.

7.2 Microgrid management system

A crucial unit that controls the operation of the MG is the MG management system (MGMS) that operates the system autonomously, connecting it to the utility grid appropriately for the bidirectional exchange of power and providing support to components within the MG. It enables the interplay of components and different controllers to operate the EMS in a controlled manner. This approach will allow customization of the system to enhance optimization to improve the overall efficiency without sacrificing the plug-and-play functionality. MGMS is broken down into three different subsystems, i.e., primary, secondary and tertiary control layers that manage the entire MG operation (Figure 7.1).

The MGMS controls the distribution generations (DGs) to maintain the balance between generation and load demand during islanded mode, grid-connected mode or the transition period between the two modes. The three control layers are described next [6]:

- *Primary control layer:* This is the base layer that has the fastest response time (typically, in the region of milliseconds to minutes) and is responsible for the control of devices that respond to system dynamics and transients. It is also known as the local or internal controller. This control is based exclusively on local measurements and requires no communications. The functions of this control include islanding detection, converter output control, frequency regulation, voltage regulation and power-sharing control. In the MG, the voltage source inverters (VSIs) are the common interface between the DERs and the MG. VSI controller requires a specially designed control to simulate the inertia characteristic of synchronous generators and provide appropriate frequency regulation. The VSI has two stages of control: inverter output and power-sharing.

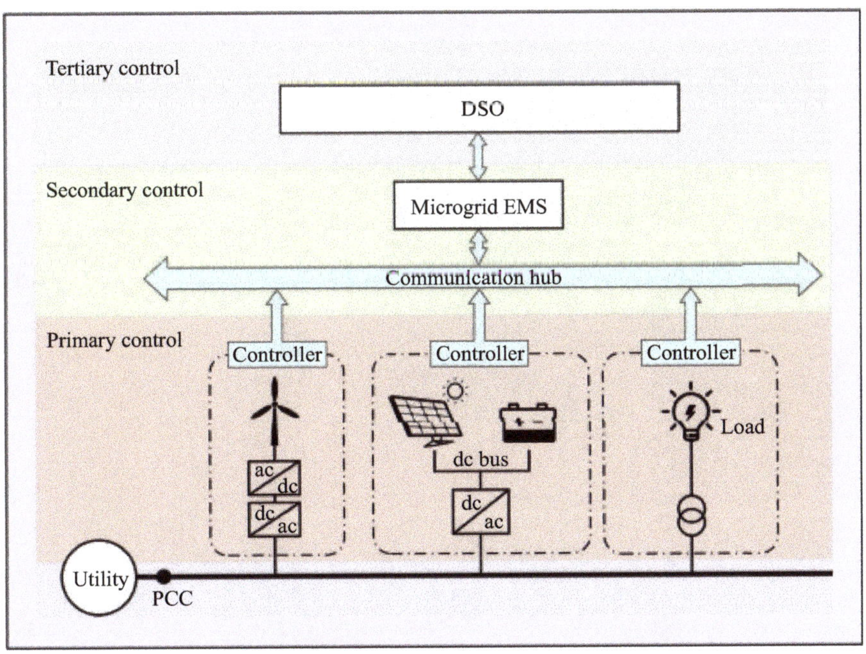

Figure 7.1 Microgrid management system (MGMS) [9]

The inverter typically consists of an outer control loop for voltage control and an inner loop for current regulation. The power-sharing control is used for sharing the active and reactive power in the system.

- *Secondary control layer:* This is the central layer (Figure 7.2) and is responsible for the reliable and economical operation of the MG. Its main function includes an EMS and automatic generation control. The secondary control also helps one to reset the frequency and voltage deviations of the droop-controlled VSIs and generators then assigns to them new optimal long-term set points calculated from the MG EMS. The EMS minimizes the MG's operation cost and maximizes its reliability in grid-connected or islanded modes of operation. The objective of the EMS consists of finding the optimal unit commitment (UC) and economical dispatch of the available DER units, to achieve load and power balance in the system. The cost function is designed in terms of economic tolls such as fuel cost, power bill, maintenance cost, shutdown and start-up cost, emissions and social welfare and battery degradation cost and cost of loss load. The reliability indices are formulated as constraints such as load forecast, forecasted power availability RES, the generation and demand balance, energy storage capacity limits and power limits for all controllable generations. The EMS resolves a multi-objective optimization problem with complex constraints and falls under mixed-integer linear programing (MILP)/nonlinear programing. The output of the optimizer is the schedule energy import/export and DGs power output. The EMS system is illustrated in Figure 7.2.
- *Tertiary control layer:* This is the highest control layer that provides intelligence for the whole system. It is responsible for buying and selling of energy

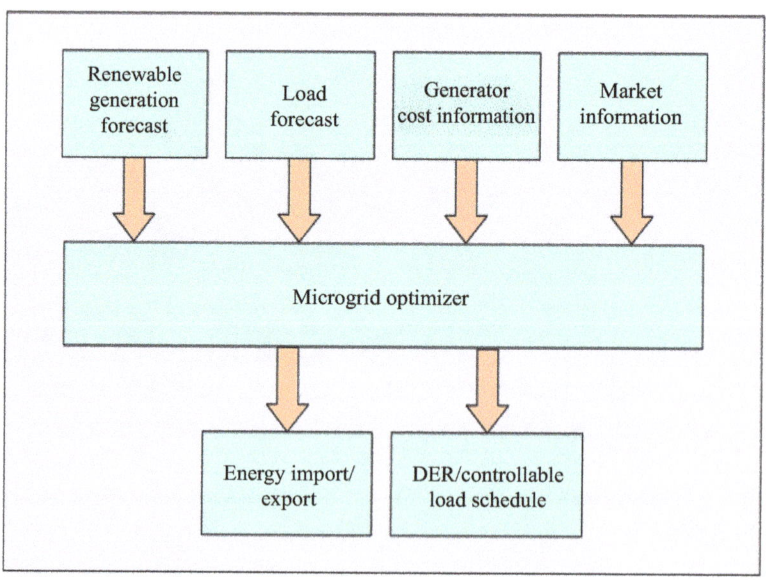

Figure 7.2 Secondary control layer [9]

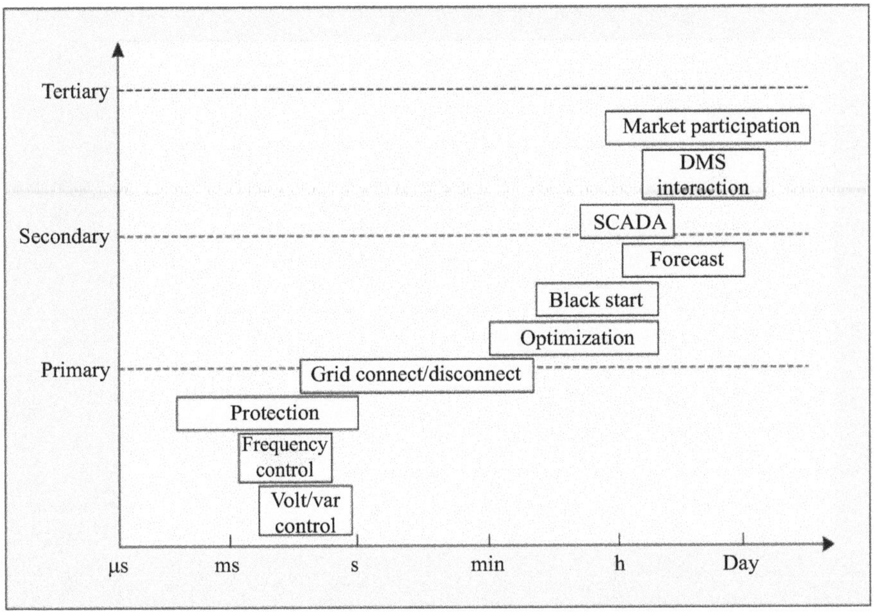

Figure 7.3 Response times of various layers in MGMS [9]

between consumers and transmission system, as well as providing active and reactive power support for the whole distribution system. Tertiary control layer is not a part of MGMS as it is recognized as a subsystem of the utility distribution system operator (DSO). But, due to the increase of MGs in the distribution system, the tertiary control layer is evolving in a concept called virtual power plant (VPP). The objective of a VPP is to coordinate the operation of multiple MGs interacting with one another within the system and communicate needs and requirements from the main grid. The VPP can provide transmission-system primary-frequency support, reactive power support and energy market participation. The control layer response time is typically of the order of several minutes to hours, providing signals to secondary controls at MGs and other subsystems that form the full grid.

Figure 7.3 shows the timescales of the three MGMS control layers in which the lower (primary) layer controls devices with fastest response times, whereas higher (secondary and tertiary) layers controls tend to have slower response times.

7.3 Microgrid energy management systems

The International Electrotechnical Commission (IEC) in the standard IEC 61970 defines an EMS as "a computer system comprising a software platform providing basic support services and a set of applications providing the functionality needed for the effective operation of electrical generation and transmission facilities so as to assure adequate security of energy supply at minimum cost" [7]. Hence, the MG EMS is a product of these features. It is usually equipped with decision-making algorithms, load and power forecasting, human machine interfaces (HMI) and supervisory control and data acquisition (SCADA) system. These functions help the EMS in optimizing MG operation while satisfying the technical constraints.

MG EMS supervisory control can be subdivided into two types, namely, centralized and decentralized EMSs. This section will be going more in depth about the two different types of control methods and also the advantages and disadvantages of both.

7.3.1 Centralized energy management

In the centralized EMS system, data information and collection are usually required from the tertiary control and primary control layers such as operating cost, weather forecast, load demand, voltage and current readings from each component. Based on the gathered information, an appropriate UC and dispatch optimization algorithm is executed to achieve efficient, economical operation and maintain power quality as well as match generation with load demand. The MG relies on the secondary layer in which a master controller with a high computing performance and a dedicated communication network is utilized for the operation. Usually the master controller supports EMS and SCADA functions with an HMI, which allows the system operator to control and monitor the MG. A centralized configuration is shown in Figure 7.4.

In Figure 7.4, the centralized configuration requires a two-way communication channel between the primary control (local controllers) and secondary control (EMS) for the exchange of information. This configuration is called a star

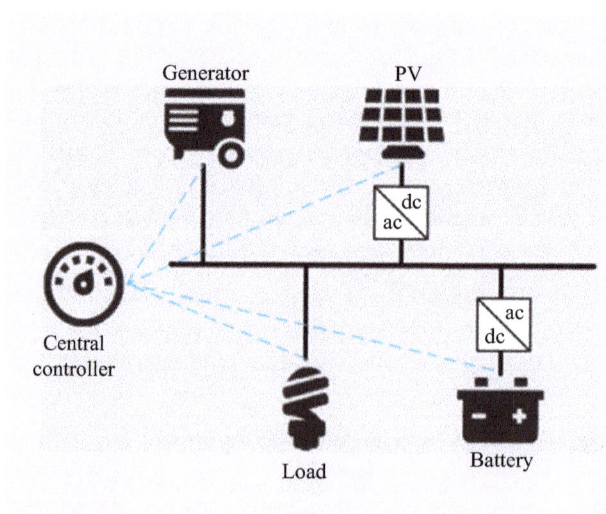

Figure 7.4 Centralized EMS configuration [9]

connection topology and a master/slave technique is established. The communication channels can be either wired or wireless depending on their requirements. Some of the available technologies are based on power line carriers, telephone lines or a wireless medium.

In centralized EMS, operation is in real time when the secondary controller frequently observes the entire system and samples the critical generation/demand information from each component. This frequent communication may cause a computational burden; therefore, a high-performance computing unit is required where the EMS can execute accurate decisions by processing the data read through the communication channel. Moreover, a high-bandwidth communication is required to meet the growing demands of EMSs.

Centralized EMS is a comparatively straightforward implementation but can also endanger the overall performance as any single-point failure or any fault at a unit can cause the entire system to breakdown. Therefore, it is considered to have a low expandability and flexibility. Considering its structure and functionality, the following options can be more desirable for the MG cases [8]:

- A small-scale MG with low communication and computational cost in which centralized information can be processed.
- Unity between the components is required that can operate the MG with a common goal for generation/demand balances.
- Must operate with a high security that keeps the data information secure.

The EMS optimizes the MGs power flow distribution, resulting in maximizing the DG's production depending on various parameters, constraints, variables and market prices provided as an input to the EMS controller. Some of the most commonly used data provided as input information to the controller to process and provide the reference values to the primary control layer may include the following:

- forecasting of the grid electricity prices,
- status of interconnection of utility grid,
- reliability and security constraints of the MG,
- operational limits of DG to be discharged,
- state of charge (SOC) of the battery ESS,
- load day-ahead forecast values and
- RESs generation forecasted power output.

Centralized EMS enables all the relevant information to be gathered at a single point for the controller to perform its function. The following steps are involved in a centralized framework for the EMS to perform its tasks [9] (Figure 7.5):

1. performing a RESs generation and load day-ahead forecast;
2. performing a day-ahead energy scheduling calculation by collecting information from all the components;
3. executing an optimization algorithm to calculate the optimal day-ahead schedules;
4. assigning optimal day-ahead schedules to the corresponding components;

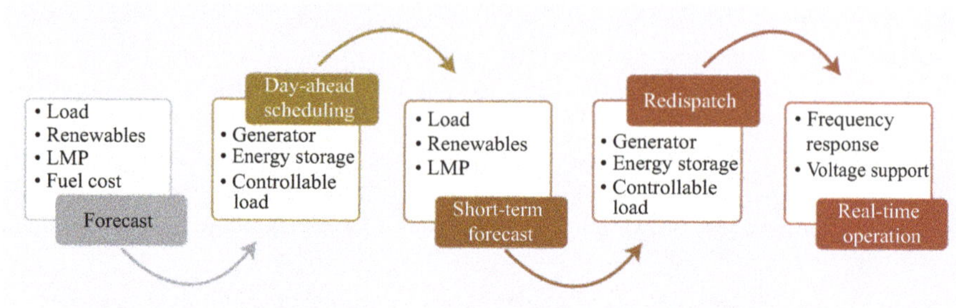

Figure 7.5 Centralized framework for the EMS [9]

5. acquiring real-time system information, as there might be unexpected events or forecast errors;
6. generating short-term forecasts during the operation;
7. reexecuting the optimization algorithm and rescheduling the dispatch of RESs (15 min);
8. finally, sending the EMS, the most updated and optimal set points to the primary control.

When the EMS is executed, a set of information is dispatched to the local controllers at the primary control level to operate the DGs in a cost-effective manner and simultaneously maximize the reliability of the MG. The set of information dispatched is as follows [10]:

- set points for DGs to dispatch the production of power,
- set points for local loads to be shed or to be served,
- market prices to serve as the input for EMS.

7.3.2 Decentralized energy management

In the decentralized EMS scheme (Figure 7.6), local controllers are interfaced with each DG unit to communicate among each other through a communication channel within the MG. Each unit is controlled by its local controller where data for each of the DG controllers is exchanged. Local controllers communicate with each other to request/offer a service, exchange information, communicate expectations and share knowledge, which is relevant to the MG operation. These controllers have advanced algorithms to make their own decisions or to process information and execute commands from the upper level. This EMS scheme is known for its intelligence as it is not fully aware of the decision made by controllers or the system-wide variables.

In Figure 7.6, the decentralized configuration illustrates a two-way communication channel between the local controllers for the exchange of information. This configuration is called a peer-to-peer (P2P) communication topology and is established within the primary control layer. The EMS is implemented locally in each of its local controllers connected to either DGs or the loads within the MG to

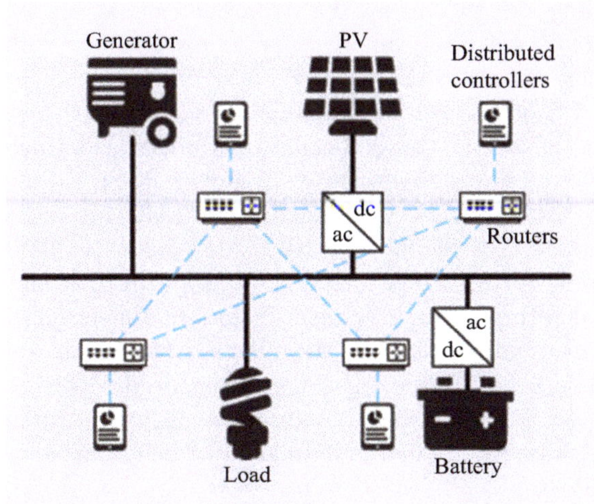

Figure 7.6 Decentralized EMS configuration [9]

allow the interaction of each unit to enable a decision-making process to optimally solve the energy management problem while providing flexibility within the MG to provide autonomy for all DGs and loads.

In a decentralized EMS operation, the need for secondary control layer is eliminated as the collaboration at the primary control layer between the local controllers that work jointly to achieve local goals to meet generation and demand of the entire MG. This can reduce the computational burden to some degree, as the customers no longer need to report their current or historical generation and demand data to EMS at the secondary control layer. Processing of information such as weather forecast, operating cost, load demand can be optimized by the local controllers, which reduces the use of hierarchical levels in the MG.

Implementation of EMS in a decentralized control architecture can increase the overall complexity of the entire MG although this can be overcome when looking at other perspectives in terms of its flexible operation and avoidance of single-point failure, which can still maintain the normal operation of EMS. Another advantage is that it can allow interaction of various other DG units, such as a plug-and-play functionality, without the need to make continuous changes to the local controller settings. Considering its structure and functionality, the following options can be more desirable for the MG cases [8]:

- Large-scale MGs, or the consumption, storage and generation are widely isolated, which can make data acquisition costly or difficult when using centralized EMS.
- Requirement of local decision-making, when the resources are owned by multiple owners.
- Adding or removing of DG.

When modeling the MG using the decentralized approach, the concept of multi-agent system (MAS) has been primarily addressed in the literature. MAS has evolved from a classical distribution control system that is specially designed for automated control systems with capabilities of controlling large and complex entities or groups of entities by dedicated controllers. Distributed control and MAS have a similar structure but what distinguishes them is the level of intelligence with which the agents are embedded. The MAS relies on a framework to achieve multiple local and global objectives autonomously, where two or more agents or intelligent agents are provided with local information. The characteristics of the local information, responsibilities and functionalities assigned to each agent and information shared by the agents between each other play a vital role in the overall performance of the system for the enhanced robustness, reliability and flexibility of MG. An intelligent agent is distinguished from a hardware or a software-automated system and can be described as an agent that possesses the following characteristics [6]:

- *Reactivity:* Ability to show reaction and reach to the changes in the environment in a timely manner.
- *Pro-activeness:* Seeking initiatives to achieve goals.
- *Social-ability:* Interaction with other agents through a communication channel.

These characteristics in local controllers work toward improving the performance of the system and do not have the main objective to maximize the revenue of the corresponding unit. This means that the intelligent agents can interact with other intelligent agents to react to environmental changes and establish a goal-oriented behavior.

Overall operation of MAS is to control objectives such as economy, reliability, energy market participation and MG operation. Although this is an overall global goal for the MG to operate in a reliable manner, in MAS only local goals are defined and not global goals. When intelligent agents cooperate among themselves and work toward local goals, the targeted global goal may be achieved with local goals responding to subparts of the global goal. The design of MAS algorithms is a complex process that requires a great deal of expertise to decompose global goal task by modeling the agent's interactions and classifying agents. Intelligent agents working together to achieve various local goals is a multi-objective problem, where the complexity of the MAS algorithm is structured in a rigorous manner for the agents to communicate autonomously.

An agent communication language is an environment for knowledge and information exchange and is required within the MG to enable agents to process only allowable or necessary knowledge for the agents to communicate among each other and eventually agree upon to determine the power mismatch between the demand and generation, as well as the estimated incremental cost. In decentralized EMS, MAS algorithms rely on a modern communication infrastructure that supports P2P communication between the agents for the exchange of data. IEC 61850 standards are utilized that support P2P data exchange. Communication can be either wired or wireless, depending on requirements.

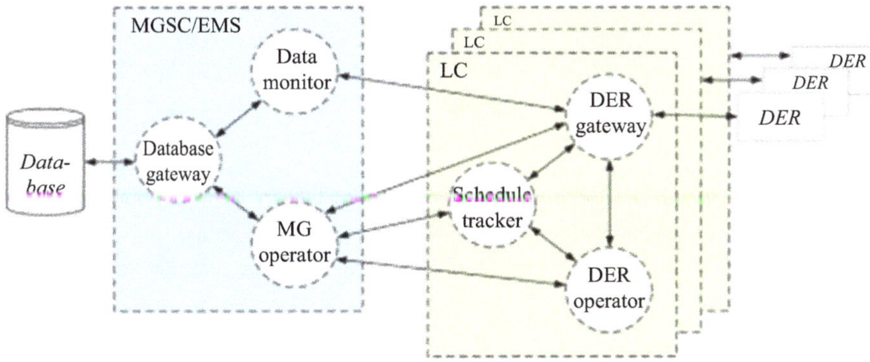

Figure 7.7 A general architecture of MAS-based MG EMS [8]

The general architecture of an MG EMS based on MAS is described in [11] and shown in Figure 7.7. Several different types of agents are defined in MGSC and LCs, including the following:

- Database Gateway agent for storing and retrieving information with databases.
- Data Monitor agent for obtaining real-time operation data from DER and feeding essential information to databases.
- MG Operator agent for optimizing the operation of the whole system.
- DER Gateway agent for interfacing other agents and control system with DER devices.
- Schedule Tracker agent for following the schedule from MG Operator agent and sending set point to DER.
- DER Operator agent for locally optimizing the operation of DER and response with MG Operator agent. Based on this scheme, a secondary control is implemented for regulating MG frequency and power flow in islanded and grid-connected modes.

7.3.3 Comparison between centralized and decentralized EMS

In addition to the comparison shown in Table 7.1, the decentralized framework has the following benefits:

- Decentralized controllers require less computing power that is more cost effective.
- Decentralized control systems have greater controller redundancy and are robust to a single point of failure. Controller failures will not cause system blackout.
- Local decision-making reduces network use, relaxing the communication bandwidth requirement.
- System maintenance and upgrades can be done without shutting down the entire system.
- The decentralized control framework is more flexible and scalable for future modifications and expansions.

Table 7.1 Comparison of centralized and decentralized EMS

Controls	Advantages	Disadvantages
Centralized	• Simple to implement • Easy to maintain • Relatively low cost • Widely used and operated • Wide control over the entire system	• Computational burden • Require high-bandwidth links • Single point of failure • Not easy to expand • Weak plug-and-play functionality
Decentralized	• Easier plug and play (easy to expand) • Low computational cost • Avoid single point of failure • Suitable for large-scale complex, heterogeneous systems	• Need synchronization • May be time-consuming for local agents to reach consensus • Convergence rates may be affected by the communication network topology • Upgrading cost on the existing control and communication facility • Need new communication structure

7.4 MG EMS solution

Many researchers have used different approaches to achieve optimal and efficient operation of MGs. This section outlines some popular methods used by researchers to solve the EMS.

7.4.1 EMS based on linear and nonlinear programming methods

In [12], the authors present an optimal energy management of residential MG to help in minimizing the operation cost. The cost function consists of energy trading cost, penalty cost on adjustable load shedding, electric vehicles (EVs) batteries wearing out cost and the range anxiety term for EVs. Three test cases have been defined using three different range of anxiety levels and have been studied to analyze the trade-off between operational cost of MG and average SOC of EVs battery. Also, three different types of loads are considered, namely, critical, adjustable and shiftable loads. This optimal EMS is formulated using MILP. Similarly, in [13] the authors proposed MILP optimization model for an optimal MG EMS to maximize daily revenue with main grid peak shaving application by introducing demand responsive loads. For the system, it was assumed that the load demand of the MG will always be more than generation. Two different systems were used to test and analyze the performance of the algorithm: first, a 1-bus MG and second, a 14-bus MG system.

The authors in [14] proposed a centralized architecture for EMS of a grid-connected MG using sequential quadratic programing method. The EMS aims to optimize the operation of the MG during interconnected operation, i.e., maximize its value by optimizing the production of the local DGs and power exchanges with the main distribution grid. Two policies are used: the first policy is operational cost minimization, while the second policy aims to maximize its profit considering energy transactions with the main grid. Likewise, authors in [15] introduced strategic EMS of a grid-connected MG, constrained by an operation window of transformer nominal operation and voltage security. The developed model minimizes MG operational cost using modified gradient descent solution method. The forward/backward sweep algorithm determines the power flow solution of MG. Three scenarios are considered in the objective function with respect to customer benefits, network losses and load leveling.

7.4.2 EMS based on dynamic programming and rule-based methods

Another solution of MG EMS is presented in [16], which is an approximate dynamic programming approach to overcome the curse of dimensionality in the proposed EMS model of a grid-connected MG. The cost-function is computed using receptive field weight regression and lookup table. Various scenarios of wind speed, load demand and ambient temperature are generated to consider their uncertainty in the proposed model that uses economic dispatch and UC operations to optimize the energy scheduling of MG. The proposed system is compared to myopic optimization and dynamic programming methods. In comparison with myopic optimization, the proposed system had lower operation cost, but higher computational time. But, with dynamic programming, the system had faster computational time and higher operation cost for MG.

7.4.3 EMS based on meta-heuristic approaches

Various authors have used meta-heuristic approaches to solve the MG EMS. In [17], the authors proposed a genetic algorithm (GA) and rule-based approach to solve an economic load dispatch and battery degradation cost-based multi-objective EMS for a remote MG. The system comprising both day-ahead and real-time operations considers diesel generator supply, battery supply and load shedding options in a sequential order to maintain load generation balance. An optimal EMS for grid-connected MG considers uncertainties of RESs, load demand and electricity price using PSO [18]. The efficiency of PSO in finding the best solution is shown to be better in comparison with GA, combinatorial PSO, fuzzy self-adaptive PSO and adaptive modified PSO. Similarly, a differential evolution (DE)–based EMS for a grid-connected MG is presented in [19]. The objectives of the proposed EMS are minimization of operational and emission costs of the MG. Operational cost of MG includes bidding cost of DERs, DR incentives and energy trading cost with main grid. The results obtained using DE algorithm are compared with the PSO-based results. In comparison, the proposed DE algorithm was found to be superior in terms of solution quality and convergence speed. In [20], an ant colony optimization-based multilayer EMS model for an islanded MG is proposed to minimize its operational cost. The objective function comprises bidding cost of RERs, DGs and battery, penalty cost on load shedding and DR incentives in both day-ahead scheduling and 5-min interval real-time scheduling layers. Three case studies were used to analyze the system: operation, sudden high requirement of load demand and plug-and-play ability. The proposed approach reduces operational cost of MG by almost 20% and 5% more than the modified conventional EMS and PSO-based EMS, respectively.

7.4.4 EMS based on artificial intelligent methods

In [21], the authors presented a fuzzy-based MG EMS. The algorithm utilized two GAs to optimize its day-ahead MG scheduling and built a fuzzy expert system to control the power output of the storage system. The first GA determined MG energy scheduling and fuzzy rules, while the second GA tuned fuzzy membership functions. Reference [22] proposed an intelligent adaptive dynamic EMS for a grid-connected MG. It maximized the utilization of RERs and minimized carbon emissions to achieve a reliable and self-sustainable system. It also improved battery lifetime. The proposed EMS was modeled using evolutionary adaptive dynamic programming and reinforcement learning concepts and solved by using two neural networks (NNs). An active NN is used to solve the proposed EMS strategy, while a critical NN checks its performance with respect to optimality. The new defined performance index evaluates the performance of dynamic EMS in terms of battery lifetime, utilization of renewable energy and minimum curtailment of controllable load. The performance of the proposed approach is better as compared to decision-tree-approach-based dynamic EMS.

7.4.5 EMS based on multi-agent systems (MAS)

A decentralized MAS-based MG EMS for a grid-connected MG is presented in [23]. For optimal operation of MG, all the consumers, storage units, generation units and grid are considered as agents. Also, the consumer consumption preference has been considered as an important factor for decision-making. The algorithm helps one to reduce the power imbalance cost while considering consumer consumption preference as an important factor in the decision-making process. In [24], multi-objective hierarchical MAS-based EMS for a grid-connected MG system is presented to minimize its operational cost, emission cost and line losses. The hierarchical MAS is divided into three levels. The upper level agent in a centralized control is responsible for energy management of the whole system to maximize economical and environmental benefits. The middle level agent is responsible for operational mode switching of this region to maintain secure voltages. The lower level agent is responsible for *f*/*V* and *PQ*-based control strategies for unit agents to manage real-time operation of DERs.

7.5 MG EMS-wireless communication

Traditional power grids that only supported one-way communication are now evolving into smart grids with two-way communication infrastructure to support intelligent mechanisms for predicting failures and monitoring the condition of the network. Communication will play a vital role to transform the smart grids into a set of MGs. Important aspects of concern in MG are the security, load sharing and the variability of DGs. The variability of power generation by the DGs influences two aspects that are a matter of concern, i.e., power flow management and voltage control in the MG. Therefore, effective communication combined with advanced control techniques is needed to maintain the stability and safe operation during islanded mode operation of the MG [25].

The communication network can be treated as the spinal cord of an MG. It connects the power generating sources, transmission, distribution and consumption systems to the management block in order to evaluate the real-time data that reflects the stability of the entire grid [26]. Information is exchanged bidirectionally among operators, energy generating sources and consumers. MG communication networks can be divided into the following categories:

1. *Home area networks:* Provide low bandwidth, two-way communication between the home appliances and equipment such as smart meters to collect the real-time data.
2. *Field area networks:* Provide low bandwidth, two-way communication between the MG control station and customer premises.
3. *Wide area networks:* Provide high bandwidth, two-way communication between the MG and utility grid.

To establish a reliable, secure and effective two-way communication for the information exchange, a communication subsystem in an MG must consist of the following requirements:

- Must support the quality of service of data. This is because the critical data must be delivered promptly.
- Must be highly reliable since a large number of devices will be connected and will be communicating with different devices.
- Must have a high coverage, so it can respond to any event in the MG.
- Must guarantee security and privacy.

Communication technologies can be wired or wireless. Wireless technologies offer significant benefits over wired technologies such as rapid deployment, low installation cost and mobility. Most commonly used wireless technologies are wireless mesh network, cellular communications, cognitive radio, IEEE 802.15, satellite communications and microwave communications, whereas some of the most commonly used wired technologies are fiber-optic communications and powerline communications.

7.6 MG EMS-test case

As a case study, an MG EMS based on PSO is presented for grid-connected and islanded modes of operation while considering a full-day load demand profile. The proposed EMS is required to verify load demand during 24-h operation with the lowest operating cost for the DGs while taking into account the grid tariff and various system constraints [27].

7.6.1 Modeling of the microgrid

The schematic diagram of the MG model used for the EMS is shown in Figure 7.8. For this study, a typical MG model is used that includes different DGs such as combined heat and power (CHP) plant, diesel generator, natural gas-fired generator, photovoltaic (PV) generator, wind generator and ESS. The DGs are connected and integrated at the point of common coupling (PCC) to provide power to a cluster of loads. The rated power for DGs is shown in Table 7.2. The MG is capable of operating in two different modes:

- *Grid-connected mode* in which the MG is able to buy power from the grid if demand exceeds available power or sell power back to the grid if production exceeds demand. The load demand in this mode is 4.35 MW.
- *Islanded mode* in which it supplies power to only the critical loads. The load demand in this mode is 2.50 MW.

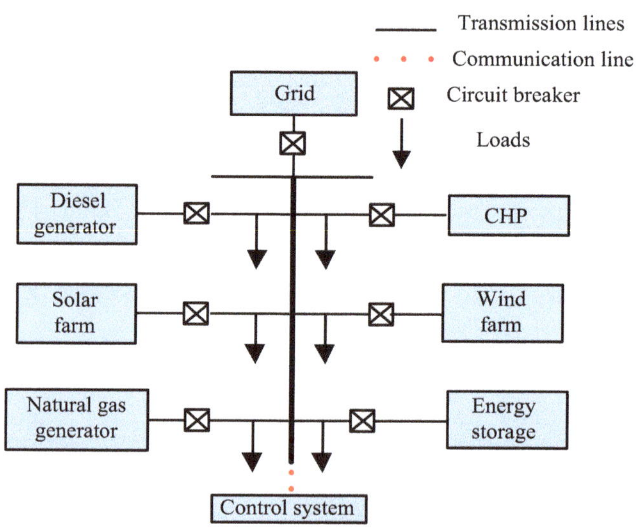

Figure 7.8 Overall system [27]

Table 7.2 Rating of all the DGs in the system

Distribution generations	Minimum power (MW)	Maximum power (MW)
Combined heat and power	0.0	1.5
Diesel generation	0.0	1.0
Natural gas generation	0.0	1.0
Solar farm	0.0	0.5
Wind farm	0.0	0.5
Energy storage	0.1	0.4
Grid	−1.0	1.0

7.6.2 Mathematical model of the system

In this section, the mathematical modeling of the system is presented.

7.6.2.1 Generators cost functions

The fuel cost function for the CHP, diesel generator and natural gas generator is typically approximated by a quadratic function, as stated in the following equation:

$$F_j(P_j) = \alpha_j + \beta_j P_j + \gamma_j P_j^2 \tag{7.1}$$

where j is the generating source; P is the power output of the source j; F is the operation cost of the source j in \$/h; and α, β and γ are the cost coefficients in \$/h (shown in Table 7.3) [28].

7.6.2.2 Solar generation cost function

The solar generation cost function is given by

$$F(P_s) = aP_s + G^e P_s \tag{7.2}$$

$$a = \frac{r}{[1 - (1 + r)^{-N}]} \tag{7.3}$$

where P_s is the solar generation (kW), a is the annuitization coefficient (dimensionless), r is the interest rate, N is the investment lifetime (taken as $N = 20$ years), I^P is the investment costs, per unit installed power (\$/kW) and G^E is the operation and maintenance (O&M) costs per unit generated energy (\$/kW).

Equations (7.2) and (7.3) are used to calculate the total generating cost of the solar energy considering the depreciation of all the equipment for generation. In this system, the values for the investment costs per unit of installed power (I^P) and O&M costs per unit of generated energy (G^E) are assumed to be equal to \$5,000 and 1.6 cents/kW, respectively. Therefore, the final cost function can be derived, represented in (7.4) [29].

$$F(P_s) = 545.016P_s \tag{7.4}$$

Figure 7.9 depicts the forecasted power for the solar farm for this study over an aggregated 24-h period. This data is not based on any one particular season or geographical area; however, it is a typical curve and is used here for discussion purposes. The seasonality over the year of particular geographical regions can also be accommodated, if desired.

Table 7.3 Cost figures for various generators

	CHP	Diesel generator	Natural gas
α (\$/h)	15.30	14.88	9.00
β (\$/h)	0.210	0.300	0.306
γ (\$/h)	0.000240	0.000435	0.000315

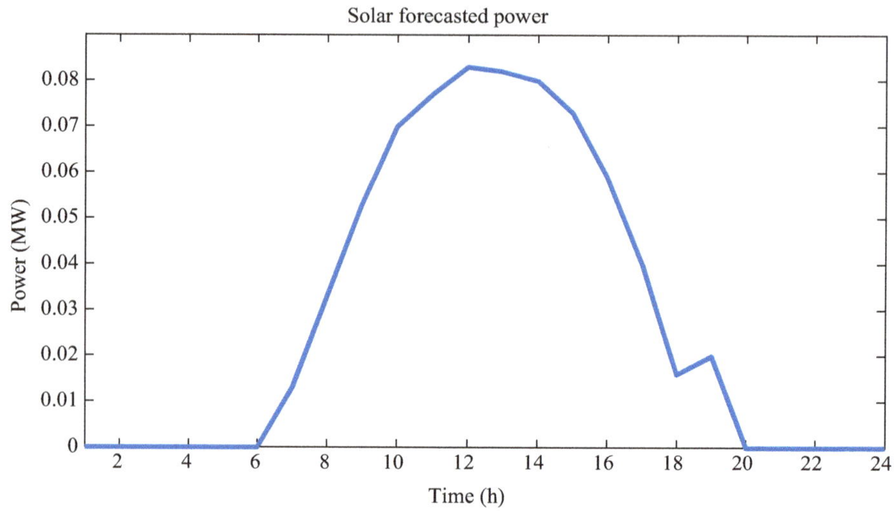

Figure 7.9 Forecasted power for solar farm [27]

7.6.2.3 Wind generation cost function

The cost function for wind generation by (7.2) and (7.3) is similar to solar generation. But, the investment costs per unit installed power (I^P) and O&M costs per unit generated energy (G^E) are assumed to be equal to 1,400 \$/kW and 1.6 cents/kW. Therefore, the final cost function can be derived and is represented in (7.5) [29].

$$F(P_w) = 152.616P_w \tag{7.5}$$

Figure 7.10 depicts the forecasted aggregated power for the wind farm for this study over a 24-h period. Again, this data is not based on any season or geographical area; however, it is typical, and it is just assumed here for test purposes.

7.6.2.4 Energy storage cost function

A 500 kWh battery is used for this study. The battery charges at a rate of capacity-divided-by-two (C/2) and discharges at capacity-divided-by-three (C/3). The charging mode of the battery is considered as a load of 3 MW. Thus, the cost function for battery is derived from (7.2) and (7.3), similar to the previous two DGs. The investment costs per unit installed power (I^P) and O&M costs per unit generated energy (G^E) are assumed to be equal to 1,000 \$/kW and 1.6 cents/kW. Therefore, the final cost function can be derived, as represented in (7.6) [30].

$$F(P_b) = 119P_b \tag{7.6}$$

For this study, the battery charging is not considered and the discharge time and rate are predetermined for testing purposes.

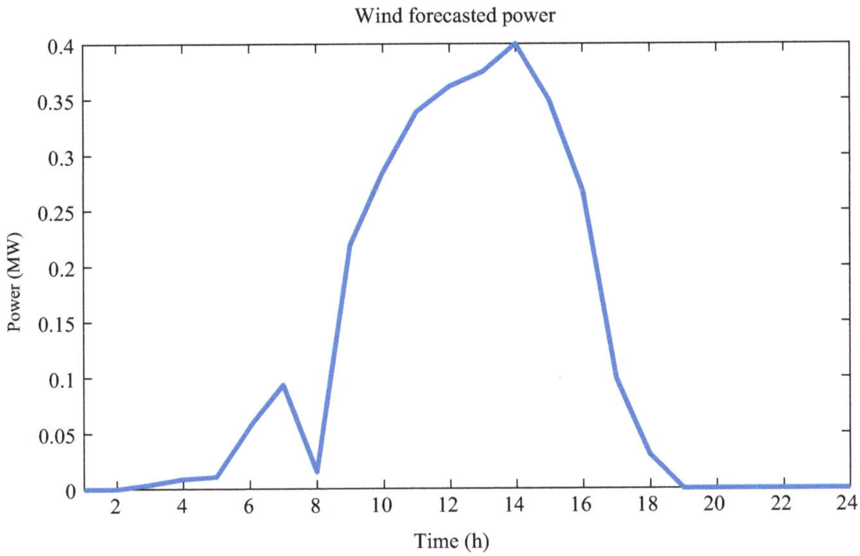

Figure 7.10 Forecasted power for wind farm [27]

7.6.2.5 Constraint functions

The constraint functions are used to help one to guide the system to achieve the desired results.

Grid-connected mode

In grid-connected mode, the MG is able to buy/sell power from/to the main grid depending on the load demand.

$$P_{generated} \neq P_{Load} \tag{7.7}$$

If (7.7) is true then

$$P_{grid} = P_{generated} - P_{Load} \tag{7.8}$$

Therefore, if P_{grid} is positive, then MG is purchasing power from the grid and if P_{grid} is negative then MG is selling power to the grid.

Islanded mode

Since in islanded mode, the MG is disconnected from the main grid, there cannot be any buying/selling of power.

$$P_{generated} = P_{Load} \tag{7.9}$$

Thus, (7.9) must always be true.

Power generation limits

Each DG has a power rating as shown in Table 7.1.

$$P_j^{min} \leq P_j \leq P_j^{max} \tag{7.10}$$

7.6.3 Results of PSO

The load demand for a 24-h period as used for both case studies is presented in Figure 7.11. It can be seen in the figure that load demand for the islanded mode is significantly lower than load demand during grid-connected mode. That is because when the system goes into islanded condition, then load shedding takes place and the system only supplies power to the critical loads.

7.6.3.1 First case study: grid-connected mode

Figure 7.12 depicts the best output achieved from the EMS for grid-connected mode. The algorithm was able to find the optimal solution to dispatch DGs to satisfy the given load demand for the time period. It can be observed that the system is able to buy and sell power from the grid during off-peak and peak hours. The total cost of operation during the 24-h period achieved is \$2,297.96.

7.6.3.2 Second case study: islanded mode

Figure 7.13 depicts the best output achieved from the EMS for islanded mode of operation. The algorithm was able to find the optimal solution to dispatch every generator to satisfy the given load demand for each time interval. As can be observed, the three generators, CHP (Gen1), diesel (Gen2) and natural gas (Gen3), are being rapidly ramped up and down to help for each time interval to help one to reduce or increase the generation to help one to satisfy the load. This ramping in the system has to be controlled by adding a ramp rate constraint in the algorithm, but for this study the ramp rate has been ignored due to the use of a 1-h time step. The total operation cost achieved during the 24-h period for the case is \$2,277.92.

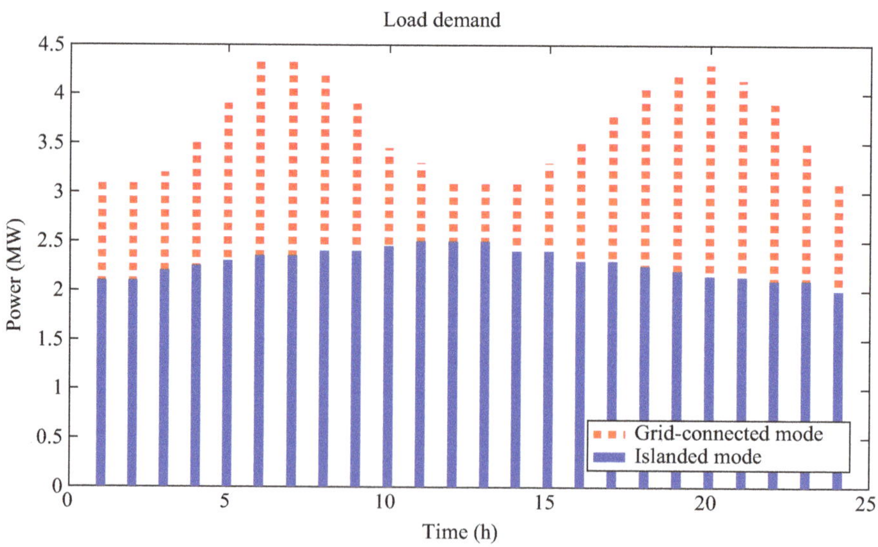

Figure 7.11 Load demand [27]

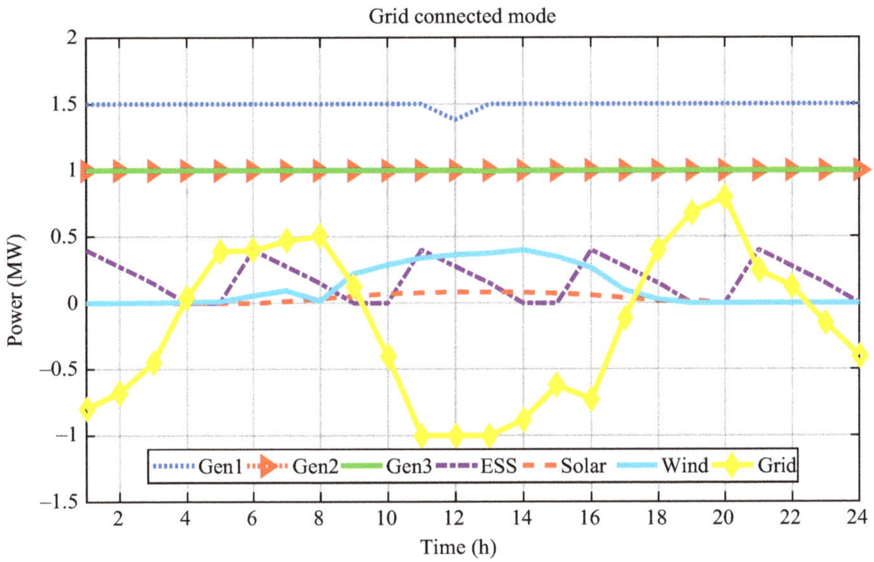

Figure 7.12 Output of PSO (first case study) [27]

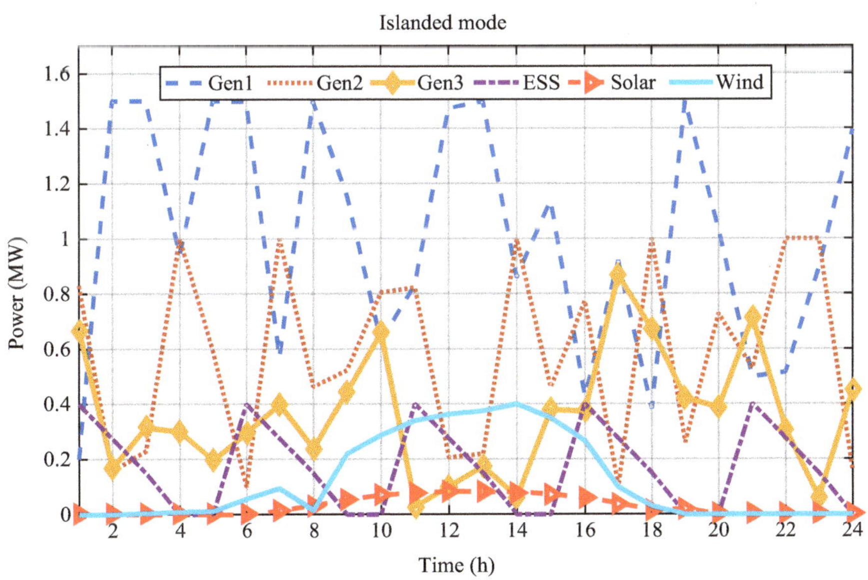

Figure 7.13 Output of PSO (second case study) [27]

7.7 Conclusion

This chapter provides insight into the role of the EMS system for an MG. The hierarchy of the various controllers utilized in the EMS is presented. Various programming methods can be used for optimizing the behavior of the EMS such that the MG can operate in a safe and reliable manner to match the demanded load with the available energy sources. The requirements for a communication system within the MG are discussed. A test case with a PSO technique is provided.

References

[1] N. I. Nwulu and X. Xia, "Optimal dispatch for a microgrid incorporating renewables and demand response," *Renewable Energy*, vol. 101, pp. 16–28, 2017. *Electron. Comput. Sci.*, 2016.
[2] P. Basak, S. Chowdhury, S. Halder Nee Dey, and S. P. Chowdhury, "A literature review on integration of distributed energy resources in the perspective of control, protection and stability of microgrid," *Renewable and Sustainable Energy Reviews*, vol. 16, no. 8, pp. 5545–5556, 2012.
[3] S. Ezhilarasan, P. Palanivel, and S. Sambath, "Design and development of energy management system for DG source allocation in a micro grid with energy storage system," *Indian Journal of Science and Technology*, vol. 8, no. 13, pp. 1–9, 2015.
[4] Y. S. Bhavsar, P. V. Joshi, and S. M. Akolkar, "Simulation of Microgrid with energy management system," *Int. Conf. Energy Syst. Appl. ICESA 2015*, ICESA, Pune, India, pp. 592–596, 2016.
[5] S. Monesha, S. G. Kumar, and M. Rivera, "Microgrid energy management and control: Technical review," *2016 IEEE Int. Conf. Autom.*, 2016, pp. 1–7.
[6] D. E. Olivares, A. Mehrizi-Sani, A. H. Etemadi, *et al.*, "Trends in microgrid control," *IEEE Transactions on Smart Grid*, vol. 5, no. 4, pp. 1905–1919, IEEE, 2014.
[7] M. F. Zia, E. Elbouchikhi, and M. Benbouzid, "Microgridsenergy management systems: A critical review on methods, solutions, and prospects," *Applied Energy*, vol. 222, pp. 1033–1055, 2018.
[8] L. Meng, E. R. Sanseverino, A. Luna, *et al.*, "Microgrid supervisory controllers and energy management systems: A literature review," *Renewable and Sustainable Energy Reviews*, vol. 60, pp. 1263–1273, 2016.
[9] Z. Cheng, J. Duan, and M. Chow, "To centralize or to distribute: That is the question: A comparison of advanced microgrid management systems," *IEEE Industrial Electronics Magazine*, vol. 12, no. 1, pp. 6–24, 2018.
[10] F. Katiraei, R. Iravani, N. Hatziargyriou, and A. Dimeas, "Microgrids management," *IEEE Power and Energy Magazine*, vol. 6, no. 3, pp. 54–65, 2008.

[11] J. Jimeno, J. Anduaga, J. Oyarzabal, and A. G. De Muro, "Architecture of a microgrid energy management system," *European Transactions on Electrical Power*, vol. 21, pp. 1142–1158, 2011.

[12] C. Corchero, M. Cruz-Zambrano, F.-J. Heredia, *et al.*, "Optimal energy management for a residential microgrid including a vehicle-to-grid system," *IEEE Transactions on Smart Grid*, vol. 5, no. 4, pp. 2163–2172, 2014.

[13] J. Shen, C. Jiang, Y. Liu, and J. Qian, "A microgrid energy management system with demand response for providing grid peak shaving," *Electric Power Components and Systems*, vol. 44, no. 8, pp. 843–852, 2016.

[14] A. G. Tsikalakis and N. D. Hatziargyriou, "Centralized control for optimizing microgrids operation," *Power and Energy Society General Meeting, 2011 IEEE*, Detroit, Michigan, pp. 1–8, IEEE, 2011.

[15] L. K. Panwar, S. R. Konda, A. Verma, B. K. Panigrahi, and R. Kumar, "Operation window constrained strategic energy management of microgrid with electric vehicle and distributed resources," *IET Generation, Transmission & Distribution*, vol. 11, no. 3, pp. 615–626, 2017.

[16] M. Strelec and J. Berka, "Microgrid energy management based on approximate dynamic programming," *Innovative Smart Grid Technologies Europe (ISGT EUROPE), 2013 4th IEEE/PES*, Lyngby, pp. 1–5, IEEE, 2013.

[17] S. Chalise, J. Sternhagen, T. M. Hansen, and R. Tonkoski, "Energy management of remote microgrids considering battery lifetime," *The Electricity Journal*, vol. 29, no. 6, pp. 1–10, 2016.

[18] J. Radosavljevic, M. Jevtić, and D. Klimenta, "Energy and operation management of a microgrid using particle swarm optimization," *Engineering Optimization*, vol. 48, no. 5, pp. 811–830, 2016.

[19] N. Tiwari and L. Srivastava, "Generation scheduling and micro-grid energy management using differential evolution algorithm," *Circuit, Power and Computing Technologies (ICCPCT), 2016 International Conference on*, Nagercoil, pp. 1–7, IEEE, 2016.

[20] M. Marzband, E. Yousefnejad, A. Sumper, and J. L. Domínguez-García, "Real time experimental implementation of optimum energy management system in standalone microgrid by using multi-layer ant colony optimization," *International Journal of Electrical Power & Energy Systems*, vol. 75, pp. 265–274, 2016.

[21] J. P. Fossati, A. Galarza, A. Martín-Villate, J. M. Echeverría, and L. Fontan, "Optimal scheduling of a microgrid with a fuzzy logic-controlled storage system," *International Journal of Electrical Power & Energy Systems*, vol. 68, pp. 61–70, 2015.

[22] G. K. Venayagamoorthy, R. K. Sharma, P. K. Gautam, and A. Ahmadi, "Dynamic energy management system for a smart microgrid," *IEEE Transactions on Neural Networks and Learning Systems*, vol. 27, no. 8, pp. 1643–1656, 2016.

[23] S. Ghorbani, R. Rahmani, and R. Unland, "Multi-agent autonomous decision making in smart micro-grids energy management: A decentralized

approach," *German Conference on Multiagent System Technologies*, Springer, Munich, pp. 223–237, 2017.

[24] C.-X. Dou and B. Liu, "Multi-agent based hierarchical hybrid control for smart microgrid," *IEEE Transactions on Smart Grid*, vol. 4, no. 2, pp. 771–778, 2013.

[25] S. Safdar, B. Hamdaoui, E. Cotilla-Sanchez, and M. Guizani, "A survey on communication infrastructure for micro-grids," *2013 9th International Wireless Communications and Mobile Computing Conference (IWCMC)*, Sardinia, pp. 545–550, IEEE, 2013.

[26] X. Fang, S. Misra, G. Xue, and D. Yang, "Smart grid—The new and improved power grid: A survey," *IEEE Communications Surveys & Tutorials*, vol. 14, no. 4, pp. 944–980, Fourth Quarter 2012.

[27] M. Y. Ali, F. Khan, and V. K. Sood, "Energy management system of a microgrid using particle swarm optimization and wireless communication system," *2018 IEEE Electrical Power and Energy Conference (EPEC)*, Toronto, ON, pp. 1–7, IEEE, 2018.

[28] Z.-L. Gaing, "Particle swarm optimization to solving the economic dispatch considering the generator constraints," *IEEE Transactions on Power Systems*, vol. 18, no. 3, pp. 1187–1195, 2003.

[29] N. Augustine, S. Suresh, P. Moghe, and K. Sheikh, "Economic dispatch for a microgrid considering renewable energy cost functions," *2012 IEEE PES Innov. Smart Grid Technol.*, Washington, DC, pp. 1–7, IEEE, 2012.

[30] M. Bhoye, S. N. Purohit, I. N. Trivedi, M. H. Pandya, P. Jangir, and N. Jangir, "Energy management of renewable energy sources in a microgrid using cuckoo search algorithm," *2016 IEEE Students' Conference on Electrical, Electronics and Computer Science (SCEECS)*, Bhopal, pp. 1–6, IEEE, 2016.

Chapter 8

Reconfiguration of distribution network using different optimization techniques

Hanan Hamour[1], Salah Kamel[1], Juan Yu[2] and Francisco Jurado[3]

Modern distribution systems consist of a number of interconnected radial circuits, which include buses, distribution lines and numerous numbers of switches: tie switches (normally open) and sectionalizing switches (normally close). These switches are used to change the status of feeders to meet a specific objective. During transferring power from generating stations to reach customers, a part of power losses occurs at transmission system and another part is at distribution system. A huge amount of power loss occurs in distribution network because the distribution system operates at very less voltages than transmission systems, so that the load currents that flow through the long distribution lines are much greater than those in transmission system. As the loads need highly safe and dependable power supply, distribution network becomes very important and active part in power system that can be optimized and controlled to improve its performance. Estimating the amount of total power losses in distribution system is considered a main tool for evaluating the system. One of the several control methods for distribution power loss reduction is *reconfiguration process*. Distribution network reconfiguration (DNR) is an effective process for power loss reduction. It is done by changing the topological structure of the network by switches on/off status correction to transfer loads from heavily loaded feeder to lightly loaded one until the minimum value of summation of all feeder power losses in the network is obtained while satisfying the network operating constraints without isolation of any load from the system. Recently, meta-heuristic methods are used for distribution system reconfiguration such as genetic algorithms (GA), ant colony algorithms, enhanced genetic algorithm (EGA), simulated annealing (SA), selective particle swarm optimization (SPSO) algorithm, grasshopper optimization algorithm and grey wolf optimization (GWO) technique and others. This chapter presents a concept of using augmented grey wolf optimization (AGWO) algorithm, comparing with other optimization techniques, to

[1]Department of Electrical Engineering, Faculty of Engineering, Aswan University, Aswan, Egypt
[2]College of Electrical Engineering, Chongqing University, Chongqing, China
[3]Department of Electrical Engineering, University of Jaén, Jaén, Spain

reconfigure the distribution system aiming to reduce the power losses. The developed algorithm is successfully applied on IEEE 33-bus distribution system that has proved its ability to find optimal solutions through lower computational time.

Nomenclature

GA	Genetic Algorithm
SA	Simulated Annealing
SPSO	Selective Particle Swarm Optimization algorithm
GWO	Grey Wolf Optimization algorithm
EGA	Enhanced Genetic Algorithm
DNR	Distribution Network Reconfiguration
AGWO	Augmented Grey Wolf Optimization algorithm
FEP	Fuzzy Adaptation of Evolution Programming

8.1 Introduction

The power distribution system is a very important part in the electrical power system, in which it receives the electrical power from transmission system to supply the loads with the required power. As distribution network has a high R/X ratio, very low voltage levels and high current flow compared with transmission system, most of power losses occur in distribution networks. Loads need to be a reliable supplying source for assuring the quality and continuity of service, so the distribution system losses must be minimized as lower as possible. Distribution network consists operationally of a number of interconnected radial circuits. Now, distribution system has become active where the operating conditions can be improved.

One of the most efficient and well-known methods for reducing real power losses of the distribution system is network reconfiguration, which can reduce system power losses and improve the voltage profile. Network reconfiguration aims to obtain the optimum construction of the distribution network with a combination of optimum total network power losses and optimum switches that are opened in the new topological structure. By maintaining system operating constraints, reconfiguration process is performed by relieving the loads from heavily loaded feeders to lower loaded ones. This is done by changing the status of on/off sectionalizing and tie switches where a number of the tie switches are closed, and at the same instant a corresponding number of sectionalizing switches are opened, so that the total number of opened switches in the new structure equals the original number of tie switches in the original configuration of the system to satisfy the network operating constraints.

Reconfiguration process consists of two stages: first one uses an efficient optimization algorithm that is responsible for performing switching processes, creating a new construction of the network and testing the satisfaction of network operating constraints. The second stage uses suitable load flow algorithm to perform load flow calculations and to provide the total system power loss calculation for each new topological structure resulting from reconfiguration process by isolating the worst obtained solutions and keeping the best ones. This is computed and called as local minimum power loss; the best value of power losses among the whole local values is determined and considered as the global optimum power losses of the system corresponding to an optimum combination of opened switches [1]. Reconfiguration problem is an optimization problem because it aims to find the best solution among all possible solutions, and it is considered as a combinatorial one because its variables take a finite set of values, so the reconfiguration problem is a combinatorial optimization problem.

In this chapter, reconfiguration problem is solved to find the best value of total system power losses that can be considered as the optimum solution. Based on overloading conditions, reconfiguration is necessary to isolate the overcurrent from the component in the distribution system such as feeders or transformers. An algorithm is required to quickly solve reconfiguration problem to minimize real power losses under maintaining network constraints, where the algorithm is a method that describes a set of actions that must be done to solve a problem [2]. There are several optimization algorithms for solving DNR problem with different levels of success. Traditionally, reconfiguration problem for power loss minimization was termed 'multiobjective optimization problem' when system reliability is considered.

To solve radial distribution system reconfiguration problem, many heuristic and meta-heuristic optimization techniques are used. Merlin and Back are the first who suggest a branch method that solves reconfiguration problem in which the system included n sectionalizing line, switches and $2n$ number of new topological structures produced by reconfiguration process, but this was quite time consuming [1]. Heuristic optimization techniques introduced a suboptimum solution for distribution system reconfiguration problem, but they introduce very low quality solutions in which they can introduce the local best solutions for each configuration of system and cannot give the global optimum solution among all the best local solutions. Second, heuristic methods depend on the problem itself, i.e. we can define a specific heuristic algorithm for a specific problem.

Heuristics were first used by Sheromohammadi and Hong [3] in which they presented heuristic algorithm to search for the best solution for DNR. In [4], the optimal reconfiguration of radial distribution systems has been achieved using fuzzy adaptation of evolutionary programming. Many of these approaches that are based on heuristic optimizations are not sufficiently accurate to solve DNR optimization problem as simultaneous switching operations are not performed due to the combinatorial distribution system switching options. Civanlar *et al.* proposed a heuristic optimization method while developing a simple model for branch exchange system power losses computation depending on performing just two pairs of switching operation at each time [5]. Naveen *et al.* used modified bacterial foraging optimization algorithm as a recent evolutionary technique to solve reconfiguration of distribution system [6]. Sathish Kumar and Jayabarathi proposed using chemo toxins on bacterial growth in bacterial foraging optimization technique [7]. Nara suggested using of GA for solving reconfiguration problem with the minimum value of power losses [8]. Zhu [9] suggested refining the GA by doing competition in mutation and crossover. Mohamed Imran [10] used the concept of fireworks in the fireworks optimization algorithm to reconfigure radial distribution network for power loss minimization.

All the previously mentioned methods just perform reconfiguration which is opening and closing sectionalizing and tie switches in distribution system. As modern electrical power distribution system becomes more complex with nonlinear operational characteristics and electrical behaviour, reconfiguration problem is a complex combinatorial problem. To solve this type of problem, recent meta-heuristic optimization techniques are based on the principle of population, and expanding the search process over all the search space is used to minimize power losses by calculating the local best solutions and to enable defining the global best solution for problem as the optimal solution.

Meta-heuristic algorithms are problem-independent techniques that can solve nearly all optimization problems. They proceed iteratively trying to improve the obtained results until reaching the required optimum solution. Modified bacterial foraging optimization algorithm was applied to solve reconfiguration problem for power loss minimizing [1]. Modified honey bee mating optimization algorithm in [11] was proposed for performing network reconfiguration to reduce power losses. EGA is proposed to reach the optimal power losses as an optimal solution for DNR problem [12]. Bacterial foraging optimization algorithm was proposed by Passino in 2002. This optimization technique is applied for many various applications where it can isolate the worst solutions and scatter the best solutions.

SA optimization algorithm was introduced to get optimal solution for distribution system reconfiguration problem by authors in [13]. Modified Tabu search algorithm is used to evaluate the optimum value of total distribution power losses as in [14]. SPSO is employed to obtain optimal solution of DNR problem and compared with binary particle swarm algorithm [15]. GWO algorithm was used to obtain the optimum power losses for distribution system [16,17]. Ant colony search algorithm is employed for getting the optimum solution for DNR problem.

In this chapter, an AGWO algorithm is presented to solve DNR problem to minimize real power loss based on the natural behaviour of alpha and beta grey wolves. The obtained results proved that AGWO [18] approach is more accurate algorithm with much less computational time compared with SPSO and other optimization methods when it is tested on 33-bus distribution system. The obtained results from AGWO are compared with the obtained results from SPSO.

8.2 Optimal network reconfiguration

Network reconfiguration concept is applied in the distribution system in two states: planning and operation of the system. In planning of the system, reconfiguration process is required for obtaining the optimal topological structure by altering the on/off state of sectionalizing and tie switches. While doing this, the overloading at a feeder is transferred into another feeder with lower loading, so that the total power losses of the system are minimized. In operation of the system, reconfiguration process is very important to automatically and rapidly restore the power supplying service after isolating the fault from the system. Applying the reconfiguration in distribution system is much simpler and less costlier compared to other methods of control distribution system's performance. Generally, reconfiguration process has two important benefits: providing maximum possible amount of the electrical power required for the loads and automatically reconfiguring the network when any problems such as 'faults' are removed. Distribution system has two types of switches that are used for both protection and reconfiguring the network, sectionalizing switches (normally close switches) within loaded feeders and tie switches (normally open switches).

Distribution system is reconstructed for minimizing total power losses, reducing the overloading from network components such as feeders and transformers, automatically restoring the power supplying to network after isolating the problem such as faults and due to this, the operating conditions are improved. Reconfiguration of electrical distribution system can be considered as a combinatorial, complex and nonlinear optimization problem.

8.2.1 Problem formulation

The power flow equations can be evaluated from the single-line diagram in Figure 8.1. From this branch model, voltage phase at bus m is $V_m \angle\delta_m$ whereas voltage at bus $m+1$ is $V_{m+1}\angle\delta_{m+1}$.

Current flow through the line is defined from the following equation:

$$\tilde{I}_{m+1} = \frac{V_m \angle\delta_m - V_{m+1}\angle\delta_{m+1}}{R_{m+1} + jX_{m+1}} \tag{8.1}$$

where R_{m+1} is the resistance of branch, X_{m+1} is the reactance of branch. The consuming power by loads is given by

$$P_{m+1} - jQ_{m+1} = \tilde{V}_{m+1} \times \tilde{I}_{m+1} \tag{8.2}$$

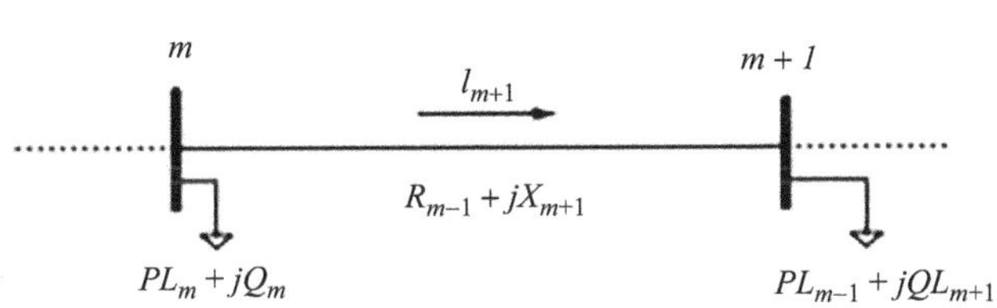

Figure 8.1 Single-line diagram for radial distribution system

From (8.1) and (8.2), the voltage magnitude at bus $m+1$ is given by

$$V_{m+1} = \left\{ \left[\left(P_{m+1}R_{m+1} + Q_{m+1}X_{m+1} - \frac{|V_m|^2}{2} \right)^2 - \left(R^2{}_{m+1} + X^2{}_{m+1} \right)\left(P^2{}_{m+1} + Q^2{}_{m+1} \right) \right]^{1/2} - \left(P_{m+1}R_{m+1} + Q_{m+1}X_{m+1} - \frac{|V_m|^2}{2} \right) \right\}^{1/2} \quad (8.3)$$

where P_{m+1} and Q_{m+1} represent the total active and reactive power losses, respectively, as follows:

$$P_{m+1} = \sum_{j=m+1}^{N-1} P_{L,j} + \sum_{j=m+1}^{N-1} P_{loss,j} \quad (8.4)$$

$$Q_{m+1} = \sum_{j=m+1}^{N-1} Q_{L,j} + \sum_{j=m+1}^{N-1} Q_{loss,j} \quad (8.5)$$

where N is the total number of buses; $P_{L,j}$ and $Q_{L,j}$ are the active and reactive load powers at bus j, respectively; $P_{loss,j}$ and $Q_{loss,j}$ are the active and reactive power losses through the branch, respectively.

$$P_{loss,m} = \frac{R_m(P^2{}_m + Q^2{}_m)}{|V_m|^2} \quad (8.6)$$

$$Q_{loss,m} = \frac{X_m(P^2{}_m + Q^2{}_m)}{|V_m|^2} \quad (8.7)$$

When summing all the power losses of all lines in the system, we will obtain total system real power loss $P_{T,loss}$ as

$$P_{T,loss} = \sum_{m=1}^{N_b b} K_m P_{loss,m} \quad (8.8)$$

where N_b is the total number of branches of system; K_m is the binary variable which represents the topological status of the feeder in radial distribution system.

Problem can be identified as

$$\text{Minimize } F = \min P_{T,loss} \quad (8.9)$$

Subjected to system constraints,

1. Radiality constraint: It assures that no closed loops are included through the network. This means that each load must be supplied from only one bath and therefore the number of branches = the number of buses − 1.

2. Feasibility constraint: This constraint means that no loads are isolated during reconfiguring the network.
3. Voltage constraints: Bus voltage must lie between min and max limits.
4. Current constraints: Branch current constraints must be in its allowable range.
5. $V_{m,min} < |V_m| < V_{m,max}$ (8.10)
6. $K_m I_m < I_{m,max}, m\epsilon[1, 2, \ldots, N_b]$ (8.11)
7. Kirchhoff's current and voltage laws, respectively.

$$g_i(I, K) = 0 \quad (8.12)$$

$$g_v(V, K) = 0 \quad (8.13)$$

8.3 Augmented grey wolf optimization algorithm

AGWO algorithm involves two major processes: the first process is exploration process that increases the diversity of search agents over the search space that makes the solutions change stochastically for quick reaching optimal solution and try to achieve a good coverage of the search space to avoid stagnation state of generated solutions and reduce the computation time. Divergence of search agents must not be very high because very large divergence does not lead to reasonable solutions. The second process is exploitation that is responsible for improving the solution quality by updating the solutions locally around the best solution obtained by exploration [19] in each iteration. AGWO algorithm is inspired from grey wolves actions that are searching, encircling, attacking and hunting the prey relying on only the strongest first two groups of grey wolves that are alpha (α) and beta (β) grey wolves as shown in Figure 8.2. Alpha and beta groups are the first two groups in the hierarchy of grey wolves.

AGWO algorithm [18] is a simple augmentation for exploration and exploitation processes of GWO [20]. Exploration process is augmented by making the value of parameter *a* nonlinearly, which varies from 2 to 1 as in Figure 8.3 instead

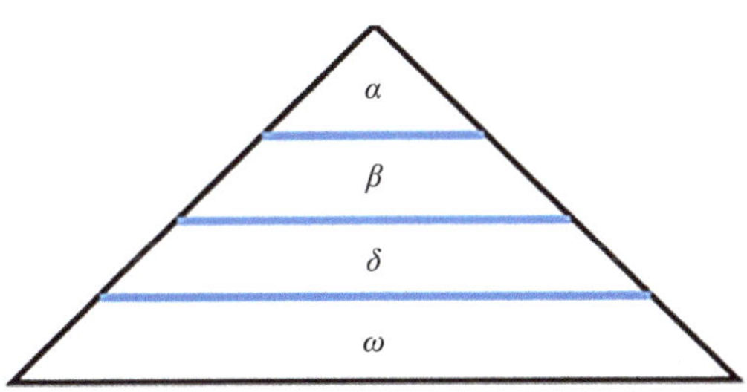

Figure 8.2 Grey wolves hierarchy

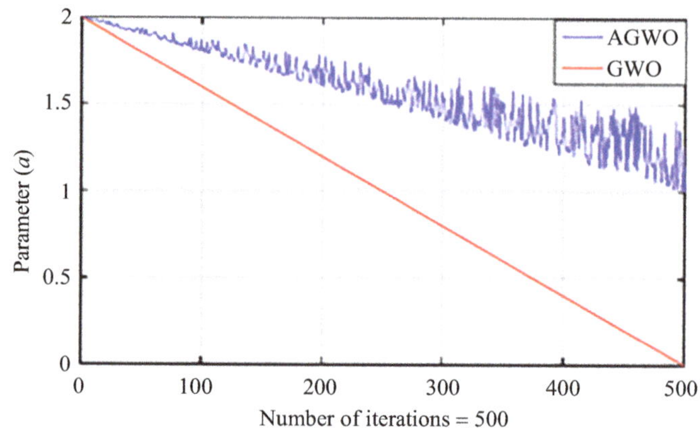

Figure 8.3 Behaviour of parameter a in AGWO and GWO optimization techniques

of varying linearly from 2 to 0 as in GWO algorithm in which parameter a is defined by (8.1). In the AGWO, hunting and decision-making depends on only α and β. The process of hunting the prey by grey wolves is divided into searching for the prey, encircling the prey, hunting and attacking the prey. Encircling the prey can be modelled as

$$\vec{D}_\alpha = \left|\vec{C}_1 \cdot \vec{X}_{\alpha i} - \vec{X}_i\right| \tag{8.14}$$

$$\vec{D}_\beta = \left|\vec{C}_2 \cdot \vec{X}_{\beta i} - \vec{X}_i\right| \tag{8.15}$$

$$\vec{X}_1 = \vec{X}_{\alpha i} - \vec{A}_1 \cdot \vec{D}_\alpha \tag{8.16}$$

$$\vec{X}_2 = \vec{X}_{\beta i} - \vec{A}_2 \cdot \vec{D}_\beta \tag{8.17}$$

where C_1 and C_2 are the random parameters, $X_{\alpha i}$ is the position of alpha at iteration i, and $\vec{X}_{\beta i}$ is the position of beta at iteration i.

Alpha and beta search agents update their positions in each iteration by the following equation:

$$\vec{X}_{i+1} = \frac{\vec{X}_1 + \vec{X}_2}{2} \tag{8.18}$$

Varying the parameter nonlinearly and randomly to augment the exploration process is expressed by the following equation:

$$\vec{a} = 2 - \cos(rand) \times \frac{t}{max\text{-}iter} \tag{8.19}$$

$$\vec{A} = 2\vec{a} \cdot \vec{r}_1 - \vec{a} \tag{8.20}$$

$$\vec{C} = 2 \cdot \vec{r}_2 \tag{8.21}$$

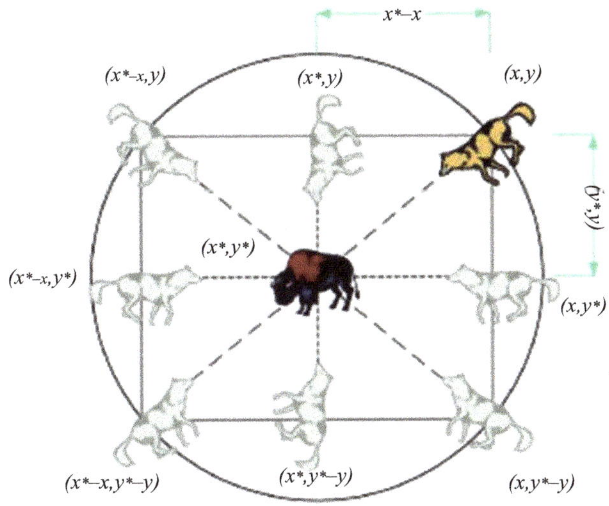

Figure 8.4 Behaviour of grey wolves to hunt the prey

where r_1 and r_2 are the uniformly distributed random vectors between 0 and 1 and parameter a is nonlinearly decreased from 2 to 1 with iteration (t) increased until reaching maximum iteration. Exploration (searching for the prey location) is achieved by divergence of search agents through the search space that is achieved when $|A| > 1$. Exploitation (attacking on the prey) by the divergence of the search agents that is $|A| < 1$ as in Figure 8.4 that shows exploration and exploitation process of grey wolves as search agents for hunting prey.

AGWO algorithm can be described step-by-step as shown in the flowchart in Figure 8.5.

8.4 Test results of IEEE 33-bus distribution system

The AGWO for minimizing power loss uses the standard IEEE 33-bus radial distribution system. A programmable code is performed to develop an effective methodology for the AGWO for achieving the objective function. The AGWO is carried out by using MATLAB®. The test system has a base voltage of 12.66 kV; total active power load of system is 3,715 kW and total reactive power load is 2,300 kVAR [4]. Initial total active power losses before reconfiguration are 211.0194 kW. At initial state of the network, it has five tie switches that are indicated by doted branches and 32 sectionalizing switches that are indicated by solid branches as in Figure 8.6. Table 8.1 shows the line and load data of the system [21].

To show the effectiveness of AGWO, an estimation is carried out by running MATLAB programming code for 50 times with the proposed parameters as shown in Table 8.2. Convergence characteristics for the AGWO are shown in Figure 8.7.

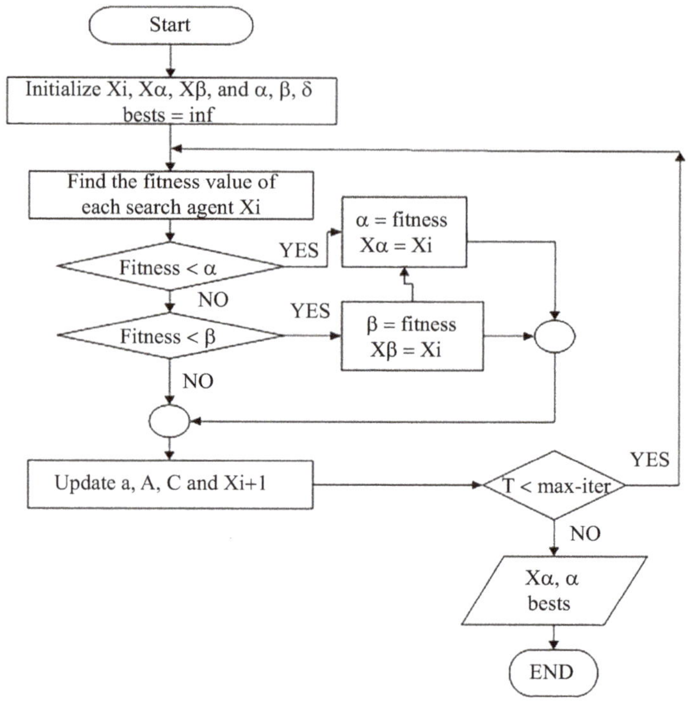

Figure 8.5 Flowchart of AGWO algorithm

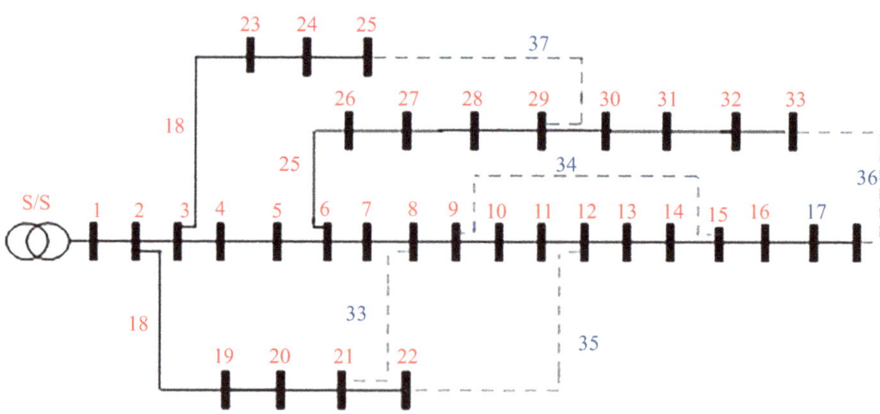

Figure 8.6 IEEE 33-bus distribution system before reconfiguration

From this figure, it can be observed that the AGWO reaches the best value of power loss at minimum simulation time compared with the SPSO technique [15]. In 33-bus system, the optimal active power loss value after reconfiguration using the AGWO is 130.8164 kW when switches S4, S14, S15, S22 and S33 are opened and

Table 8.1 Line data and bus data of IEEE 33-bus system

From	To	*P* (kW)	*Q* (kW)	*R* (Ω)	*X* (Ω)	I_{max}
1	2	100	60	0.0922	0.0470	400
2	3	90	40	0.4930	0.2510	400
3	4	120	80	0.3661	0.1864	400
4	5	60	30	0.3811	0.1941	400
5	6	60	20	0.8190	0.7070	400
6	7	200	100	0.1872	0.6188	300
7	8	200	100	1.7117	1.2357	300
8	9	60	20	1.0299	0.7400	200
9	10	60	20	1.0440	0.7400	200
10	11	45	30	0.1967	0.0651	200
11	12	60	35	0.3744	0.1237	200
12	13	60	35	1.4680	1.1549	200
13	14	120	80	0.5416	0.7129	200
14	15	60	10	0.5909	0.5260	200
15	16	60	20	0.7462	0.5449	200
16	17	60	20	1.2889	1.7210	200
17	18	90	40	0.7320	0.5739	200
2	19	90	40	0.1640	0.1564	200
19	20	90	40	1.5042	1.3555	200
20	21	90	40	0.4095	0.4784	200
21	22	90	40	0.7089	0.9373	200
3	23	90	50	0.4512	0.3084	200
23	24	420	200	0.8980	0.7091	200
24	25	420	20	0.8959	0.7010	200
6	26	60	25	0.2031	0.1034	300
26	27	60	25	0.2842	0.1447	300
27	28	60	20	1.0589	0.9338	300
28	29	120	70	0.8043	0.7006	200
29	30	200	600	0.5074	0.2585	200
30	31	150	70	0.9745	0.9629	200
31	32	210	100	0.3105	0.3619	200
32	33	60	40	0.3411	0.5302	200

Table 8.2 Used parameters in the AGWO algorithm

Population	50
Maximum number of solutions	100
Dimensions	5
Voltage limits	$0.9 \le V \le 1$

form the optimal system configuration as shown in Figure 8.8, whereas it is smaller than those obtained by the other algorithms shown in Table 8.3.

In this table, the AGWO is compared with in terms of tie switches, total system power losses, power losses reduction. Changing the construction

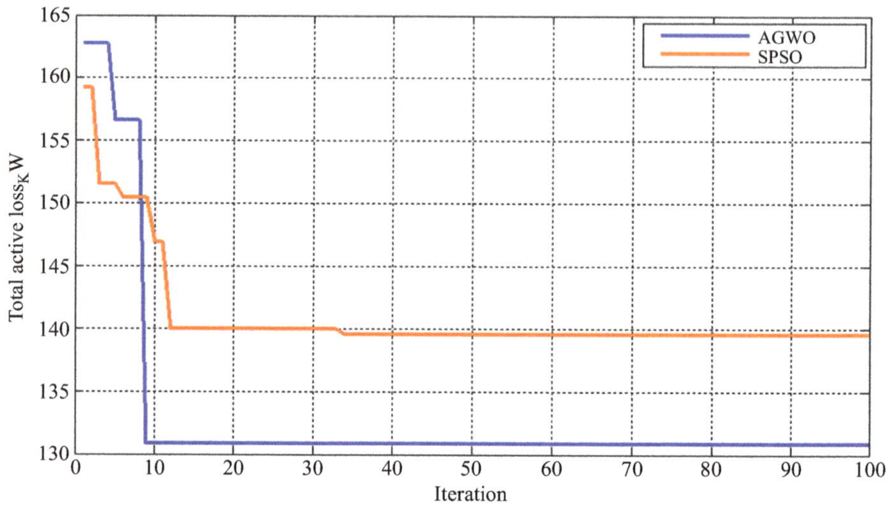

Figure 8.7 Convergence characteristics of the AGWO compared with convergence characteristics with SPSO

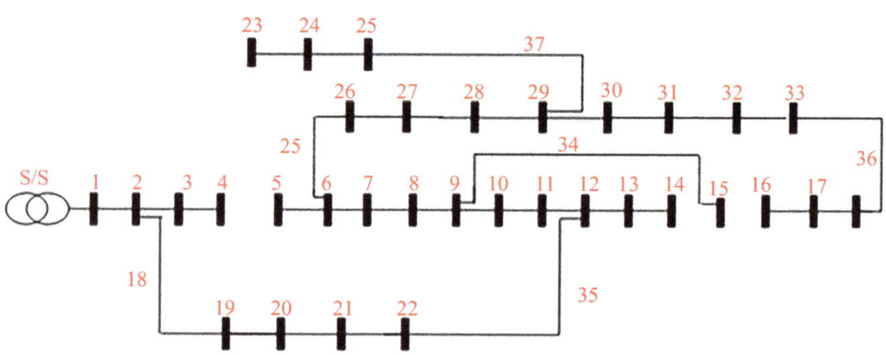

Figure 8.8 IEEE 33-bus distribution system after reconfiguration using AGWO algorithm

Table 8.3 Power loss results produced by AGWO compared with other algorithms

	AGWO	**EGA [12]**	**SPSO [15]**	**FEP [4]**
Power loss (kW)	130.8164	139.55	138.92	139.83
Open switches	4, 14, 15, 22, 33	7, 9, 14, 32, 37	7, 9, 14, 32, 37	7, 9, 14, 28, 32
Reduction (%)	38.0074	33.86	33.36	33.79

of network by reconfiguration process effects on the system voltage profile at each node is shown in Figure 8.9. Various switching operations for different constructions of system using AGWO with different total system power losses are presented

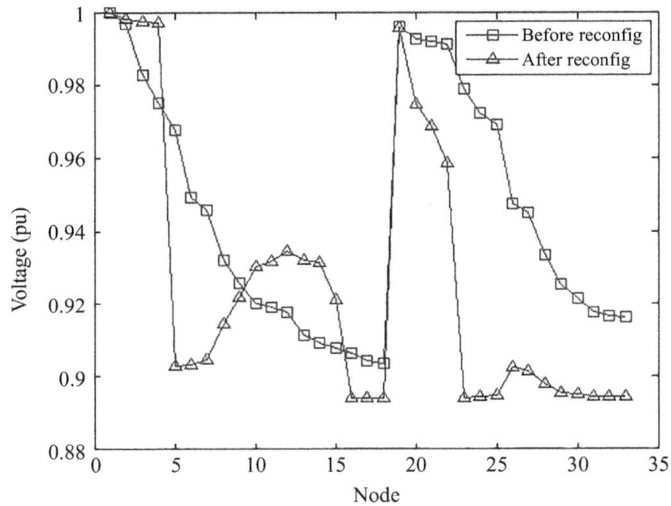

Figure 8.9 Node voltages after network reconfiguration by AGWO

Table 8.4 The obtained results of system power losses using AGWO at different switching operations

Closed switches	**Open switches (kW)**	**Power losses (kW)**	**Reduction (%)**
1,2,3,4,5,6,8,9,10,11,12,13, 15,16,18,19,20,21,22,23,24,25, 26,27,28,29,30,31,32,33,34,36	156.0594	7, 14, 17, 35, 37	26.045
1,2,3,4,5,6,8,9,10,11,12,13,15,16,17, 18,19,20,21,22,23,24,25,26,28, 29,30,31,33,34,35,36,37	155.2794	7, 14, 27, 32, 35	26.41
1,2,3,4,5,6,7,8,10,11,12,13,14,15,16, 17,18,19,20,21,22,23,24,25, 26,27,29,30,31,35,36,37	145.3723	9, 28, 32, 33, 34	31.11
1,2,3,4,5,6,8,10,11,12,13,15,16,17, 18,19,20,21,22,23,24,25,26, 28,29,30,31,33,34,35,36,37	143.3196	7, 9, 14, 27, 32	32.08
1,2,3,4,5,6,8,9,10,12,13,14,15,16, 17,18,19,20,21,22,23,24,25,26, 27,29,30,31,32,33,35,36,37	143.2075	7, 11, 28, 32, 34	32.14
1,2,3,4,5,6,8,9,11,12,13,15,16,17, 18,19,20,21,22,23,24,25,26,27, 29,30,31,33,34,35,36,37	140.7275	7, 10, 14, 28, 32	33.31
1,2,3,4,5,6,8,10,11,12,13,15,16,17, 18,19,20,21,22,23,24,25,26,27, 29,30,31,33,34,35,36,37	139.9998	7, 9, 14, 28, 32	33.66

in Table 8.4. This proves effectiveness and ability of the AGWO to solve this combinatorial reconfiguration problem.

8.5 Conclusion

This chapter proposes a new efficient concept of optimal reconfiguration that uses the AGWO algorithm to study the optimal reconfiguration of radial distribution system for power losses reduction. The AGWO is used to determine the optimal switches that can be opened to reconfigure the distribution system and to obtain the minimum system power loss. The simulation results of the IEEE 33-bus show that after reconfiguration, system power loss reduction is 38.0074%. From simulation results, it is observed that the AGWO achieves superior results when compared with other optimization techniques. The convergence characteristic of the AGWO proved that it is able to solve difficult optimization problems in different fields.

References

[1] A. Merlin and H. Back, "Search for a minimum loss operating spanning tree configuration for urban power distribution system," *Proc. 5th Power Syst. Comput. Conf. (PSCC)*, Cambridge, 1–18, 1975.
[2] S. K. Goswami and S. K. Basu, "A new algorithm for the reconfiguration of distribution feeders for loss minimization," *IEEE Trans. Power Delivery*, vol. 7, no. 3, pp. 1484–1491, 1992.
[3] D. Shirmohammadi and H. W. Hong, "Reconfiguration of electric distribution networks for resistive line loss reduction," *IEEE Trans. Power Del.*, vol. 4, no. 1, pp. 1492–1498, 1989.
[4] B. Venkatesh and R. Ranjan, "Optimal radial distribution system reconfiguration using fuzzy adaptation of evolutionary programming," *Int. J. Electr. Power Energy Syst.*, vol. 25, no. 10, pp. 775–780, 2003.
[5] S. Civanlar, J. J. Grainger, H. Yin, and S. S. H. Lee, "Distribution feeder reconfiguration for loss reduction," *IEEE Trans. Power Del.*, vol. 3, no. 3, pp. 1217–1223, 1988.
[6] S. Naveen, K. Sathish Kumar, and K. Rajalakshmi, "Distribution system reconfiguration for loss minimization using modified bacterial foraging optimization algorithm," *Int. J. Electr. Power Energy Syst.*, vol. 69, pp. 90–97, 2015.
[7] K. Sathish Kumar and T. Jayabarathi, "Electrical power and energy systems power system reconfiguration and loss minimization for an distribution systems using bacterial foraging optimization algorithm," *Int. J. Electr. Power Energy Syst.*, vol. 36, no. 1, pp. 13–17, 2012.
[8] K. Nara, A. Shiose, M. Kitagawa, and T. Ishihara, "Implementation of genetic algorithm for distribution systems loss minimum re-configuration," *IEEE Trans. Power Syst.*, vol. 7, no. 3, pp. 1044–105, 1992.

[9] J. Z. Zhu, "Optimal reconfiguration of electrical distribution network using the refined genetic algorithm," *Electr. Power Syst. Res.*, vol. 62, no. 1, pp. 37–42, 2002.

[10] A. Mohamed Imran and M. Kowsalya, "A new power system reconfiguration scheme for power loss minimization and voltage profile enhancement using Fireworks Algorithm," *Int. J. Electr. Power Energy Syst.*, vol. 62, pp. 312–322, 2014.

[11] J. Olamaei, T. Niknam, and S. B. Arefi, "Distribution feeder reconfiguration for loss minimization based on modified honey bee mating optimization algorithm," *Energy Procedia*, vol. 14, pp. 304–311, 2012.

[12] D.-L. Duan, X.-D. Ling, X.-Y. Wu, and B. Zhong, "Reconfiguration of distribution network for loss reduction and reliability improvement based on an enhanced genetic algorithm," *Int. J. Electr. Power Energy Syst.*, vol. 64, pp. 88–95, 2015.

[13] H.-C. Chang and C.-C. Kuo, "Network reconfiguration in distribution systems using simulated annealing," *Electr. Power Syst. Res.*, vol. 29, no. 3, pp. 227–238, 1994.

[14] A. Y. Abdelaziz, F. M. Mohamed, S. F. Mekhamer, and M. A. L. Badr, "Distribution system reconfiguration using a modified Tabu Search algorithm," *Electr. Power Syst. Res.*, vol. 80, no. 8, pp. 943–953, 2010.

[15] A. Tandon and D. Saxena, "A comparative analysis of SPSO and BPSO for power loss minimization in distribution system using network reconfiguration," in *Innovative Applications of Computational Intelligence on Power, Energy and Controls With Their Impact on Humanity (CIPECH)*, pp. 226–232, IEEE, 2014.

[16] A. V. Sudhakara Reddy, M. Damodar Reddy, and M. Satish Kumar Reddy, "Network reconfiguration of Primary distribution system using GWO algorithm," *International Journal of Electrical and Computer Engineering*, vol. 7, no. 6, pp. 3226–3234, 2017.

[17] C.-T. Su, C.-F. Chang, and J.-P. Chiou, "Distribution network reconfiguration for loss reduction by ant colony search algorithm," *Electr. Power Syst. Res.*, vol. 75, no. 2–3, pp. 190–199, 2005.

[18] M. H. Qais, H. M. Hasanien, and S. Alghuwainem, "Augmented grey wolf optimizer for grid-connected PMSG-based wind energy conversion systems," *Appl. Soft Comput.*, vol. 69, pp. 504–515, 2018.

[19] N. Mittal, U. Singh, and B. S. Sohi, "Modified grey wolf optimizer for global engineering optimization," *Applied Computational Intelligence and Soft Computing*, vol. 2016, article ID 7950348, pp. 1–16.

[20] S. Mirjalili, S. M. Mirjalili, A. Ewis, "Grey wolf optimizer," *Advances in Engineering Software*, vol. 69, pp. 46–61, 2014.

[21] M. M. Aman, G. B. Jasmon, A. H. A. Bakar, and H. Mokhlis, "A new approach for optimum simultaneous multi-DG distributed generation Units placement and sizing based on maximization of system load ability using HPSO (hybrid particle swarm optimization) algorithm," *Energy*, vol. 66, pp. 202–215, 2014.

Chapter 9

Intelligent optimization scheduling of isolated microgrids considering energy storage integration, traditional generation, and renewable energy uncertainty costs

Erik Carvajal[1], Gustavo Muñoz[1], David Romero-Quete[1] and Sergio Rivera[1]

Economic and technical issues have contributed to difficulties experienced in accessing or supply of electricity in areas far away from the main electrical grids, such as islands. However, the implementation of the microgrids, integrated with the renewable energy sources, has contributed to ensuring that these areas are supplied with electricity. Besides, systems incorporating renewable energy and energy storage systems (ESSs) have and continue to play a significant role as an energy supply solution for these areas, through improved reliability and efficiency of the system. In full appreciation of the impacts of ESS, a novel multi-objective intelligent scheduling formulation is proposed that considers batteries life loss cost, operation cost, and uncertainty costs associated with the variability of renewable sources. Mitigation of issues due to intermittency of renewable energy sources through ESSs is demonstrated, using a non-dominated genetic algorithm (NSGA II). The test system used corresponds to a microgrid in an island with similar features of a real microgrid in Dongfushan Island in China.

9.1 Introduction

Due to economic and technical issues, most areas with difficult access to electricity, such as islands, use diesel generators for their electrical supply. The main drawbacks of this type of energy generation include the high cost of fuel and the emission of polluting gases [1]. Therefore, the implementation of energy systems that make use renewable sources is essential, since these regions are generally rich in natural resources [2]. Microgrids that are small-scale power systems capable of operating in stand-alone mode or connected to the power network [3,4] provide an

[1]Engineering Faculty, Universidad Nacional de Colombia, Bogota, Colombia

effective and widely accepted solution to integrate renewable energies in isolated areas [5].

An important challenge for the implementation of microgrids integrating renewable generation is the intermittent and fluctuant nature of the sources [2,3]. In fact, these sources of energy are considered in the economic dispatch problem as non-dispatchable, so they have to be addressed differently from conventional sources—being modelled, for instance, by probability distribution functions (PDFs) that adequately describe their stochastic behaviour [3,4]. It is important for the economic operation of the system to adequately use these energy sources while maintaining the system reliability and stability. For this reason, some of the points analysed in this chapter are mitigation of these issues due to intermittency of renewable energy sources through ESSs. The intelligent scheduling proposal handles the minimization of fuel consumption, the loading and unloading times of the batteries, and the uncertainty cost of renewable energy, due to the variations in climatic factors such as changes in wind speed, temperature, or solar irradiation on a surface [3].

A large number of studies aimed at promoting the penetration of renewable energies in response to energy needs have been carried out in recent years, especially in areas not connected to the conventional electricity grid [2–4]. Such is the case of the islands of the territory of Croatia in the Adriatic Sea [6]. Given the large number of islands and islets of the Republic of Croatia (49), with a population of almost 125,000 people, this is an adequate scenario for the implementation and use of solar energy, with great acceptance among the localities and the government. In the study by [6], four steps were proposed for the analysis of the optimization of the microgrid of the Island of Vis. Notably, in the software HOMER [6] the energy was used to analyse the energy strategies for both economic and environmental impacts.

Another study [7] with a similar methodology to the present study presents a microgrid typical of an island. In the microgrid, photovoltaic (PV) panels, wind turbines (WT), diesel generator, a system of batteries, and a system of desalination of water for the consumption of the residents were incorporated. In order to minimize the operation cost of the microgrid, differential evolution algorithms were used to find the optimal operating parameters. Importantly, the researchers in [7] found that the respective simulation after 500 iterations provides favourable results in terms of reducing the cost of generation.

A similar formulation is carried out in [8]. A multi-objective optimization in a DC microgrid was also proposed for the optimal dispatch of energy of a system that integrates PV generation and charging stations for electric vehicles. In this work, a mathematical development was presented; their objective functions are the purchase cost of electricity and the cycle life of batteries; solving the problem through the non-dominant genetic algorithm NSGA II. As in many mathematical models that seek to approach reality [2,5], this model also takes into account some restrictions such as state of charge (SOC) of batteries, the charge and discharge range of the ESS, the nominal power demanded by the network of distribution, and the balance of the power system, among others.

Lastly, the optimization in the dispatch of a microgrid includes functions that have the characteristic of a non-linear problem and with multiple restrictions [3].

This study [3] presented the appropriate scenario to use the solution algorithm that reduces the complexity of the calculations and provides a very accurate optimal result through uncertainty cost functions (UCFs).

Another fundamental element of microgrids is ESSs that are able to improve the efficiency and reliability of the system [2,9,10]. Since batteries represent an important part of the total investment that has a very short life cycle [10], they become economic hosts for the analysis and optimization of the system. In this way, this work proposes to extend (using the UCFs [3]) the analysis presented in [2] that uses two main objectives: (1) the maximization of the useful life of the batteries by means of the inclusion of a complete analysis of the characteristics of service of these and (2) the minimization of the cost of power generation.

9.2 Background of the study

This research is applied to a microgrid with similar features of a real microgrid that includes a seawater desalination plant and the load of the island [2]. In the past, the island generated its energy through diesel generators, something that changed since the implementation of renewable sources [2].

The microgrid has the following components [2]: 7 WT of 30 kW (each turbine), PV panels (100 kW) dispersed in 556 solar panels, a 200 kW diesel generator, and finally 480 batteries for a total of 960 kW, as shown in Table 9.1, and a schematic representation of the components of the power system in Figure 9.1 [1], for a total of 960 kW.

The control strategy that governs this project is master–slave. The diesel generator and the battery system are used as backup of the system, are the master control units which ensure that voltage and frequency levels are controlled at all times.

Some of the most important parameters of the microgrid control system are the SOC_{min} (state of charge minimum charge) and the SOC_{stp} (state of stopping battery charge). SOC_{stp} is the parameter for adjustment to which the SOC reaches a value at which the diesel generator is turned off. Other important parameters include P_{ren}, P_{load}, P_{excess}, and P_{charge}, described in the next paragraph.

The control logic works as follows: when the SOC parameter is larger than SOC_{min}, the master control unit is the batteries [2]. This energy storage is in different modes, one is in 'charge', other in 'standby', and finally, in 'discharge'. States can be determined by P_{ren}, P_{load}, and P_{excess}. When P_{ren} (output power of renewable resources) exceeds P_{load} (total power demanded) and the power in excess is superior to P_{excess}, the batteries are

Table 9.1 Characteristics of the system [2]

Name	Eolic turbines	Solar panels	Diesel generator	Lead–acid batteries
Type	30 kW	180 W	200 kW	2 V/1,000 Ah
Quantity	7	556	1	480
Capacity	210 kW	100 kW	200 kW	960 kWh

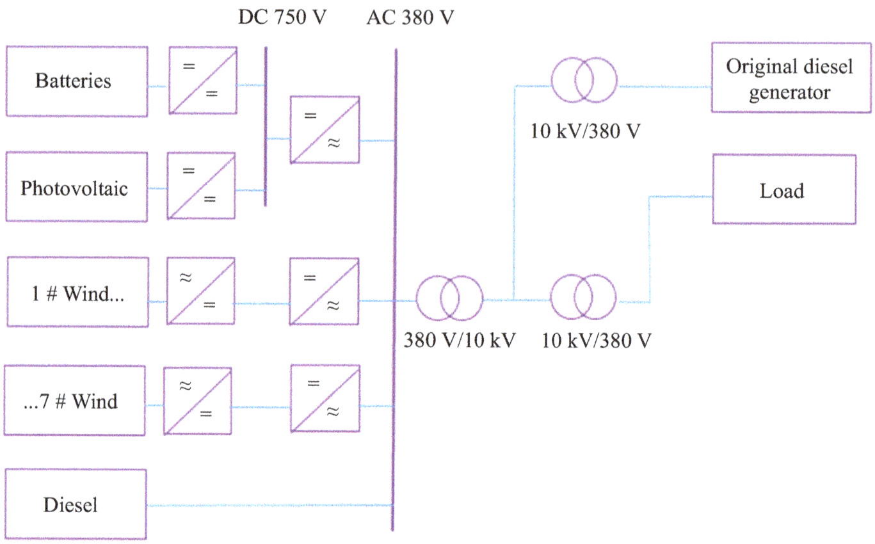

Figure 9.1 Representation of the microgrid. Modified, with permission, from Reference [2]

placed in charge mode. If P_{excess} is lower, the batteries are placed in standby mode, and when P_{ren} is lesser than P_{load} clearly, the power demand is greater than what is possible to generate; at this moment, the batteries are in discharge mode.

At no time the charge of the batteries should be lower than P_{charge}, because when the SOC of the battery is lower than SOC_{min}, the diesel generator is turned on and behaves as a master control unit. At this moment, the charge state of the batteries increases until it gets the SOC_{stp} and the generator stops. The batteries are the master control unit. The scheme of the mentioned algorithm is shown in Figure 9.2.

The non-dominant genetic classification algorithm (NSGA II) was employed in optimizing the multi-objective problem (Figure 9.3).

9.2.1 Generation resources modelling

9.2.1.1 Wind generation

It is deduced from the relationship between the input and output power of WT that use the power of the wind to generate electrical energy, as suggested by the approximate function that describes this behaviour in [3].

$$P_{wt}(v) = \begin{cases} 0 & v \leq v_{ci} \text{ or } v \geq v_{co} \\ \dfrac{P_{rated\text{-}wt}(v - v_{ci})}{v_r - v_{ci}}, & v_{ci} \leq v \leq v_r \\ P_{rated\text{-}wt} & v_r \leq v \leq v_{co} \end{cases} \tag{9.1}$$

where v_{ci} is the cut-in speed, v_{co} is the cut-off speed, v_r is the nominal wind speed, and $P_{rated\text{-}wt}$ is the nominal output power of WT.

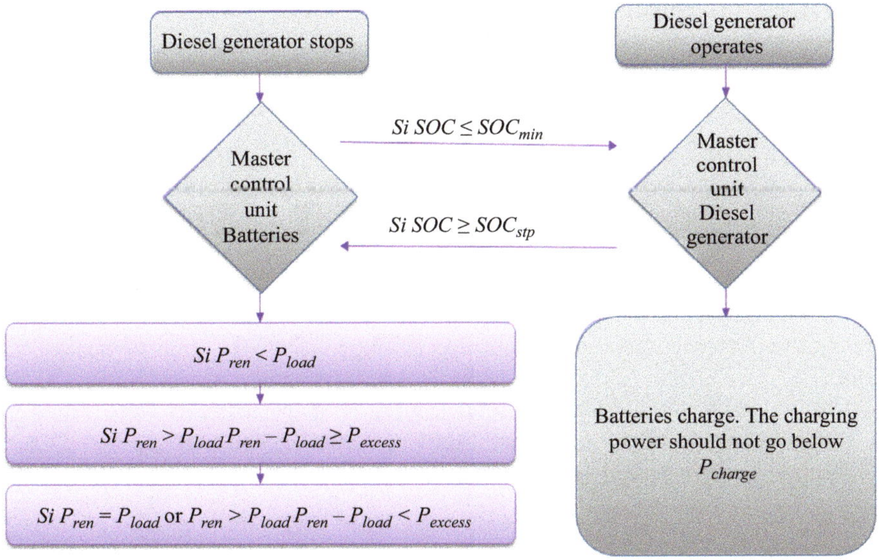

Figure 9.2 Problem formulation. Modified, with permission, from Reference [2]

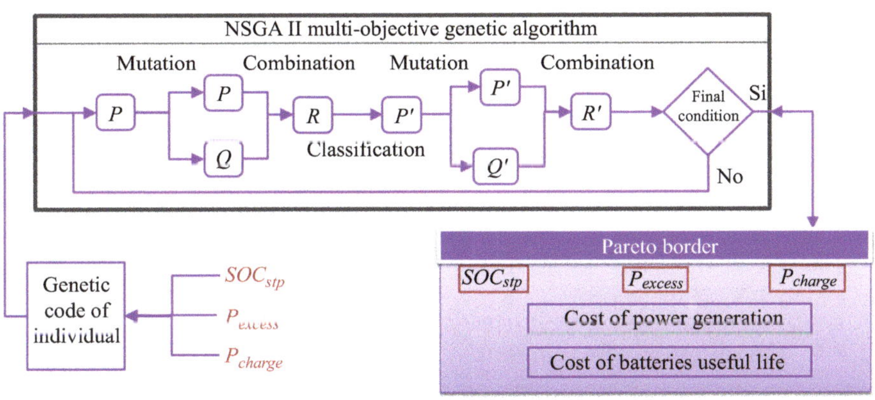

Figure 9.3 Multi-objective optimization algorithm

9.2.1.2 Photovoltaic generation

For its modelling, the electrical output power in standard conditions, the intensity of the ambient light, and its temperature are taken into account: solar radiation G_{stc} is 100 W/m and temperature PV T_{stc} is 25 °C. The following function is obtained [3]:

$$P_{pv} = P_{stc}\frac{G_c}{G_{stc}}[1 + k(T_c + T_{stc})] \tag{9.2}$$

where P_{pv} is the output power of the PV generator; G_c, the irradiation of the operation point; k, the coefficient of temperature; P_{stc}, the nominal output power under standard conditions; and T_c, the PV temperature of the operation point. Equation (9.3) shows that the output power of the PV generator can be manipulated by lowering its power.

$$P_{out} - pv \leq P_{pv} \tag{9.3}$$

9.2.1.3 Batteries model

The ESS developed on the island is one of the main parts of the microgrid [2]. It is a matter of importance that the optimization strategy affects the performance of the system.

It must be considered that the operating mode of the batteries is governed by the SOC that must remain within two limits: the SOC_{max} and SOC_{min}.

Then the output power of the batteries should have the same behaviour:

$$P_{cha_{max}} \leq P_{bat} \leq P_{discha_{max}} \tag{9.4}$$

where $P_{cha_{max}}$ and $P_{discha_{max}}$ are the maximum allowable load power and the maximum allowable discharge power, respectively [2]. P_{bat} is positive when it is discharged and negative for the charge. Now, the SOC in a period of time $t + \Delta t$ is determined by the value at that time, as can be seen in (9.5).

$$SOC_{t+\Delta t} = SOC_t - P_{bat\text{-}t} \times \frac{\Delta t}{C_{bat}} \tag{9.5}$$

where $P_{bat\text{-}t}$ is the power of the battery between t and $t + \Delta$, and C_{bat} is the capacity of the battery. It must be considered that the SOC cannot exceed or be inferior to its established maximum and minimum values for any reason.

The SOC is strongly associated with the useful life of the batteries. It is known that a high SOC must be operated to give efficiency to the useful life [1].

9.2.1.4 Diesel generation model

This type of generators is a reserve type; they are in the system to supply the needs of power when necessary for the stability and continuous operation of the system. The energy is supplied from a combustible material, which will have a linear volume relation to the output power given by

$$F = F_0 \times P_{rated\text{-}gen} + F_1 \times P_{gen} \tag{9.6}$$

where $P_{rated\text{-}gen}$ is the nominal power output, P_{gen} is the real output power, and F_0 and F_1 are the adjustment coefficients of the fuel consumption curve. In [8], F_0 is set to 0.08415 and F_1 is 0.246. Diesel generators have power limits that can be formulated in the following equation [2]:

$$K_{gen} P_{rated\text{-}gen} \leq P_{gen} \leq P_{rated\text{-}gen} \tag{9.7}$$

9.2.2 Economic analysis

9.2.2.1 Modified generation cost

Basically, the modified generation has costs of two types: traditional (diesel) and renewables-related costs.

$$\begin{aligned} C_{gen} &= C_{die} + C_{ren} \\ C_{die} &= C_{loss\text{-}die} + C_{om-die} + C_{fuel} + C_{po\text{-}die} \\ C_{ren} &= C_{loss\text{-}ren} + C_{om\text{-}ren} - C_{sub} \end{aligned} \tag{9.8}$$

where C_{die} is the diesel generator cost; $C_{loss\text{-}die}$ is the life loss cost; $C_{om\text{-}die}$ is the operation and maintenance cost; C_{fuel} is the traditional cost of fuel; $C_{po\text{-}die}$ is the cost of pollution; C_{ren} is the renewable energy generating cost; $C_{loss\text{-}ren}$ and $C_{om\text{-}ren}$ are used in life cost and operation and maintenance; and C_{sub} is the subsidy for renewable generation.

The diesel generation is composed of costs of the useful life of the machine, its operation and maintenance, and fuel, as well as the economic impacts of the pollution it generates. According to (9.8), the cost of renewables also has important effects on the objective 1, the generation costs: their useful life cost and operation and maintenance costs. However, the generation of clean energy also produces benefits that are a subsidy for generation in this case. The mentioned costs of renewable resources cannot be constant data that are programmed before the optimization.

9.2.2.2 Batteries life loss cost

The losses in the useful life of the batteries can be measured by a considerable factor to assess its capacity in Ah. This is expressed in the following equation:

$$L_{loss} = \frac{A_c}{A_{total}} \tag{9.9}$$

where L_{loss} is the loss of life of the batteries, A_c is the cumulative yield Ah effective in a determined period, and A_{total} is the cumulative total of Ah in the life cycle. Different studies carried out define that manufacturers usually give the useful life of batteries, where the variables can be obtained for the relationship [2,11–13]. Batteries also have costs in terms of their initial investment that can be high and the operation that depends on several factors, including the SOC.

9.2.2.3 Uncertainty cost of renewable resources

The elements of solar and wind generation are considered as non-controllable in the operation optimization of power systems. This is because these resources or charges in the state mode of the batteries have stochastic behaviour that will be modelled with a PDF in this research [3,14]. A mathematical model designed to calculate an expected cost associated with the uncertainty of these energy elements (renewable generators and batteries), that is to say a UCF, will be used in this study.

The analytical calculation was verified in [14] through Monte Carlo simulations to be appropriate in the simulation of the variability of solar and wind energy sources, since they can be modelled by knowing the PDFs [3].

Given the uncertainty of the sources, the operation scheduling of the energy systems that make up these resources must include a reasonable term to consider the aforementioned uncertainty. This term is made up of two parts:

1. Penalty cost due to underestimation [14]
   This case occurs in the dispatch model. Notably, it occurs when an amount of energy coming from a generator is planned but is less than the power available in it. The power available is, therefore, not totally used. It is represented as

$$W_{s,i} < W_{Av,i} \tag{9.10}$$

   where $W_{s,i}$ is the power that is programmed in dispatch from generator i and $W_{Av,i}$ is the real power available in generator i.
2. Penalty cost due to overestimation [14]
   This case occurs when in the dispatch a quantity of power of a generator greater than the available power in the source is planned, that means, power not available at a given moment. In order to supply the load, the model requires another power source activated. In this case, it is a penalty for using another source of energy or for energy not supplied.

$$W_{Av,i} < W_{s,i} \tag{9.11}$$

The proposed UCF will be given by adding costs due to underestimation and overestimation:

$$UCF = C(u,i)(W(s,i), W(Av,i)) + C(o,i)(W(s,i), W(Av,i)) \tag{9.12}$$

where $C(u, i)$ refers to the cost for underestimation and $C(o, i)$ to the cost for overestimation.

In this way, the expected penalty cost due to underestimation is given by the equations in [3]. These new costs should be taken into account in the economic analysis (9.8), resulting in

$$C_{ren} = C_{loss\text{-}ren} + C_{om\text{-}ren} - C_{sub} + C_{u,i} + C_{o,i} \tag{9.13}$$

where $C_{u,i}$ and $C_{o,i}$ are the costs associated with underestimation and overestimation, respectively. Therefore, these costs will be associated with the power generation of the system, updating the values of (9.8):

$$C_{gen} = C_{die} + C_{ren} \tag{9.14}$$

9.3 Intelligent scheduling without consideration of uncertainty cost

Data such as solar irradiation, temperature, and wind speed are the data historically measured for Dongfushan Island [2]. The data obtained for the same time periods are used to estimate the optimal parameters of the three decision variables

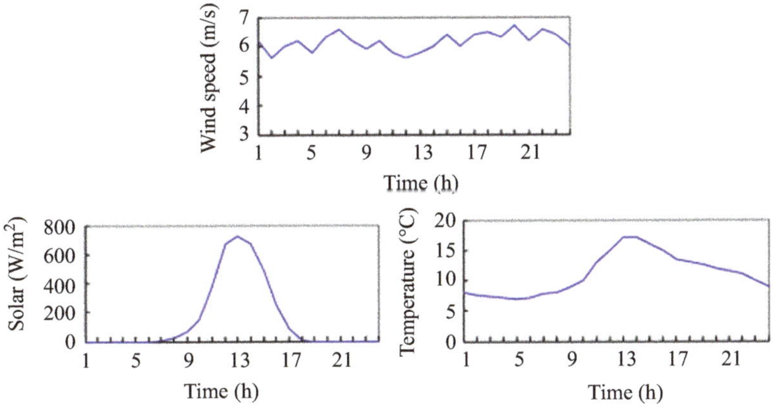

Figure 9.4 Forecast of primary sources and temperature in the case of plentiful renewable sources. Modified, with permission, from Reference [2]

calculated for different climatic conditions. To make a correct comparison and to do an analysis of the data, two different climatic conditions are proposed, as well as the plentiful and few renewable resources. In this way, two renewable energy scenarios are taken into account, the plentiful and sparse of renewable primary energy sources (solar radiation and wind speed), respectively. Figure 9.4 shows the characteristics of the case with plentiful renewable sources.

9.3.1 Plentiful renewable primary energy sources

For this case, the solar irradiation and wind speed are high. The power output is therefore greater, as shown in Figure 9.5. Therefore, the optimal result is the one shown in Figure 9.6. The first Pareto diagram is obtained from the simulation, objective 1 is in the x-axis and objective 2 is in the y-axis, in the condition of plentiful renewable energy sources.

Three cases are to be analysed from the scenario with plentiful renewable resources. In this way, for each of the following graphs three optimal points from the Pareto in Figure 9.6 will be analysed. Where in the case of scheme operation no. 1 the point taken is dedicated to the lower cost of diesel power generation, the second figure refers to a central point in the Pareto, and the third one from the second objective where loss of life and cost of the batteries are the lowest.

Figure 9.7 Case 1, where the operation of the batteries is more continuous throughout the day, shows how batteries provide energy to the system until batteries need to be charged. This operation profile corresponds to a profile where the objective is to minimize the generation power cost: the lesser use of the diesel generator.

For the scheme of operation no. 2, an average Pareto point is taken where the costs of generation and life loss in the batteries are average between both. Figure 9.8 shows this case.

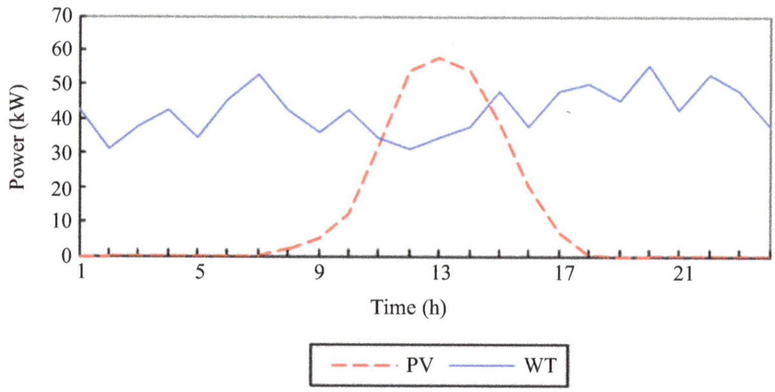

Figure 9.5 Power available in the case of plentiful renewable sources

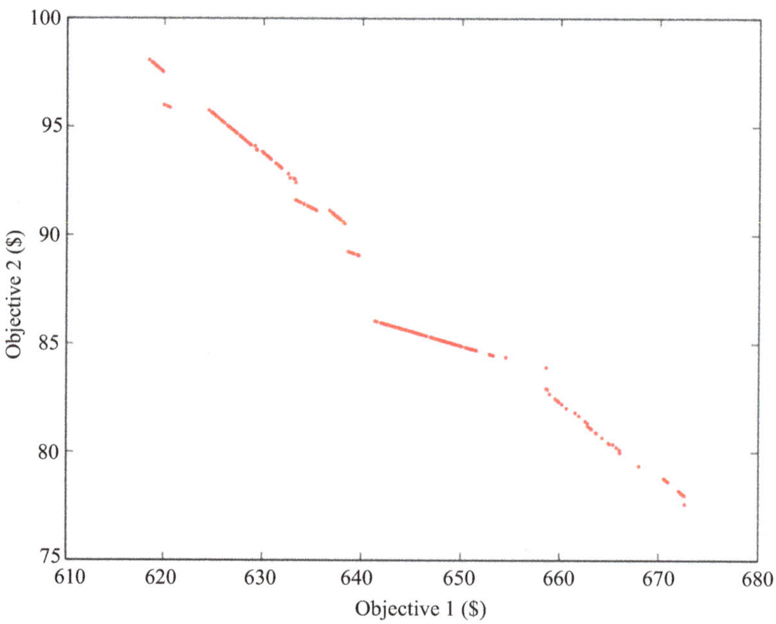

Figure 9.6 Possible optimal solutions (last iteration of the algorithm), plentiful natural resources case

Figure 9.9 shows the third case, when the diesel generator is used mainly. In this case, the generator is continuously used almost the whole day; and then, the batteries will be used throughout the day but without so many variations between charging and discharging.

It is observed that in order to reduce the life loss cost, the operating time of the diesel generator and the generation cost must be increased. Due to the plentiful

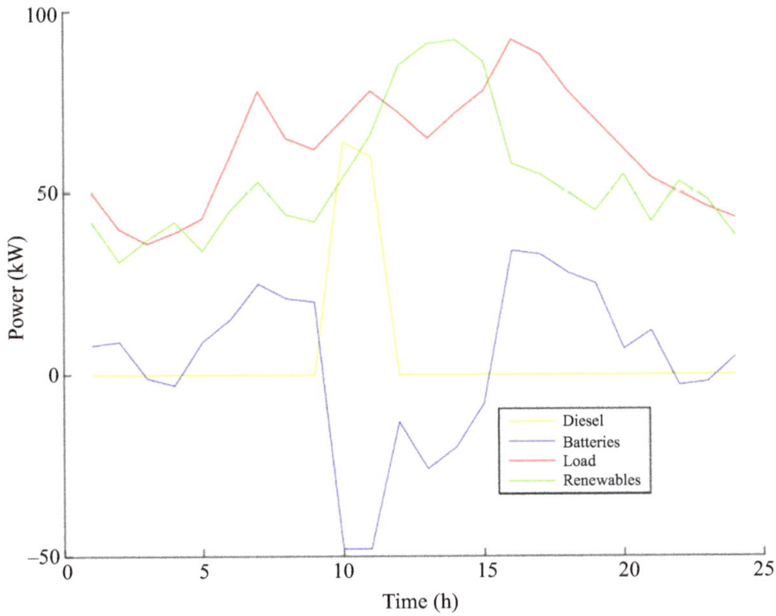

Figure 9.7 Profiles of operation of the system under the scheme of operation no. 1 case of plentiful natural resources

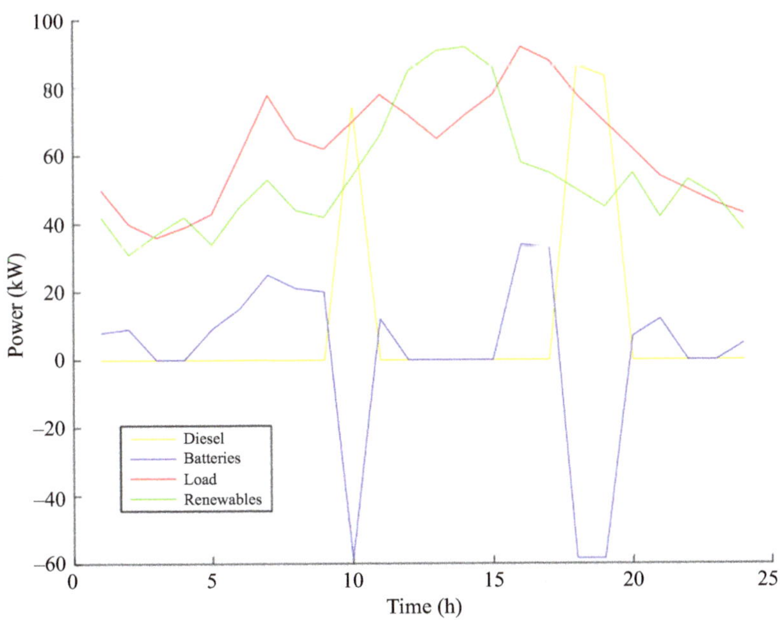

Figure 9.8 Profiles of operation of the system under the scheme of operation no. 2 case of plentiful natural resources

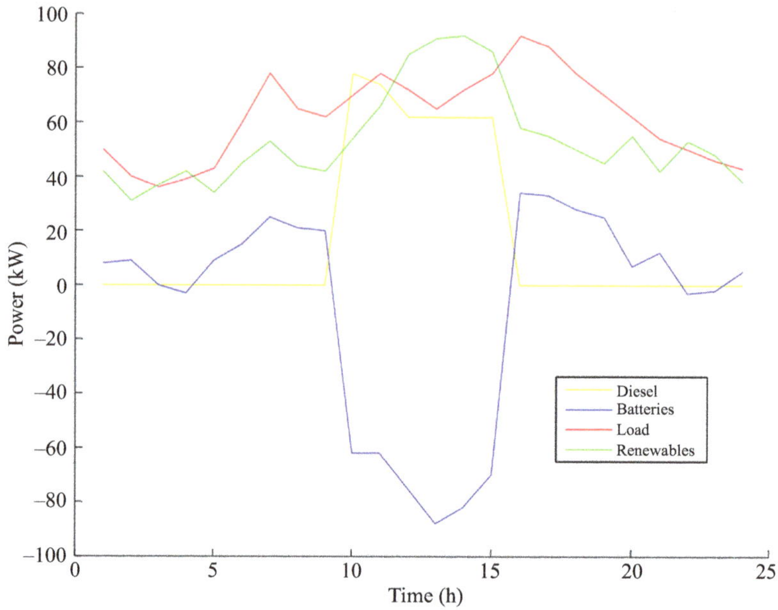

Figure 9.9 Operation profiles of the system under the scheme of operation no. 3 case of plentiful natural resources

Table 9.2 Results from natural resources combination of objectives

No.	SOC_{stp}	P_{charge} (kW)	P_{excess} (kW)	Objective
1	0.54	48.0	0.78	358.24
2	0.53	58.44	30.62	362.86
3	0.88	61.89	1.55	375.10

resources of renewable energy, the reduction in the cost of batteries is related to the increase in the generator operation, but it may not be the greatest option to reduce costs.

To find an optimal operation, a ponderation method is used; the two objectives are integrated in one (Table 9.2).

$$Objective = a_1 \times Objective_1 + a_2 \times Objective_2 \tag{9.15}$$

where a_1 and a_2 are weightiness coefficients and their sum is equal to 1.

It is possible to see that in the case of plentiful sources of renewable energy, the scheme of operation no. 1 should be selected, since it has a relatively high value of SOC_{stp}, can reduce the generation cost, and protects the batteries from a reduced useful life [2].

9.3.2 Sparse of renewable primary energy sources

For this case, the wind speed and solar radiation are in relatively weak conditions, as shown in Figure 9.10.

The corresponding renewable output WT and PV are shown next, followed by the result of the optimization, in Figures 9.11 and 9.12, respectively.

Similar to the case of plentiful renewable sources, the search to maximize the use of renewable energy can cause batteries to work longer, thereby increasing the life loss.

In contrast, an increase in the operating time of the traditional generator can deliver a better condition for the batteries in order to reduce the life loss of the aforementioned ones by paying the cost of the generation, which is more expensive.

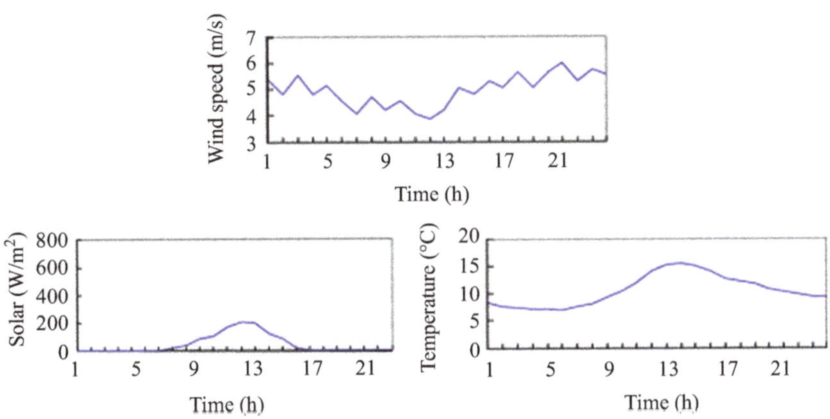

Figure 9.10 Profiles of primary sources and temperature in the case of a few renewable resources

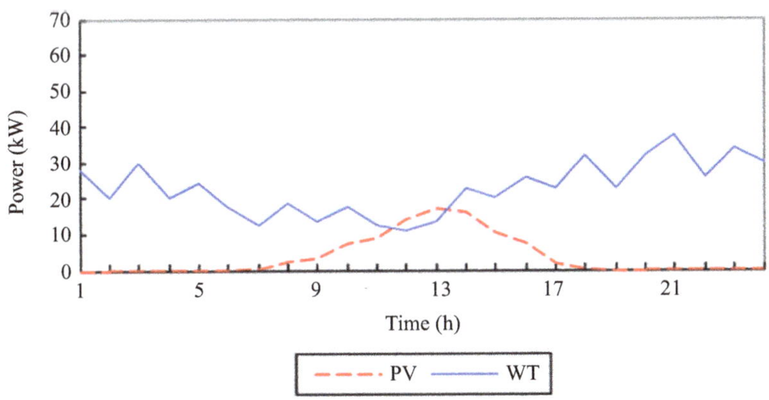

Figure 9.11 WT and PV output power profiles, short of renewable resources case

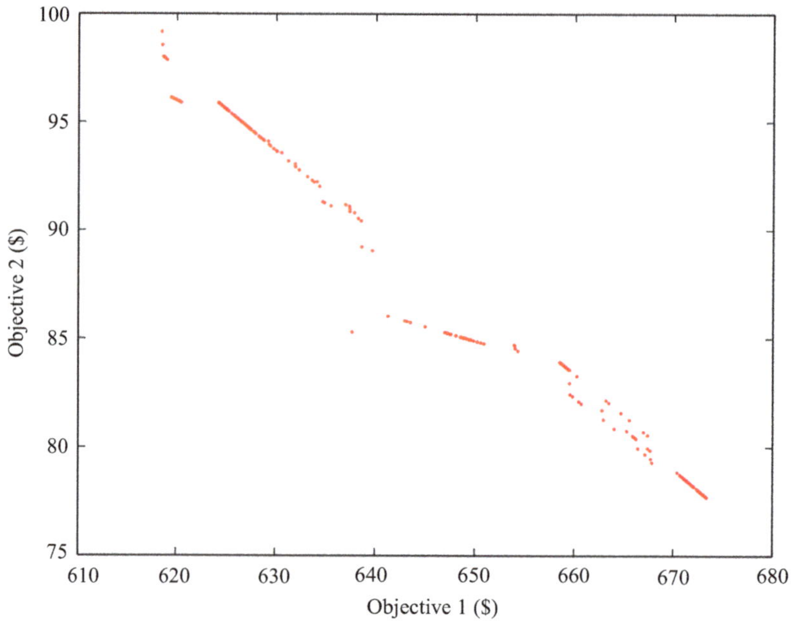

Figure 9.12 Possible optimal solutions (last iteration of the algorithm), a few renewable resources case

The three operating profiles of the system can be observed in Figures 9.13–9.15. In this first case (Figure 9.13), the operation of the batteries is more continuous throughout the day. It can be seen how they supply the system to the point that the batteries have to be charged. Therefore, this operation profile corresponds to the one whose objective is to improve the generation cost; on the contrary, that is to say, raise the cost of life loss in the batteries.

If a balance between objectives is sough, a medium point in the Pareto must be selected; the life loss cost of the batteries and the generation cost are apparently equitable. Figure 9.14 shows the mentioned operation profile.

Figure 9.15 shows how the contributions of the diesel generator and the batteries work fully during the 24-h journey. However, when this generation scheme is compared with the profile mentioned earlier, it can be noted how the batteries lower their time of operation and the maximum peak of power that they supply. In contrast, in this case the diesel generator generates a little more. The batteries still have a long operation time and the diesel generator provides enough energy to reach the balance of costs. In this case, the most important matter is to minimize the battery life loss cost. On the contrary, the diesel generator must generate enough power to supply the system, causing a greater cost of energy. In the previous profiles, it can be noted that batteries practically have a single daily charge mode, reducing their operating time and saving useful life.

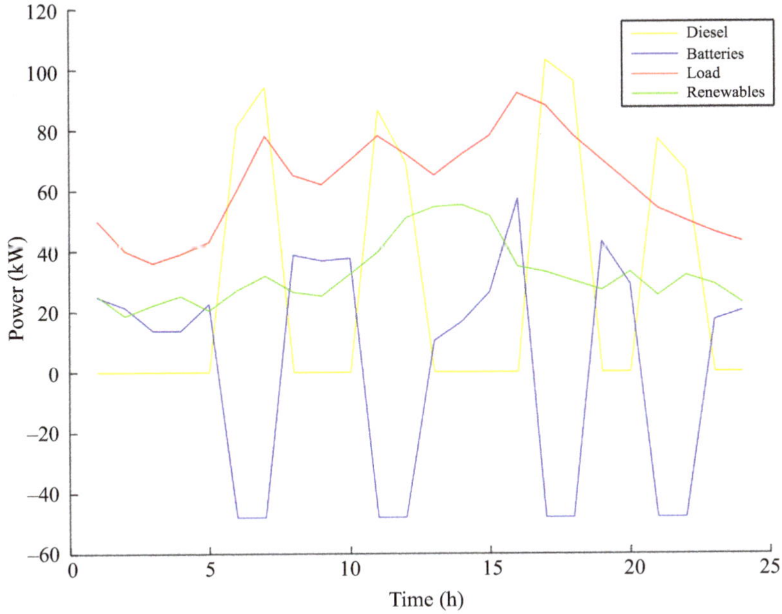

Figure 9.13   Operation profile of the system under optimization 1, a few renewable resources case

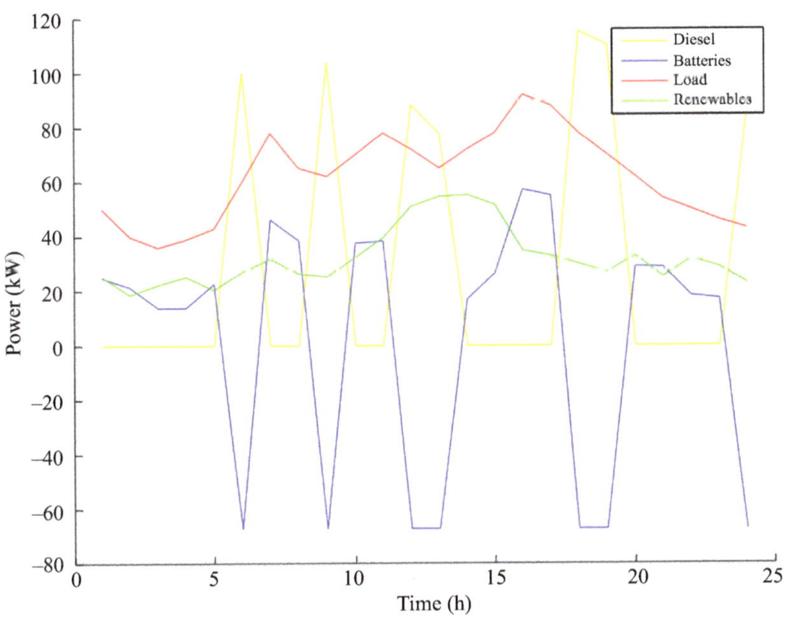

Figure 9.14   Operation profile of the system under optimization 2, a few renewable resources case

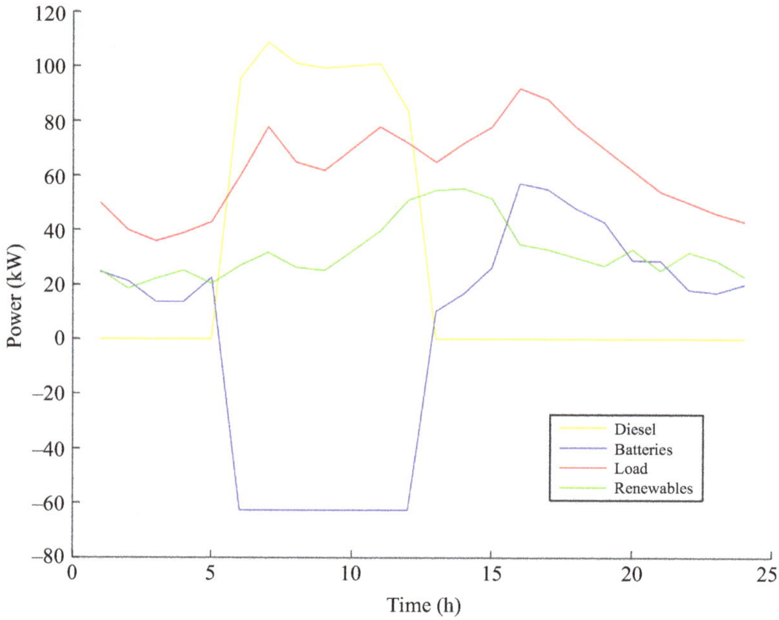

Figure 9.15 Operation profile of the system under optimization 3, a few renewable resources case

Table 9.3 Results of the optimization in the case of short of renewable sources with a combined objective

No.	SOC_{stp}	P_{charge} (kW)	P_{excess} (kW)	Objective
1	0.56	48.02	8.76	358.79
2	0.55	67.13	52.82	368.16
3	0.92	62.55	76.41	375.54

Table 9.3 shows the results when the two objectives are combined together with the equal ponderation factors, i.e. $a_1 = a_2 = 0.5$, in the optimization figures.

9.4 Intelligent scheduling considering the uncertainty cost functions

9.4.1 Optimization results with uncertainty cost analysis of wind energy

In order to make a more accurate assessment of the situation that occurs when the energy is delivered in the power system, the inclusion of the costs associated with the uncertainty is made in this case. The values of solar irradiation and the speed of the wind at a future moment are obtained.

As mentioned earlier, since it is possible to model the sources of solar energy (solar radiation) and wind (wind speed, by means of PDFs), some authors have conducted studies where those sources are modelled through log-normal, beta, or Weibull distributions [3].

When considering the cost of uncertainty of renewable energies, there is a penalty cost due to the underestimation and overestimation of the dispatch. For the case of energy from the wind component of the generation park of the microgrid in Dongfushan Island, the uncertainty cost of wind energy is implemented in the model, maintaining solar generation constant for this analysis. Then, the simulation is carried out again using the genetic algorithm NSGA II, obtaining a Pareto diagram, shown in Figure 9.16; it presents all the possible optimized values for this scenario. Thus, the operation of the system will be analysed for three specific cases: first, the cost of operating batteries is the minimum; second, the generation cost is the minimum; and third, a balance of the costs is reached.

According to the previous figure, the closer the point is taken to the horizontal axis of objective 1, the higher the generation cost is; however, the lower the battery life loss cost is. The loss of the useful life of the batteries will be more expensive than the cost of generating energy. Table 9.4 shows the optimization results, where listing of the different values for the three decision variables for three optimal

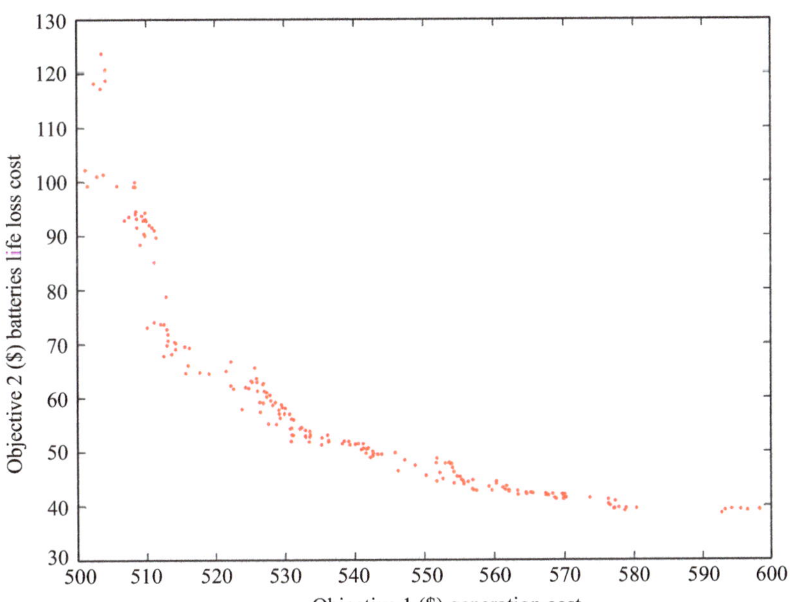

Figure 9.16 Possible optimal solutions (last iteration of the algorithm), a few renewable resources case

Table 9.4 Results of optimization, including the uncertainty cost of wind energy

No.	SOC_{stp}	P_{charge} (kW)	P_{excess} (kW)	Objective 1 ($)	Objective 2 ($)	Objective ($)
1	0.63	69.76	75.95	501.23	123.76	312.5
2	0.64	82.05	89.53	529.78	61.22	295.5
3	0.86	83.09	82.46	598.33	38.68	318.50

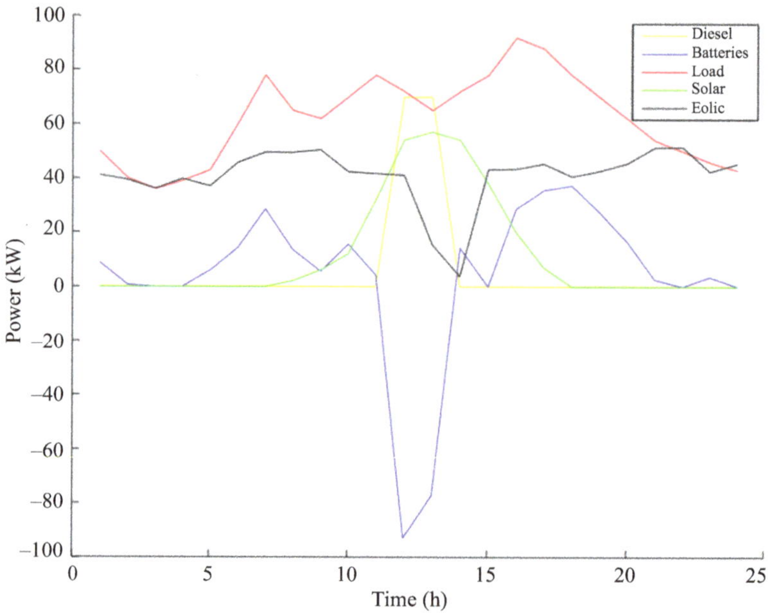

Figure 9.17 Operation profile minimizing the power generation cost

points was obtained from Pareto in Figure 9.15, resulting in three different operating strategies that can be obtained.

Case 1. Minimize the cost of power generation

In this case, the cost of generating power is minimized, but the life loss cost of the batteries is higher in return. Therefore, the batteries must be placed in operation for most of the time, as can be seen in Figure 9.17.

According to Figure 9.16, the minimization of the generation cost occurs because the diesel generator is placed in operation only for a few hours a day, specifically at the moment when the batteries are in charge mode, that is to say, when they are in constant switching between charge and discharge in five periods of time during the day.

According to the information in Table 9.3, for this generation scheme the value of SOC_{stp} is 0.63, indicating that at a relatively lower value the diesel generator stops. Therefore, the master control unit of the power system is the batteries.

A value that allows Case 1 to be the lowest generation cost is the lowest value of P_{charge} of 69.76 compared to the high values of the other generation schemes. It should be remembered that this value is the value of battery charge adjustment when the diesel generator operates.

Case 2. Balance between costs

For this case, a midpoint is taken from the Pareto in Figure 9.15, where the balance between the life loss cost of the batteries and the generation cost is in equilibrium. The simulation scheme is shown next.

Figure 9.18 shows how the value of power delivered by the diesel generator increases, with an output power greater than 100 kW. The batteries decrease their operating time in comparison to the previous case, where there was a valley between 11 and 15 h, and the batteries were in standby mode. Therefore, for this period of time the power coming from the renewable resources is greater than the power required by the load.

Since in this generation scheme the value of SOC_{stp} is 0.64, this value is more beneficial for the useful life of the batteries than in the previous case and the cost is reduced by almost a half. Although the diesel operation and the power delivered are greater, the operation scheme of the microgrid turns out to be the most economical.

Case 3. Minimize the cost in useful life of the batteries

This case intends to use the Pareto to minimize the cost of life loss. Therefore, a point close to the axis of objective 1 is selected, whose decision variables' values

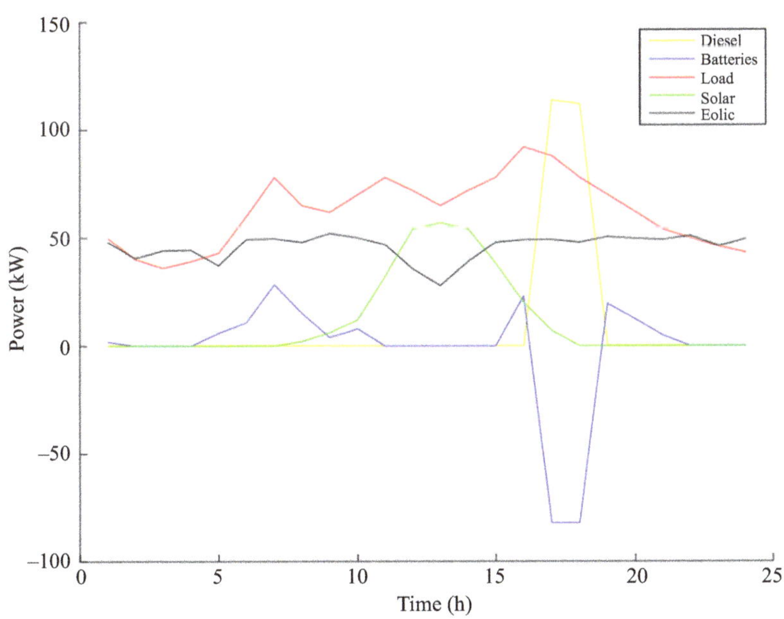

Figure 9.18 Operation profile with balance between objectives

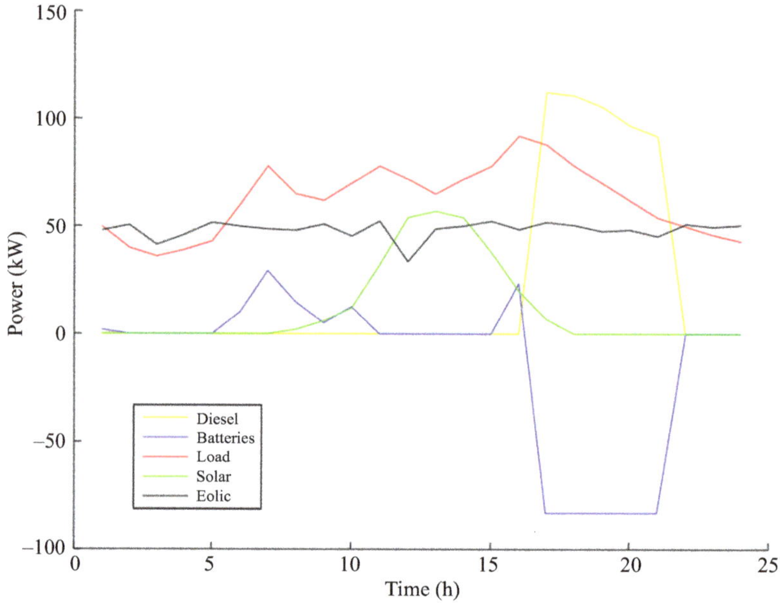

Figure 9.19 Operation profile minimizing the life loss cost of the batteries

correspond to those shown for Case 3 in Table 9.3. Figure 9.19 shows the operation scheme for this case.

In this case, the effects of SOC_{stp} can be noticed because it is high – 0.86. This is detrimental to the generation cost but it makes this scheme the one with the lowest value of battery life. According to Figure 9.18, the batteries are used relatively little throughout the day, and from 2 p.m. to 10 p.m. they are charged. The value of P_{charge} is 83.09, the highest value of battery charge adjustment, while the diesel generator remains in service, increasing its use. Therefore, scheme 3 turns out to be the most expensive of the three schemes analysed for the inclusion of uncertainty of wind power.

9.4.2 Uncertainty costs for wind and solar generation case

In this case, the codes that model the uncertainty costs due to penalization for underestimation and overestimation were modified. The solar generation is, therefore, no longer assumed as constant; instead, the multi-objective optimization is performed on the basis of the PDF that models the solar irradiation as shown in Section 9.3. The radiation function is therefore transferred to the expected value with its respective uncertainty [3], obtaining the UCF for renewable solar and wind energy for the optimization.

Similar to the previous case, where only the probabilistic model for wind power generation was included, the analysis of three important cases where relevant information can be found was based on the Pareto presented in Figure 9.20.

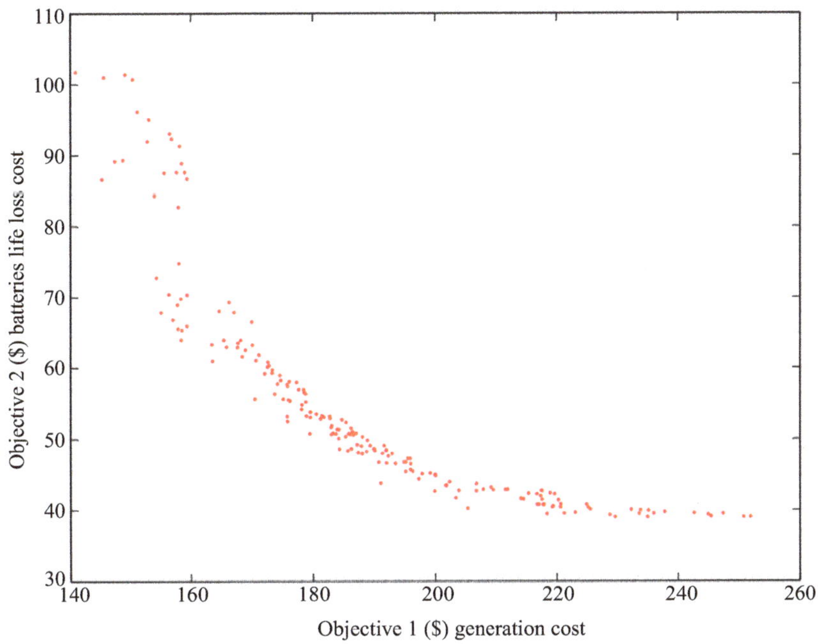

Figure 9.20 Possible optimal solutions (last iteration of the algorithm)

Table 9.5 Results of optimization, including uncertainty cost of wind and solar energy

No.	SOC_{stp}	P_{charge} (kW)	P_{excess} (kW)	Objective 1 ($)	Objective 2 ($)	Objective ($)
1	0.60	67.59	80.46	140.87	101.74	121.31
2	0.64	80.89	83.83	186.00	50.00	118.00
3	0.89	82.61	82.12	251.7	39.03	145.39

The results are shown in Table 9.5; different values of SOC_{stp}, P_{excess}, and P_{charge} are listed for the three different operating cases, each with independent characteristics.

The information in Table 9.5 demonstrates that the operation profile no. 2 is the best option, since the value of the target cell that is the combined target has the lowest value. Next, the three cases are shown graphically in terms of power and time and an analysis of each case is done.

Case 1. Improving the cost of power generation

In the first operating profile, in which the lowest generation costs are intended by reducing the operation of the traditional generator to the maximum, the value of SOC_{stp} is adjusted to 0.61. Therefore, as shown in Figure 9.21, this generator only operates between 12 and 14 h; the rest of the day the system operates with the

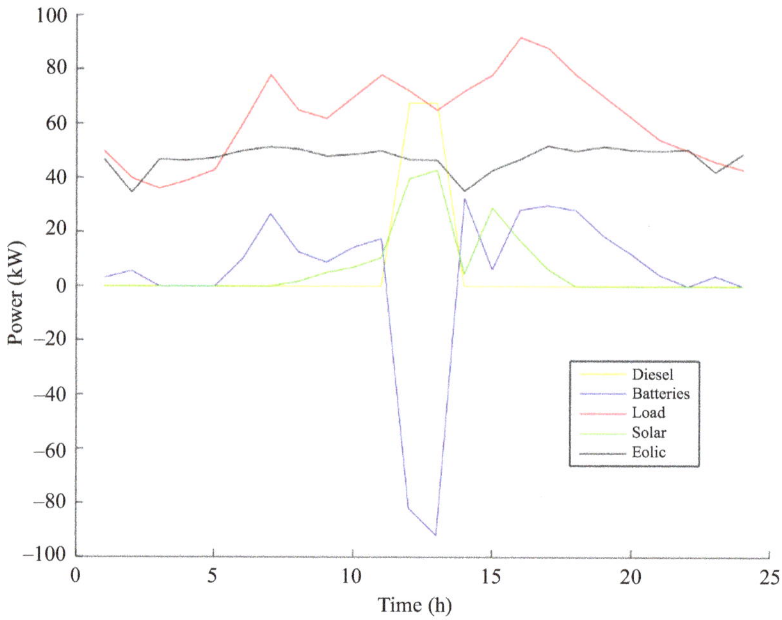

Figure 9.21 Operation profile minimizing the cost of power generation

energy from renewable sources. However, this makes the cycle of batteries something more resounding – they are in constant charge and discharge mode, so the cost of life loss in batteries is higher than all generation schemes. Moreover, P_{excess} set as 80.46 causes a waste of renewable resources. Additionally, it should be noted that in these results the PV power profile has a more real daily dispatch, both uncertainty and the possible reality of the system are simulated.

Case 2. Balance between costs

The second profile aims to achieve balance of costs between diesel generation and the life cost of the batteries. The value of SOC_{stp} is set at 0.64, so the switching between the states of charge and discharge of the batteries is lesser than in the previous case. Figure 9.22 also shows how the stress of the battery cycle is reduced, resulting in lower costs in the useful life. However, the diesel generator slightly increases its delivered energy as well as the expected value of the PV energy. In this case, P_{excess} is set at 84, resulting in more renewable power that is not used more properly. These circumstances make objective 1 increase by 25% compared to the previous case. Despite this situation, the cost of the combined objectives is the best, and the generation scheme of Case 2 is the most economical.

Case 3. Minimize the cost in battery life

The latter case is intended to minimize the cost in the useful life of the batteries, as shown in Figure 9.23. The batteries have a long charging time, which is beneficial for the useful life of the batteries, a situation consistent with this case.

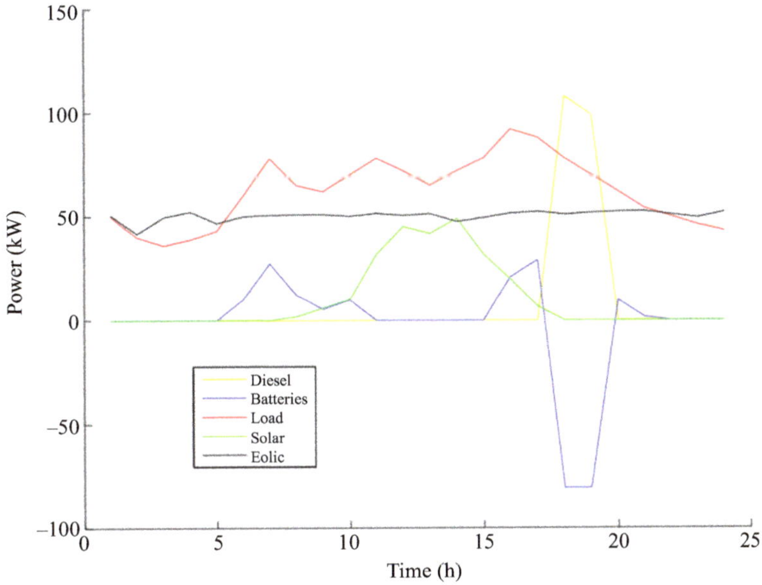

Figure 9.22 Operation profile with balance between objectives

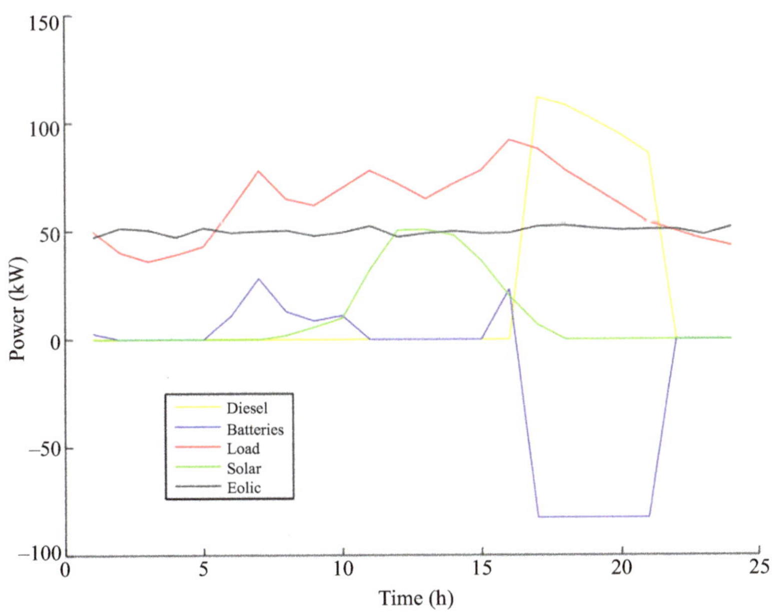

Figure 9.23 Operation profile minimizing the life loss cost of the batteries

According to the information in Table 9.5, SOC_{stp} was set at 0.89. This value is higher than the other schemes. Thus, the generator remains longer in service. Besides, it justifies the higher value of objective 1 of $234 for this scheme.

As the objectives to be optimized are contradictory, the lowest value of the life loss cost results in the highest generation cost in all schemes. It is also possible to note that the amount of the energy delivered by the diesel generator increases approximately twice compared to the previous graph.

In this last analysis, where the uncertainty of renewable PV is added and the uncertainty of the wind was also considered, the results are the most similar to those that actually occur in the system; they are, for obvious reasons, taken as the most important results of this study. With these results, it is possible to make decisions on the dispatch of the system in an optimal way, considering life loss cost, generation costs, and the uncertainty cost of renewable resources. Figure 9.23 presents a comparison of results. The optimal costs are lower in the latter case where the stochastic behaviour of renewable solar and wind energy is taken into account.

9.5 Conclusion

The multi-objective genetic algorithm used in the present work, NSGA II, plays a fundamental role in the development of optimization. It makes it possible to find different optimal values for the operation of the power system, reaching the two central objectives of the research: the minimization in the operating costs and in the life loss costs of the battery.

Once the Pareto is obtained with the possible optimal solutions, it can be used for the benefit of the dispatch, taking into account the existing variables and the characteristics of the system, as follows: when the system has batteries with a worn-out life, a Pareto point must be chosen in which the state of the batteries aims to minimize the cost of life loss; although the cost of the power generation is increased, it will be possible to have healthy conditions of batteries for a longer time.

The results of the evaluation of the four proposed scenarios are optimization of the microgrid with plentiful renewable sources and a few resources with the inclusion of uncertainty costs for wind generation; and finally, the most complete evaluation, where uncertainty is included with both wind and solar component. It is, therefore, possible to make decisions based on the information tool regarding the environmental, political, and economical characteristics of the region where the power system is located.

Moreover, it is possible to extrapolate and apply the processes and the different scenarios proposed here to different isolated systems that fulfil similar conditions to those of the power system developed in China. There is also a possibility for extension of the results of this work to different types of batteries taking into account the values provided by the manufacturers. Otherwise, further studies should be done on the hyper volume analysis, which should be considered for possible improvement of the results in the Paretos.

References

[1] Arévalo J., Santos F., and Rivera S. Uncertainty cost functions for solar photovoltaic generation, wind energy generation, and plug-in electric vehicles: Mathematical expected value and verification by Monte Carlo simulation. *International Journal of Power and Energy Conversion*. 2019; **10**(2): 171–207.

[2] Zhao B., Zhang X., Chen J., Wang C., and Guo L. Operation optimization of standalone microgrids considering lifetime characteristics of battery energy storage system. *IEEE Transactions on Sustainable Energy*. 2013; **4**(4): 934–943.

[3] Mojica-Nava E., Rivera S., and Quijano N. Game-theoretic dispatch control in microgrids considering network losses and renewable distributed energy resources integration. *IET Generation, Transmission & Distribution*. 2017; **11**(6): 1583–1590.

[4] Lezama F., Soares J., Vale Z., Rueda J., Rivera S., and Erlich I. 2017 IEEE competition on modern heuristic optimizers for smart grid operation: Testbeds and results. *Swarm and Evolutionary Computation*. 2019; **44**: 420–427.

[5] Vargas S., Rodriguez D., and Rivera S. Mathematical formulation and numerical validation of uncertainty costs for controllable loads. *Revista Internacional de Métodos Numéricos para Cálculo y Diseño en Ingeniería*. 2019; **35**(1): 1–17.

[6] Bernal-Rubiano J., Neira J., and Rivera S. Mathematical uncertainty cost functions for controllable photo-voltaic generators considering uniform distributions. *WSEAS Transactions on Mathematics*. 2019; **18**: 137–142.

[7] Peña A., Romero D., and Rivera S. Generation and demand scheduling in a micro-grid with battery-based storage systems, hybrid renewable systems and electric vehicle aggregators. *WSEAS Transactions on Power Systems*. 2019; **14**(2): 8–23.

[8] Arango D., Urrego R., and Rivera S. Robust loss coefficients: Application to power systems with solar and wind energy. *International Journal of Power and Energy Conversion*. 2018; **9**(4): 351–383.

[9] Mejia W., Rodriguez D., Rivera S., and Rosero J. Heuristic estimation of parameters in high-frequency models of induction motors for bearing currents simulation. *International Review of Automatic Control*. 2016; **9**(6): 355–364.

[10] Vargas A., Saavedra O., Samper M., Rivera S., and Rodriguez R. Latin American energy markets: Investment opportunities in nonconventional renewables. *IEEE Power and Energy Magazine*. 2016; **14**(5): 38–47.

[11] Xu G., Shang C., Fan S., Hu X., and Cheng H. A hierarchical energy scheduling framework of microgrids with hybrid energy storage systems. *IEEE Access*. 2018; **6**: 2472–2483.

[12] Khani H., Dadash M. R., and Varma R. Optimal scheduling of independently operated, locally controlled energy storage systems as dispatchable assets

in a competitive electricity market. *IET Generation, Transmission & Distribution*. 2017; **11**(6): 1360–1369.

[13] Kazemi M., Zareipour H., Amjady N., Rosehart W. D., and Ehsan M. Operation scheduling of battery storage systems in joint energy and ancillary services markets. *IEEE Transactions on Sustainable Energy*. 2017; **8**(4): 1726–1735.

[14] Molina F., Pérez S., and Rivera S. Formulación de Funciones de Costo de Incertidumbre en Pequeñas Centrales Hidroeléctricas dentro de una Microgrid. *Ingenierías USBmed*. 2017; **8**(1): 29–36.

Chapter 10

ADMM-based consensus droop control and distributed pinning droop control of isolated AC microgrids

Chia-Chi Chu[1], Lin-Yu Lu[2], and Nelson Fabian Avila[3]

Microgrid (MG) systems have been demonstrated as an effective technology to integrate the increasing distributed energy resources (DERs) into the existing AC power grid in the past decade. In order to enable the interoperability of various controllers and components in MG operations, the hierarchical control of DER units electronically coupled by DERs interface converters (DICs) plays a critical role for MG operations and control. According to the time frame of operations and infrastructure requirements, the conventional hierarchical control consists of three distinct levels: (i) primary, (ii) secondary, and (iii) tertiary.

In this chapter, we will focus on the primary and secondary controls for proper power sharing and set-point restoring operations of MGs. Traditionally, the power sharing is achieved by the decentralized droop scheme of each DIC. Although proper real power sharing can be achieved in most existing methods, incorrect reactive power sharing has been observed. In recent years, advocates of distributed multiagent systems (MAS) have been broadly adopted as promising solutions for MG control and operations. Under this framework, the primary control and the secondary control can be efficiently integrated into a single distributed task.

This work aims to provide a tutorial of distributed droop control for accurate active and reactive power sharing in isolated AC MGs. Two recently developed distributed droop control mechanisms will be studied for improving the slow convergence rate of the conventional consensus-based droop control method. First, the alternating direction multipliers method (ADMM)–based consensus control is proposed. The advantages of ADMM-based consensus control include (i) faster convergence rate, (ii) resilience to quantization errors and communication noises, and (iii) more convex update design. Two implementations of consensus-based distributed droop control by the ADMM will be investigated. It can be shown that the

[1]Department of Electrical Engineering, National Tsing Hua University, Hsinchu, Republic of China
[2]Industrial Automation BG, Delta Electronics, Taoyuan, Republic of China
[3]Power System Limits/Engineering Projects, Independent Electricity System Operator (IESO), Toronto, Ontario, Canada

closed-loop system is described by the second-order vector differential equation, and the stability of every trajectory can be analyzed by the energy function (EF) approach under certain mild conditions. Real-time digital simulations are performed to validate the effectiveness of the proposed distributed droop control mechanism.

Next, a distributed pinning droop control is investigated. Under this framework, both DICs and loads in the MG are treated as agents in a MAS. Since only a fraction of the agents in the network have access to critical information, the required communication bandwidth and control cost can be significantly reduced without degrading the dynamical performance of the power sharing compared to consensus-based droop control techniques. Theoretical studies of this pinning-based distributed droop control, including pinning localization, grid partition, and convergence criteria, will be explored in detail. In order to validate the feasibility and the correctness of the proposed distributed droop control method, a MAS platform is developed for a test-bed MG system in Taiwan. Simulation results demonstrate that the proposed pinning-based distributed droop control can indeed improve autonomous operations of an isolated AC MG.

10.1 Introduction

MG systems have been demonstrated as an effective technology to integrate the increasing DERs into the existing AC power grid in the past decade [1,2]. MGs encompass several distributed generators (DGs), energy storage units, and loads. DERs are integrated into the existing power grid through DICs. Generally speaking, an MG can operate either in grid-connected mode or in isolated mode. This work focuses on the isolated mode of operation. In order to enable the interoperability of various controllers and components in MG operations, the hierarchical control of DER units electronically coupled by DERs DICs plays a critical role for MG operations and control [2]. According to the time frame of operations and infrastructure requirements, the conventional hierarchical control, as shown in Figure 10.1, consists of the following three distinct levels [3,4]:

1. Primary control utilizes local measurements to synthesize local control actions to meet the following two objectives: (i) regulate the output voltages and currents, and (ii) achieve adequate active and reactive power sharing [5–7].
2. Secondary control is responsible for reliable, secure, and economical operations of MGs [8]. Since the primary control may introduce steady-state errors in frequency deviations and voltage magnitudes, these errors will be restored by secondary control actions.
3. Tertiary control will set the optimal operational conditions and is responsible for coordinating the operation of multiple DICs and communicating needs or requirements from the main power grid.

Traditionally, the secondary and tertiary levels rely on a centralized architecture [9], where the MG central controller concentrates critical information and periodically calculates and communicates the optimal reference values to the primary

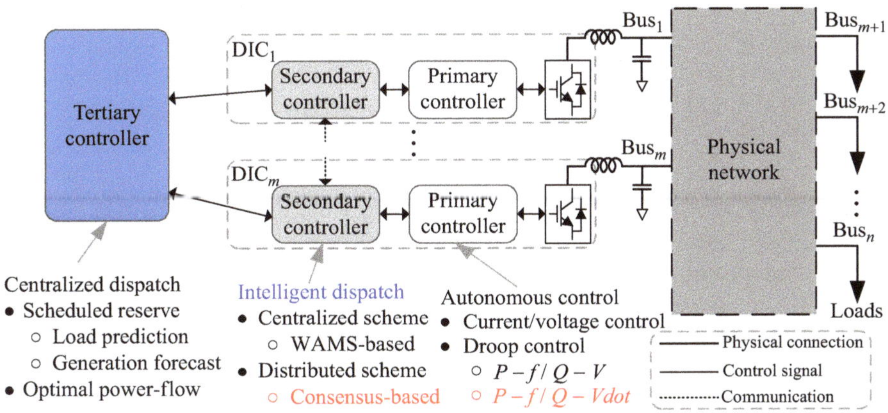

Figure 10.1 DICs in an isolated AC MG with various control levels

controllers [10–12]. Although the centralized control may provide the optimal solution, this approach possesses the following disadvantages:

1. tremendous amount of data processing when there are several small DGs in the MG,
2. high-speed communication requirement, and
3. degraded reliability and security due to a single point of failure of the central computer.

In order to overcome these shortcomings, various decentralized droop control schemes have been proposed. Although proper real power sharing can be achieved in most existing primary and secondary control methods, incorrect reactive power sharing has been observed [6,8]. Despite the potential benefits of DIC-based MGs, one major difficulty comes from the fact that unlike bulk power systems, where a large number of conventional synchronous generators (SGs) ensure a relatively large inertia, these DICs in MG are rather inertialess [13,14]. This low inertia characteristic may lead to severe frequency deviations specifically in isolated MGs, where frequency stability raises much more concerns than traditional power systems [14]. In order to overcome this difficulty, the concept of the synchronverter, also known as virtual synchronous generators (VSGs), in which designing DIC control to mimic the dynamics of conventional SGs, was proposed recently [13,15].

In recent years, advocates of distributed MAS [16,17] have been broadly adopted as promising solutions for MG control and operations. Various distributed MAS-based strategies have been proposed in the past decade [11,12,18–24]. Under this framework, the primary control and the secondary control can be efficiently integrated into a single distributed task. Among all existing MAS techniques, the class of consensus-based algorithms is particularly important since this class of algorithms can overcome the problem of improper power sharing and operating set points by the conventional droop control with only sparse communication network deployed [25]. The state-of-the-art MAS approaches for distributed secondary and tertiary control can be essentially categorized into the following two categories:

1. distributed average consensus [25–30] and
2. pinning-based distributed control [31–37].

In [28], a distributed averaging integral controller is proposed to regulate the system frequency and provide active power sharing among DICs. Extended approaches have been conducted in [25,29] for reactive power sharing using the first-order consensus algorithm. Similarly, [27] devices distributed load-sharing mechanism for DC–DC converters using consensus-voting protocols. In [30], a distributed subgradient coordination scheme is proposed to coordinate the operation of various renewable generators in MG. This last approach is extended to coordinate multiple-battery energy storage systems in [38]. The optimal economic operation of MGs has also been targeted through similar consensus algorithms and their variants, for instance, consensus+innovations [24] and incremental cost consensus [39,40]. More recently, researchers in [41] developed an optimal control system for MG via fully distributed diffusion strategy, which outperforms the convergence rate of the traditional consensus algorithm. Although these approaches are effective, a two-way communication system is still required and the convergence rate highly depends on the underlying communication topology [16]. Conversely, by devising an appropriate pinning control scheme, the control objective can be achieved with unidirectional communication, the required number of controllers can be reduced [42], the communication bandwidth can be indeed diminished, and the control objective can be accomplished even under arbitrary given communication topologies [43].

In the following, we will focus on recent developments of the average consensus-based droop control and distributed pinning-based droop control.

10.1.1 Average consensus-based droop control

More recently, the average consensus-based $P-f/Q-\dot{V}$ droop control was extended with considering the effect of the virtual inertia [29]. The stability of every trajectory in the closed-loop system has been analyzed by the advanced EF approach under mild conditions. Since the kinematic energy (KE) is included in composing EFs, the system damping can be indeed improved. The dynamics of virtual inertia also make the consensus-based control scheme in P–f loop become a second-order system [44].

Nevertheless, the convergence rate of this average consensus-based droop control method still highly depends on the structure of the underlying communication network [16,44]. In order to speed up the convergence rate, several alternative methods have been proposed in the past. For example, the ADMM, in which solutions of small local subproblems are properly coordinated such that the global optimal solution can be obtained, has been investigated lately [45–47]. It has been adopted for optimized power systems monitoring and operation in recent years [48,49]. For example, the consensus algorithm has been reformulated as a special distributed convex optimization problem. In order to solve convex optimization problems by the so-called decomposition-coordination (or splitting) procedure, the ADMM has been investigated recently [45]. The advantages of ADMM-based implementations include (i) faster convergence rate, (ii) resilience to quantization errors and communication noises, and (iii) more convex update design. More recently, it has been shown that the fast consensus can be achieved by average consensus through the ADMM [50]. As a consequence, the aim of this work is to investigate the consensus-based droop control by ADMM implementations.

In summary, distinct from our previous works in [25,29], the major improvement of the proposed ADMM implementations of consensus-based droop control in isolated MGs can be outlined as follows [51]:

1. Unlike conventional consensus-based droop control in isolated MGs [52,53], these objectives of accurate power sharing with restored operating set points for both the active and reactive power control are formulated as *consensus optimization* problems. Accordingly, a distributed ADMM implementation of the consensus-based droop control is proposed. This novel implementation provides better convergence rate and resilience to noises. Moreover, due to the improved convexness in the secondary control update design, it also allows more flexibility in the primary control settings. Two variants, including double message exchange and unique message exchange, are studied in this work.
2. Theoretical investigations about asymptotic behavior of system trajectories of the closed-loop systems under this new consensus-based distributed droop control are analyzed by the EF under mild conditions. Compared to our recent work in [54], since the KE is further included in the EF, real power dissipation will be involved during the transient period, and the system damping can be effectively improved.
3. A more flexible scheme for setting the droop gain ratios of VSGs under ADMM-based control algorithms is proposed. In contrast to our latest work in [55], such flexible setting scheme will provide some robustness for accurate real and reactive power sharing even when droop gains of all VSGs are different. As a consequence, this flexible scheme would be more appropriate for practical industrial applications.

10.1.2 Distributed pining-based droop control

Lately, a significant amount of research has been devoted toward distributed pinning cooperative control for MG applications. In [31,32], by applying input–output feedback linearization, the secondary voltage and frequency controls of the MG were represented as first- and second-order MAS, respectively. Then, synchronization among DGs was achieved by pinning a single-leader agent, which sets the control reference to synchronize all DGs to a common value of the reference voltage and frequency. However, accurate reactive power sharing among DGs was not considered in their work. This cooperative control approach was further extended as a multiobjective framework in [33]. Although active power sharing and reactive power sharing were deliberated, there was no mechanism to refine the power control reference of the pinning node when the MG experiences large disturbances. More recently, a novel cooperative control approach for MG has been presented in [34], where only one DG participates in the droop control and the remaining DGs operate in the PQ mode. Then, upon the occurrence of a load disturbance, a distributed pinning scheme is employed for obtaining the active/reactive power references of the converters operating in the PQ mode. Nevertheless, no power-sharing mechanism among various droop-controlled DICs was presented. Innovative results on the upper and lower bounds to accomplish pinning control for MG with weakly connected communication topologies are reported in [35]. More recently, applications of [35] have been extended to devise intelligent single and multiple pinning mechanism for distributed cooperative secondary control of DGs in isolated MGs [36]. This last approach confirms that the selection of a pinning set based on the degree of connectivity and distance of leader DGs from the rest of the system can indeed improve the transient performance of the MG.

These previously mentioned approaches have pioneered the application of pinning-based cooperative control in MG. An unavoidable assumption in these schemes is that the secondary control will always be achieved by selecting a unique leader agent in the entire MG. However, this assumption may be affected by the communication topology among DICs. More specifically, for these approaches to work properly, the directed graph (diagraph) that represents the communication network among the DICs must contain a directed spanning tree (DST) [56]. Unfortunately, as more DGs are integrated into the MG, the communication topology may change significantly, and, under the worst-case scenario, the communication diagraph may not always contain the required DST. Consequently, a solitary pinning agent may not be able to accomplish the global synchronization for secondary control. It is therefore necessary to investigate alternative MAS schemes that can guarantee the global secondary control regardless of the infrastructure of the communication network.

By further extending our preliminary work in [57], we propose a new effective distributed pinning droop control scheme for secondary control in autonomous operations of isolated AC MGs. The proposed framework aims to satisfy the following objectives: (i) provide accurate active/reactive power sharing among several droop-controlled DICs in the MG and (ii) achieve frequency restoration after the occurrence of small and large disturbances. In order to achieve these objectives, the following new concepts have been adopted into the proposed framework [58]:

1. A new scheme for pinning localization and grid partition is proposed for separating the MG into zones. Under this framework, the required communication bandwidth can indeed be reduced and the control performance can be enhanced by appending more pinning agents into the network-controlled system. Moreover, this partition scheme ensures that each leader agent has a direct connection to the remaining nonleader agents even when the communication topology is disconnected.

2. The convergence criteria of the proposed approach are presented through both analytical and numerical simulations.
3. Comparison with the conventional average consensus technique is developed and presented.

In order to validate the feasibility and the correctness of the proposed distributed droop control method, a MAS platform is developed for a test bed MG system in Taiwan. Two types of agents are developed in the proposed MAS: (i) DG agents (DGAs) and (ii) and load agents (LAs). DGAs are implemented in order to craft the pinning control algorithm. The main purpose of these types of agents is to execute simple mathematical operations required to reach a consensus via distributed pinning control. The LAs are integrated at the load side and their main purpose is to monitor load requirements and to trigger the secondary pinning droop control after the occurrence of large disturbances. Simulation tests under normal conditions, heavily loading conditions, and plug-in and plug-out scenarios are investigated in MATLAB®/Simulink™ and the Java Agent Development Framework (JADE) [59].

This chapter will provide a comprehensive treatment of both ADMM-based droop control and distributed pinning-based droop control. The problem formulation of droop control in isolated AC MGs will be discussed first. Technical backgrounds and implementations of both methods in droop control will also be explored. Comprehensive analysis of the stability of the closed-loop system is considered in detail. Real-time simulations of these two distributed droop control schemes have been performed to validate the feasibility and the correctness of these two methods. Finally, some conclusions and future research directions will be outlined.

10.2 Droop control in isolated AC MGs

In this section, the background of the conventional droop control for DICs in an isolated MG will be reviewed first. The control mechanism of VSGs will also be addressed later.

10.2.1 Problem formulation

Consider an isolated MG with n buses as shown in Figure 10.1. The subset of buses $\mathcal{N} := \{1,\ldots,m\}$ is power sources embedded with DICs and the subset of load buses is denoted by $\mathcal{N}_L := \{m+1,\ldots,n\}$. Let the bus voltage magnitude and the angle be represented by V_i and θ_i, respectively. P_i and Q_i are active and reactive power injections to the MG in each bus. P_i^r and Q_i^r are nominal active and reactive power ratings of DIC_i. $P_{L,i}$ and $Q_{L,i}$ are load demands in each load bus. In order to develop the analysis framework for the proposed control, we also make the following mild assumptions:

(A1) All transmission lines are lossless.
(A2) All possible load variations in the MG can be supported by DICs without violating the rating limits of DICs or transmission lines.

(A3) All loads in the MG are with $\dot{\theta}/\dot{V}$ dependencies.

Indeed, to cohere with the analysis employed by the EF approach, (A1) can be further relaxed to consider that all transmission lines are with a uniform R/X ratio in our previous studies [25]. For simplicity, (A1) is considered. (A2) can be guaranteed in the planning phase of the MG development, while (A3) is based on the field measurement studies in the literature [60–62].

The main objectives of DIC control in isolated MGs are maintaining the following two operating conditions at the desired postdisturbed stable equilibrium point (SEP):

1. Maintaining stable operating points of DICs such that in the postdisturbed steady state,

$$\dot{\theta}_i := \omega_i - \omega_b = 0 \quad \text{and} \quad \dot{V}_i = 0, \forall i \tag{10.1}$$

where ω_i and ω_b are the output frequency of each DIC$_i$ and the nominal grid frequency, respectively.
2. Autonomous power sharing forces all DICs to share total loads according to their nominal ratings such that

$$\frac{P_i}{P_i^r} = \frac{P_j}{P_j^r}, \quad \text{and} \quad \frac{Q_i}{Q_i^r} = \frac{Q_j}{Q_j^r}, \quad \forall i,j \in \mathcal{N}. \tag{10.2}$$

10.2.2 Model descriptions

Since the variations in amplitude and/or phase of phasor variables in MGs are slow enough in comparison with fast transient dynamics of power electronics, the pseudo steady-state analysis is sufficient for these MG droop-control problems. In order to preserve the identity of all grid components, including DICs, loads, and lines, and provide more physical interpretations among them, we will use the *network-preserving model* [25,60,63–66]. Thus, the following model descriptions are considered:

1. *Power injections at each bus*: Under (A1), the active power output P_i and reactive power output Q_i of each bus must satisfy the following power-flow equations [67]. For $i \in \mathcal{N} \cup \mathcal{N}_L$,

$$P_i = \sum_{j=1}^{n} V_i V_j B_{ij} \sin \theta_{ij}, \tag{10.3a}$$

$$Q_i = -\sum_{j=1}^{n} V_i V_j B_{ij} \cos \theta_{ij} \tag{10.3b}$$

where B_{ij} and $\theta_{ij} := \theta_i - \theta_j$ are the transmission line admittance and the angle difference between bus i and j, respectively.

2. *Dynamical load models*: Based on the field measurement studies in literature the [60–62], the load dynamics can be represented by the following singular perturbed equations for $i \in \mathcal{N}_L$ [29,68]:

$$P_{Li} = -P_i - D_{Lp,i}\dot{\theta}_i, \tag{10.4a}$$

$$Q_{Li} = -Q_i - D_{Lq,i}\dot{V}_i. \tag{10.4b}$$

3. *DIC and VSG models*: In order to increase the inertia of conventional DICs, the concept of the VSGs was proposed recently in which the DIC mimics conventional SGs by designing proper parameters J_i of the SG into each droop-control mechanism [29]. Note that if $J_i = 0$, the VSG will be reduced to the conventional DIC. The $P - f/Q - \dot{V}$ droop control for DICs and VSGs can be implemented by modifying the conventional $P - f/Q - \dot{V}$ droop control proposed in [6]. That is, for $i \in \mathcal{N}$,

$$\dot{\theta}_i = \omega_i - \omega_b, \tag{10.5a}$$

$$J_i\omega_b\dot{\omega}_i + D_{p,i}\dot{\theta}_i = P_i^r - P_i - p_i, \tag{10.5b}$$

$$k_{p,i}\dot{p}_i = J_i\omega_b\dot{\omega}_i + D_{p,i}\dot{\theta}_i, \tag{10.5c}$$

$$D_{q,i}V_i\dot{v}_i = Q_i^r - Q_i - q_i, \tag{10.5d}$$

$$k_{q,i}\dot{q}_i = D_{q,i}V_i\dot{v}_i \tag{10.5e}$$

where $\dot{v}_i$ is the transformed bus voltage magnitude defined by $v_i := \ln V_i$. $D_{p,i}$ and $D_{q,i}$ are droop gains for active and reactive power control. p_i and q_i describe $\dot{\theta}$ and $\dot{V}$ restoration mechanism for regulating the frequency variable and the voltage magnitude, respectively.

The control block diagram for each individual VSG is shown in Figure 10.2 with the control inputs from consensus ADMM network turned off. It should be pointed out that in order to execute real-time digital simulations, detailed power

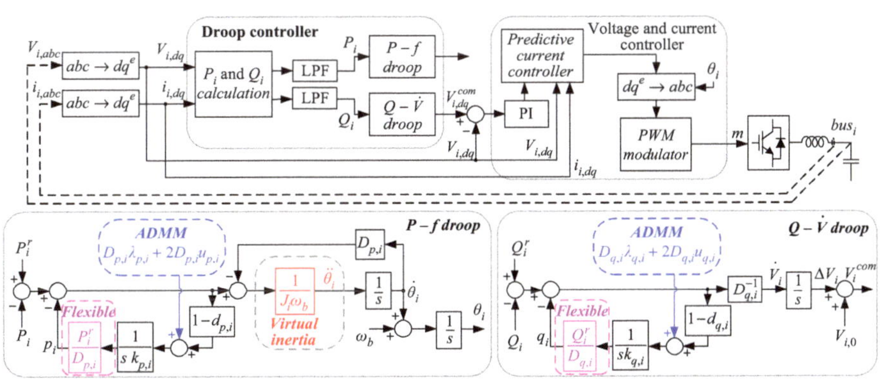

Figure 10.2 ADMM-based droop control with virtual inertia for each DIC_i

electronics control blocks are also included. In fact, this droop control mechanism can be considered as an enhancement of the so-called synchronverter proposed by [15]. However, the above two control objectives (i) and (ii) cannot be achieved simultaneously [29]. Fortunately, two new control mechanisms can overcome this obstacle. We will discuss this issue in the following sections.

10.3 Average consensus through ADMM

In this section, basic concepts of ADMM implementations for fast consensus will be reviewed first. Then, two variants of ADMMs for solving average consensus problems will be considered.

10.3.1 Basic concepts

Consider the consensus problem with n nodes in a connected network $\mathcal{G}$. Let $\phi_i,\ i = 1,\ldots,n$ be the local measurements of each node, the goal of the consensus algorithm is to extract the following average value:

$$\bar{\phi} = \frac{1}{n}\sum_{i=1}^{n}\phi_i. \tag{10.6}$$

The ADMM implementations of the consensus algorithm can be proceeded by the so-called decomposition–coordination scheme. First, the original problem (10.6) is decomposed by several subproblems, which can be solved concurrently. Then, the global optimal solution can be reached iteratively by coordinating solutions of each subproblem with proper control actions [45,46]. Depending on the existence of the centralized agent, this problem can be solved by the following two possible implementations:

1. One additional centralized agent is appended. As illustrated in Figure 10.3(a), x_i is the solution of each subproblem. The auxiliary variable z defined in the centralized agent is utilized for information exchange among all nodes. Mathematically, the original problem (10.6) can be rephrased as the following

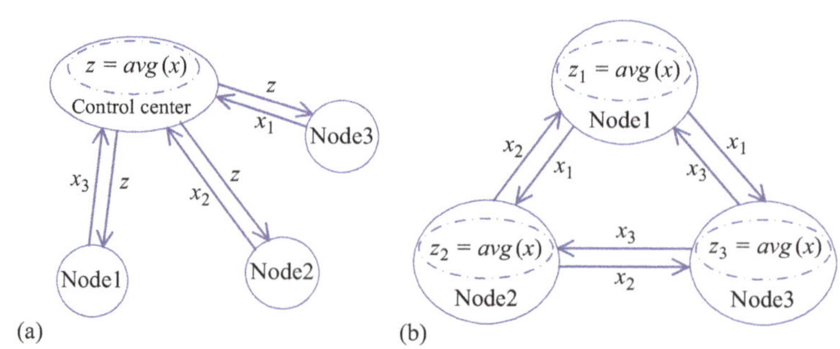

Figure 10.3 ADMM architectures: (a) centralized agent and (b) fully distributed

constrained optimization problem [48]:

$$\begin{aligned}\bar{\phi} = \quad & \arg\min_{\mathbf{x}} \sum\nolimits_{i=1}^{n} (x_i - \phi_i)^2 \\ & \text{subject to } x_i = z, \quad \forall i\end{aligned} \tag{10.7}$$

where the auxiliary variable z, which is defined as the average of all x_i, $z = avg(x) = 1/n \sum_{i=1}^{N} x_i$, will be updated in each iteration of the coordination process. Since this implementation requires one centralized agent for coordination of the global optimal solution among all subtasks in each node, more comprehensive communication topologies are needed. Therefore, this implementation seems not to be suitable for MG operations since the plug-and-play feature of MGs cannot be easily demonstrated.

2. An alternative approach is the so-called fully distributed ADMM implementation. This implementation can be achieved by reformulation (10.6) as the following constrained optimization problem [50]:

$$\begin{aligned}\bar{\phi} = \quad & \arg\min_{\mathbf{x}} \sum\nolimits_{i=1}^{n} (x_i - \phi_i)^2 \\ & \text{subject to } x_i = x_j, \forall i, j\end{aligned} \tag{10.8}$$

as illustrated in Figure 10.3(b). The optimal solution of (10.8) can be obtained by taking the gradient with respect to x_i

$$\sum_{i=1}^{n} (x_i - \phi_i) = 0 \underset{x_i = x_j}{\Rightarrow} x^* = \frac{1}{n}\sum_{i=1}^{n} x_i = \frac{1}{n}\sum_{i=1}^{n} \phi_i \tag{10.9}$$

which indeed implies the average consensus.

Note that the major difference between (10.7) and (10.8) lies on their associated equality constraints. Since no centralized variable z is introduced in (10.8), its implementation can be achieved in a fully distributed manner. Therefore, it seems to be more appropriate for demonstrating the plug-and-play feature of MGs. We will further explore this distributed implementation later.

10.3.2 Distributed implementations

The following two variants of distributed ADMM-based consensus methods are considered [69,70].

10.3.2.1 Method A

The constrained convex optimization problem (10.8) can be stated equivalently as

$$\begin{aligned}\bar{\phi} = \quad & \arg\min_{\{x_i\},\{z_j\}} \sum\nolimits_{i=1}^{n} (x_i - \phi_i)^2 \\ & \text{subject to } x_i = z_j, \quad \forall i \text{ and } j \in \mathcal{N}_i\end{aligned} \tag{10.10}$$

where x_i being the state of each node and $\mathcal{N}_i$ denotes the one-hop-neighbors of node i, which also include node i itself. By introducing auxiliary variables z_j, the set of equality constraints $x_i = z_j$ ensures that all state variables x_i coincide across the given network. As proposed in [50], by defining the corresponding augmented Lagrangian of (10.10), the update equations for the prime variable $x(k)$ and the dual variable $\overline{\lambda}(k)$ of this ADMM implementation of (10.10) can be derived as follows,

for $k \geq 1$:

$$x_i(k+1) = (1-d_i)\phi_i + d_i x_i(k) + \overline{\lambda}_i(k) + 2u_i(k), \tag{10.11a}$$

$$\overline{\lambda}_i(k+1) = \overline{\lambda}_i(k) + u_i(k). \tag{10.11b}$$

The control input $u_i(k)$ is defined by

$$u_i(k) = \sum_{j\in\mathcal{N}_i} b_{i,j}\Big(\sum_{\ell\in\mathcal{N}_j} a_{\ell,j}x_\ell(k)\Big) - d_i x_i(k) \tag{10.12}$$

where

$$a_{i,j} = \frac{l_{i,j}}{\sum_{\ell\in\mathcal{N}_j} l_{\ell,j}}, \quad b_{i,j} = \frac{l_{i,j}}{\left(1+\sum_{\ell\in\mathcal{N}_i} l_{i,\ell}\right)}, \quad d_i = \sum_{j\in\mathcal{N}_i} b_{i,j}. \tag{10.13}$$

Each entry $l_{i,j} > 0$ comes from the Laplacian matrix of the communication network. It represents the communication strength between agent i and j. The initial conditions are

$$x_i(1) = (1-d_i)\phi_i, \overline{\lambda}_i(1) = 0. \tag{10.14}$$

One distinguishing feature of Method A described by (10.12) is that $u_i(k)$ can be evaluated in a distributed manner by two message exchanges: the first for the summation in $\mathcal{N}_j$ and the second for the summation in $\mathcal{N}_i$, which also means that the information from two-hop-neighbors is required as shown in Figure 10.4(a).

10.3.2.2 Method B

The optimization problem (10.8) can also be written as

$$\begin{aligned} \bar{\phi} = \; & \underset{\{x_i\},\{z_{i,j}\}}{\arg\min} \sum\nolimits_{i=1}^{n} (x_i - \phi_i)^2 \\ & \text{subject to } x_i = z_{i,j}, \\ & \text{and to } z_{i,j} = z_{j,i}, \forall i \text{ and } j \in \mathcal{N}_i. \end{aligned} \tag{10.15}$$

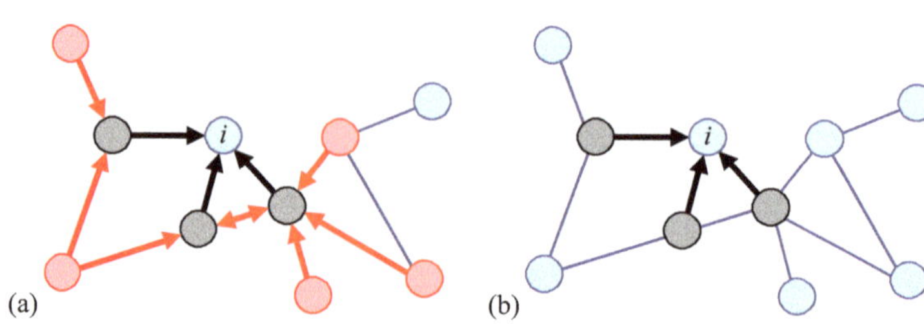

Figure 10.4 Distributed ADMMs: (a) Method A and (b) Method B

Again, the use of auxiliary variables $z_{i,j}$ accompanied with equality constraints in (10.15) would force all x_i and $z_{i,j}$ to coincide across the network. Similar to Method A, by defining the corresponding augmented Lagrangian of (10.15), another distributed ADMM implementation of the average consensus algorithm can be depicted by (10.11a) and (10.11b) and (10.14), where (10.12) becomes

$$u_i(k) = \sum_{j \in \mathcal{N}_i} e_{i,j} x_j(k) - \left(\sum_{j \in \mathcal{N}_i} e_{i,j} \right) x_i(k) \tag{10.16}$$

where

$$e_{i,j} = \frac{(1 - d_i) l_{i,j} l_{j,i}}{(l_{i,j} + l_{j,i})}. \tag{10.17}$$

As opposed to the double message exchange used in Method A, now only a unique message exchange is used in Method B. This means that only the information from one-hop-neighbors of each node i is required as shown in Figure 10.4(b). Therefore, less influence from noisy environments in comparison with Method A since less information are required in Method B.

Similar to those developments of the basic consensus algorithm with two agents, the update equations for the prime variable $x(k)$ and the dual variable $\lambda(k)$ in (10.11a) and (10.11b) can be easily extended to the continuous system. Here we omit the details. This continuous-time version of ADMM for the average consensus algorithm will be further utilized for subsequent analysis.

10.4 Consensus-based droop control

In this section, theoretical foundations of consensus-based droop control of VSGs via ADMM will be investigated in detail. First, we will investigate the desired SEP fulfilling objectives (10.1) and (10.2). It will then be shown that the control problem for achieving these objectives can therefore be cast into a *consensus optimization* one. Finally, ADMM implementations of this consensus-based droop control will be considered.

10.4.1 Optimal stable equilibrium point

For notational simplicity, let X/x represent either the notions for $P-f$ or those for $Q-\dot{V}$ droop control, namely, P/p and Q/q. The equilibrium points of $P-f$ and $Q-\dot{V}$ droop control fulfilling the objectives in Section 10.2.1 can be depicted as

Proposition 1. *A SEP* $(\dot{\theta}^*, P^*, p^*, \dot{V}^*, Q^*, q^*)$ *that fulfills objectives* (10.1) *and* (10.2) *must also satisfy*

$$D_{x,i}^{-1}(X_i^r - X_i^* - x_i^*) = 0, \quad \forall i \in \mathcal{N}, \tag{10.18a}$$

$$\frac{x_i^*}{D_{x,i}} = \frac{x_j^*}{D_{x,j}}, \quad \forall i,j \in \mathcal{N}. \tag{10.18b}$$

Proof: Assuming the fast dynamics in $\dot{\omega}$ already diminished, replacing $\dot{\theta}_i = 0$ and $\dot{V}_i = 0$, respectively, into (10.5b) and (10.5d) gives (10.18a). As to (10.18b), dividing (10.5b) and (10.5d), respectively, with P_i^r and Q_i^r gives

$$\frac{P_i}{P_i^r} = 1 - \frac{D_{p,i}}{P_i^r}\frac{p_i^*}{D_{p,i}} - \frac{D_{p,i}}{P_i^r}\dot{\theta}_i^*, \tag{10.19a}$$

$$\frac{Q_i}{Q_i^r} = 1 - \frac{D_{q,i}}{Q_i^r}\frac{q_i^*}{D_{q,i}} - \frac{D_{q,i}}{Q_i^r}\dot{V}_i^*. \tag{10.19b}$$

Since at a SEP, $\dot{\theta}_i^* = \dot{\theta}_j^*$ and $\dot{V}_i^* = \dot{V}_j^*, \forall i$ can be assured, uniform $p_i^*/D_{p,i}$ and $q_i^*/D_{q,i}$ in (10.19a) and (10.19b) can guarantee the achieving of objective (10.2). It is however worth noting that uniform droop gain ratio settings, namely,

$$\frac{D_{p,i}}{P_i^r} = \frac{D_{p,j}}{P_j^r}, \quad \text{and} \quad \frac{D_{q,i}}{Q_i^r} = \frac{D_{q,j}}{Q_j^r}, \quad \forall i,j \in \mathcal{N} \tag{10.20}$$

are required to achieve accurate active and reactive power sharing.

10.4.2 Consensus optimization

Based on the conditions (10.18a) and (10.18b) for the optimal SEP, and assuming that the communication network among VSGs forms an undirected and connected graph $\mathcal{G} = (\mathcal{V}, \mathcal{E})$, the control problems of achieving objectives (10.1) and (10.2) can be formulated as the following *consensus optimization* problems:

$$\begin{aligned} &\min_x \sum\nolimits_{i=1}^{m} D_{x,i}^{-2}[(X_i^r - X_i) - x_i]^2 \\ &\text{subject to } \frac{x_i}{D_{x,i}} = \frac{x_j}{D_{x,j}} \quad \forall j \in \mathcal{N}_i. \end{aligned} \tag{10.21}$$

Compared with (10.8), the optimizers for (10.21) can be found as

$$\bar{\phi}_x = \frac{1}{m}\sum_{i=1}^{m}\left(\frac{X_i^r - X_i}{D_{x,i}}\right). \tag{10.22}$$

Substitute $p_i/D_{p,i} = \bar{\phi}_p$ and $q_i/D_{q,i} = \bar{\phi}_q$ into (10.5b) and (10.5d), the power sharing ratios can be found as

$$\frac{X_i}{X_i^r} = 1 - \frac{D_{x,i}}{mX_i^r}\sum_{i=1}^{m}\left(\frac{X_i^r - X_i}{D_{x,i}}\right) \tag{10.23}$$

which indeed achieve both the control objectives of ensuring accurate power sharing and maintaining restored operating points.

10.4.3 ADMM implementations

In our previous work [25], the consensus problems described by (10.21) are solved by implementing the *first-order consensus* algorithm [16]. In order to achieve better convergence rate and resilience to communication noises, two ADMM variants of consensus-based droop control are investigated in this work.

By integrating the discrete-time version of the ADMM described by (10.11a) and (10.11b) into the continuous-time $P-f$ and $Q-\dot{V}$ droop control law (10.2), the dynamics of p_i and q_i of two ADMMs can both be described as

$$k_{x,i}\dot{x}_i = (1-d_i)(X_i^r - X_i - x_i) + D_{x,i}\lambda_{x,i} + 2D_{x,i}u_{x,i} \tag{10.24}$$

while the dynamics of $\lambda_{pq,i}$ and $u_{pq,i}$ can be depicted as follows.

10.4.3.1 Method A

$$\dot{\lambda}_{x,i} = u_{x,i} = \sum_{j\in\mathcal{N}_i} b_{i,j} \sum_{k\in\mathcal{N}_j} a_{k,j}\frac{x_k}{D_{x,k}} - d_i\frac{x_i}{D_{x,i}} \tag{10.25}$$

where notions $a_{i,j}$, $b_{i,j}$, and d_i are defined by (10.13).

10.4.3.2 Method B

$$\dot{\lambda}_{x,i} = u_{x,i} = \sum_{j\in\mathcal{N}_i} e_{i,j}\frac{x_j}{D_{x,j}} - \left(\sum_{j\in\mathcal{N}_i} e_{i,j}\right)\frac{x_i}{D_{x,i}} \tag{10.26}$$

where $e_{i,j}$ is defined in (10.17).

Proposition 2. *Given the proper stepsize designs* (10.13) *and* (10.17), *the proposed control algorithm with* (10.24) *replaced into* (10.5a)–(10.5e), *while accompanied by the two variants* (10.25) *and* (10.26), *converges to the optimal solution of* (10.21).

First, consider that the optimization problem (10.21) is a linear quadratic one with strong convex objective function. Moreover, albeit the nonlinearity in (10.3a) and (10.3b) and (10.5a)–(10.5e), the existence and feasibility of the desired SEP can always be guaranteed by (A2) [71]. Therefore, the optimum of (10.21) can indeed be achieved by the proposed control algorithms.

Note that since the proposed ADMM implementations of consensus-based droop control are described by the continuous-time systems, the original discrete versions of the distributed ADMM implementations for consensus problems need to be slightly modified. Meanwhile, compared to Method A described in (10.25), Method B requires less-complicated communication links and is thus expected to be able to achieve better resilience to communication noises.

10.4.4 ADMM implementations with flexible droop gain ratios

Although the ADMM-based droop control described by (10.24) can achieve accurate power sharing among VSGs, the precise settings (10.20) are also required. This condition will restrict the applications for practical large-scale MG since damping variations will destroy the perfect power sharing. In order to avoid this restriction and provide more flexibility for individual droop gain settings, the $P-f/Q-\dot{V}$ droop controls in (10.5b) and (10.5d) are modified as

$$J_i\omega_b\dot{\omega}_i + D_{p,i}\dot{\theta}_i = P_i^r - P_i - D_{p,i}^{-1}P_i^r p_i, \tag{10.27a}$$

$$D_{q,i}V_i\dot{v}_i = Q_i^r - Q_i - D_{q,i}^{-1}Q_i^r q_i. \tag{10.27b}$$

Under this framework, the two ADMM implementations of consensus-based droop control can also be extended by replacing the corresponding $\lambda_{pq,i}$ and $u_{pq,i}$ into (10.24). Compared to (10.5b) and (10.5d), the optimization problems in (10.21) become

$$\begin{aligned}&\min_x \sum\nolimits_{i=1}^m D_{x,i}^{-2}\left[(X_i^r - X_i) - D_{x,i}^{-1}X_i^r x_i\right]^2\\ &\text{subject to } x_i/D_{x,i} = x_j/D_{x,j}, \quad \forall j \in \mathcal{N}_i.\end{aligned} \tag{10.28}$$

Consequently, the optimizers for (10.28) should be modified to

$$\bar{\phi}_{x,f} = \frac{1}{m}\sum_{i=1}^m \left(\frac{X_i^r - X_i}{X_i^r}\right). \tag{10.29}$$

Substitute $p_i/D_{p,i} = \bar{\phi}_{p,f}$ and $q_i/D_{q,i} = \bar{\phi}_{q,f}$, respectively, into (10.27a) and (10.27b), the power-sharing ratios can be found as

$$\frac{X_i}{X_i^r} = 1 - \frac{1}{m}\sum_{i=1}^m \left(\frac{X_i^r - X_i}{X_i^r}\right). \tag{10.30}$$

In comparison with (10.23), it is obvious that the ratios are now independent of each individual droop gain setting and the restriction (10.20) is eliminated. Therefore, the above formula suggests that flexible individual droop gain setting and precise power sharing can be achieved simultaneously. Moreover, since the proposed modifications do not alter the properties of underlying objective functions, the same conclusions of achieving the optimum illustrated in Proposition 2 can still be made on (10.27a) and (10.27b).

10.5 Stability analysis of closed-loop MG systems with consensus-based droop-controlled VSGs

Similar to our previous work in [29], the stability of the closed-loop MG system with the consensus-based droop-controlled VSGs will be investigated by the EF approach. First, define the state variables for individual DIC$_i$ as $y_i := [\theta_i, p_i, \lambda_{p,i}, v_i, q_i, \lambda_{q,i}]^T$ and $z_i := dy_i/dt$, $\forall i \in \mathcal{N}$. Next, for load buses, since no droop controllers are embedded, the state variables are defined as $y_{L,i} := [\theta_i, v_i]^T$ and $z_{L,i} := dy_{L,i}/dt$, $\forall i \in \mathcal{N}_L$. Under this framework, the state variables of the overall system are represented by $y := [y_1^T, y_2^T, \ldots, y_m^T, y_{L,m+1}^T, \ldots, y_{L,n}^T]^T$ and $z := dy/dt$. Now define the corresponding potential energy (PE) of (10.5a)–(10.5e) as follows:

$$U(y) = U_1(y) + U_2(y) + U_3(y) + U_4(p, \lambda_p, q, \lambda_q). \tag{10.31}$$

Analytical expressions of each component in PE can be derived as follows:

$$\begin{aligned} U_1(y) &= -\sum_{i=1}^{n}\sum_{j=1}^{n} V_i V_j B_{ij} \cos\,\theta_{ij}, \\ U_2(y) &= -\sum_{j=1}^{m} P_j^r \theta_j + \sum_{j=m+1}^{n} P_{L,j}\theta_j, \\ U_3(y) &= -\sum_{j=1}^{m} Q_j^r v_j + \sum_{j=m+1}^{n} Q_{L,j} v_j, \\ U_4(p, \lambda_p, q, \lambda_q) &= \frac{1}{2}\sum_{j=1}^{m}\left(p_j^2 + \lambda_{p,j}^2 + q_j^2 + \lambda_{q,j}^2\right). \end{aligned} \tag{10.32}$$

Each term can be related to a physical quantity of the associated closed-loop MG systems. For each load bus, by (10.4a) and (10.4b), dynamical equations can be expressed as

$$D_{L,i}(y_{L,i})\frac{d}{dt}y_{L,i} = -A_{L,i}\frac{\partial U(y)}{\partial y_{L,i}} \tag{10.33}$$

where

$$D_{L,i}(y_{L,i}) = \begin{bmatrix} D_{Lp,i} & 0 \\ 0 & V_i D_{Lq,i} \end{bmatrix}, \quad A_{L,i} = \begin{bmatrix} 1 & 0 \\ 0 & 1 \end{bmatrix}. \tag{10.34}$$

Depending on the various droop control actions for VSGs, the overall closed-loop system can be described in different representations.

10.5.1 $P - f/Q - \dot{V}$ droop

For each DIC bus under the $P - f/Q - \dot{V}$ droop control with virtual inertia, (10.5a)–(10.5e) can be rewritten as the following compact representation:

$$\dot{y}_i = z_i, \quad M_i \dot{z}_i = -D_i z_i - A_i \frac{\partial U(y)}{\partial y_i} \tag{10.35}$$

where $M_i = diag[J_i\omega_b, 0, 0, 0, 0, 0], D_i = diag[D_{p,i}, k_{p,i}, 1, D_{q,i}V_i, k_{q,i}, 1]$, and

$$A_i = \begin{bmatrix} N_A & 0 \\ 0 & N_A \end{bmatrix}, \quad N_A = \begin{bmatrix} 1 & 1 & 0 \\ 1 & 1 & 0 \\ 0 & 0 & 0 \end{bmatrix}.$$

Derivations of (10.35) can be easily realized by taking (10.5b) as an example. Each term of (10.5b) can be replaced by part of $\partial U(y)/\partial y_i$ as follows:

$$J_i\omega_b\frac{d^2\theta_i}{dt^2}+D_{p,i}\frac{d\theta_i}{dt}=-\frac{\partial U_2}{\partial\theta_i}-\frac{\partial U_1}{\partial\theta_i}-\frac{\partial U_4}{\partial p_i}$$

and the rest of derivations follow. Obviously, both M_i and A_i are positive semi-definite, while D_i is positive definite. Now the overall closed-loop system can be expressed as the following compact form:

$$\dot{y}=z,\quad M\dot{z}=-Dz-A\frac{\partial U(y)}{\partial y},\tag{10.36}$$

$$M=\begin{bmatrix}N_{M1}&0\\0&0\end{bmatrix},\quad D(y)=\begin{bmatrix}N_{D1}&0\\0&N_{D4}\end{bmatrix},\quad A=\begin{bmatrix}N_{A1}&0\\0&N_{A4}\end{bmatrix},$$

where

$$\begin{aligned}&N_{M1}=diag[M_i],\quad \forall i\in\mathcal{N},\\&N_{D1}=diag[D_i],\quad N_{A1}=diag[A_i],\quad \forall i\in\mathcal{N},\\&N_{D4}=diag[D_{L,i}],\quad N_{A4}=diag[A_{L,i}],\quad \forall i\in\mathcal{N}_L.\end{aligned}$$

Since A is singular, (10.36) cannot be transformed into the following compact representation (10.37) for constructing analytical EFs described in [65,72,73],

$$\dot{y}=z,\quad \widetilde{M}\dot{z}=-\widetilde{D}z-\frac{\partial U(y)}{\partial y}.\tag{10.37}$$

Thus, the stability cannot be guaranteed by $P-f/Q-\dot{V}$ droop control of VSGs and the corresponding EF is not well defined. In fact, the singularity of A may lead to multiple nonisolated SEPs of (10.36). As shown in simulation studies later, when this class of droop control is applied, the frequency and voltage magnitude may be restored to their nominal values. However, accurate active and reactive power sharing may not be reached since an undesired nonisolated EP is reached.

10.5.2 ADMM implementations

Indeed, the closed-loop systems of ADMM-based droop-controlled VSGs can be reformulated as the following compact form:

$$\widehat{M}\ddot{\widehat{y}}=-\widehat{D}\dot{\widehat{y}}-A_a\frac{\partial U}{\partial\widehat{y}}\tag{10.38}$$

where $\widehat{y}^T=[\theta^T,p^T,\lambda_p{}^T,v^T,q^T,\lambda_q{}^T]$. The rearrangement from y to $\widehat{y}$ comes from including the communication protocol introduced by the ADMM mechanism. Also,

M, D, and A are modified to $\widehat{M}$, $\widehat{D}$, and $\widehat{A}$, respectively. Detailed formulations of $\widehat{M}$, $\widehat{D}$, and $\widehat{A}$ are illustrated in Appendix A. $A_a = (\widehat{A} + \widehat{A}_a)$, where

$$\widehat{A}_a = \begin{bmatrix} N_p & 0 \\ 0 & N_q \end{bmatrix}, \quad N_p = \begin{bmatrix} 0 & 0 & 0 \\ 0 & 2D_pL & 0 \\ 0 & L & 0 \end{bmatrix}, \quad N_q = \begin{bmatrix} 0 & 0 & 0 \\ 0 & 2D_qL & 0 \\ 0 & L & 0 \end{bmatrix}. \tag{10.39}$$

D_p and D_q are the diagonal matrices consisting of $D_{p,i}$ and $D_{q,i}$ for $\forall i \in \mathcal{N}$, respectively. The matrix L represents the communication protocol of ADMMs described by (10.25) or (10.26). Either can be substituted into the closed-loop system with corresponding distributed control mechanism. Since $2n$ zero-eigenvalues of A in (10.36) are eliminated by the ADMM-based control mechanism, A_a will not be singular.

Now, by defining $\widehat{z} := d\widehat{y}/dt$, (10.38) can be rewritten into the following compact representation:

$$\dot{\widehat{y}} = \widehat{z}, \quad M_a\dot{\widehat{z}} = -D_a\widehat{z} - \frac{\partial U(\widehat{y})}{\partial \widehat{y}} \tag{10.40}$$

where $M_a = A_a^{-1}\widehat{M}$ and $D_a = A_a^{-1}\widehat{D}$. Note that the major difference between system (10.40) and that described in [72] is that M_a is always singular since the closed-loop system (10.40) is a hybrid of second-order systems and first-order systems. Equation (10.40) is indeed differential-algebraic equations (DAE) and the analytical expressions of EF in [72] cannot be directly applied. The corresponding EF can be constructed by singular perturbation techniques [68,74]. The singular perturbation approach treats the algebraic equation as a limit of the fast dynamics and the singular perturbed system formulated as *boundary-layer equations* (BLE) [68,74]. Therefore, the analysis of the DAE system can then be achieved by analyzing the BLE system.

For our applications in (10.40), the state variable $\widehat{y}$ can be split into $y_u = [\theta_{DIC}]$ and y_d containing the rest of $\widehat{y}$. Then by introducing a diagonal matrix ε with sufficiently small positive entries, (10.40) can be modified as

$$\begin{bmatrix} \dot{y}_u \\ \dot{y}_d \end{bmatrix} = \begin{bmatrix} z_u \\ z_d \end{bmatrix}, \tag{10.41a}$$

$$\begin{bmatrix} M_{a,u} & 0 \\ 0 & \varepsilon \end{bmatrix}\begin{bmatrix} \dot{z}_u \\ \dot{z}_d \end{bmatrix} = -\begin{bmatrix} D_{a,u} & 0 \\ 0 & D_{a,d} \end{bmatrix}\begin{bmatrix} z_u \\ z_d \end{bmatrix} - \begin{bmatrix} \dfrac{\partial U(\widehat{y})}{\partial y_u} \\ \dfrac{\partial U(\widehat{y})}{\partial y_d} \end{bmatrix}. \tag{10.41b}$$

If we denote the matrix M_a after this modification as $M_{a,\varepsilon}$, it is easy to see that $M_{a,\varepsilon}$ is now positive definite. Also, since the conditions for the *Tikhonov theorem* are valid for the underlying structure-preserving MG model (10.40), the BLE system described by (10.41a) and (10.41b) indeed possesses the same SEP as the DAE system (10.40).

Under this situation, the corresponding EF of the closed-loop system can be constructed as follows:

$$W(\hat{y},\hat{z}) = \frac{1}{2}\hat{z}^T M_{a,\varepsilon}\hat{z} + U(\hat{y}) = KE(\hat{z}) + PE(\hat{y}). \tag{10.42}$$

If (10.42) is indeed the EF of (10.40), the stability of every trajectory in the closed-loop system (10.40) can be ensured. The ADMM-based droop-control mechanism of VSGs will then be an effective strategy for controlling isolated AC MGs. Theorem 1 summarizes the main results.

Theorem 1 (existence of an analytical EF for ADMM-based droop control of VSGs). *Consider the closed-loop system* (10.40) *under the ADMM-based droop control of VSGs.* $W(\hat{y},\hat{z})$ *defined in* (10.42) *is an analytical representation of the EF for system* (10.40).

Proof: See Appendix B.

Since the EF of the closed-loop system (10.40) can be derived analytically, the behavior of the trajectories is very simple in the sense that any trajectory whether converges to one of the EPs or goes to be unbounded [63,64]. In particular, if an EF exists for system (10.40), which has an asymptotically SEP, then the stability boundary is contained in the set, which is the union of stable manifolds of all unstable equilibrium points on the stability boundary [63,64]. In order to apply the EF approach for stability analysis of (10.40), empirically it is assumed that the SEP of the predisturbed system and that of the droop-control stage are sufficiently close to each other. Thus, the predisturbed system is operated in its SEP which is located inside the stability region of the droop-control system. Since the droop-control system (10.40) admits an EF and all of its EPs are isolated, this trajectory will eventually approach the desired SEP under the proper power sharing.

10.5.3 ADMM implementations with flexible droop gain ratios

With the modifications for flexible droop gains made in (10.27a) and (10.27b), the closed-loop system in this case can still be reformulated as the compact form similar to (10.38), with only A_a being replaced by $A_f = (\widehat{A}_f + \widehat{A}_a)$ and detailed $\widehat{A}_f$ being shown in Appendix A. It is obvious that the properties of $\widehat{A}$ can still be preserved after the modifications made in $\widehat{A}_f$, and the overall system can still be rewritten as the following compact representation:

$$\dot{\hat{y}} = \hat{z}, \quad M_f\dot{\hat{z}} = -D_f\hat{z} - \frac{\partial U(\hat{y})}{\partial \hat{y}} \tag{10.43}$$

where $M_f = A_f^{-1}\widehat{M}$ and $D_f = A_f^{-1}\widehat{D}$. With the same singular perturbation technique used in developing (10.41a) and (10.41b), the corresponding transient energy function (TEF) of the closed-loop system can be modified from (10.42) as follows:

$$W_f(\widehat{y},\widehat{z}) = \frac{1}{2}\widehat{z}^T M_{f,\varepsilon}\widehat{z} + U(\widehat{y}) = KE_f(\widehat{z}) + PE(\widehat{y}). \tag{10.44}$$

Since Theorem 1 can also be applied to $W_f(\widehat{y},\widehat{z})$, stability of the closed-loop system can be ensured and the flexible ADMM-based droop control is indeed an effective strategy for controlling isolated AC MGs, while providing extra damping for this class of VSGs.

10.6 Real-time simulations of ADMM-based consensus droop control

In order to validate the performance of ADMM-based $P-f/Q-\dot{V}$ droop control for VSGs, simulations of a 14-bus/6-VSG MG depicted in Figure 10.5 have been conducted by real-time simulators OPAL-RT [75]. Details of the MG line parameters can be found in [76]. The use of real-time simulators enables the inclusion of details such as PWM switching transient in simulations under real-time scale. The phase and the magnitude of the bus voltage for inner loop voltage controllers are determined through $P-f$ and $Q-\dot{V}$ droop controllers, respectively. Parameters of VSGs are listed as follows: (i) system voltage: three-phase 220 V, 60 Hz; (ii) VSGs: rated 2 kV A, two-level PWM converters, switching frequency $f_s = 10$ kHz, output filter inductor $L_f = 2$ mH, output filter capacitor $C_f = 4.7$ μF. Notice that the same rating among all VSGs is only set to ease the observation of accurate power sharing. The power sharing under diverse ratings can be easily ensured as long as the uniform $D_{p,i}/P_i^r$ and $D_{q,i}/Q_i^r$ being maintained.

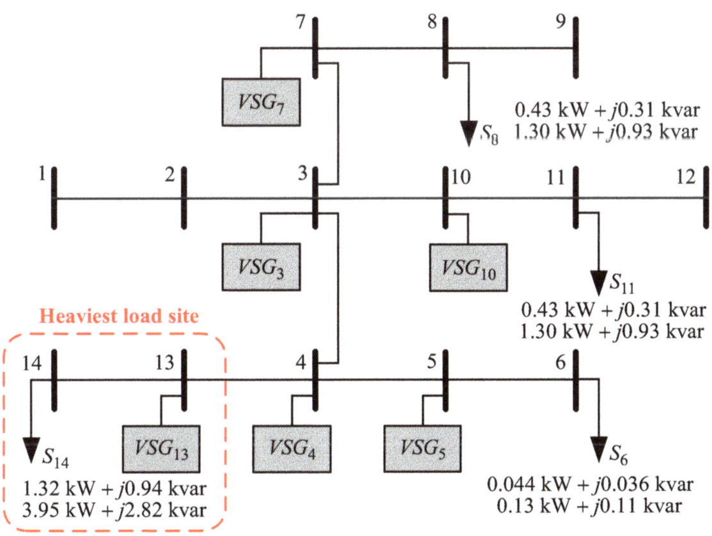

Figure 10.5 One-line diagram of the 14-bus/6-VSG microgrid

For simplicity, since the convergence rates of Methods A and B of ADMMs are similar, which will be illustrated in Part 3, only the results of Method B will be shown in other parts. The corresponding droop-control coefficients under base case are listed as (i) virtual inertia: $J_i = 1.061$ kg m^2; (ii) droop coefficients: $D_{p,i}^{-1} = 2 \times 10^{-5}$ rad/(s W) and $D_{q,i}^{-1} = 6 \times 10^{-3}$ V/(s var); (iii) restoration coefficients: $k_{p,i} = k_{q,i} = 1 \times 10^{-3}$s; and (iv) ADMM coefficients: $d_i = 0.9524$, for $i = 3, 4, 5, 7, 10, 13$. Worth noticing that the common droop coefficients are set among the VSGs in the base cases to account for the same rating of converters. Such restriction will be eliminated whenever the proposed flexible droop gain scheme being deployed. The proper coefficients can be determined by including all the fast transient dynamics of power electronics into the stability analysis framework [6], which will provide satisfactory convergence rate of droop operations, while ensuring the stability of VSG inner loop controllers.

1. *With and without ADMM implementations of consensus-based droop-control mechanism*: The droop-control mechanism of each VSG is studied with and without the ADMM-based consensus control. Two cases start from identical initial condition with light loads, and then both switched to heavily loading conditions. Simulation results of $P - f$ droop are shown in Figure 10.6, and those of $Q - \dot{V}$ are illustrated in Figure 10.7. Although the $P - f/Q - \dot{V}$ droop control can reach one SEP, the $\dot{\theta}/\dot{V}$ restoration mechanisms (10.5c) and (10.5e) only guarantee the achieving of objective (10.1). Consequently, each VSG varies its control variables p/q locally and results in the diverged active and reactive power outputs after load variations. This problem is then resolved by activating the ADMM-based consensus control mechanism. The secondary

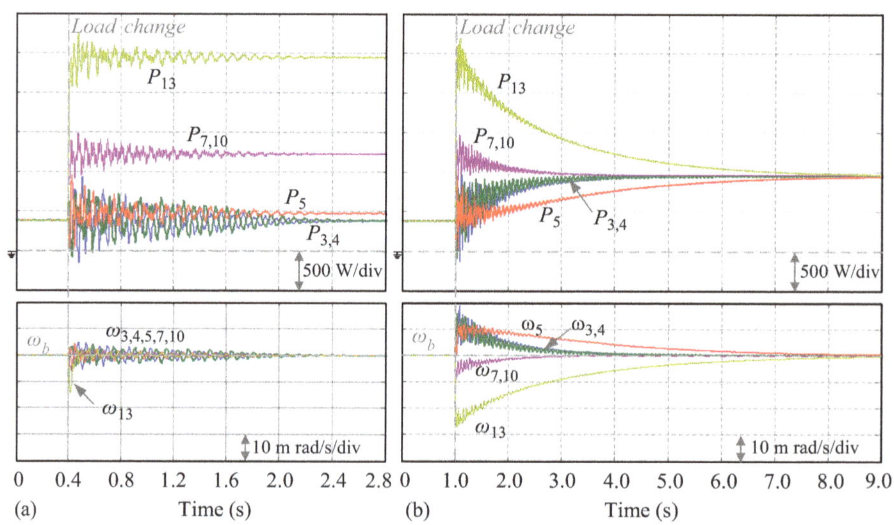

Figure 10.6 Droop control of VSGs: (a) $P - f$ droop and (b) ADMM-based $P - f$ droop

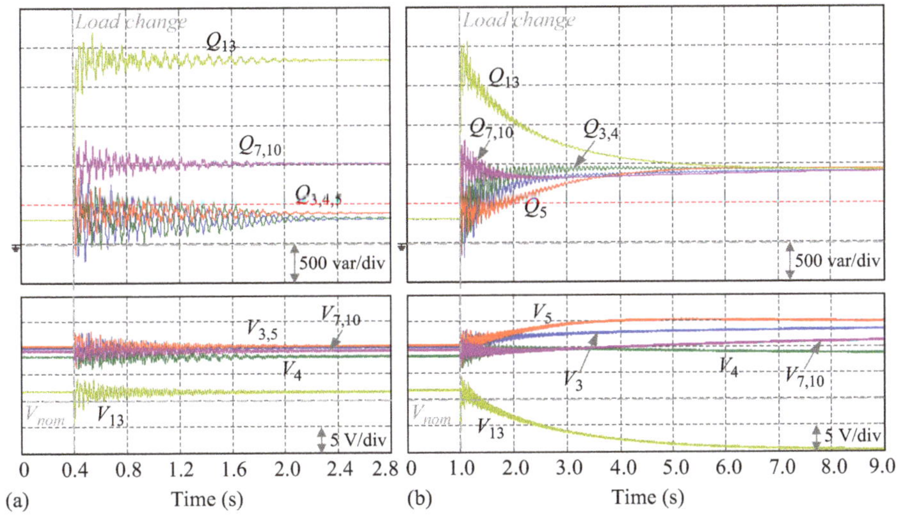

Figure 10.7 Droop control of VSGs: (a) $Q - \dot{V}$ droop and (b) ADMM-based $Q - \dot{V}$ droop

ADMM-based control soon dominates VSGs' outputs and achieves accurate power sharing only after a few seconds, which results in the convergence of P_i and Q_i shown in these figures.

2. *Frequency damping of virtual inertia*: The effects of virtual inertia are also investigated with the ADMM-based consensus control mechanism activated. Load variations from 95% of the total load to the full loading condition are studied. Since the major load of the MG is located at Bus_{14}, the response of the nearby VSG_{13} is investigated. Effects of varying the virtual inertia J are shown in Figure 10.8(a) and those of varying the virtual damping D_p are illustrated in Figure 10.8(b). As can be seen from the results, increasing D_p also contributes to the damping of frequency. However, one can also realize from (10.5b), larger D_p is also equal to smaller droop ratio, which eventually slows down the convergence of power sharing as shown in the zoomed scene. The larger virtual inertia on the other hand reduces dips and oscillations of frequency without degrading the convergence rate. This makes the virtual inertia more preferable for extra damping of DICs in isolated MGs.
3. *Performance evaluations of consensus-based droop control via ADMM implementations*: Convergence rates of two versions of ADMMs on proper active power and reactive power sharing are studied in this section. The consensus-based droop control for VSGs proposed in [29] is also developed for comparative studies. Again the active and reactive power outputs of VSG_{13} are specifically illustrated in Figure 10.9. The results indicate that the ADMM-based consensus is indeed better than simple first-order consensus in speaking of the convergence rate. For the ADMM algorithms, the speeds of Methods A

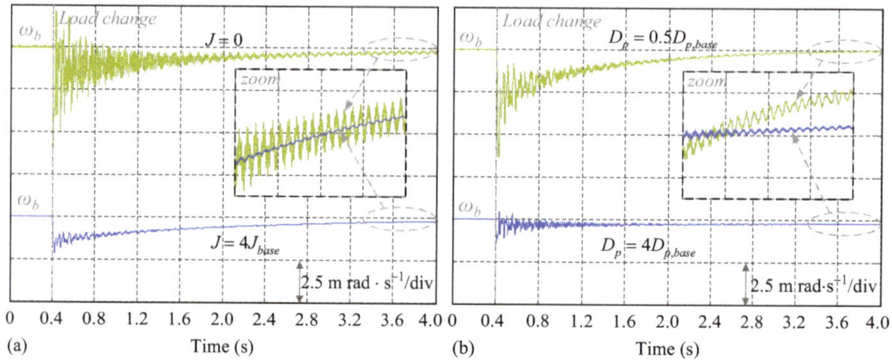

Figure 10.8 Frequency response of DIC_{13}: (a) different virtual inertia J, $J_{base} = 1.061$ kg m^2 and (b) different virtual damping D_p, $D_{p,base}^{-1} = 2 \times 10^{-5}$ rad/(s W)

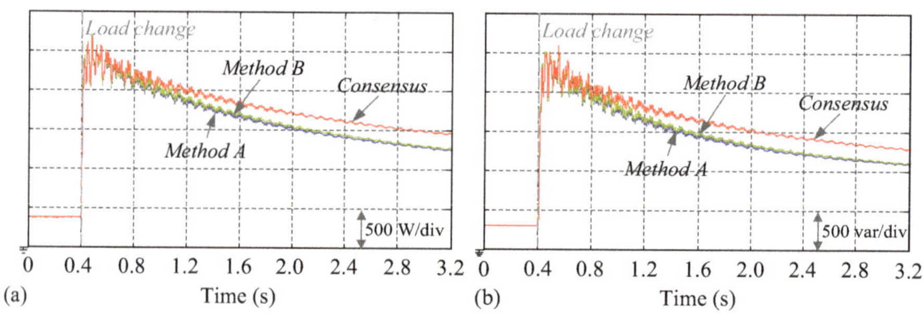

Figure 10.9 Performance comparisons of consensus-based and ADMM-based droop control methods: (a) convergence of P_{13} and (b) convergence of Q_{13}

and B are almost the same, which has also been investigated analytically in [50].

4. *ADMM implementations with flexible droop gain ratios*: The performance of flexible scheme for consensus control is illustrated by varying the droop gain of VSG_{13}. Three test scenarios are carried out as follows:
   (i) Case I: ADMM-based droop control with uniform droop gain ratio settings;
   (ii) Case II: ADMM-based droop control with nonuniform droop gain ratio settings: $D_{p,13}^{-1} = 4 \times 10^{-5}$ rad/(s W) and $D_{q,13}^{-1} = 12 \times 10^{-3}$ V/(s var);
   (iii) Case III: ADMM-based droop control with flexible droop gain ratio settings: $D_{p,13}^{-1} = 4 \times 10^{-5}$ rad/(s W) and $D_{q,13}^{-1} = 12 \times 10^{-3}$ V/(s var).

It is worth mentioning that Case I is developed to demonstrate the desired power sharing and nominal damping. For all of the three cases, the system loads are

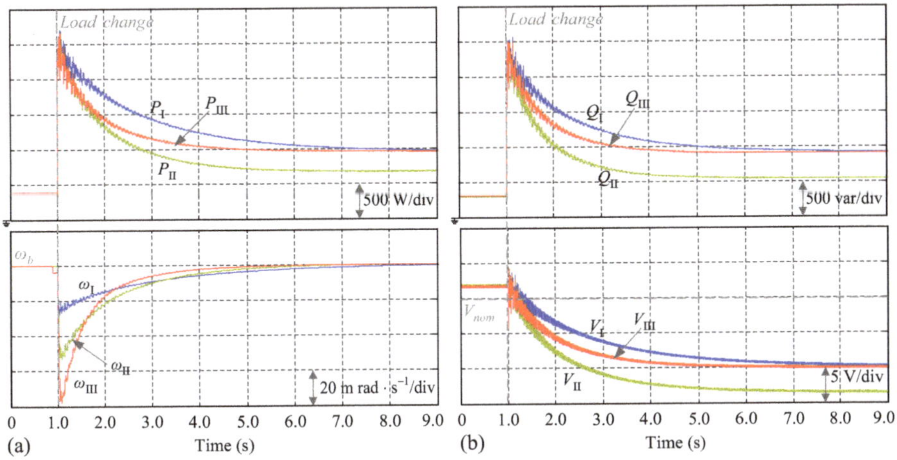

Figure 10.10 Flexible droop control of VSG_{13}: (a) $P - f$ droop and (b) $Q - \dot{V}$ droop

varied from S_{light} to S_{heavy} at $t = 1$ s, and the simulation results are shown in Figure 10.10. The outputs of VSG_{13} for each case are indicated by corresponding subscripts. When imperfectly droop setting is applied in Case II, though with faster converging rate, the original ADMM-based droop results in undesired power sharing under the steady state. This is then resolved by the proposed flexible ADMM-based droop control that provides much faster power output responses and achieves accurate power sharing as well.

10.7 Distributed pinning control

The complex dynamical behavior of a distributed MAS can be examined from graph theory analysis. In order to make this manuscript self-contained, notions of graph theory and pinning control will be presented first. Theoretical developments will also be explored in detail in this section.

10.7.1 Basic concepts

A *directed graph*, or called *digraph*, $\vec{G}$ is defined as an ordered pair $\vec{G} = (V(\vec{G}),\ E(\vec{G}))$. $V(\vec{G}) = \{v_1, v_2, \ldots, v_N\}$ is a set of vertices and $E(\vec{G}) = \{(v_i, v_j) \cdots (v_m, v_N)\}$ is a set of edges. The *adjacency matrix* of digraph $\vec{G}$, denoted as $\mathbf{A}$, is a matrix whose rows and columns are indexed in the order of $V(\vec{G})$. Entries of $\mathbf{A}$ are obtained from the sets of vertices and edges. More specifically, $\mathbf{A}[i][j] = (a_{i,j}) > 0$ if the pair of vertices $(v_i, v_j) \in E(\vec{G})$, otherwise $A[i][j] = 0$. The *in-degree* of vertex $v_i \in V(\vec{G})$, denoted as $deg^+(v_i)$, is the number of elements in $E(\vec{G})$ of the form (v_k, v_i) $(v_k \in V(\vec{G}), k \neq i)$. The *Laplacian matrix* $\mathbf{L} = (l_{i,j})_{N\times N}$ of digraph $\vec{G}$ is defined as $l_{i,j} = -a_{i,j}$ if $i \neq j$, otherwise

$l_{i,j} = deg^+(v_i)$. Digraph $\vec{G}$ contains a DST if there exists a root vertex $v_r \in V(\vec{G})$ with a directed path to every other vertex in $V(\vec{G})$ [77].

10.7.2 Distributed pinning control

Consider a MAS of N agents $S_a = \{S_1, \ldots, S_N\}$, each $S_i \in S_a$ can be interpreted as a vertex in a digraph. Each agent S_i has its own state L-dimensional state-variable vector $x_i(t) = [x_i^1(t) \cdots x_i^L(t)]^T$ with the following dynamic equation:

$$\frac{dx_i(t)}{dt} = f(t, x_i(t)) + c\sum_{j=1}^{N} a_{ij}(x_j(t) - x_i(t)) \tag{10.45}$$

where constant c represents a coupling strength and $a_{i,j}$ is the entry (i,j) of the network adjacency matrix $\mathbf{A}$ of diagraph $\vec{G}$, which represents the communication topology of the MAS. Now assume that our objective is to achieve a homogeneous stationary state $\bar{x}$ for all agents $S_i \in S_a$ as follows:

$$\bar{x} = x_1(t) = x_2(t) = \cdots = x_N(t). \tag{10.46}$$

In order to reduce the number of controllers required to move the system to the desired stationary state $\bar{x}$, one can apply the idea of pinning control [42,78,79] where a small fraction $S_a^{pin} \subset S_a$ of the total agents is selected to move the coupled MAS to the required equilibrium state in (10.46). This set of selected agents is also referred to as *pinning* or *leader* agents, to which control inputs are to be assigned. Thus, the dynamic equation of $S_k \in S_a^{pin}$ takes the following form:

$$\frac{dx_k(t)}{dt} = f(t, x_k(t)) + c\sum_{j=1}^{N} a_{kj}(x_j(t) - x_k(t)) \underbrace{-g_k(x_k(t) - \bar{x})}_{\mathbf{u_k}} \tag{10.47}$$

where the control input is chosen as $\mathbf{u_k} = -g_k(x_k(t) - \bar{x})$ and g_k is referred to as the *pinning gain* of S_k. The pinning gains can be arranged as $\mathbf{G} = diag(g_1, \ldots, g_N)$. Note that $g_i > 0$ if $S_i \in S_a^{pin}$, otherwise $g_i = 0$.

In order to ensure that all agents can reach a consensus state $\bar{x}$ in (10.46), several attempts about pinning control haven been developed. For example, if S_a^{pin} contains only one agent, Chen *et al.* [80] stated that, if $\vec{G}$ has a DST and the chosen pinning agent is located at its root vertex, the coupled MAS in (10.45) can be synchronized to the consensus state (10.46) asymptotically. However, if this condition is not fulfilled, more pinning agents need be appended into S_a^{pin}. Theoretically speaking, the MAS (10.45) can be synchronized to (10.46) if and only if digraph $\vec{G}$ is partitioned into strongly connected components and the pinning agents S_a^{pin} are located at the root of the DST of every created component [43]. Here, we apply the DST-based partition technique for appending additional agents in the pinning-based droop control to achieve the goal of accurate real power and reactive power sharing in the isolated AC MG.

10.8 Distributed pinning-based droop control

A control block diagram of reactive power droop under the $P-f$ and $Q-\dot{V}$ control framework can be illustrated in Figure 10.11(a). It can be observed that this reactive droop controller incorporates a voltage restoration mechanism aiming to regulate $\dot{V}_i$ to zero in steady state. Although the closed-loop system is stable, accurate reactive power sharing may not always be accomplished. The reason comes from the fact that the closed-loop system may possess nonisolated SEPs [25].

In order to prevent this drawback, the distributed secondary communication-based control strategies for MG will be considered. These distributed secondary controllers are normally designed through neighbor-to-neighbor communication by

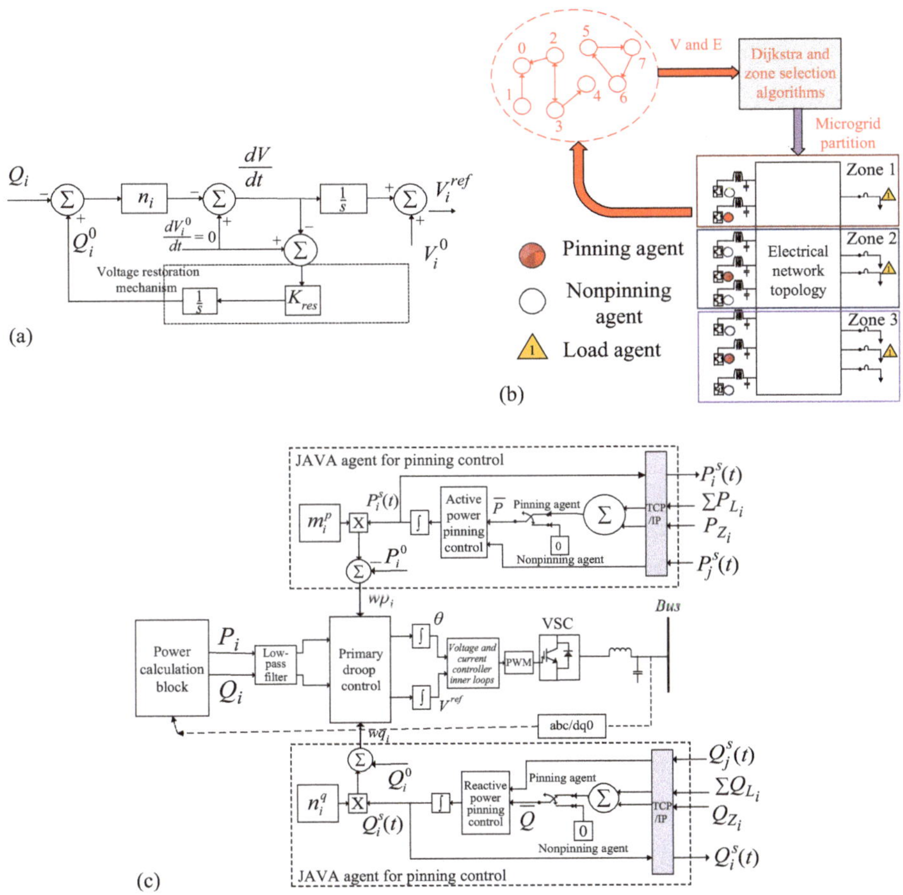

Figure 10.11 Proposed distributed framework: (a) block diagram of the primary reactive power droop control under the $P-f$ and $Q-\dot{V}$ framework; (b) example of an MG partition; and (c) block diagram of the proposed pinning-based distributed secondary controller

integrating various average consensus algorithms [25,28,30]. The performance of these techniques depends on the underlying communication topology. One possible solution is to apply the aforementioned pinning control [31,32] since, as previously stated, the stability of the closed-loop system can be achieved regardless of the given communication infrastructure. Moreover, by properly locating the pinning nodes, the overall communication bandwidth may be reduced compared to the so-called average consensus algorithm.

10.8.1 MAS development

Since the distributed secondary control algorithm requires communication among DICs and loads in the MG, it is necessary to implement software agents with communication and reaction capabilities. These software agents are to be attached to the DICs and the loads participating in the MG. To this end, two different types of agents are proposed:

- *DGA*: This agent is assigned into each DIC of the MG. The main purpose of this agent is to reach equilibrium state via distributed pinning control. As discussed earlier, the concept behind pinning control is to reach a homogeneous stationary state by selecting a small fraction of the agents S_a^{pin} of the total agents S_a in the MG. Therefore, the DGAs can be categorized into (i) pinning agents and (ii) nonpinning agents. The pinning agents are assigned to the DICs carrying the reference of the required stationary state.
- *LA*: This agent is attached to the loads in the MG and its main purpose is to monitor and broadcast active/reactive power requirements to the selected set of pinning DGA.

Conventionally, in a multiagent environment, all agents are registered in a so-called directory facilitator agent (DFA) [81]. Roughly speaking, the DFA contains a database with unique agent identification numbers and the Internet Protocol address of each node. Additionally, agents can register (or deregister) their services to the DFA. Likewise, agents in our framework are registered into the DFA, and the service they provide is the manufactured power rating.

10.8.2 Pinning localization and grid partition

In order to develop an effective pinning scheme for secondary control of MGs, an appropriate localization of the pinning agents must be considered. The reasons for adopting the grid partition and selection of pinning agents are 3-fold:

1. to guarantee distributed control regardless of the given communication topology [43],
2. to improve pinning control performance compared to that of a single controller [56], and
3. to reduce the communication bandwidth compared to the conventional average consensus mechanisms.

The adopted pinning-based control can be achieved by the following two phases. In the first phase, the controllability of the MG using a single pinning agent is analyzed by verifying the existence of a DST. This is conducted by executing Dijkstra's algorithm [82]. Dijkstra's algorithm is used to compute a distance matrix $\mathbf{M} = (m_{i,j})_{N\times N}$ with the minimum communication path lengths $m_{i,j}$ between two distinct vertices v_i and v_j $(\forall v_i, v_j \in V)$. If $\vec{G}$ does not have a DST, additional pinning agents should be considered. Then, the entire MG is partitioned into several nonoverlapping zones $\{Z_k\}_{k=1}^m$, such that, for every Z_k, it is possible to select a pinning node with directed paths to the remaining nodes in Z_k. Since Z_k is either a connected subgraph or a connected component of $\vec{G}$, the localization of the pinning nodes can be determined by repeatedly executing Dijkstra's algorithm into Z_k. This allows us to find the minimum DST and its root vertex in every zone. In the second phase, pinning agents are appended into the root vertices so that they have a directed path to all the followers in Z_k.

For illustration purposes, consider the digraph depicted in Figure 10.11(b). Executing Dijkstra's algorithm for every pair of vertices of this network, the distance matrix **M** can be obtained. By inspecting matrix M, we have $Z_1 = (v_0, v_1)$, $Z_2 = (v_2, v_3, v_4)$, and $Z_3 = (v_5, v_6, v_7)$. Note that vertex v_1 could have been part of either Z_1 or Z_2, however, since Z_1 has a less number of agents compared to Z_2, vertex v_1 was appended into Z_1. Now, the pinning agents are selected by obtaining the DST for every created Z_i. For instance, the distance matrix using Dijkstra's algorithm into Z_2 can be obtained as follows:

$$\mathbf{M}_{\mathbf{Z_2}} = \begin{bmatrix} 0 & 1 & 2 \\ 1 & 0 & 1 \\ +\infty & +\infty & 0 \end{bmatrix}.$$

Since $\mathbf{M}_{\mathbf{Z_2}}$ has bounded row-sum for the first and second rows (corresponding to v_2 and v_3), it can be concluded that all agents in zone 2 can be reached through both v_2 and v_3. Therefore, v_2 and v_3 are pinning agent candidates in Z_2. In this particular case, v_3 is selected as a pinning node because $\min(M_{Z_2}[2][:]) < \min(M_{Z_2}[1][:])$, where $\min(M_{Z_2}[i][:])$ represents the minimum element of all the entries in row i. The physical meaning of this is that v_3 is closer to the remaining nonpinning agents compared to v_2. An identical procedure is followed to determine the location of the pinning agents for Z_1 and Z_3. More detailed descriptions of this pinning selection algorithm are shown in Algorithm 1, which can be explained in the following steps S:

- **S1** (Line 1): The selected pinning agents are to be stored in a **pinAgents** list.
- **S2** (Line 2): The function *Localize*() returns a list of two items: (i) **root** and (ii) **Zones**. These two outputs are explained as follows:
  - **root**: If the input digraph contains a DST, the function *Localize*() returns its root node. In this case, the MG's partition is not required and the two-dimensional list **Zones** is left empty.
  - **Zones**: If the input digraph is disconnected or does not contain a DST, function *Localize*() returns an empty variable **root**. On the other hand, the two-dimensional list **Zones** will be composed of sets of vertices grouped into their corresponding zone number. Each $\mathbf{Z}_k \in$**Zones** must have at least one node with directed paths to the remaining nodes in $\mathbf{Z}_k$.
- **S3** (Lines 3–10): If **Zones** is empty, a unique pinning node is required and it must be located at **root**. Thus, **pinAgents** = [**root**]. Otherwise, the function *Localize*() must be used for each $\mathbf{Z}_k$ separately, and the localization of the

pinning nodes is given by **pinAgents** = [**root**$_1$,... **root**$_k$]. It should be kept in mind that each $\mathbf{Z}_k$ is either a connected subgraph or a connected component of the original communication graph.

The detailed algorithm for function *Localize* used in Algorithm 1 is shown in Algorithm 2, and each required step **S** can be described as follows:

- **S1** (Line 1–3): Definition of function *Localize*() and its local variables.
- **S2** (Line 4–8) For each node v_i, Dijkstra's algorithm is executed between v_i and all the remaining nodes v_j. This is done by the function *DijkstraAlgorithmDistance*(), which returns the list d_i = [distance(v_i, v_1), . . . , distance(v_i, v_n)]. The maximum distance between node v_i and any other arbitrary node v_j (max(d_i)) is obtained and temporarily stored in the list **f**. The one-dimensional list d_i is also appended as an element of a two-dimensional list **M** (the so-called distance matrix).
- **S3** (Line 9–11): Since the *i*th element of **f** contains the distance between v_i and the furthest node from v_i, then, this list can be used to verify the existence of a DST. If the minimum element of **f** is bounded (min(**f**)<+∞), then we can conclude the existence of a DST and the function must return the root node as the index of the minimum element in the list **f** (line 10). In this case, **Zones** will be left empty, as shown in Line 11 (**Zones**=*null*).
- **S4** (Line 13–23) If the minimum element in **f** is not bounded (min(**f**)=∞), then the MG must be partitioned into zones. For every node v_i we must identify and group all nodes v_j for which there is a bounded element $m_{i,j}$ of matrix **M**. This group of nodes can create a "possible zone" named $zone_k$. This possible zone is appended into a bigger two-dimensional list named **Zones**.
- **S5** (Line 24): After all nodes have been analyzed in **S4** and all possible zones have been obtained and appended into **Zones** (Line 23), the function *superSets()* is used to obtain all the supersets in **Zones**. The main purpose of function *superSets()* is to eliminate the possibility that any zone $Z_i \in$ **Zones** is contained into another zone $Z_k \in$ **Zones**.

Algorithm 1 MG partition and pinning localization algorithm.

```
pinAgents=[]
[rootm, Zones]=Localize(V(G⃗), E(G⃗))
if Zones.isNull( ) then
  pinAgent.append(root)
else
  for Z_k ∈Zones do
    root_k=Localize(V(Z⃗_k), E(Z⃗_k))[0]
    pinAgent.append(root_k)
  end for
end if
```

Algorithm 2 Localize function: Used in Algorithm 1.

```
function Localize(V, E)
    A = adjacencyMatrix(V, E), M = [][], Zones=[][]
    f = [], d_i = [], zone_k=[]
    for v_i ∈ V do
        d_i = dijkstraAlgorithmDistances(v_i, A)
        f.append(max(d_i))
        M.append(d_i)
    end for
    if min(f).isBounded() then
        root=argmin(f)
        return[root, null]
    else
        for v_i ∈ V do
            for v_j ∈ V do
            if m_{i,j}.isBounded() then
            zone_k.append(v_j)
            end if
        end for
        if ¬ (zone_k ∈ Zones) then
            Zones.append(zone_k)
        end if
        zone_k=null
    end for
    Zones=superSets(Zones)
    return [null, Zones]
end if
end function
```

10.8.3 Pinning-based droop control

The proposed MAS operates in the secondary control level and adopts the $P-f$ and $Q-\dot{V}$ droop control as a primary controller for all the DICs inside the MG. That being said, when the pinning control framework is activated, two additional terms, wp_i and wq_i, are appended into the primary controllers of equations as follows:

$$\omega_i(t) - \omega^0 = m_i(P_i^0 - P_i) + \underbrace{m_i(m_i^p P_i^s(t) - P_i^0)}_{wp_i} = m_i(m_i^p P_i^s(t) - P_i(t)), \tag{10.48a}$$

$$\frac{dV_i^0}{dt} - \frac{dV_i}{dt}(t) = n_i(Q_i^0 - Q_i) + \underbrace{n_i(n_i^q Q_i^s(t) - Q_i^0)}_{wq_i} = n_i(n_i^q Q_i^s(t) - Q_i) \tag{10.48b}$$

where

$$m_i^p = \frac{P_i^*}{\sum_{i\in C} P_i^*}, \quad \text{and} \quad n_i^q = \frac{Q_i^*}{\sum_{i\in C} Q_i^*} \tag{10.49}$$

and

$$\frac{dP_i^s(t)}{dt} = \sum_{j\in\aleph_i} a_{ij}(P_j^s(t) - P_i^s(t)) - g_i(P_i^s(t) - \bar{P}), \tag{10.50}$$

$$\frac{dQ_i^s(t)}{dt} = \sum_{j\in\aleph_i} a_{ij}(Q_j^s(t) - Q_i^s(t)) - g_i(Q_i^s(t) - \bar{Q}). \tag{10.51}$$

As can be observed in (10.49), m_i^p and $n_i^q \in [0, 1]$ represent the normalized active and reactive power capacity of each DIC, respectively. The denominator and numerator of these variables are preassigned to every agent in the MAS. The total generation capacity of the MG shown in the denominator of (10.49) is assumed to remain constant except under outage scenarios requiring the plug-in or plug-out state of related DGs. Under this situation, the total active and reactive power generation capacity of the MG is updated using the DFA. For instance, when a DG is plugged out from the system, it is immediately deregistered from the DFA, next, the power rating of the failed DG is communicated to the remaining DGs through the DFA, aiming to update the total capacity of the MG. It is worthwhile to mention that the DFA has no role in the pinning-based distributed droop-control algorithm, but only to update total generation capacity through agent communication with the DGAs. The set of DICs in the MG is denoted as C. State variables $P_i^s(t)$ and $Q_i^s(t)$ in (10.48a) and (10.48b) are referred to as *synchronization functions*. These two variables are derived in a completely decentralized manner from the proposed pinning control shown in (10.50) and (10.51). Note that (10.50) and (10.51) are equivalent to the MAS dynamics in (10.45) and the dynamics of the selected agents in (10.47), where $x_i(t) = [P_i^s(t), Q_i^s(t)]^T$, $\bar{x} = [\bar{P}, \bar{Q}]^T$, α_{ij} is a coupling strength defined by $\alpha_{ij} = c \times a_{i,j}$ if $i \neq j$, else $\alpha_{ij} = -\sum_{j\in\aleph_i} c \times a_{i,j}$. The variable g_i is the called *pinning gain* as previously indicated. Constants $\bar{P}$ and $\bar{Q}$ are the postdisturbed desired consensus values. That is, these values are defined as the sum of active and reactive power of all DICs in the MG. These prescribed consensuses are determined through TCP/IP communication when the MG reaches a postdisturbed SEP after a disturbance. More specifically, once a postdisturbed SEP is attained, the pinning agent located at zone k calculates the desired consensus state as follows:

$$\bar{P} = \sum_{i=1}^{N_L^k} P_{L_i} + \sum_{i=1, i\neq k}^{N_Z} P_{Z_i}, \tag{10.52a}$$

$$\bar{Q} = \sum_{i=1}^{N_L^k} Q_{L_i} + \sum_{i=1, i\neq k}^{N_Z} Q_{Z_i} \tag{10.52b}$$

where N_L^k is the number of loads in zone k. P_{L_i} and Q_{L_i} are the active and reactive power of load i in zone k, respectively. The pinning agent obtains this information through communication with the LAs located in zone k. P_{Z_i} and Q_{Z_i} are, respectively, the active and reactive power demand in zone i. The pinning agent in zone k obtains this information through communication with other LAs located in zone i $(i \neq k)$. To make this possible, all LAs k must broadcast $P_{Z_k} = \sum_{i=1}^{N_L^k} P_{L_i}$ and $Q_{Z_k} = \sum_{i=1}^{N_L^k} Q_{L_i}$ to the selected pinning agents in the network.

A control block diagram of the proposed pinning controllers is illustrated in Figure 10.11(c) where the primary controllers are the $P-f$ and $Q-\dot{V}$ shown in Figure 10.11(a). Equations (10.48a) and (10.48b) are graphically illustrated in Figure 10.11(c).

10.8.4 Stability criteria

Let $E_{Pi} = P_i^s(t) - \bar{P}$ and $E_{Qi} = Q_i^s(t) - \bar{Q}$ denote the error between the synchronization functions and the desired consensus state. Clearly, (10.50) and (10.51) can also be expressed as

$$\frac{dE_{Pi}}{dt} = \sum_{j\in\aleph_i} a_{ij}(E_{Pj} - E_{Pi}) - g_i(E_{Pi}), \tag{10.53}$$

$$\frac{dE_{Qi}}{dt} = \sum_{j\in\aleph_i} a_{ij}(E_{Qj} - E_{Qi}) - g_i(E_{Qi}). \tag{10.54}$$

The error can also be represented in the vector form as $\xi_i = [E_{Pi}, E_{Qi}]^T$ and reformulate (10.53) and (10.54) as

$$\frac{d\xi_i}{dt} = \sum_{j\in\aleph_i} a_{ij}(\xi_j - \xi_i) - g_i\xi_i \tag{10.55}$$

By denoting the total error in vector form as $\xi = [\xi_1, \xi_2, \ldots, \xi_n]^T$, the stability properties of the pinning control system are obtained by using the following Lyapunov candidate:

$$V = \frac{1}{2}\sum_{i=1}^{n} \xi_i^T \xi_i. \tag{10.56}$$

It then follows that

$$\begin{aligned}
\frac{dV}{dt} &= \sum_{i=1}^{n} \xi_i^T \frac{d\xi_i}{dt} \\
&= \sum_{i=1}^{n} \xi_i^T \left\{ \sum_{j\in\aleph_i} a_{ij}(\xi_j - \xi_i) - g_i\xi_i \right\} \\
&\leq \sum_{i=1}^{n} \xi_i^T \sum_{j\in\aleph_i} a_{ij}(\xi_j - \xi_i) - \sum_{i=1}^{n} g_i\|\xi_i\|^2 \qquad (10.57a) \\
&\leq \sum_{i=1}^{n} \|\xi_i^T\| \sum_{j\in\aleph_i} a_{ij}\|\xi_j - \xi_i\| - \sum_{i=1}^{n} g_i\|\xi_i\|^2 \\
&\leq \sum_{i=1}^{n} \|\xi_i^T\| \sum_{j\in\aleph_i} a_{ij}(\|\xi_j\| + \|\xi_i\|) - \sum_{i=1}^{n} g_i\|\xi_i\|^2
\end{aligned}$$

$$= |\xi|^T(\alpha - \mathbf{G})|\xi| \tag{10.57b}$$

where $\mathbf{G} = diag(g_1, g_2, \ldots, g_n)$ is the pinning matrix introduced in Section 10.7.2 and $\alpha = \{a_{i,j}\}_{N\times N}$ is the communication coupling matrix previously defined.

Based on (10.57b), one can guarantee the asymptotic stability of the closed-loop system by selecting $\alpha - \mathbf{G}$ as negative definite. This can be achieved by selecting a proper pinning matrix $\mathbf{G}$ such that $\alpha - \mathbf{G} < 0$. Since $\alpha - \mathbf{G}$ becomes a Hurwitz matrix, the maximum real part of all its eigenvalues, denoted as $\vartheta(\alpha - G)$, is less than zero. Consequently $lim_{t\to\infty}|\xi_i| = 0$, $lim_{t\to\infty}P_i^s(t) = \bar{P}$, and $\lim_{t\to\infty}Q_i^s(t) = \bar{Q}$ asymptotically. Now, with a careful examination of (10.48a) and (10.48b), it can be observed that both right-sides of (10.48a) and (10.48b) will approach zero since $\lim_{t\to\infty}m_i^pP_i^s(t) = m_i^p\bar{P}$, which is identical to P_i. Meanwhile, $\lim_{t\to\infty}n_i^qQ_i^s(t) = n_i^q\bar{Q}$ that is identical to Q_i. The above two statements confirm that both frequency and the voltage magnitude will restore to the desired post-disturbed SEP asymptotically. This analysis confirms that the proper power sharing can be ensured by using the proposed pinning control.

10.9 Simulation verifications of the pinning-based droop control

10.9.1 Test-bed descriptions and partition

The selected test-bed system is located in Taiwan. At initial stage, the proposed distributed pinning droop control is only verified through numerical simulations of the test-bed MG. Real-time hardware implementations of the system will be discussed in the near future. The one-line diagram of the system consists of ten DGs and it is depicted in Figure 10.12(a). The communication digraph of the test-bed is given as $V = \{DG1, DG2, DG3, DG4, DG5, DG6, DG7, DG8, DG9, DG10\}$ and $E = \{(DG2, DG1), (DG1, DG3), (DG2, DG3), (DG4, DG6), (DG6, DG7), (DG7, DG5), (DG5, DG4), (DG8, DG9), (DG9, DG10), (DG10, DG8)\}$. Both the set of nodes V and edges E are used in the MG partition and selection of pinning agents described in Algorithm 1. After executing the algorithm, the MG was partitioned into three nonoverlapping zones, in this case, the obtained zones are the connected components of the given communication digraph. This can be clearly observed from the communication topology in Figure 10.12(b). The Laplacian matrix $\mathbf{L}$ of the digraph in Figure 10.12(b) is given by

$$\mathbf{L} = \begin{bmatrix} 1 & 0 & 0 & -1 & 0 & 0 & 0 & 0 & 0 & 0 \\ -1 & 1 & 0 & 0 & 0 & 0 & 0 & 0 & 0 & 0 \\ 0 & -1 & 1 & 0 & 0 & 0 & 0 & 0 & 0 & 0 \\ 0 & 0 & -1 & 1 & 0 & 0 & 0 & 0 & 0 & 0 \\ 0 & 0 & 0 & 0 & 1 & -1 & 0 & 0 & 0 & 0 \\ 0 & 0 & 0 & 0 & 0 & 1 & -1 & 0 & 0 & 0 \\ 0 & 0 & 0 & 0 & -1 & 0 & 1 & 0 & 0 & 0 \\ 0 & 0 & 0 & 0 & 0 & 0 & 0 & 1 & -1 & 0 \\ 0 & 0 & 0 & 0 & 0 & 0 & 0 & 0 & 1 & -1 \\ 0 & 0 & 0 & 0 & 0 & 0 & 0 & -1 & 0 & 1 \end{bmatrix}.$$

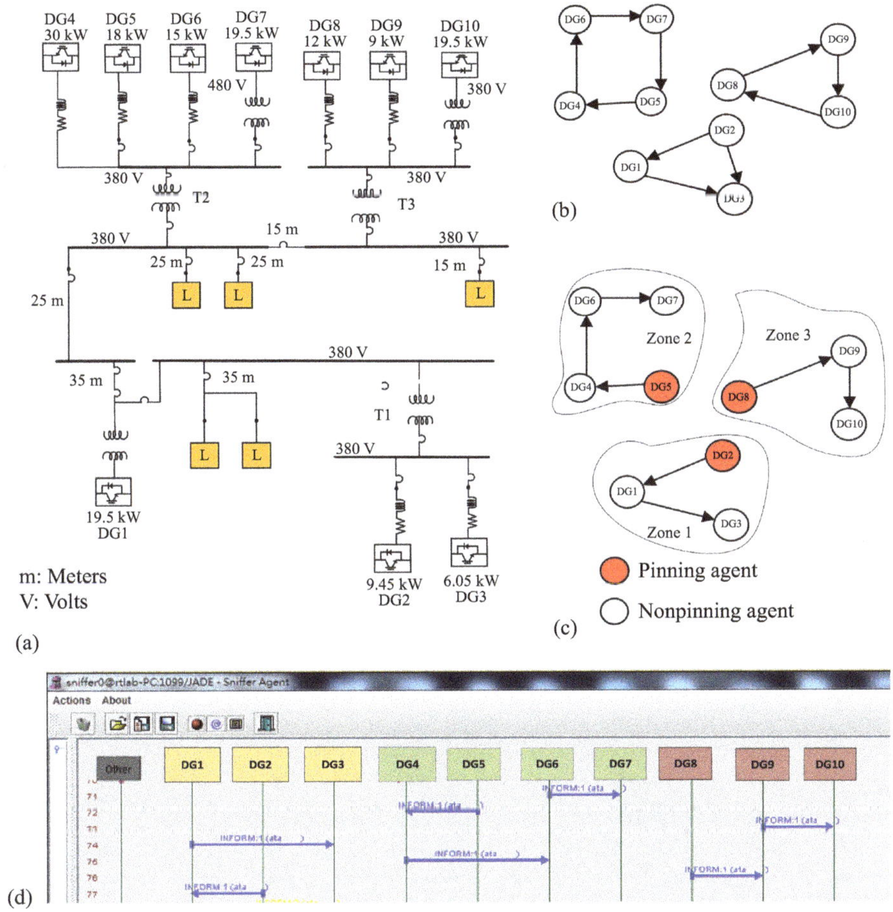

Figure 10.12 System implementation: (a) one-line diagram of the test-bed system; (b) MG partition of the communication topology; (c) appropriate location of pinning-agents to reduce the overall bandwidth; and (d) tracked messages among parallel DICs using the SA tool in JADE

Since matrix **L** has a zero-eigenvalue with multiplicity $m = 3$, it can be concluded that the digraph does not contain a DST and the MG cannot be controlled using a single pinning agent [80]. Furthermore, since the collective dynamics of the conventional average consensus algorithm can be expressed in general form as $\dot{\mathbf{x}} = -\mathbf{L}\mathbf{x}$ [16,25], the multiplicity of the zero eigenvalue will jeopardize the convergence of the average consensus technique. If we aimed to accomplish the secondary control using a single pinning agent, it would be necessary to create more communication links such that the graph has the desired DST.

Subsequently, the communication bandwidth can be reduced by obtaining the DST in each created zone Z_i, appending a pinning agent in its root vertex and

Table 10.1 Eigenvalue analysis for different number of pinning agents

Selected pinning agents	$\vartheta(\alpha - G)$
DG2, DG5, DG8	−1.674
DG2, DG5, DG4, DG8	−2.269
DG2, DG5, DG7, DG8, DG9	−2.269
DG1, DG2, DG5, DG8, DG9, DG10	−1.674

removing communication links that are not part of the DST, as shown in Figure 10.12(c). It is important to point out that selecting more than one pinning agent per zone would not negatively affect the overall control performance, since the maximum real part of the eigenvalues in (10.57b) would not increase [56, Proposition 3]. To clarify, Table 10.1 shows the maximum real part of the eigenvalues of (10.57b), represented as $\vartheta(\alpha - G)$, for different selection of pinning agents into every zone. To obtain these eigenvalues, we have used the communication network in Figure 10.12(b) with a coupling strength $c = 10$ and a pinning gain $g = 0.9 \times c$. As seen in Table 10.1, although selecting more pinning agents may in some cases improve the performance by moving $\vartheta(\alpha - G)$ further to the left half plane, this would involve a trade-off between control cost and system performance. To simplify our control objective, the aim of this work is limited to minimizing the number of pinning agents, that is, one pinning agent per zone.

10.9.1.1 System development and testing

The MAS has been developed under the JADE. This framework is a Java agent-oriented programing paradigm distributed by *Telecom Italia* [59]. The standard Agent Communication Language required by the Foundation of Physical Agents can be straightforwardly established using available Java libraries. By default, JADE incorporates the following two agents:

- Sniffer agent (SA): This agent has the capability to track the communication messages exchanged in the MAS. The communication among several parallel DICs can be observed in Figure 10.12(d).
- DFA: This provides a "Yellow Pages" service [81]. As stated in Section 10.8.1, all agents in our framework are registered into the DFA.

For more information about the JADE platform, we refer the reader to [59]. Extensive simulations of the MG in Figure 10.12(a) have been conducted under the SimPowerSystem toolbox of MATLAB/Simulink. The TCP-IP communication between JADE and MATLAB is established by the open source Java libraries developed in [83].

10.9.1.2 Simulation results

The correctness of the proposed pinning-based droop control has been assessed by simulation under both (i) normal MG operations and (ii) load and outage disturbances.

Pinning control convergence

After executing the distributed pinning control of (10.50) and (10.51) assigning aleatory initial conditions to the state variables, numerical experiments confirm that both distributed controllers reach convergence after approximately 114 iterations within the JADE platform, as can be observed in Figure 10.13.

Normal MG operation

At first, the MG operates using the primary droop controllers. The power allocation, the system frequency, and the terminal voltages can be observed in Figure 10.14. At $t \approx 0.66$ s, the distributed secondary control is activated in order to move the system to the desired operating condition. As observed in Figure 10.14(a), the activation of the pinning control has little effect on the active power output of each DIC. On the contrary, Figure 10.14(b) manifests that most DICs experience a change in reactive power output aiming to deliver reactive power according to their ratings. The MAS activation has a positive impact on the steady-state frequency of the whole MG and this can be observed in Figure 10.14(c). When the MG operates with the droop control method, the system reaches a steady-state frequency below the nominal set points. Upon the activation of the pinning control, a new steady-state frequency of about 60 Hz is attained. The voltages at the terminals of each DIC are shown in Figure 10.14(d). It is worth pointing out that the transition between the conventional control and the secondary control seems to be smooth to the connected loads.

Heavy loading condition

Initially, as marked using a dotted vertical line in Figure 10.15(a) and (b), the pinning control algorithm is deactivated at $t \approx 0.8$ s and the MG operates with the primary controllers. At $t \approx 0.9$ s, a large load disturbance is added to the system and a new steady-state condition is quickly attained using the droop control. Nevertheless, the new steady-state condition is characterized by erroneous reactive power dispatch as pointed out using a red circle in Figure 10.15(b). In order to counteract these errors, the reaction mechanism of the LA reactivates the pinning control at $t \approx 1.3$ s by informing the pinning agents about the load variations,

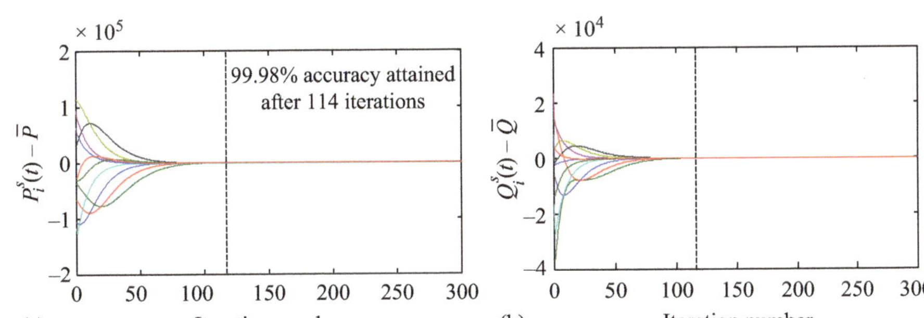

Figure 10.13 Distributed pinning control convergence: (a) active power consensus and (b) reactive power consensus

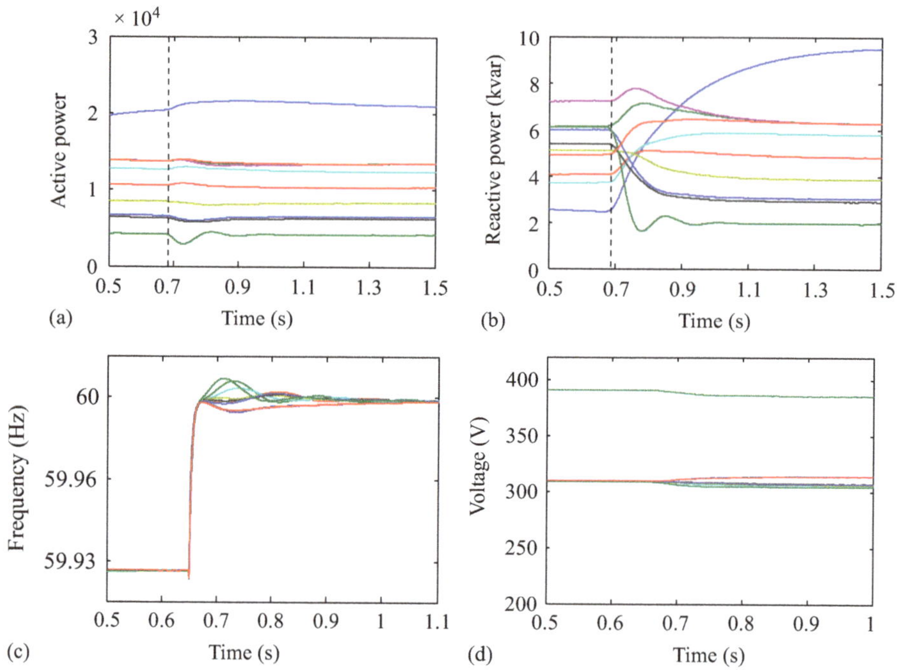

Figure 10.14 Microgrid under normal operations: (a) active power allocation in the microgrid; (b) reactive power allocation in the microgrid; (c) output frequency of the DICs; and (d) phase-to-ground voltages

aiming to update the postdisturbed SEP ($\bar{P}$ and $\bar{Q}$) according to (10.52a) and (10.52b). In this manner, proper reactive power dispatch is obtained, as marked using a green circle in Figure 10.15(b). It is important to note that the added disturbance has an adverse impact into the entire system frequency, as depicted in Figure 10.15(c). Fortunately, the pinning control in (10.48a) auspiciously restores the frequency to its nominal value as shown in Figure 10.15(c) for $t > 1.3$ s.

Plug-in and plug-out scenarios

Initially, the MG operates under normal conditions using the proposed control. At $t = 1$ s, the DG9, located in zone 3, is isolated from the system due to a temporary short circuit in its transmission line, as shown in Figure 10.16(a). The unavailable DGA in DG9 informs the DFA about its temporary unavailability. Since the DFA knows the internet address of the remaining DGAs in the MG, the DFA immediately notifies the remaining DGAs about the outage condition. This message contains the rated power of the unavailable DIC. In this manner, the remaining DGAs can update the terms in (10.49). Additionally, the out-of-service DGA informs its corresponding neighbor (DG10) about this temporary unavailability such that the neighbor can update the pinning equations in (10.50) and (10.51). Note that this procedure is equivalent to updating the network adjacency matrix. Aiming to properly achieve the secondary control, DG9 acts as a router in by interconnecting DG8 to DG10. Therefore, DG10

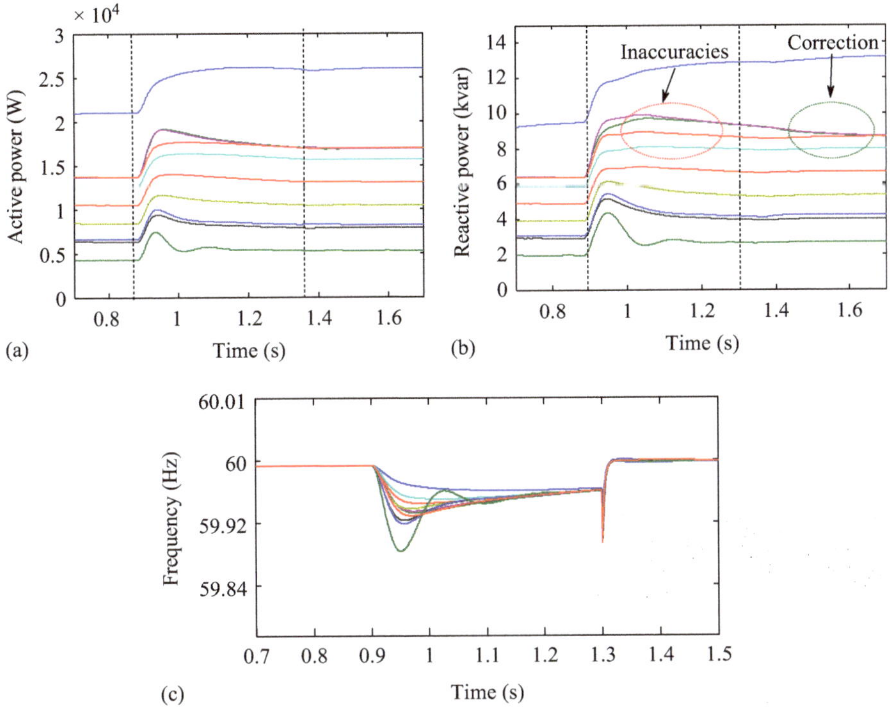

Figure 10.15 Heavy loading condition: (a) active power allocation; (b) reactive power allocation; and (c) measured frequency in rad/s

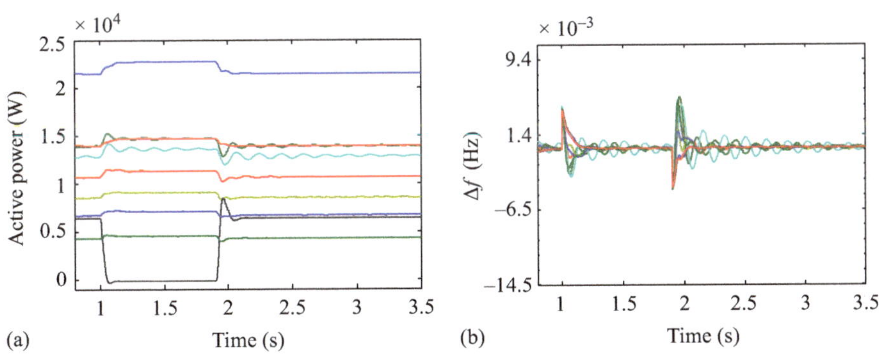

Figure 10.16 Plug-out and plug-in scenario: (a) active power allocation and (b) frequency restoration mechanism

appends DG8 as a new neighbor. In this manner, the distributed secondary control is guaranteed to be stable as shown in Figure 10.16(a) ($1\ \text{s} < t < 1.9\ \text{s}$). As depicted in Figure 10.16(a), the unavailable DGA is plugged-in at $t = 1.9\ \text{s}$ and the previously described updating procedure is repeated. With this in mind, the proposed pinning control can support the plug-and-play requirement and this can be observed from the frequency restoration mechanism depicted in Figure 10.16(b).

In this scenario, it is worthwhile noticing that the DFA is not part of the distributed secondary control technique, which is completely accomplished in a distributed manner through neighbor-to-neighbor communication. Thus, the DFA does not need to process any control or optimization functionality, and it is mainly in charge of updating the MG's capacity after outage scenarios. Reference [84] proposes a distributed tracking of the number of participating agents in the network; this however requires a double-way communication among all agents in the system. In this work, we have modeled the communication network as a directed graph, as a result, it is not possible to assume a two-way communication between every pair of neighbors.

Performance comparisons

The secondary pinning-based control has been compared with the consensus droop-control strategy in [25]. At first (see Figure 10.16), it is assumed that the system operates with the primary controllers. At $t \approx 0.7$ s, the secondary control takes place. Figure 10.16(a) shows the total reactive power error $\xi_{rp} = \sum_C |n_i^q \bar{Q} - Q_i|$ produced by all DICs in the MG under the communication network shown in Figure 10.12(c). As indicated in Section 10.9.1, the conventional consensus-based droop control fails to converge due to the repeated zero-eigenvalue of the Laplacian matrix obtained from the underlying communication topology, creating instability problems in the MG. In addition, it can also be observed that selecting one or two pinning agents leads to an incomplete synchronization since $lim_{t\rightarrow\infty}\xi_{rp} \neq 0$. This fact confirms that the appropriate secondary control in MG converges by selecting at least one pinning DGA per zone. If the communication topology in Figure 10.12(b) is modified in such a way that the digraph contains a DST (adding two edges DG2–DG5 and DG2–DG8), the consensus-based droop-control [6] algorithm and the proposed pinning scheme show a similar performance as depicted in Figure 10.17(b). Nevertheless, this requires more communication links among the DGAs.

10.9.2 Possible limitations

- Extra control efforts are appended into the pinning agents compared to the nonpinning agents. This requires additional processing into the pinning DGA.
- The pinning nodes must be secured since they are the selected leaders that carry the control reference in the MAS. A possible solution to counteract this could be selecting a backup pinning agent in the case of a serious malfunction of the selected pinning node. For instance, regarding Figure 10.12(c), DG4 can be selected as a backup pinning agent to DG5 since it has direct connections to the nonpinning agents in the zone.
- Under outage conditions, it is assumed that the communication between the DGA and other agents is still maintained. This assumption can be relaxed by separating the communication infrastructure from that of the MG. A prominent solution to this problem is the integration of wireless sensor networks [85] for agent communication.

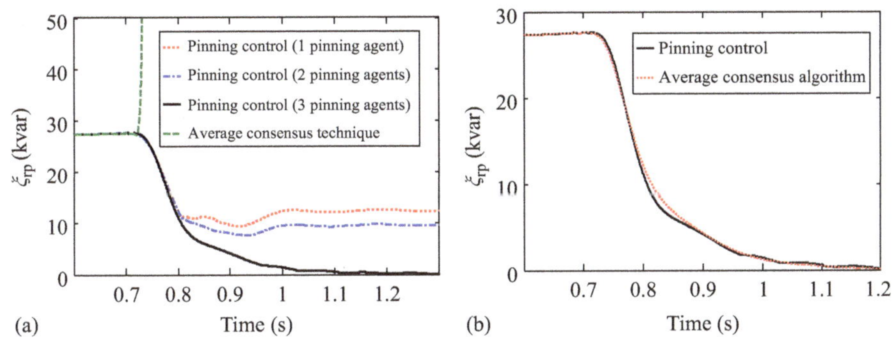

Figure 10.17 Comparisons: (a) pinning control with 1, 2, and 3 pinning agents and the traditional average consensus algorithm and (b) pinning control versus average consensus modifying the communication topology such that there exists a DST

10.10 Conclusions

This chapter provides a tutorial of consensus-based droop control for accurate active and reactive power sharing in isolated AC MGs. Two recently developed distributed droop-control mechanisms will be studied for improving the slow convergence rate of the conventional consensus-based droop-control method. First, ADMM implementations of consensus-based droop control for VSGs have been proposed to provide extra moment of inertia and damping under the transient period. Compared to the simple first-order consensus algorithms, ADMM implementations can indeed achieve faster convergence rate and resilience to noises. Two variants of ADMMs have been developed and applied to the $P-f/Q-\dot{V}$ droop-controlled VSGs. The EFs are constructed to ensure the stability of the closed-loop system. Since the KE is stored on the virtual rotating parts of VSGs, it is expected that more damping can be provided and the frequency deviation can be significantly reduced. Real-time simulations for a 14-bus/6-VSG MG are then developed on real-time simulator OPAL-RT. The droop control for VSGs with and without ADMM implementations of consensus-based control activated are examined. Performance of the ADMM implementation of consensus algorithm has also been compared to the first-order consensus algorithm, which shows faster convergence rates for both variants of the ADMMs.

In the second part, the distributed pinning droop control is investigated. Since only a fraction of the agents in the network have access to critical information, the required communication bandwidth and control cost can be significantly reduced without degrading the dynamical performance of the power sharing compared to consensus-based droop-control techniques. Theoretical studies of this pinning-based distributed droop control, including pinning localization, grid partition, and convergence criteria, have been explored comprehensively. Simulation results indicate that the MAS framework can indeed improve the autonomous operation of the MG by suppressing the errors that the conventional droop control is not able to counteract. Thus, this pinning droop-control architecture can serve as a real-time tool for energy management systems in modern MG control and operations.

Some extension work along this direction is possible. For example, we plan to investigate the feasibility of our proposed two new droop control with random time-varying communication typologies and delays in the communication network. In addition, the robustness of these two droop control under variations of network parameters is also an aspect worthy studying in the near future. Hopefully, this progress will be reported in the near future.

Appendix A Formulations of $\widehat{M}$, $\widehat{D}$, $\widehat{A}$, and $\widehat{A}_f$

In order to align with

$$\widehat{y}^T = [\theta^T, p^T, \lambda_p{}^T, v^T, q^T, \lambda_q{}^T] = [\theta_{DIC}^T, \theta_L^T, p^T, \lambda_p{}^T, v_{DIC}^T, v_L^T, q^T, \lambda_q{}^T]$$

where subscripts DIC and L represent DIC buses and load buses, matrices D and A should be rearranged as

$$\begin{aligned} \widehat{D} &= diag[D_p, D_{Lp}, p, \lambda_p, V_{DIC}D_q, V_L D_{Lq}, q, \lambda_q], \\ \widehat{M} &= diag[J\omega_b, 0],\ \widehat{A} = \begin{bmatrix} \widehat{A}_1 & 0 \\ 0 & \widehat{A}_2 \end{bmatrix}, \end{aligned} \tag{10.58}$$

$$\widehat{A}_1 = \begin{bmatrix} I & 0 & I & 0 \\ 0 & I & 0 & 0 \\ d' & 0 & d' & D_p \\ 0 & 0 & 0 & 0 \end{bmatrix}, \widehat{A}_2 = \begin{bmatrix} I & 0 & I & 0 \\ 0 & I & 0 & 0 \\ d' & 0 & d' & D_q \\ 0 & 0 & 0 & 0 \end{bmatrix} \tag{10.59}$$

where each matrix component consists of $N = diag[N_i]$, $d_i' = (1 - d_i)$, and I is the identity matrix. It is then clear that $\widehat{D}$ is positive definite, while $\widehat{M}$ and $\widehat{A}$ are only positive semidefinite. Now with the modifications made in (10.27a) and (10.27b), $\widehat{A}$ is modified as

$$\widehat{A}_f = \begin{bmatrix} \widehat{A}_{f,1} & 0 \\ 0 & \widehat{A}_{f,2} \end{bmatrix}, \tag{10.60}$$

$$\widehat{A}_{f,1} = \begin{bmatrix} I & 0 & \dfrac{P^r}{D_p} & 0 \\ 0 & I & 0 & 0 \\ d' & 0 & \dfrac{d'P^r}{D_p} & D_p \\ 0 & 0 & 0 & 0 \end{bmatrix}, \widehat{A}_{f,2} = \begin{bmatrix} I & 0 & \dfrac{Q^r}{D_q} & 0 \\ 0 & I & 0 & 0 \\ d' & 0 & \dfrac{d'Q^r}{D_q} & D_q \\ 0 & 0 & 0 & 0 \end{bmatrix} \tag{10.61}$$

which is also positive semidefinite as $\widehat{A}$.

Appendix B Validity of EF $W(\widehat{y}, \widehat{z})$

Considering the nonlinear system

$$\dot{x}(t) = f(x(t)) \tag{10.62}$$

where $x(t) \in R^n$. A state vector $\widehat{x}$ is called an EP of (10.62) if $f(\widehat{x}) = 0$. A function $E : R^n \to R$ is called an EF for the system (10.62) if the following three conditions are satisfied [63,64]:

(E1) Derivative of the EF $E(x)$ along any system trajectory $x(t)$ is nonpositive, i.e., $\dot{E}(x(t)) \le 0$.
(E2) If $x(t)$ is a nontrivial trajectory, i.e., $x(t)$ is not an EP, then, along the nontrivial trajectory $x(t)$, the set $\{t \in R : \dot{E}(x(t)) = 0\}$ has measure zero in R.
(E3) A trajectory $x(t)$ having a bounded value of $E(x(t))$ for $t \in R > 0$ implies that the trajectory $x(t)$ is also bounded.

Now we will check conditions (E1) to (E3) of the EF $W(\widehat{y}, \widehat{z})$.
(E1) Differentiating $W(\widehat{y}, \widehat{z})$ along the trajectory:

$$\dot{W}(\widehat{y}, \widehat{z}) = \frac{\partial W}{\partial \widehat{z}}\dot{\widehat{z}} + \frac{\partial W}{\partial \widehat{y}}\dot{\widehat{y}} = \widehat{z}^T M_{a,\varepsilon}\dot{\widehat{z}} + \frac{\partial U(\widehat{y})}{\partial \widehat{y}}\dot{\widehat{y}} = -\widehat{z}^T D_a \widehat{z} \le 0. \tag{10.63}$$

This inequality shows that condition (E1) of the EF is satisfied
(E2) Suppose that there is an interval $t \in [t_1, t_2]$ such that $\dot{W}(\widehat{y}, \widehat{z}) = 0$. Since D_a is positive definite, from (10.63) we have

$$\dot{W}(\widehat{y}, \widehat{z}) = 0 \Leftrightarrow \widehat{z}(t) = \dot{\widehat{y}}(t) = 0.$$

That is, $\dot{\theta} = \dot{p} = \dot{\lambda}_p = \dot{v} = \dot{q} = \dot{\lambda}_q = 0$. It then follows that the system is at an EP; hence condition (E2) of the EF also holds.

(E3) Since the closed-loop system (10.41a) and (10.41b) under the ADMM-based droop control of VSGs is a class of second-order vector differential equations described in [72], we can follow their proof with simple modifications. First, according to Theorem 3-3 in [72], if $W(\widehat{y}, \widehat{z})$ is bounded, then the trajectory $(\widehat{y}(t), \widehat{z}(t))$ of (10.41a) and (10.41b) converges to an EP. Also, by Theorem 3-1 in [72], every bounded trajectory $(\widehat{y}(t), \widehat{z}(t))$ of (10.41a) and (10.41b) converges to an EP. Therefore, it can be concluded that bounded $W(\widehat{y}, \widehat{z})$ implies that trajectory $(\widehat{y}(t), \widehat{z}(t))$ being bounded completes part (E3).

References

[1] R. H. Lasseter, J. H. Eto, B. Schenkman, *et al.*, “CERTS microgrid laboratory test bed,” *IEEE Transactions on Power Delivery*, vol. 26, pp. 325–332, 2011.

[2] D. E. Olivares, A. Mehrizi-Sani, A. H. Etemadi, *et al.*, "Trends in Microgrid Control," *IEEE Transactions on Smart Grid*, vol. 5, no. 4, pp. 1905–1919, 2014.
[3] J. Guerrero, J. Vasquez, J. Matas, L. Garcia, and M. Castilla, "Hierarchical control of droop-controlled AC and DC microgrids—A general approach toward standardization," *IEEE Transactions on Industrial Electronics*, vol. 58, no. 1, pp. 158–172, 2011.
[4] J. Guerrero, M. Chandorkar, T.-L. Lee, and P. Loh, "Advanced control architectures for intelligent microgrids—Part I: Decentralized and hierarchical control," *IEEE Transactions on Industrial Electronics*, vol. 60, no. 4, pp. 1254–1262, 2013.
[5] F. Katiraei and M. Iravani, "Power management strategies for a microgrid with multiple distributed generation units," *IEEE Transactions on Power Systems*, vol. 21, no. 4, pp. 1821–1831, 2006.
[6] C.-T. Lee, C.-C. Chu, and P.-T. Cheng, "A new droop control method for the autonomous operation of distributed energy resource interface converters," *IEEE Transactions on Power Electronics*, vol. 28, no. 4, pp. 1980–1993, 2013.
[7] A. Elrayyah, F. Cingoz, and Y. Sozer, "Construction of nonlinear droop relations to optimize islanded microgrid operation," *IEEE Transactions on Industry Applications*, vol. 51, no. 4, pp. 3404–3413, 2015.
[8] C. Sao and P. Lehn, "Control and power management of converter fed microgrids," *IEEE Transactions on Power Systems*, vol. 23, no. 3, pp. 1088–1098, 2008.
[9] M. Abdelaziz, M. Shaaban, H. Farag, and E. El-Saadany, "A multistage centralized control scheme for islanded microgrids with PEVs," *IEEE Transactions on Sustainable Energy*, vol. 5, no. 3, pp. 927–937, 2014.
[10] J. Lopez, L. Moreira, and A. Madureira, "Defining control strategies for microgrids islanded operation," *IEEE Transactions on Power Systems*, vol. 21, no. 2, pp. 916–924, 2006.
[11] J.-Y. Kim, J.-H. Jeon, S.-K. Kim, *et al.*, "Cooperative control strategy of energy storage system and microsources for stabilizing the microgrid during islanded operation," *IEEE Transactions on Power Electronics*, vol. 25, no. 12, pp. 3037–3048, 2010.
[12] Y. Zhang and H. Ma, "Theoretical and experimental investigation of networked control for parallel operation of inverters," *IEEE Transactions on Power Electronics*, vol. 59, no. 4, pp. 1961–1970, 2012.
[13] N. Soni, S. Doolla, and M. C. Chandorkar, "Improvement of transient response in microgrids using virtual inertia," *IEEE Transactions on Power Delivery*, vol. 28, pp. 1830–1838, 2013.
[14] J. O'Sullivan, A. Rogers, D. Flynn, P. Smith, A. Mullane, and M. O'Malley, "Studying the maximum instantaneous non-synchronous generation in an island system–frequency stability challenges in Ireland," *IEEE Transactions on Power Systems*, vol. 29, pp. 2943–2951, 2014.
[15] Q. C. Zhong and G. Weiss, "Synchronverters: Inverters that mimic synchronous generators," *IEEE Transactions on Industrial Electronics*, vol. 58, pp. 1259–1267, 2011.

[16] R. Olfati-Saber, J. Fax, and R. M. Murray, "Consensus and cooperation in networked multi-agent systems," *Proceedings of the IEEE*, vol. 95, pp. 215–233, 2007.

[17] M. Wooldridge, "Intelligent agents," in *Multiagent Systems: A Modern Approach to Distributed Artificial Intelligence*. Cambridge, MA: MIT Press, 1999.

[18] Q. Shafiee, J. Guerrero, and J. Vasquez, "Distributed secondary control for islanded microgrids—A novel approach," *IEEE Transactions on Power Electronics*, vol. 29, no. 2, pp. 1018–1030, 2014.

[19] Q. Shafiee, C. Stefanovic, T. Dragicevic, P. Popovski, J. Vasquez, and J. Guerrero, "Robust networked control scheme for distributed secondary control of islanded microgrids," *IEEE Transactions on Industrial Electronics*, vol. 61, no. 10, pp. 5363–5374, 2014.

[20] Q. Shafiee, J. Vasquez, and J. Guerrero, "Distributed secondary control for islanded microgrids—A networked control systems approach," in *38th Annual Conference on IEEE Industrial Electronics Society*, 2012, pp. 5637–5642.

[21] F. Guo, J. Wen, and Y. Song, "Distributed secondary voltage and frequency restoration control of droop-controlled inverter-based microgrids," *IEEE Transactions on Industrial Electronics*, vol. 62, no. 7, pp. 4355–4364, 2015.

[22] T. Logenthiran, D. Srinivasan, A. Khambadkone, and H. Aung, "Multiagent system for real-time operation of a microgrid in real-time digital simulator," *IEEE Transactions on Smart Grid*, vol. 3, no. 2, pp. 925–933, 2012.

[23] T. Logenthiran, T. Naayagi, W. Woo, V.-T. Phan, and K. Abidi, "Intelligent control system for microgrids using multiagent system," *IEEE Journal of Emerging and Selected Topics in Power Electronics*, vol. 3, no. 4, pp. 1036–1045, 2015.

[24] G. Hug, S. Kar, and C. Wu, "Consensus + innovation approach for distributed multiagent coordination in a microgrid," *IEEE Transactions on Smart Grid*, vol. 6, no. 4, pp. 1893–1903, 2015.

[25] L. Y. Lu and C. C. Chu, "Consensus-based droop control synthesis for multiple DICs in isolated micro-grids," *IEEE Transactions on Power Systems*, vol. 30, pp. 2243–2256, 2014.

[26] J. Simpson-Porco, Q. Shafiee, F. Dörfler, J. Vasquez, J. Guerrero, and F. Bullo, "Secondary frequency and voltage control of islanded microgrids via distributed averaging," *IEEE Transactions on Industrial Electronics*, vol. 62, no. 11, pp. 7025–7038, 2015.

[27] S. Moayedi, V. Nasirian, F. Lewis, and A. Davoudi, "Team-oriented load sharing in parallel DC–DC converters," *IEEE Transactions on Industry Applications*, vol. 51, no. 1, pp. 479–490, 2015.

[28] J. Simpson-Porco, F. Dörfler, and F. Bullo, "Synchronization and power sharing for droop-controlled inverters in islanded microgrids," *Automatica*, vol. 49, pp. 2603–2611, 2013.

[29] L.-Y. Lu and C.-C. Chu, "Consensus-based secondary frequency and voltage droop control of virtual synchronous generators for isolated AC microgrids," *IEEE Journal on Emerging and Selected Topics in Circuits and Systems*, vol. 5, no. 3, pp. 443–455, 2015.

[30] Y. Xu, W. Zhang, W. Liu, *et al.*, "Distributed subgradient-based coordination of multiple renewable generators in a microgrid," *IEEE Transactions on Power Systems*, vol. 29, no. 1, pp. 23–33, 2014.
[31] A. Bridam, A. Davoudi, F. Lewis, and Z. Qu, "Secondary control of microgrids based on distributed cooperative control of multi-agent systems," *IET Generation, Transmission and Distribution*, vol. 7, no. 8, pp. 822–831, 2014.
[32] A. Bridam, A. Davoudi, and J. L. Guerrero, "Distributed cooperative secondary control of microgrids using feedback linearization," *IEEE Transactions on Power Systems*, vol. 28, no. 3, pp. 3462–3470, 2013.
[33] A. Bridam, A. Davoudi, and Lewis, "Multi-objective distributed control framework for islanded AC microgrids," *IEEE Transactions on Industrial Informatics*, vol. 10, no. 3, pp. 1785–1798, 2014.
[34] W. Liu, W. Gu, W. Sheng, X. Meng, and M. Chen, "Pinning-based distributed cooperative control for autonomous microgrids under uncertain communication topologies," *IEEE Transactions on Power Systems*, vol. 31, no. 2, pp. 1320–1329, 2016.
[35] S. Manaffan, M. Talebi, A. Jain, and A. Behal, "Synchronization in networks of identical systems via pinning: Applications to distributed secondary control of microgrids," *IEEE Transactions on Control Systems Technology*, vol. 25, no. 6, pp. 2227–2234, 2017.
[36] M. Telebi, S. Mannaffam, A. Jain, and A. Behal, "Intelligent pinning-based cooperative secondary control of distributed generators for microgrid in islanding operation," *IEEE Transactions on Power Systems*, vol. 33, no. 2, pp. 1364–1373, 2018.
[37] J. Lai, H. Zhou, X. Lu, and Z. Liu, "Distributed power control of DERs based on networked multiagent systems with communication delays," *Neurocomputing*, vol. 179, pp. 135–143, 2016.
[38] Y. Xu, W. Zhang, G. Hug, S. Kar, and Z. Li, "Cooperative control of distributed energy storage systems in a microgrid," *IEEE Transactions on Smart Grid*, vol. 6, no. 1, pp. 238–248, 2015.
[39] Y. Xu and Z. Li, "Distributed optimal resource management based on the consensus algorithm in a microgrid," *IEEE Transactions on Industrial Electronics*, vol. 62, no. 4, pp. 2584–2592, 2015.
[40] C. Li, M. Savaghebi, J. Vasquez, and J. Guerrero, "Multiagent based distributed control for operation cost minimization of droop controlled AC microgrid using incremental cost consensus," in *17th European Conference on Power Electronics and Applications (EPE'15 ECCE-Europe)*, Geneva, Switzerland, 2015, pp. 1–9.
[41] R. Azevedo, H. Mehmet, T. Ma, and O. Mohammed, "Multiagent-based optimal microgrid control using fully distributed diffusion strategy," *IEEE Transactions on Smart Grid*, vol. 8, no. 4, pp. 1997–2008, 2017.
[42] F. Chen, Z. Chen, L. Xiang, Z. Liu, and Z. Yuan, "Reaching a consensus via pinning control," *Automatica*, vol. 45, no. 5, pp. 1215–1220, 2009.

[43] C. W. Wu, "Localization of effective pinning control in complex networks of dynamical systems," in *IEEE Int. Symposium on Circuits and Systems (ISCAS)*, Seattle, WA, 2008, pp. 2530–2533.

[44] Y. Cao, W. Yu, W. Ren, and G. Chen, "An overview of recent progress in the study of distributed multi-agent coordination," *IEEE Transactions on Industrial Informatics*, vol. 9, pp. 427–438, 2013.

[45] D. P. Bertseka and J. N. Tsitsiklis, *Parallel and Distributed Computation: Numerical Methods*. New York, NY: Athena Scientific, 2003.

[46] S. Boyd, N. Parikh, E. Chu, B. Peleato, and J. Eckstein, "Distributed optimization and statistical learning via the alternating direction method of multipliers," *Foundations and Trends in Machine Learning*, vol. 3, no. 1, pp. 1–122, 2011.

[47] J. F. C. Mota, J. M. F. Xavier, P. M. Q. Aguiar, and M. Püschel, "D-ADMM: A communication-efficient distributed algorithm for separable optimization," *IEEE Transactions on Signal Processing*, vol. 61, no. 10, pp. 2718–2723, 2013.

[48] J. Zhang, S. Nabavi, A. Chakrabortty, and Y. Xin, "ADMM optimization strategies for wide-area oscillation monitoring in power systems under asynchronous communication delays," *IEEE Transactions on Smart Grid*, vol. 7, pp. 2123–2133, 2016.

[49] Y. Wang, L. Wu, and S. Wang, "A fully-decentralized consensus-based ADMM approach for DC-OPF with demand response," *IEEE Transactions on Smart Grid*, vol. 8, no. 6, pp. 2637–2647, 2017.

[50] T. Erseghe, D. Zennaro, E. Dall'Anese, and L. Vangelista, "Fast consensus by the alternating direction multipliers method," *IEEE Transactions on Signal Processing*, vol. 59, pp. 5523–5537, 2011.

[51] L.-Y. Lu and C.-C. Chu, "Consensus-based droop control of isolated micro-grids by ADMM implementations," *IEEE Transactions on Smart Grid*, vol. 9, no. 5, pp. 5101–5112, 2018.

[52] S. T. Cady, A. D. Domínguez-García, and C. N. Hadjicostis, "A distributed generation control architecture for islanded AC microgrids," *IEEE Transactions on Control Systems Technology*, vol. 23, no. 5, pp. 1717–1735, 2015.

[53] F. Dörfler, J. W. Simpson-Porco, and F. Bullo, "Breaking the hierarchy: Distributed control and economic optimality in microgrids," *IEEE Transactions on Control of Network Systems*, vol. 3, no. 3, pp. 241–253, 2016.

[54] L.-Y. Lu and C.-C. Chu, "Consensus-based distributed droop control of isolated micro-grids by alternating direction multipliers methods," in *IEEE Region 10 Conference (TENCON2015)*, Macao, 2015, pp. 1–5.

[55] L.-Y. Lu and C.-C. Chu, "Consensus-based distributed droop control of VSGs for isolated AC micro-grids by ADMMs," *IEEE Int. Symposium on Circuits and Systems (ISCAS)*, Montreal, QC, 2016, pp. 2138–2141.

[56] Q. Song, J. Cao, and W. Yu, "M-matrix strategies for pinning-controlled leader-following consensus in multiagent systems with nonlinear dynamics," *IEEE Transactions on Cybernetics*, vol. 43, no. 6, pp. 1688–1697, 2013.

[57] N. Avila and C. Chu, "Pinning-based distributed droop control of power converters in isolated AC microgrids," in *IEEE 8th International Power Electronics and Motion Control Conference (IPEMC-ECCE Asia)*, Hefei, 2016, pp. 171–178.

[58] N. Avila and C. Chu, "Distributed pinning droop control in isolated AC microgrids," *IEEE Transactions on Industry Applications*, vol. 53, no. 4, pp. 3237–3249, 2017.

[59] F. Bellifemine, G. Caire, and D. Greenwood, *Developing Multi-Agent Systems with JADE*. John Wiley & Sons Ltd, Chichester, 2007.

[60] A. R. Bergen and D. J. Hill, "A structure preserving model for power system stability analysis," *IEEE Transactions on Power Apparatus and Systems*, vol. PAS-100, pp. 25–35, 1981.

[61] P. Varaiya, F. Wu, and R.-L. Chen, "Direct methods for transient stability analysis of power systems: Recent results," *Proceedings of the IEEE*, vol. 73, pp. 1703–1715, 1985.

[62] A. Bergen, D. Hill, and C. L. D. Marco, "Lyapunov function for multi-machine power systems with generator flux decay and voltage dependent loads," *International Journal of Electrical Power & Energy Systems*, vol. 8, pp. 2–10, 1986.

[63] H.-D. Chiang, Direct Methods for Stability Analysis of Electric Power Systems: Theoretical Foundation, BCU Methodologies, and Applications. John Wiley & Sons Ltd, Hoboken, NJ, 2011.

[64] H. D. Chiang, C. C. Chu, and G. Cauley, "Direct stability analysis of electric power systems using energy functions: Theory, applications, and perspective," *Proceedings of the IEEE*, vol. 83, pp. 1497–1529, 1995.

[65] C.-C. Chu and H.-D. Chiang, "Constructing analytical energy functions for network-preserving power system models," *Circuits, Systems, and Signal Processing*, vol. 24, pp. 363–383, 2005.

[66] N. A. Tsolas, A. Arapostathis, and P. P. Varaiya, "A structure preserving energy function for power system transient stability analysis," *IEEE Transactions on Circuits and Systems*, vol. CAS-32, pp. 1041–1049, 1985.

[67] A. Bergen and V. Vittal, *Power Systems Analysis*, 2nd ed. Prentice Hall, Upper Saddle River, NJ, 2000.

[68] Y. Zou, M.-H. Yin, and H.-D. Chiang, "Theoretical foundation of the controlling UEP method for direct transient-stability analysis of network-preserving power system models," *IEEE Transactions on Circuits and Systems—I: Fundamental Theory and Applications*, vol. 50, pp. 1324–1336, 2003.

[69] I. Schizas, A. Ribeiro, and G. Giannakis, "Consensus in *ad hoc* WSNs with noisy links—Part I: Distributed estimation of deterministic signals," *IEEE Transactions on Signal Processing*, vol. 56, pp. 350–364, 2008.

[70] H. Zhu, G. Giannakis, and A. Cano, "Distributed in-network channel decoding," *IEEE Transactions on Signal Processing*, vol. 57, pp. 3970–3983, 2009.

[71] N. Ainsworth and S. Grijalva, "A structure-preserving model and sufficient condition for frequency synchronization of lossless droop inverter-based AC

networks," *IEEE Transactions on Power Systems*, vol. 28, no. 4, pp. 4310–4319, 2013.

[72] H.-D. Chiang and F. Wu, "Stability of nonlinear systems described by a second-order vector differential equation," *IEEE Transactions on Circuits and Systems*, vol. 35, pp. 703–711, 1988.

[73] C.-C. Chu and H.-D. Chiang, "Constructing analytical energy functions for lossless network-reduction power system models: Framework and new developments," *Circuits, Systems, and Signal Processing*, vol. 18, pp. 1–16, 1999.

[74] H. K. Khail, *Nonlinear Systems*, 3rd ed. Englewood Cliffs, NJ: Prentice Hall, 2002.

[75] C. Dufour and J. Belanger, "On the use of real-time simulation technology in smart grid research and development," *IEEE Transactions on Industry Applications*, vol. 50, pp. 3963–3970, 2014.

[76] L. Y. Lu and C. C. Chu, "Consensus-based secondary frequency and voltage droop control of virtual synchronous generators for isolated AC micro-grids," *IEEE Journal on Emerging and Selected Topics in Circuits and Systems*, vol. 5, pp. 443–455, 2015.

[77] J. Bang-Jensen and G. Gutin, *Digraphs: Theory, Algorithms and Applications*, 2nd ed. Springer-Verlag, London, 2008.

[78] X. Wang and H. Su, "Pinning control of complex networked systems: A decade after and beyond," *Annual Reviews in Control*, vol. 38, no. 1, pp. 103–111, 2014.

[79] X. Liu and T. Chen, "Cluster synchronization in directed networks via intermittent pinning control," *IEEE Transactions on Neural Networks*, vol. 22, no. 7, pp. 1009–1019, 2011.

[80] T. Chen, X. Liu, and W. Liu, "Pinning complex networks by a single controller," *IEEE Transactions on Circuits and Systems I: Regular Papers*, vol. 54, no. 6, pp. 1317–1326, 2007.

[81] G.-H. Lee, Y.-J. Yoon, S.-H. Lee, K.-J. Choi, and D.-R. Shin, "Design of directory facilitator for agent-based service discovery in ubiquitous computing environments," in *IEEE Int. Workshop on Semantic Computing and Applications*, Incheon, 2008, pp. 412–419.

[82] R. Ahuja, R. Magnanti, and J. Orlin, *Network Flows: Theory, Algorithms and Applications*, 1st ed., Englewood Cliffs, NJ: Prentice-Hall, 1992.

[83] C. Robinson, P. Mendham, and T. Clarke, "MACSimJX: A tool for enabling agent modelling with Simulink using JADE," *Journal of Physical Agents*, vol. 4, no. 3, pp. 1–7, 2010.

[84] G. Binetti, A. Davoudi, F. L. Lewis, D. Naso and B. Turchiano, "Distributed consensus-based economic dispatch with transmission losses," *IEEE Transactions on Power Systems*, vol. 29, no. 4, pp. 1711–1720, 2014.

[85] R. Majumder, G. Bag, and K. Kim, "Power sharing and control in distributed generation with wireless sensor networks," *IEEE Transactions on Smart Grid*, vol. 3, no. 2, pp. 618–634, 2012.

Part III

Converters

Chapter 11

Extendable multiple outputs hybrid converter for AC/DC microgrid

Anish Ahmad[1] *and Rajeev Kumar Singh*[1]

High gain converters are very popular nowadays due to large demand in the area of microgrid-based renewable energy resources. The renewable energy resources have low-voltage generation sources that require high voltage gain converters to step up the input DC voltage. Also, multi-output converters are demanded in the present scenario due to a requirement of different AC and DC load demands. The conventional converters used for this purpose have a few limitations such as duty cycle is limited and they have to operate on extreme duty cycle to achieve high gain. To encounter these problems, hybrid multi-output converters are designed and proposed for high gain in less duty cycle. In this chapter, an extendable multiple-output hybrid converter is presented for AC/DC microgrid applications. The proposed converter is derived from single-switch-derived quadratic boost converter topology. In the proposed hybrid converter, both the filter inductors of the converter are magnetically coupled and the main switch of the converter is replaced by an H-bridge inverter circuit. This inverter bridge can be arranged by series or parallel and depending upon the need of output voltage and current, an *n*-number of bridges can be connected in series or parallel. Thus, the proposed hybrid converter is capable of giving single DC and *n*-number of AC outputs simultaneously. This type of converter topologies is mostly suited for rural-based microgrid application where requirements are of less maintenance, high power density and less cost. Detailed steady-state analysis and PWM control algorithm for both series and parallel connected topologies have been carried out to demonstrate its advantages. In this chapter, the proposed topology is validated for two simultaneous AC and single DC outputs on a 500 W prototype through simulation and experimental results.

11.1 Introduction

Owing to the benefits of higher power-processing density, nowadays lots of researches are being done on hybrid AC/DC microgrids. They offer simultaneous

[1]Department of Electrical Engineering, Indian Institute of Technology (Banaras Hindu University) Varanasi, India

DC and AC outputs and can power both DC and AC loads at the same time, e.g. a hybrid AC/DC microgrid can power both AC fan and LED lamp at the same time [1–9], especially for rural applications. Figure 11.1(a) shows the architecture of a conventional microgrid in which it can be observed that separate power converters for each power conversion (DC/DC and DC/AC) are required. However, it can be observed from the architecture shown in Figure 11.1(b) that a hybrid AC/DC microgrid using the proposed extendable multiple-output hybrid converter utilizes single stage to perform DC/DC and DC/AC conversion.

It can be observed from Figure 11.1 that the classical microgrid uses mainly two different converters: DC–DC and DC–AC. The input to these converters is generally given from the renewable energy resources. The solar PV-based renewable energy is freely available and causes zero pollution to the environment. However, a solar PV renewable resource-based microgrid has drawbacks of low-voltage generation, a more number of conversion stages and reduced efficiency. Thus, in order to step up the voltage of the renewable sources, a high step-up converter is an essential part in the microgrid [10,11]. Moreover, in microgrid application, multi-output converters are needed for different power requirements. These converters have higher power density, and they require improved reliability due to multi-output system. In conventional multi-port system, several converters are used in parallel to achieve multiple outputs. In view of a more number of conversion stages, such systems have poor performance and reduced overall efficiency.

In this chapter, first of all a brief review of a hybrid multi-output converter is given in Section 11.2. The proposed extendable multiple-output hybrid converter is described in Section 11.3. Section 11.4 explains the steady-state behaviour of the proposed hybrid multi-output converters and Section 11.5 deals with the details of pulse width modulation (PWM) control techniques. Design criteria is discussed in Section 11.6. Verification and discussion about the results of the proposed topology are discussed in Section 11.7. Conclusions are drawn in Section 11.8.

11.2 State of the art of hybrid multiple-output converters

The converter used for the microgrid applications is characterized as shown in Figure 11.2. Generally, the converter used in microgrid applications can be divided into the DC–DC converter, DC–AC converter (inverter) and hybrid converters. The DC–DC converter is broadly classified in bidirectional, coupled inductor/transformer, interleaved, soft switching, multilevel, switched inductor/capacitor, impedance source converters and multi-outputs converters [12]. Similar topologies are also found in the DC–AC converter, i.e. inverter. However, in addition the inverter bridge circuits are used after the DC–DC converters for high gain. The hybrid converter topologies are used nowadays. This is due to the simultaneous AC and DC output capability, increase power-processing capability, compact size, reliability and flexibility. Due to these advantages, it is most suited for microgrid application.

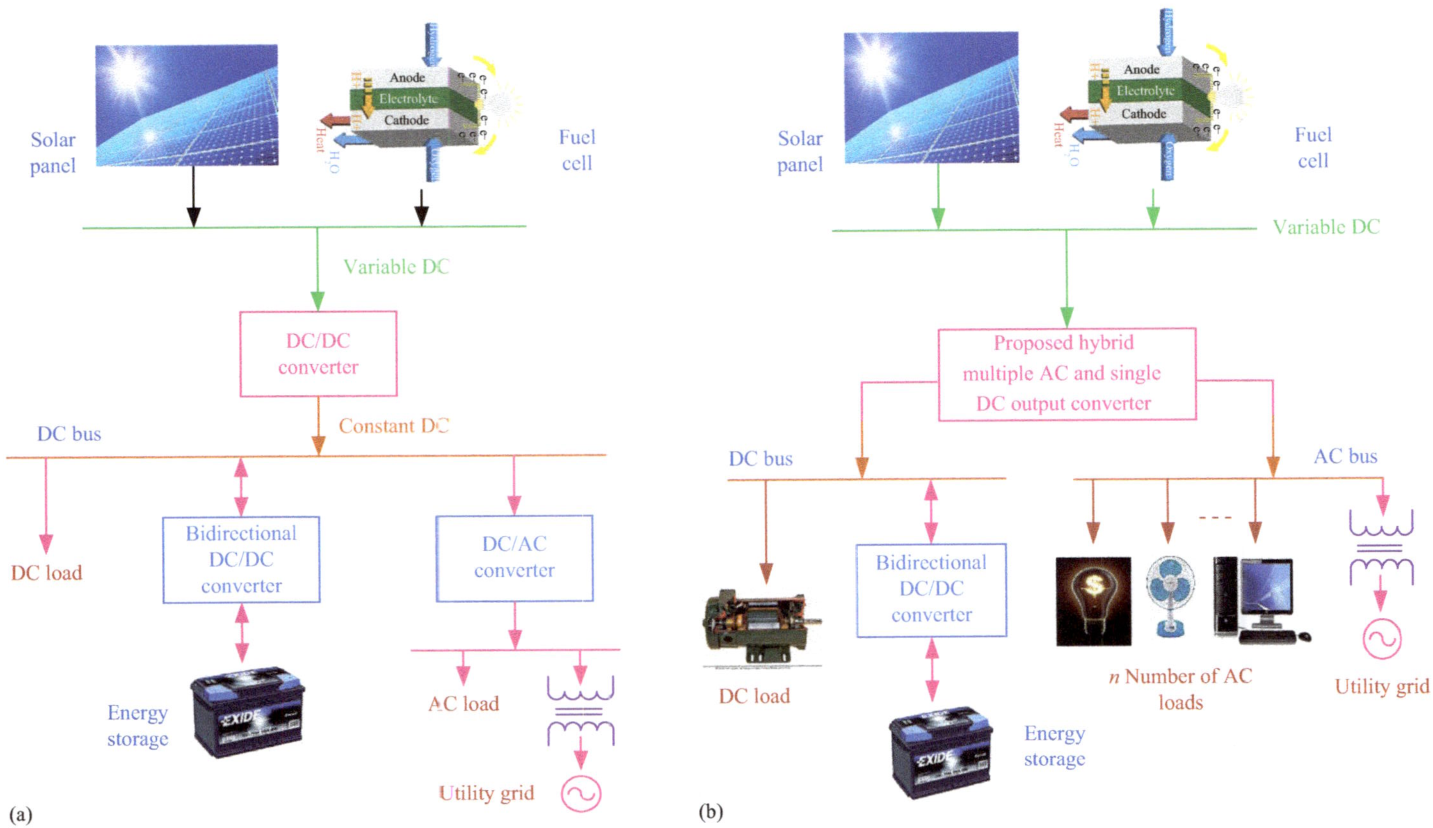

Figure 11.1 AC/DC microgrid: (a) conventional and (b) proposed microgrid

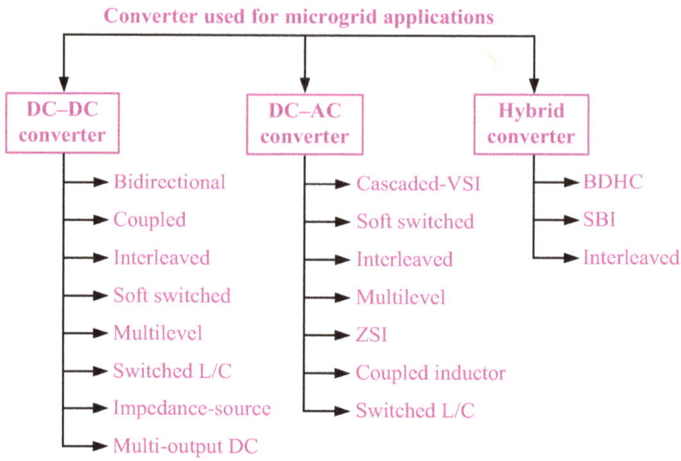

Figure 11.2 Converters classification for microgrid applications

In the recent years different multi-output converters have been proposed for microgrid [13–16]. The isolated-based converters contain a more number of circuit components and involve complex control. Also, the multiple-output architecture-based converter systems' weight, volume and size increase due to the presence of the magnetic elements. The isolated topologies are required where galvanic isolation between the different ports is necessary.

However, conventional hybrid multiple-output converters also give simultaneously single-DC and single-AC outputs [17–22]. Decoupled single DC and AC outputs by a hybrid interleaved boost converter are also reported in [23]. It may be observed that the existing hybrid multiple-output converters give either multiple DC outputs or one AC and one DC outputs simultaneously.

The nonisolation-based multiple-output converter is reported in [24–27]. Figure 11.3(a) shows basic multiple-output-based DC–DC converters. The problem of this basic single-input dual-output (SIDO) converter is that it has cross voltage regulation problem. Figure 11.3(b) shows the synchronous buck multi-output converter. The limitation of this converter is that it has low-voltage gain outputs [28,29]. Literature [30–33] discuss the different multiple-output converters for various applications, but all these are basically DC/DC converters giving multiple DC outputs. Single-stage high-gain inverters are also used in microgrid applications [34–41] but have only single AC outputs.

An impedance source-based inverter such as a switched boost inverter (SBI) as shown in Figure 11.4(a) and a current fed switched inverter (CFSI) also give simultaneous AC and DC outputs as shown in Figure 11.4(b). A multi-output hybrid interleaved boost converter discussed in [23] also gives single AC and DC outputs. In SBI, DC output is received at the V_c terminal and in CFSI AC output is received across the capacitor C. Literature [42] discusses a hybrid topology with multilevel AC and one DC outputs simultaneously. A split-source-based hybrid

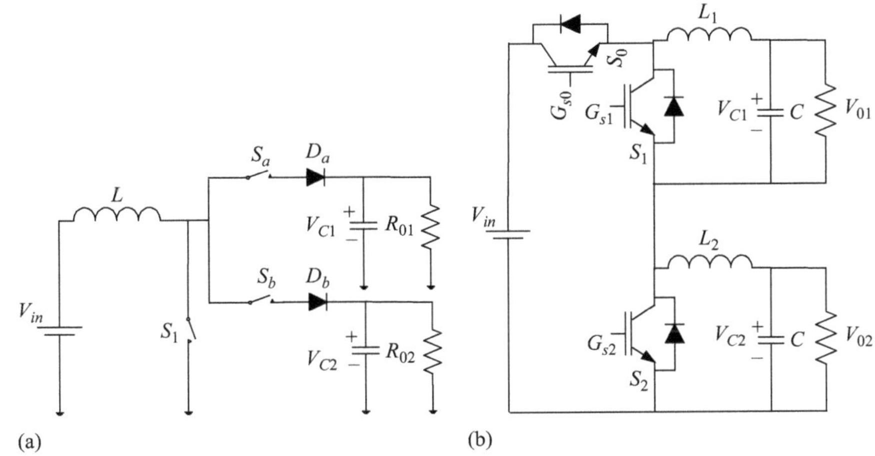

Figure 11.3 Single-input multi-output DC–DC converter: (a) SIDO converter and (b) synchronous buck

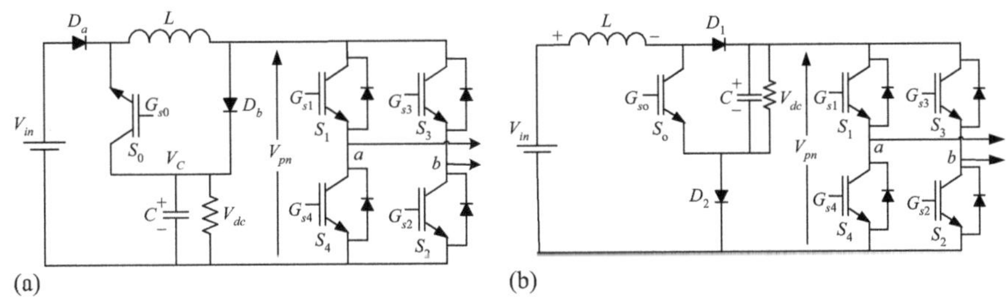

Figure 11.4 Hybrid output ZSI: (a) switched boost inverter [22] and (b) current fed switched inverter [19]

converter gives only single AC output [43]. Overall, it may be observed that the existing multi-output converters give either multiple DC outputs or only one AC and one DC outputs simultaneously.

Based on the literature review and challenges associated with the conventional high-gain converters, some desired characteristics of high-gain converters for microgrid applications are also identified and listed as follows:

1. The power converter used in microgrid should have high boost inversion capability, as the renewable energy resources such as solar PV has less voltage generation capability.
2. The continuous input current should be drawn from the input DC power supply, which is suitable for renewable energy application. The continuous input current-based converter does not require any input filter.
3. The converter should not operate at extreme shoot-through duty cycle to achieve high-voltage gain. Operating on extreme shoot-through duty cycle

leads to lower values of modulation index and thereby increases total harmonics distortions.
4. The converter should have higher power-processing density and should have inherent shoot-through protection for increased reliability.
5. The converter should not require dead band for the switching so that waveform distortion is avoided for DC–AC conversion. The number of passive components should be least.

11.3 Proposed generalized hybrid multi-output converters

The hybrid multiple-output system has higher power-processing density and higher reliability (due to inherent shoot-through protection), suitable for compact size system. The proposed quadratic boost derived magnetically coupled extendable multiple-output hybrid converter for AC/DC microgrid is shown in Figure 11.5. The proposed converter is capable of giving n-number of simultaneous AC along with single DC outputs. The proposed converter is developed from single switch derived quadratic converter by magnetically coupling both the filter inductors of the converter and replacing the main switch of the converter by an H-bridge inverter circuit. The inverter bridge can be arranged in series or parallel and depending upon the output requirement, n-number of bridges can be connected in series or parallel as shown in Figure 11.5. The proposed topology in this chapter has higher power density and improved reliability due to inherent shoot-through protection property. These properties help the proposed hybrid converter, useful for compact systems with multiple AC and one DC loads. The salient points of the works are listed next:

1. The proposed converter is capable of giving n-simultaneous AC and single DC outputs.
2. It is derived from a single switch quadratic boost converter by coupling the filter inductors of the converter and replacing the switch with n-number of H-bridge networks. Due to magnetic coupling, the size of the inductor reduces and thereby overall weight and volume of the system also reduce.
3. Steady state and dynamic modelling of the proposed converter are carried out to verify its steady state and transient characteristics.
4. The concept is validated with simulation and experimental results on 500 W prototype for two simultaneous AC and single DC outputs.

Operating modes in both the topology inverter brides are assumed as a single switch. The proposed converter topology consists of two states: (1) nonshoot-through state $(1-D_s)T_s$ and (2) shoot-through state (D_sT_s), and equivalent circuits are presented in Figure 11.6(a) and (b), respectively. All the switches and elements are assumed to be ideal. The operating waveforms of the proposed hybrid multiple-output topology converters are shown in Figure 11.6(c).

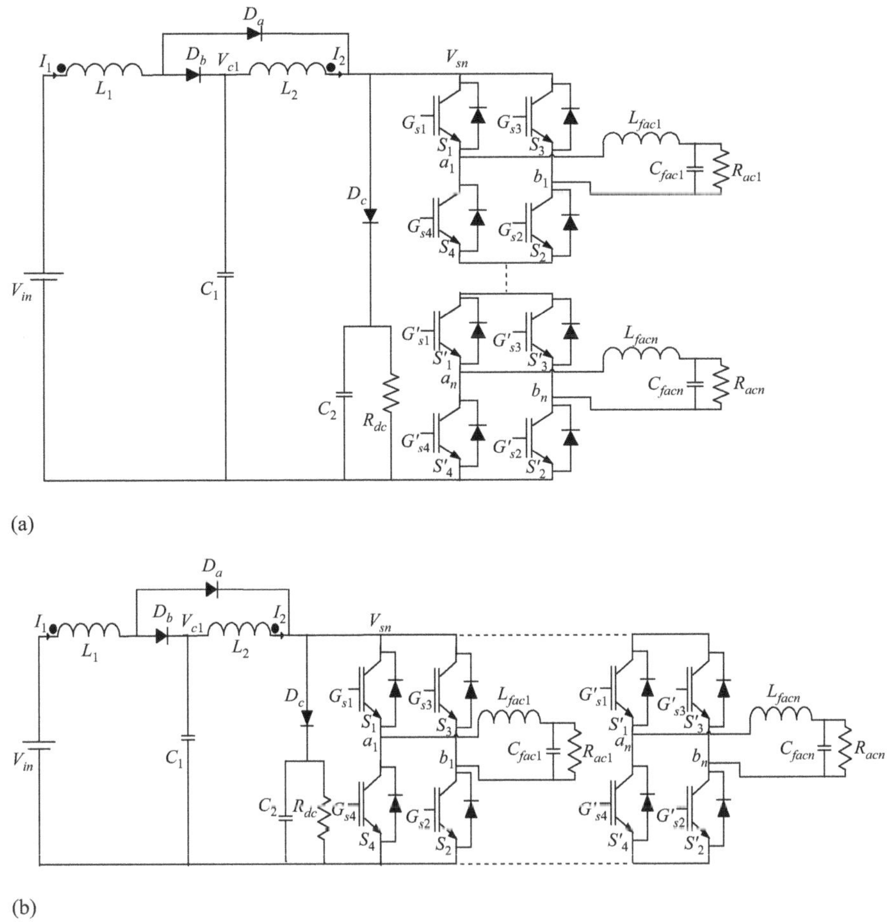

Figure 11.5 Proposed extendable multiple-output hybrid converter topology: (a) in series mode and (b) in parallel mode

11.3.1 Nonshoot-through interval ($D_sT_s \leq t \leq (1 - D_s)T_s$)

In this nonshoot-through mode, the diode D_a is reverse biased. The inverter bridge switches S_1–S_2 or S_3–S_4 of H-bridge network are operating and diodes D_b–D_c are conducting. If n inverter bridges are connected in series or parallel topology then the same PWM switches (S_1–S_2 or S_3–S_4) operate simultaneously. The inductors and input power are transferred to the AC and DC loads.

11.3.2 Shoot-through interval ($0 \leq t \leq D_sT_s$)

In this shoot-through mode, the diodes D_b and D_c are reverse biased. The inverter bridge switches S_1–S_4 or S_2–S_3 of H-bridge network are operating and diode D_a is conducting. The inductors (L_1 and L_2) store the energy from the DC input supply.

11.4 Steady-state behaviour of hybrid multi-output converters

The equivalent circuit during shoot-through interval is shown in Figure 11.6(b). The current and voltage expressions of the inductors and capacitors during the shoot-through interval are as follows:

$$L_1\frac{di_1}{dt} - M\frac{di_2}{dt} = v_{in} \tag{11.1}$$

$$L_2\frac{di_2}{dt} - M\frac{di_1}{dt} = v_{C1} \tag{11.2}$$

$$C_1\frac{dv_{C1}}{dt} = -i_2 \tag{11.3}$$

$$C_2\frac{dv_{C2}}{dt} = -i_{dco} = -\frac{v_{C2}}{R} \tag{11.4}$$

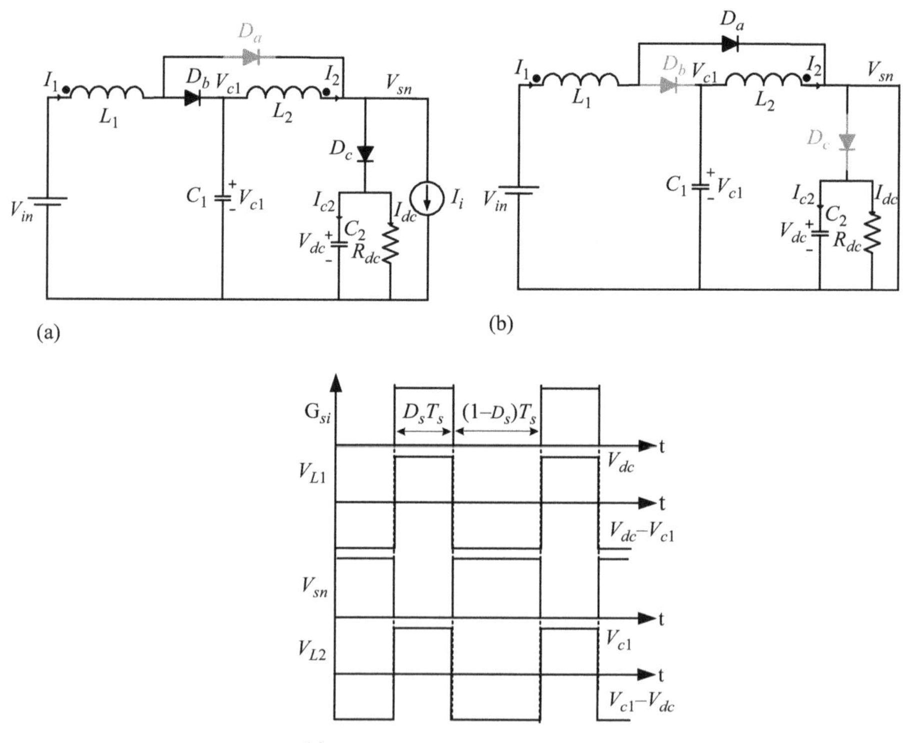

Figure 11.6 Steady-state performance: (a) nonshoot-through interval, (b) shoot-through interval and (c) operating waveforms

where M is the mutual inductance between the inductors (L_1 and L_2). The current and voltage expressions of the inductors and capacitors during the nonshoot-through state following equations are derived from Figure 11.6(a):

$$L_1\frac{di_1}{dt} - M\frac{di_2}{dt} = v_{in} - v_{C1} \tag{11.5}$$

$$L_2\frac{di_2}{dt} - M\frac{di_1}{dt} = v_{C1} - v_{C2} \tag{11.6}$$

$$C_1\frac{dv_{C1}}{dt} = i_1 - i_2 \tag{11.7}$$

$$C_2\frac{dv_{C2}}{dt} = i_2 - i_{dco} - i_i \tag{11.8}$$

Under steady-state condition, the average voltage across the inductor and average current through the capacitor in one switching cycle should be zero [44].

By applying voltage–second balance principle on L_1 and L_2, we have

$$V_{c1} = \frac{V_{in}}{(1 - D_s)} \tag{11.9}$$

From Figure 11.6, DC output voltage relation is as follows:

$$V_{dco} = V_{C2} = \frac{V_{c1}}{(1 - D_s)} = \frac{V_{in}}{(1 - D_s)^2} \tag{11.10}$$

By applying charge–second balance principle on C_1 and C_2,

$$I_1 = \frac{I_2}{(1 - D_s)} \tag{11.11}$$

$$I_2 = I_i + \frac{I_{dco}}{(1 - D_s)} \tag{11.12}$$

From the above two current expressions, we have

$$I_1 = \frac{I_i}{(1 - D_s)} + \frac{I_{dco}}{(1 - D_s)^2} \tag{11.13}$$

From the transfer characteristic (11.10), it can be observed that duty cycle is varied from the $0 < D_s < 1$. Voltage across the inverter bridge and current through the inverter bridge change according to the series and parallel topology.

11.4.1 Series-connected inverter bridges expression

In this, series mode of inverter topology is connected in series as shown in Figure 11.5(a). The total number of AC output voltage depends on the total switch node voltage (V_{sn}). The inverter current (I_i) is the same for the n number of connected inverter bridge. $n = 1$ is a single inverter bridge, and it is connected up to an n number of inverter bridges. In this work, all the loads are symmetrical.

Input voltage across the inverter bridges,

$$\begin{aligned}
&V_{Inac} = V_{dco} \\
&V_{Inac1} = V_{Inac2} = \frac{V_{dco}}{2} \\
&V_{Inac1} = V_{Inac2} = V_{Inac3} = \frac{V_{dco}}{3} \\
&\qquad\vdots \\
&V_{Inac1} = V_{Inac2} = V_{Inac3} = \cdots = V_{Inacn} = \frac{V_{dcon}}{n}
\end{aligned} \tag{11.14}$$

Input current through the inverter bridges,

$$I_{Inac1} = I_{Inac2} = I_{Inac3} = \ldots = I_{Inacn} = I_{inv} \tag{11.15}$$

where

$$I_{in} = \frac{M^2 V_{dco}}{nR_{acn}} \tag{11.16}$$

Fundamental AC peak output voltages of the inverter bridges,

$$\begin{aligned}
&V_{aco} = M_a V_{dco} \\
&V_{aco1} = V_{aco2} = \frac{M_a V_{dco}}{2} \\
&V_{aco1} = V_{aco2} = V_{aco3} = \frac{M_a V_{dco}}{3} \\
&\qquad\vdots \\
&V_{aco1} = V_{aco2} = V_{aco3} = \ldots = V_{acon} = \frac{M_a V_{dco}}{n}
\end{aligned} \tag{11.17}$$

Fundamental AC peak output currents of the inverter bridges,

$$\begin{aligned}
&I_{aco} = \frac{M_a V_{dco}}{R_{ac}} \\
&I_{aco1} = I_{aco2} = \frac{M_a V_{dco}}{2R_{ac}} \\
&I_{aco1} = I_{aco2} = I_{aco3} = \frac{M_a V_{dco}}{3R_{ac}} \\
&\qquad\vdots \\
&I_{aco1} = I_{aco2} = I_{aco3} = \ldots = I_{acon} = \frac{M_a V_{dco}}{nR_{ac}}
\end{aligned} \tag{11.18}$$

AC output powers of the inverter bridges,

$$
\begin{aligned}
&P_{aco} = \frac{M_a^2 V_{dco}^2}{2(1^2)R_{ac}} \\
&P_{aco1} = P_{aco2} = \frac{M_a^2 V_{dco}^2}{2(2^2)R_{ac}} \\
&P_{aco1} = P_{aco2} = P_{aco3} = \frac{M_a^2 V_{dco}^2}{2(3^2)R_{ac}} \\
&\vdots \\
&P_{aco1} = P_{aco2} = P_{aco3} = \ldots = P_{acon} = \frac{M_a^2 V_{dco}^2}{2(n^2)R_{ac}}
\end{aligned}
\tag{11.19}
$$

In this proposed one, only two inverter bridges are analysed and validated. From (11.14) and (11.15), it is clear that as the number of inverter bridge increases, the current remains the same but the voltage is equally divided for the case of symmetrical load.

11.4.2 Parallel-connected inverter bridges expression

In this, parallel mode of inverter topology connected in parallel is shown in Figure 11.5(b). The total number of AC output voltage does not depend on the switch node voltage (V_{sn}). The inverter current (I_i) is equally divided by the same n number of connected inverter bridge.

Input voltage across the inverter bridges,

$$V_{Inac1} = V_{Inac2} = V_{Inac3} = \ldots = V_{Inacn} = V_{dco} \tag{11.20}$$

Input current through the inverter bridges,

$$
\begin{aligned}
&I_{Inac} = I_{inv} \\
&I_{Inac1} = I_{Inac2} = \frac{I_{inv}}{2} \\
&I_{Inac1} = I_{Inac2} = I_{Inac3} = \frac{I_{inv}}{3} \\
&\vdots \\
&I_{Inac1} = I_{Inac2} = I_{Inac3} = \ldots = I_{Inacn} = \frac{I_{inv}}{n}
\end{aligned}
\tag{11.21}
$$

where

$$I_{in} = \frac{nM^2 V_{dco}}{R_{ac}} \tag{11.22}$$

Fundamental peak AC output voltages of the inverter bridges,

$$V_{aco1} = V_{aco2} = V_{aco3} = \ldots = V_{acon} = M_a V_{dco} \tag{11.23}$$

Fundamental peak AC output currents of the inverter bridges,

$$I_{aco1} = I_{aco2} = I_{aco3} = \ldots = I_{acon} = \frac{M_a V_{dco}}{R_{ac}} \tag{11.24}$$

AC output powers of the inverter bridges,

$$P_{aco1} = P_{aco2} = P_{aco3} = \ldots = P_{acon} = \frac{M_a^2 V_{dco}^2}{2R_{ac}} \tag{11.25}$$

In this proposed one, only two inverter bridges are analysed and validated. From (11.20) and (11.21), it is clear that as the number of inverter bridge increases the voltage across the inverter bridge is constant but the current is equally divided for the case of symmetrical load.

11.4.3 Voltage stress and current expression

During shoot-through duty cycle and nonshoot-through duty cycle, intervals of voltage stress are shown in Table 11.1. The current expressions are also listed in Table 11.2 for different intervals for the proposed magnetically coupled inverter. Inverter bridge currents are expressed in Table 11.3 for all intervals.

Table 11.1 Voltage stress of each component

Parameters	Nonshoot-through state	Shoot-through state
V_{L1}	$-D_s V_{in}/(1-D_s)$	V_{in}
V_{L2}	$-D_s V_{in}/(1-D_s)^2$	$V_{in}/(1-D_s)$
V_{C1}	$V_{in}/(1-D_s)$	$V_{in}/(1-D_s)$
V_{C2}	$V_{in}/(1-D_s)^2$	$V_{in}/(1-D_s)^2$
V_{Da}	$-D_s V_{in}/(1-D_s)^2$	0
V_{Db}	0	$-V_{in}/(1-D_s)$
V_{Dc}	0	$-V_{in}/(1-D_s)^2$
V_{Si}	$V_{in}/(1-D_s)^2$	0

Table 11.2 Current stress of each component

Components	Nonshoot-through state	Shoot-through state
I_{L1}	I_1	I_1
I_{L2}	$(1-D_s)I_1$	$(1-D_s)I_1$
I_{C1}	$D_s I_1$	$-(1-D_s)I_1$
I_{C2}	$I_2 - I_{dco} - I_i$	$-I_{dco}$
I_{Da}	0	I_1
I_{Db}	I_1	0
I_{Dc}	$I_{dco}/(1-D_s)$	0

Table 11.3 Current expression of an inverter bridge in the proposed hybrid converter

Operation intervals	Conducting switches	Current expression through switches			
		i_{s1}	i_{s2}	i_{s3}	i_{s4}
Shoot-through interval	S_1, S_4, S_2	I_2	I_{aco1}		$I_2 - I_{aco1}$
	S_1, S_4, S_3	$I_2 + I_{aco1}$		$-I_{ac1}$	I_2
	S_3, S_2, S_4		$I_2 + I_{aco1}$	I_2	$-I_{aco1}$
	S_3, S_2, S_1	I_{aco1}	I_2	$I_2 - I_{aco1}$	
Power interval	S_1, S_2	I_{aco1}	I_{aco1}		
	S_3, S_4			$-I_{aco1}$	$-I_{aco1}$
Zero interval	S_1, S_3	I_{aco1}		$-I_{aco1}$	
	S_2, S_4		I_{aco1}		$-I_{aco1}$

11.5 Hybrid PWM control techniques

In hybrid PWM techniques shoot-through, power interval (or nonshoot-through interval) and zero interval are used for hybrid multiple AC and DC outputs. For generating PWM signals shown in Figure 11.7, two shoot-through ($-V_{st}$ and V_{st}) voltages are used to compare with the high-frequency triangular waveform. Two sinusoidal reference signals ($-V_{ref}$ and V_{ref}) are used with peak voltage of V_{pref} and $-V_{pref}$. The PWM techniques given to the inverter bridge are the same switches (S_1–S_1', S_2–S_2', S_3–S_3' and S_4–S_4'). The analogue model of the PWM generation is shown in Figure 11.7. The PWM signals generated by Figure 11.7 are given to the gate terminals G_{s1} to G_{s4} (and G_{s1}' to G_{s4}') of the respective switches. Figure 11.8(a) and (b) shows the PWM waveform for $V_{ref} > 0$ and $V_{ref} < 0$, respectively. It is shown in Figure 11.8 that in shoot-through interval, three switches are turned ON simultaneously. The hybrid PWM [18] used for this proposed hybrid multiple AC outputs and single DC converter has the following constraint between D_s and modulation index (M_a) as follows:

$$D_s + M_a \leq 1 \tag{11.26}$$

The maximum AC voltage gain is satisfied by the constraint (11.26).

11.6 Design consideration

Designs of inductors and capacitors are differed as far as hybrid converters are concerned because inductors and capacitors have to handle both high-frequency and low-frequency components, which are unlike with conventional DC–DC converter. A low-pass filter ($L_{acf}C_{acf}$) is used at the output terminal of the inverter bridge to filter out the higher frequency components.

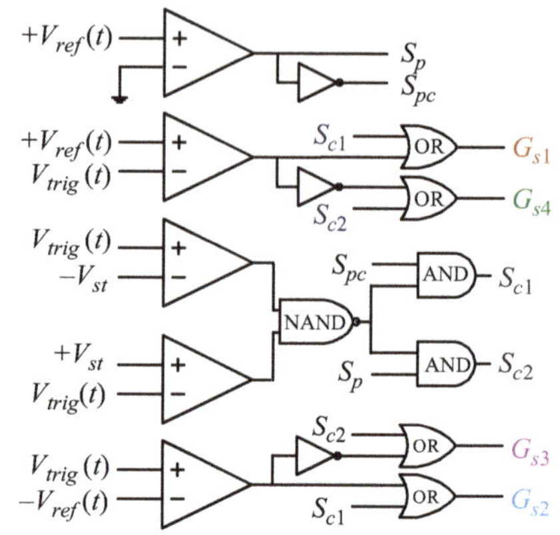

Figure 11.7 Analogue model for generating modified PWM signals

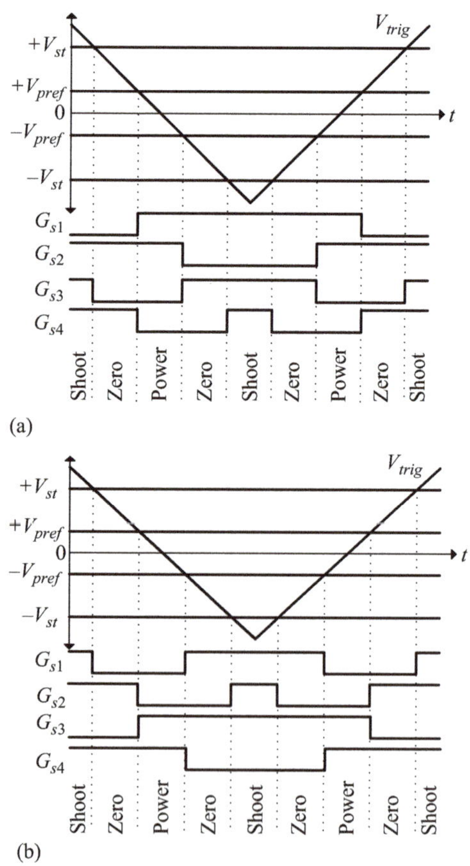

Figure 11.8 Operational PWM waveforms: (a) when $V_{ref}>0$ and (b) when $V_{ref}<0$

The instantaneous output power of the inverter bridge is given as

$$p_{ab} = P_{ab} + \widetilde{P_{ab}} = v_{ab}i_{ab} \\ 0.5V_{abm}I_{abm}\cos\theta - 0.5V_{abm}I_{abm}\cos(2\omega t - \theta) \tag{11.27}$$

where v_{ab} and i_{ab} are the instantaneous voltage and current of the inverter bridge, respectively. V_{abm} and I_{abm} are the peak values of output voltage and current of inverter bridge, respectively. Total output power (P_T) will be the addition of DC power as well as AC power, which will be written as

$$P_T = \frac{V_{odc}^2}{R_{odc}} + p_{ab} \tag{11.28}$$

From (11.27) and (11.28) after simplification, P_T will be as in the following equation:

$$P_T = \frac{V_{odc}^2}{R_{odc}} + 0.5V_{abm}I_{abm}\cos\theta - 0.5V_{abm}I_{abm}\cos(2\omega t - \theta) \tag{11.29}$$

From (11.29), it is clear that the power will consist of average components as well as double-frequency components. The instantaneous ripple power will be given by the following equation:

$$\widetilde{P_{ab}} = 0.5V_mI_m\cos(2\omega t - \theta) \tag{11.30}$$

This instantaneous ripple power should be handled by the inductors (L_1, L_2) and the capacitors (C_1, C_2) present in the circuit. The values of inductors (L_1, L_2) and the capacitors (C_1, C_2) depend on allowable current and voltage ripple, respectively. It can be estimated from the following power balance equation:

$$\begin{aligned}\widetilde{P_{ab}} &= \frac{d}{dt}(E_{L1} + E_{L2} + E_M + E_{c1} + E_{c2}) \\ &= \frac{d}{dt}\left(\frac{1}{2}L_1I_1^2 + \frac{1}{2}L_2I_2^2 - MI_1I_2 + \frac{1}{2}C_1V_{c1}^2 + \frac{1}{2}C_2V_{odc}^2\right)\end{aligned} \tag{11.31}$$

It is clear from the power balance (11.31) that mutually coupled inductance depends on the instantaneous power balance. From (11.31) it is clear that the inductors' current and DC output voltage depend upon AC output power and double frequency ripple. As the active power level increases, the magnitude of the power frequency components will be higher than the ripple due to high frequency. The inductors' currents consist of double frequency as well high frequency ripple, which is clear from (11.29) [17]. The effect of high switching frequency ripple is controlled by conventional DC–DC converters design procedures [44].

11.7 Simulation and experimental verifications

The proposed hybrid multiple outputs of AC and DC with series and parallel topologies are validated with simulation and experimental results with 500 W

Table 11.4 List of parameters and component list for prototype

Specification	
Components/parameters	**Attributes**
Inductors (L_1, L_2)	1,250 μH, 2,120 μH
Coefficient of coupling (k)	0.9
Capacitance (C_1, C_2)	147 μF, 940 μF
Carrier frequency (f_s)	10 kHz
Fundamental frequency	50 Hz
Component list	
Components	*Manufacturer*
Gate drivers	FOD3184 (Fairchild)
Inverter switches	IRGP4790D (IR Corp.)
Diodes	RURG8060 (Fairchild)

(200 W AC+300 W DC) prototype for two simultaneous AC and one DC outputs. It is important to note that the proposed concept is validated for two AC and one DC outputs that can be extended for n-number of outputs. Experimental verification of the proposed topology is validated by using DSP TMS320F28335. The parameters used to validate the concept with their attributes are listed in Table 11.4 and the parameters are designed as per Section 11.6.

11.7.1 Steady-state results of series topology-based converter

The proposed topology is verified in both simulation and experimental domains for series mode of operation. The simulation and experimental waveforms of the hybrid PWM are shown in Figures 11.9 and 11.10, respectively, based on the V_{ref} polarity. In series-connected topology, the voltage across the inverter bridges is equally divided based on the number of inverter bridges and current flowing through the inverter input terminals is the same for symmetrical AC loads. Steady-state simulation results of the proposed topology for two series AC and one DC outputs for $D_s = 0.4$ and $M_a = 0.5$ are shown in Figure 11.11. It is observed from Figure 11.11 that the proposed topology gives two AC output voltages, $V_{aco1} = 66.5$ V (pk–pk) and $V_{aco2} = 66.5$ V (pk–pk) along with DC output voltage $V_{dco} = 133$ V for the input voltage $V_{in} = 48$ V. Experimental verifications are shown in Figure 11.12(a) and (b) for $D_s = 0.4$, $M_a = 0.5$ and $D_s = 0.5$, $M_a = 0.4$, respectively. It can be observed from Figure 11.12(a) that two AC output voltages $V_{aco1} = 53.6$ V (pk–pk) and $V_{aco2} = 52$ V (pk–pk) along with DC output voltage $V_{dco} = 110$ V for the input voltage $V_{in} = 48$ V are obtained from the proposed hybrid converter. Similarly, it can be observed from Figure 11.12(b) that two AC output voltages $V_{aco1} = 61.6$ V (pk–pk) and $V_{aco2} = 59.9$ V (pk–pk) along with DC output voltage $V_{dco} = 165$ V for the input voltage $V_{in} = 48$ V are obtained from the proposed topology. The results validate the steady-state behaviour of the proposed hybrid topology in two series–connected inverter bridges for different values of D_s and M_a.

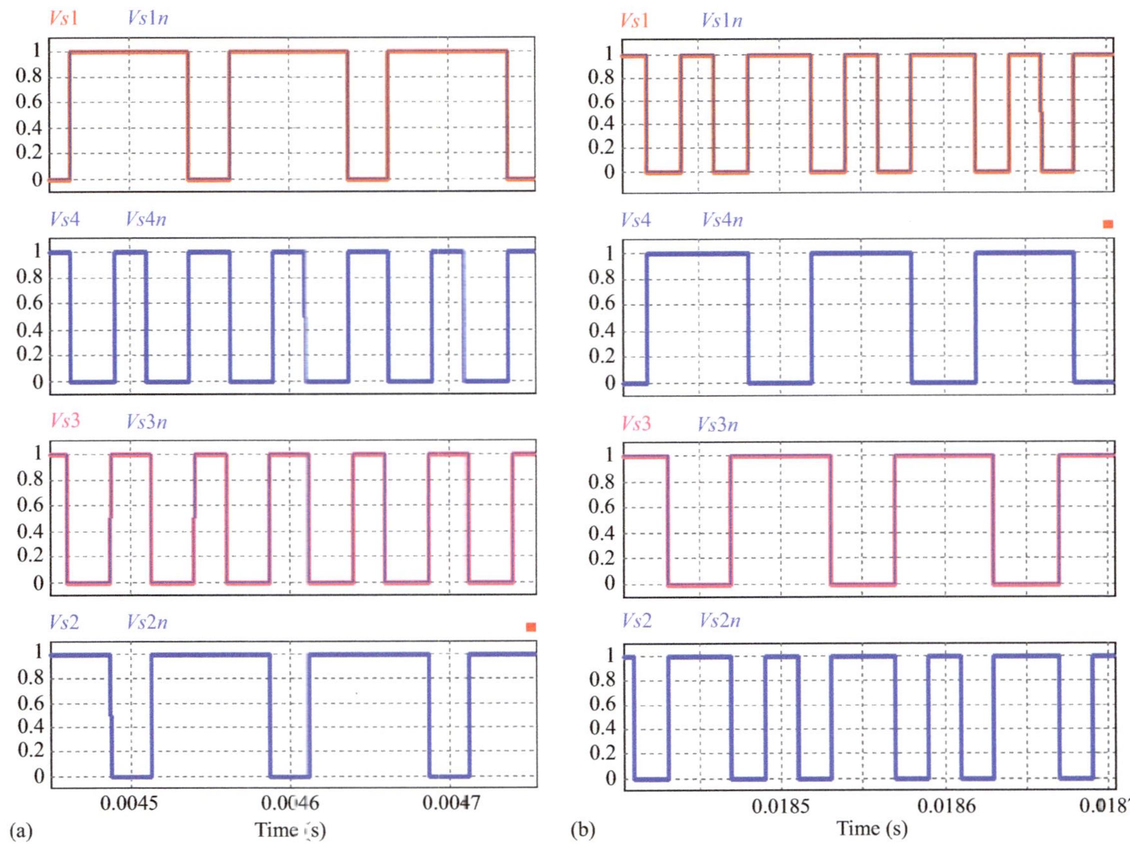

Figure 11.9 Simulation verification of PWM generation: (a) $V_{ref}>0$ and (b) $V_{ref}<0$

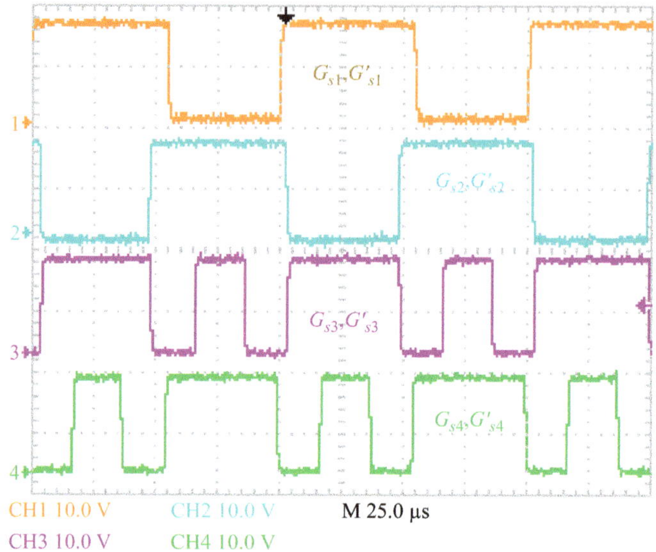

(a)

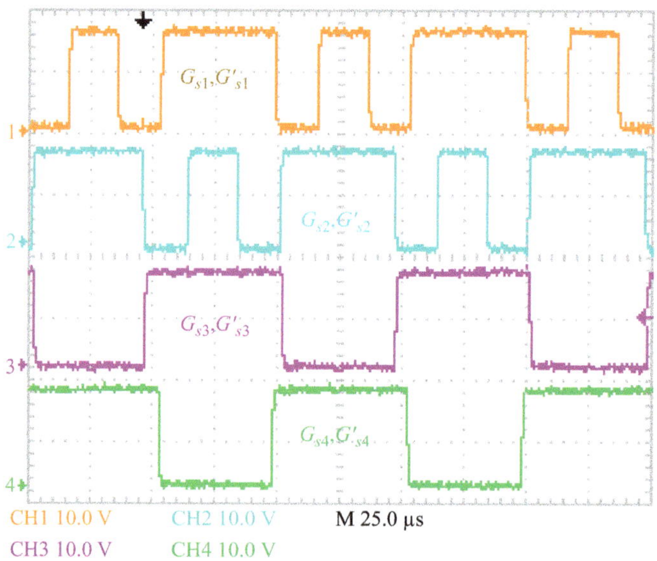

(b)

Figure 11.10 Experimental verification of PWM generation: (a) $V_{ref} > 0$ and (b) $V_{ref} < 0$

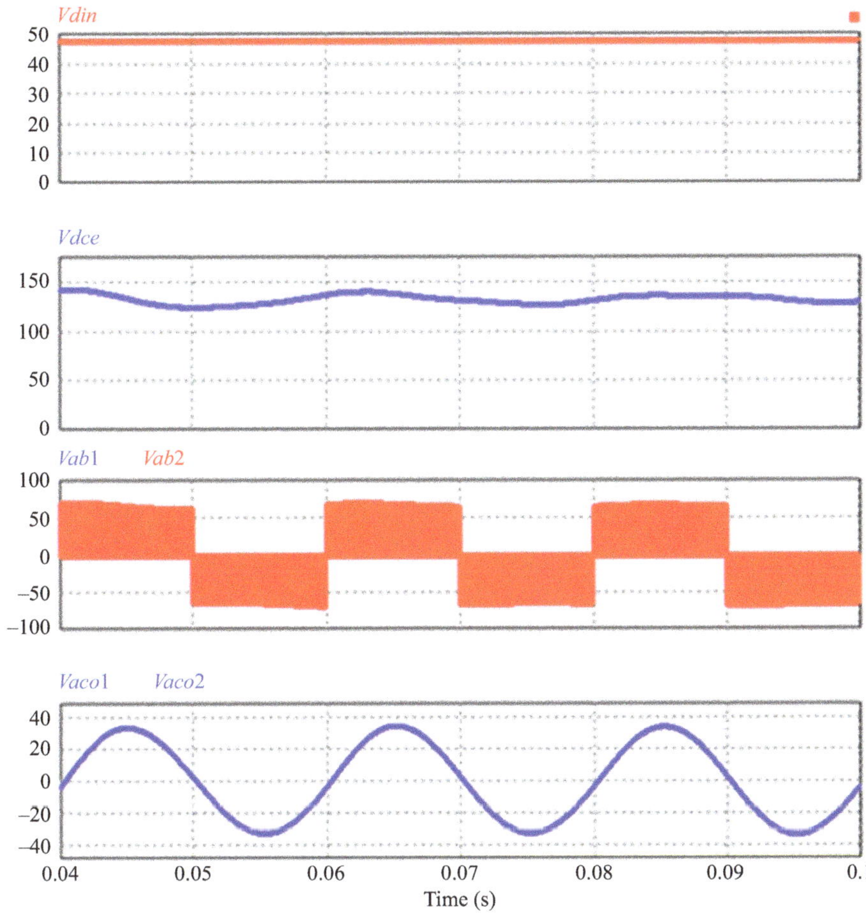

Figure 11.11 Simulation waveforms of two series-connected inverter bridges at $D_s = 0.4$ and $M_a = 0.5$

11.7.2 Steady-state results of a parallel topology-based converter

This section discusses the performance of hybrid topology for two parallel-connected inverter bridges with single DC output. The PWM signals for parallel mode of operation are the same as series mode of operation. In parallel-connected topology, the same switch node voltage appeared across the parallel inverters and current drawn by the inverters are equally divided for the symmetrical AC loads. Steady-state simulation results of the proposed topology for two parallel AC and one DC outputs for $D_s = 0.4$ and $M_a = 0.5$ are shown in Figure 11.13. It is observed from Figure 11.13 that the proposed topology gives two AC output voltages $V_{aco1} = 133$ V (pk–pk) and $V_{aco2} = 133$ V (pk–pk) along with DC output voltage

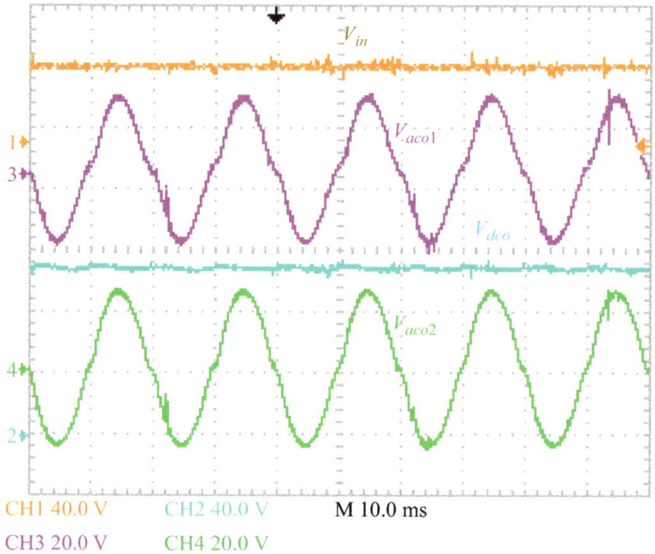

(a)

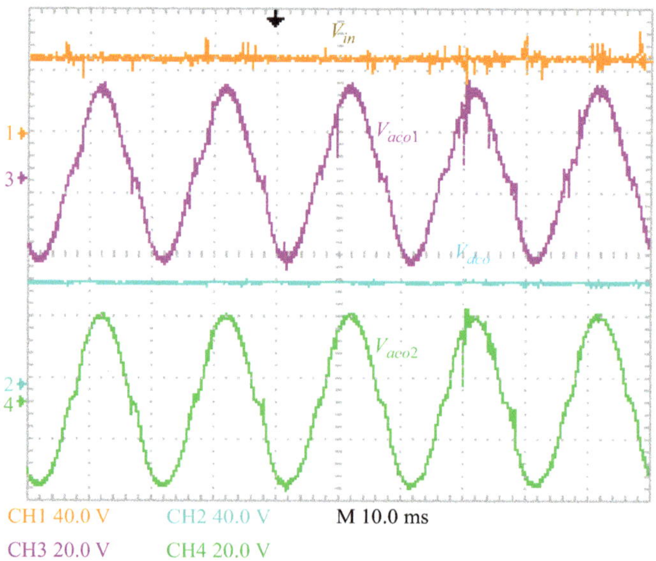

(b)

Figure 11.12 Experimental verification of two series-connected inverter bridges with single DC output: (a) at $D_s = 0.4$ and $M_a = 0.5$ and (b) at $D_s = 0.5$ and $M_a = 0.4$

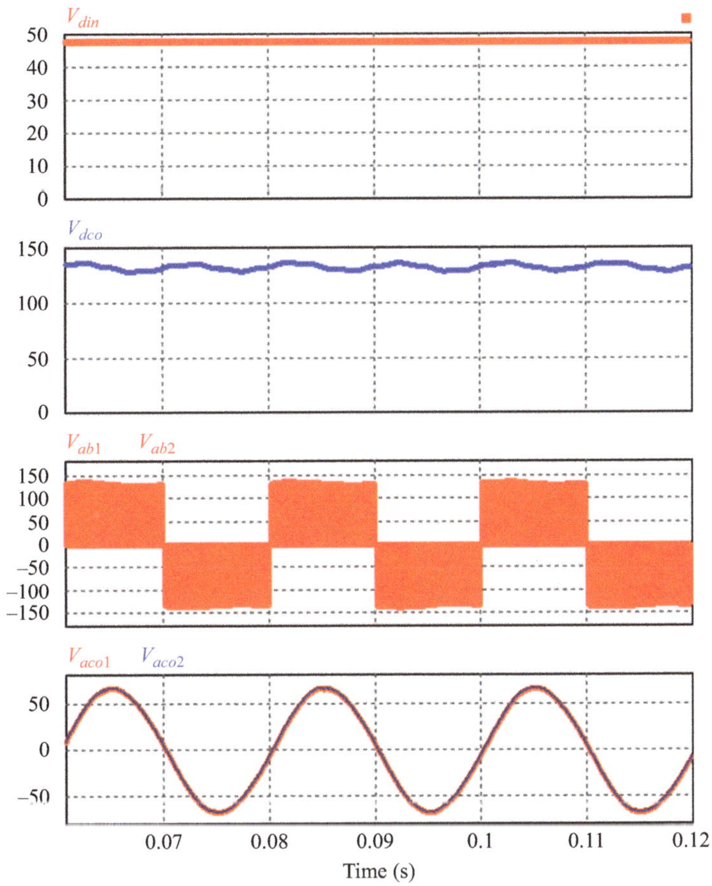

Figure 11.13 Simulation waveforms of two parallel-connected inverter bridges with single DC output at $D_s = 0.4$ and $M_a = 0.5$

$V_{dco} = 133$ V for the input voltage $V_{in} = 48$ V. Experimental verifications are shown in Figure 11.14(a) and (b) for $D_s = 0.4$, $M_a = 0.5$ and $D_s = 0.5$, $M_a = 0.4$, respectively. From Figure 11.14(a), it can be observed that two AC output voltages $V_{aco1} = 88.2$ V (pk–pk) and $V_{aco2} = 87$ V (pk–pk) along with DC output voltage $V_{dco} = 110$ V for the input voltage $V_{in} = 48$ V are obtained experimentally from the proposed topology. Similarly, it can be observed from Figure 11.14(b) that the topology gives two AC output voltages $V_{aco1} = 104$ V (pk–pk) and $V_{aco2} = 103$ V (pk–pk) along with DC output voltage $V_{dco} = 148$ V for the input voltage $V_{in} = 48$ V. Thus, experimental results verify the steady-state behaviour of the proposed hybrid topology in two parallel–connected inverter bridges for different values of D_s and M_a.

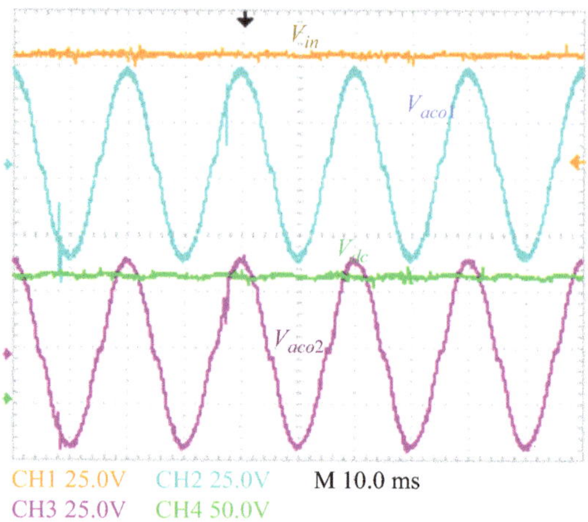

(a)

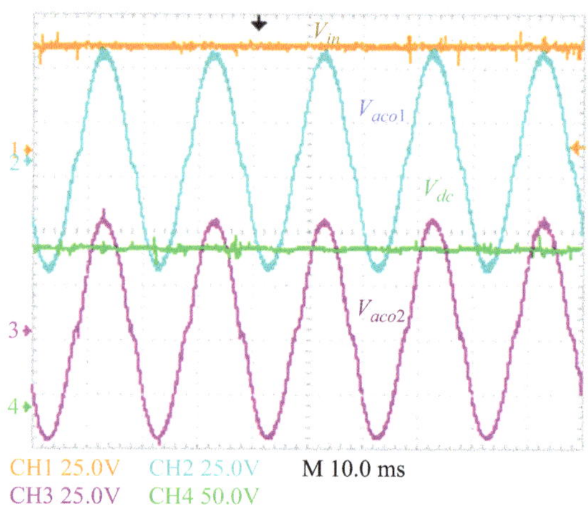

(b)

Figure 11.14 Experimental verification of two parallel-connected inverter bridges with single DC output: (a) at $D_s = 0.4$ and $M_a = 0.5$ and (b) at $D_s = 0.5$ and 0.4

11.7.3 Analysis and comparison

For the analysis of current and voltage profiles of the proposed converter, some experimental results at $D_s = 0.4$ and $M_a = 0.5$ are shown in Figure 11.15 for single AC and single DC system. Figure 11.15(a) shows the voltages and currents ripple information of the proposed converter. From Figure 11.15(a), double-frequency ripple content can be observed in input DC current ($I_{in} = I_1$) and also in intermediate capacitor voltage (V_{C1}) profiles. It is important to note that the double-frequency ripple magnitude can be reduced by selecting a proper inductor and capacitor values. Figure 11.15(b) shows the currents (I_1 and I_2) through the inductors along with voltages (V_{C1} and V_{dco}) across the DC capacitors.

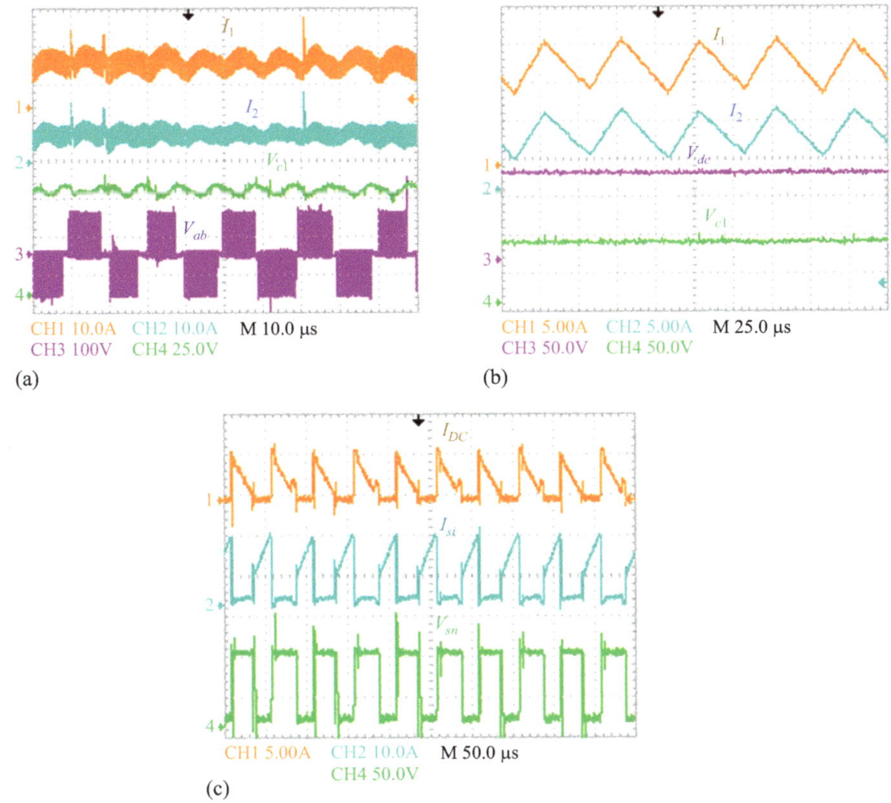

Figure 11.15 Experimental verification of operating waveforms: (a) effect of double frequency in inductor currents and capacitor voltage, (b) steady-state inductor currents and capacitor voltages and (c) switching currents and voltages

Figure 11.15(c) shows the current (I_{Dc}) through the diode D_2, switch node voltage (V_{sn}) and shoot-through current (I_{Si}).

To observe the dynamic behaviour of the proposed converter, closed loop (DC voltage control) simulation results are shown in Figure 11.16 for the step change in DC and AC loads. Figure 11.16(a) and (b) shows the step-up and step-down changes in DC load, respectively, while keeping the AC loads constant. From Figure 11.16(a), it can be observed that DC load voltage (V_{dco}) is well regulated under step-up change in DC load (from $I_{dco} = 2.3$ to 4.4 A) and the AC voltages (V_{aco1} and V_{aco2}) and AC currents (I_{aco1} and I_{aco2}) are not affected. From Figure 11.16(b), it can be observed that DC load voltage (V_{dco}) is well regulated under step-down change in DC load (from $I_{dco} = 4.4$ to 2.3 A) and the AC voltages (V_{aco1} and V_{aco2}) and AC currents (I_{aco1} and I_{aco2}) are affected during transition period. Figure 11.16(c) and (d) shows the step-down and step-up changes in one of the AC loads of the parallel-connected topology while keeping the DC load

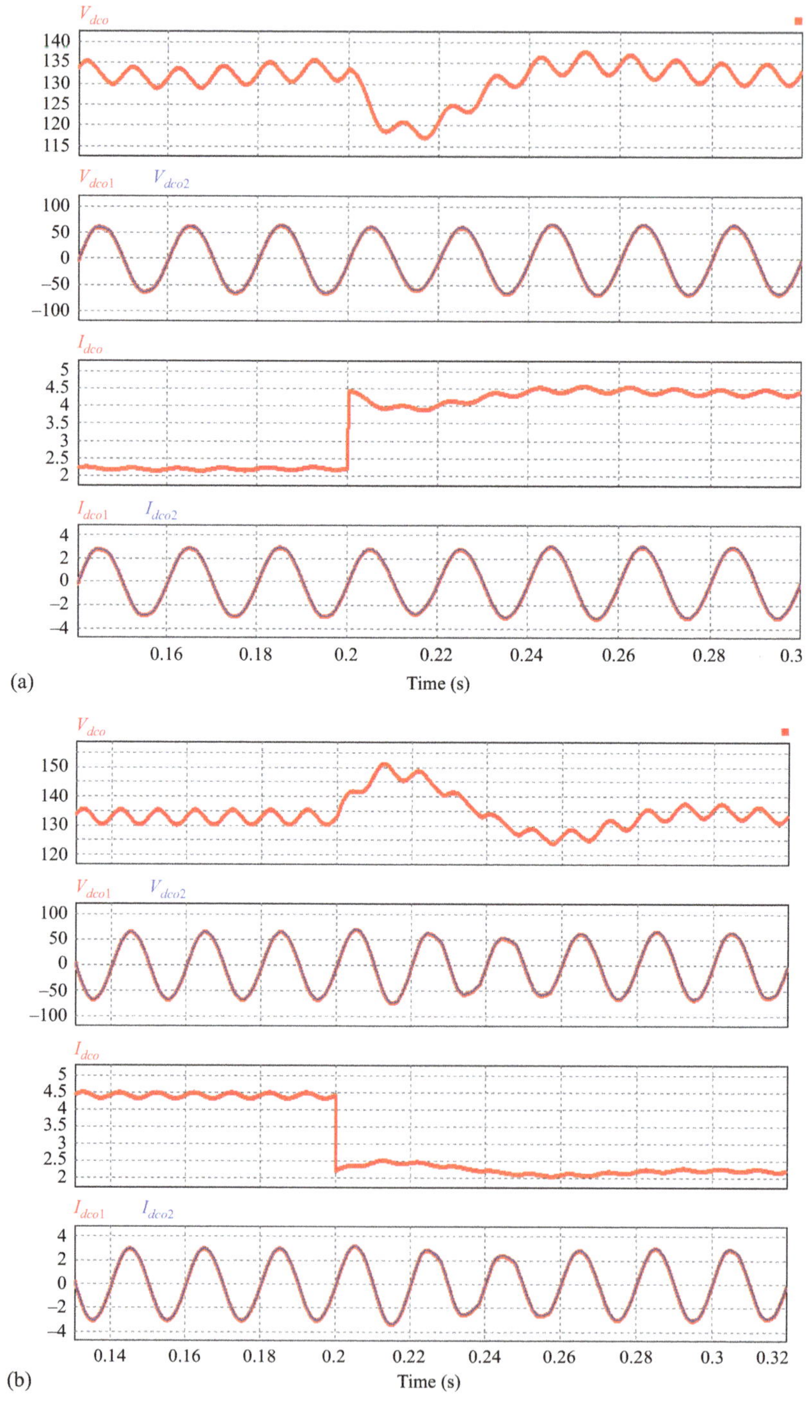

Figure 11.16 Dynamic response of the proposed converter under step change in DC and AC loads: (a) step-up (2.3 to 4.4 A) change in DC load, (b) step-down (4.4 to 2.3 A) change in DC load, (c) step-down (10 (pk–pk) to 5 A) change in one of the AC load and (d) step-up (5 (pk–pk) to 10 A) change in one of the AC load

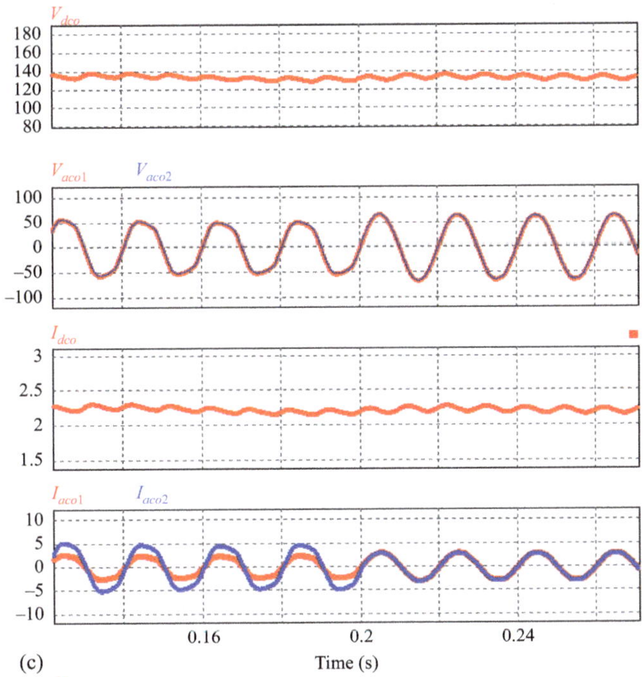

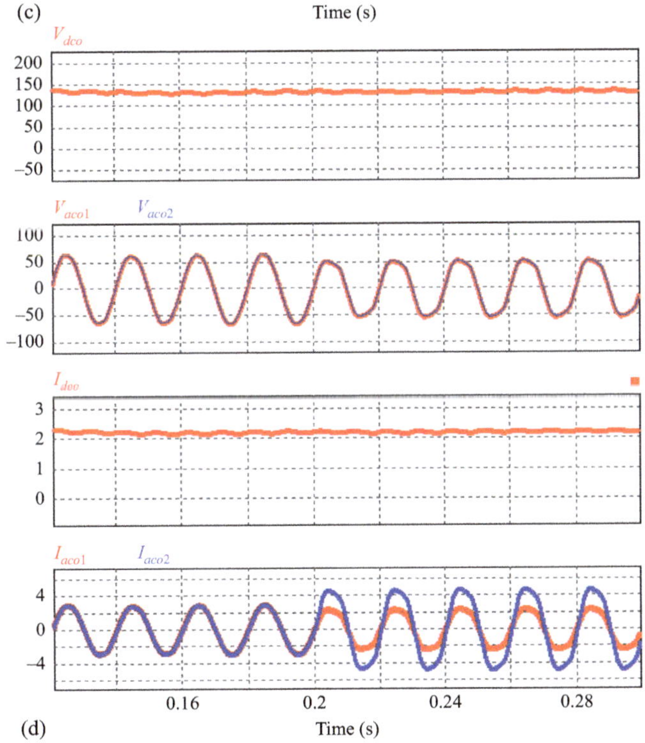

Figure 11.16 (*Continued*)

constant. From Figure 11.16(c), it can be observed that small variation in the AC load voltages (V_{aco1} and V_{aco2}) under step-down change in AC load (from I_{aco1} = 10 to 5 A) and the DC voltage (V_{dco}) and currents (I_{dco}) are not affected. From Figure 11.16(d), it can be observed that small variation in the AC load voltages (V_{aco1} and V_{aco2}) under step-up change in AC load (from $I_{aco1} = 5$ to 10 A) and the DC voltage (V_{dco}) and currents (I_{dco}) are not affected.

The variation of DC gain and AC gains (series and parallel) with respective to the shoot-through duty ratio (D_S) is shown in Figure 11.17. Figure 11.17(a) shows the DC voltage gain variation over the (D_S), and it can be observed that

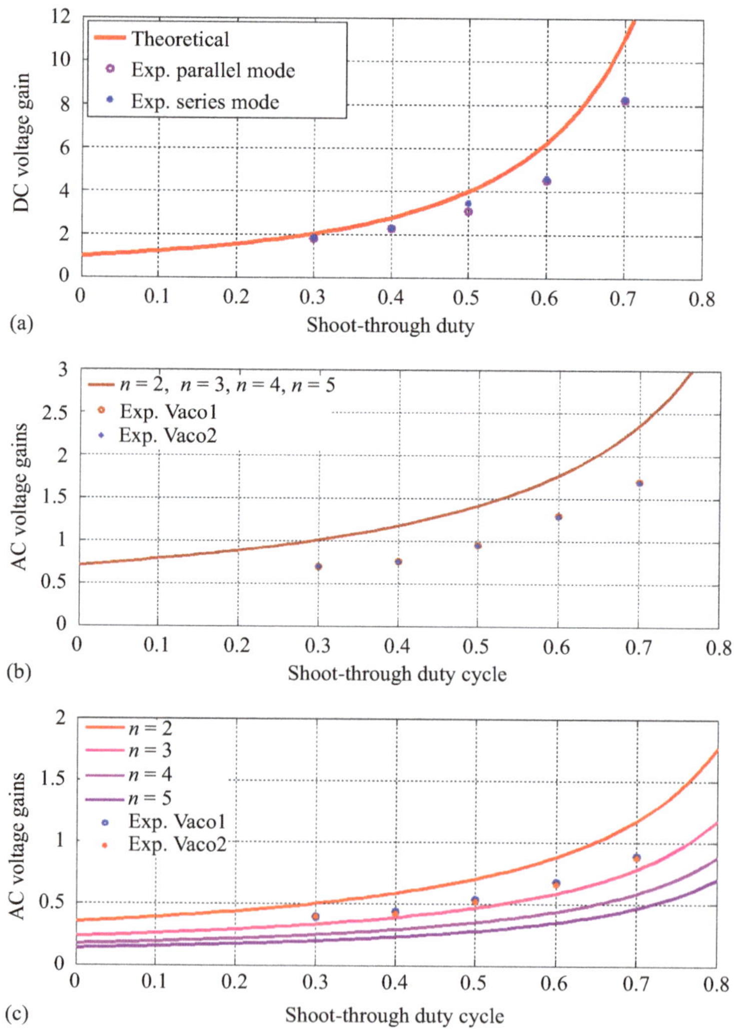

Figure 11.17 Variation of voltage gains of the proposed converter over the D_s: (a) DC voltage gain, (b) AC voltage in parallel mode of operation and (c) AC voltage in series mode of operation

experimental DC gain is close to theoretical gain in series and parallel mode of operations. Figure 11.17(b) shows the AC voltage gain variation over the (D_S) for the parallel-connected topology. Figure 11.17(b) shows that the experimental AC voltage gains of two inverter circuits are almost equal and less than the theoretical AC voltage gain when two inverter circuits are used. For an n number of inverter circuits in parallel mode, the AC voltage gain is constant for the symmetrical AC loads and the same can be observed from Figure 11.17(b). Figure11.17(c) shows the AC voltage gain variation over the (D_S) for the series-connected topology. From Figure 11.17(c), it can be observed that the experimental AC voltage gains of two inverter circuits are almost equal and less than the theoretical AC voltage gain when two inverter circuits are used. In series mode, the AC voltage gain is varied for the symmetrical AC loads according to n number of inverter circuits and the same can be observed from Figure 11.17(c).

11.8 Conclusion

This chapter has presented a magnetically coupled extendable multiple-output hybrid converter for an AC/DC microgrid, derived from single switch quadratic boost converter topology. The filter inductors of the converter are coupled magnetically and therefore, the weight and volume of the system are reduced. The main switch of the converter is replaced by n number of series or parallel–connected H-bridge inverter circuits to supply multiple AC outputs with single DC output according to requirements of a hybrid AC/DC microgrid. A comprehensive mathematical modelling of the proposed converter has been given in this chapter for the series and parallel mode of operations. The detailed steady-state and dynamic analysis are also carried for analysing the steady state and transient behaviour of the proposed converter. To validate the proposed concept, two H-bridge inverters are considered for series and parallel mode operation. Thus, the proposed concept is validated with simulation and experimental results on a 500 W laboratory prototype for two simultaneous AC and one DC outputs.

References

[1] Guerrero J. M, Blaabjerg F., Zhelev T., *et al.*, Distributed generation: Toward a new energy paradigm, *IEEE Ind. Electron. Mag.*, 2010; 4(1):52–64.

[2] Chauhan R. K., Rajpurohit B. S., Singh S. N., and Gonzalez-Longatt F. M., Intelligent energy management system for PV-battery-based microgrids in future DC homes, *Int. J. Emerg. Electr. Power Syst.*, 2016; 17(1): 173–189.

[3] Rathore A. K., Patil D. R., and Srinivasan D., Non-isolated bidirectional soft-switching current-fed LCL resonant DC/DC converter to interface energy storage in DC microgrid, *IEEE Trans. Ind. Appl.*, 2016; 52(2): 1711–1722.

[4] Chauhan R. K., Rajpurohit B. S., Hebner R. E., Singh S. N., and Gonzalez-Longatt F. M., Voltage standardization of DC distribution system for residential buildings, *J. Clean Energ Technol.*, 2016; 4(2):167–172.

[5] Loh P. C., Li D., Chai Y. K., *et al.*, Autonomous operation of hybrid microgrid with AC and DC subgrids, *IEEE Trans. Power Electron.*, 2013; 28 (5): 2214–2223.

[6] Yu X., Khambadkone A. M., Wang H., *et al.*, Control of parallel-connected power converters for low-voltage microgrid—Part I: A hybrid control architecture, *IEEE Trans. Power Electron.*, 2010; 25(12): 2962–2970.

[7] Eghtedarpour N., and Farjah E., Power control and management in a hybrid AC/DC microgrid, *IEEE Trans. Smart Grid*, 2014; 5(3): 1494–1505.

[8] Sasidharan N., and Singh J. G., A novel single-stage single-phase reconfigurable inverter topology for a solar powered hybrid AC/DC home, *IEEE Trans. Ind. Electron.*, 2017; 64(4): 2820–2828.

[9] Chandrasena R. P. S., Shahnia F., Rajakaruna S., and Ghosh A., Dynamic operation and control of a hybrid nanogrid system for future community houses, *IET Gener. Transm. Distrib.*, 2015; 9(11): 1168–1178.

[10] Carrasco J. M., Franquelo L. G., Bialasiewicz J. T., *et al.*, Power electronic systems for the grid integration of renewable energy sources: A survey, *IEEE Trans. Power Electron.*, 2006; 53(4): 1002–1016.

[11] Li W., and He X., Review of nonisolated high step up DC-DC converters in photovoltaic grid-connected applications, *IEEE Trans. Ind. Electron.*, 2011; 58: 1239–1250.

[12] Forouzesh M., Siwakoti Y. P., Gorji S. A., Blaabjerg F., and Lehman B., Step-up DC–DC converters: A comprehensive review of voltage-boosting techniques, topologies, and applications, *IEEE Trans. Power Electron.*, 2017; 32(12): 9143–9178.

[13] Patra P., Patra A., and Misra N., A single-inductor multiple-output switcher with simultaneous buck, boost and inverted outputs, *IEEE Trans. Power Electron.*, 2012; 27(4): 1936–1951.

[14] Jamil M., Hussain B., Abu-Sara M., Boltryk R. J., and Sharkh S. M., Microgrid power electronic converters: State of the art and future challenges, *44th International Universities Power Engineering Conference (UPEC)*, Glasgow, IEEE, 2009, pp. 1–5.

[15] Naqvi S. B. Q., Singh D., and Sarwar A., Hybrid multi output converter for microgrid applications, *Adv. Comput. Sci. Technol.*, 2017; 10(4): 599–611.

[16] Rehman Z., Al-Bahadly I., and Mukhopadhyay S., Multiinput DC–DC converters in renewable energy applications – An overview, *Renew. Sustain. Energy Rev.*, 2015; 41: 521–553.

[17] Ray O., and Mishra S., boost-derived hybrid converter with simultaneous DC and AC outputs, *IEEE Trans. Ind. Appl.*, 2014; 50(2): 1082–1093.

[18] Ahmad A., Bussa V. K., Singh R. K., and Mahanty R., Quadratic boost derived hybrid multi-output converter, *IET Power Electron.*, 2017; 10(15): 2042–2054.

[19] Nag S. S., and Mishra S., Current-fed switched inverter, *IEEE Trans. Ind. Electron.*, 2014; 61(9): 4680–4690.

[20] Nguyen M. K., Le T. V., Park S. J., and Lim Y. C., A class of quasi switched boost inverters, *IEEE Trans. Ind. Electron.*, 2015; 62(3): 1526–1536.

[21] Ahmad A., Singh R. K., and Mahanty R., Minimum phase hybrid coupled inductor quadratic boost inverter, *Proc. Annual Conf. of the IEEE Ind. Electron. Society (IECON),* Florence, IEEE, 2016, pp. 3727–3732.

[22] Ravindranath A., Mishra S. K., and Joshi A., Analysis and PWM control of switched boost inverter, *IEEE Trans. Ind. Electron.*, 2013; 60(12): 5593–5602.

[23] Bussa V. K., Ahmad A., Singh R. K., and Mahanty R., Interleaved hybrid converter with simultaneous DC and AC outputs for DC microgrid applications, *IEEE Trans. Ind. Appl.*, 2018; 54(3): 2763–2772.

[24] Lakshmi M., and Hemamalini S., Nonisolated high gain DC–DC converter for DC microgrids, *IEEE Trans. Ind. Electron.*, 2018; 65(2): 1205–1212.

[25] Ray O., Josyula A. P., Mishra S., and Joshi A., Integrated dual-output converter, *IEEE Trans. Ind. Electron.*, 2015; 62(1): 371–382.

[26] Chen G., Jin Z., Deng Y., He X., and Qing X., Principle and topology synthesis of integrated single-input dual-output and dual-input single-output DC–DC converters, *IEEE Trans. Ind. Electron.*, 2018; 65: 3815–3825.

[27] Boora A. A., Zare F., and Ghosh A., Multi-output buck-boost converter with enhanced dynamic response to load and input voltage changes, *IET Power Electron.*, 2011; 4(2): 194–208.

[28] Chen Y., Huang A. Q., and Yu X., A high step-up three-port DC–DC converter for stand-alone PV/battery power systems, *IEEE Trans. Power Electron.*, 2013; 28(11): 5049–5062.

[29] Cornea O., Andreescu G., Muntean N., and Hulea D., Bidirectional power flow control in a DC microgrid through a switched-capacitor cell hybrid DC–DC converter, *IEEE Trans. Ind. Electron.*, 2017; 64(4): 3012–3022.

[30] Tran V., and Choi W., Novel time division multiple control method for multiple output battery charger, *IEEE Trans. Power Electron.*, 2014; 29(10): 5102–5105.

[31] Kurokawa F., and Matsuo H., A new multiple-output hybrid power supply, *IEEE Trans. Power Electron.*, 1988; 3(4): 412–419.

[32] Huang W., Qahouq J. A. A., and Dang Z., CCM–DCM power-multiplexed control scheme for single-inductor multiple-output DC–DC power converter with no cross regulation, *IEEE Trans. Ind. Appl.*, 2017; 53(2): 1219–1231.

[33] Fan S., Xue Z., Lu H., Song Y., Li H., and Geng L., Area-efficient on-chip DC–DC converter with multiple-output for bio-medical applications, *IEEE Trans. Circuits Syst. I, Reg. Pap.*, 2014; 61(11): 3298–3308.

[34] Zhu X., Zhang B., and Qiu D., A new half-bridge impedance source inverter with high voltage gain, *IEEE Trans. Power Electron.*, 2019; 34(4): 3001–3008.

[35] Ahmad A., Bussa V. K., Singh R. K., and Mahanty R., Switched-boost modified Z-source inverter topologies with improved voltage gain capability, *IEEE J. Emerg. Sel. Top. Power Electron.*, 2018; 6(4): 2227–2244.

[36] Das M., Pal M., and Agarwal V., Novel high gain, high efficiency DC–DC converter suitable for solar PV module integration with three-phase grid tied inverters, *IEEE J. Photovoltaics*, 2019; 9(2): 528–537.

[37] Ahmad A., and Singh R. K., Embedded dual switched-capacitor based continuous input current switched-inverter for renewable energy application, *IET Power Electron.*, 2019; 12(5): 1263–1273.

[38] Kim K., Cha H., and Kim H., A new single-phase switched-coupled-inductor DC–AC inverter for photovoltaic systems, *IEEE Trans. Power Electron.*, 2017; 32(7): 5016–5022.

[39] Sreekanth T., Lakshminarasamma N., and Mishra M. K., A single-stage grid-connected high gain buck–boost inverter with maximum power point tracking, *IEEE Trans. Energy Convers.*, 2017; 32(1): 330–339.

[40] Ahmad A., and Singh R. K., Single-phase modified continuous input current switched boost inverter for high voltage gain, *2017 IEEE Transportation Electrification Conference (ITEC-India)*, Pune, IEEE, 2017, pp. 1–6.

[41] Bussa V. K., Ahmad A., Singh R. K., and Mahanty R., Single-phase high-voltage gain switched LC Z-source inverters, *IET Power Electron.*, 2018; 11 (5): 796–807.

[42] Kishore P. M., and Bhimasingu R., A simplified converter with simultaneous multi-level AC and boost DC outputs for hybrid microgrid applications, *Proc. IEEE Int. Conf. on Power Electronics, Drives and Energy Systems (PEDES)*, Trivandrum, IEEE, 2016, pp. 1–6.

[43] Lee S. S., and Heng Y. E., Improved single-phase split-source inverter with hybrid quasi-sinusoidal and constant PWM, *IEEE Trans. Ind. Electron.*, 2017; 64(3): 2024–2031.

[44] R. W. Erickson, and D. Maksimovic, *Fundamentals of Power Electronics*, 2nd ed., Springer International Edition, New Delhi, India, 2010.

Chapter 12
Multi-input converters for distributed energy resources in microgrids

Hossein Torkaman[1], Soheyl Khosrogorji[1], and Ali Keyhani[2]

In this chapter, the multi-input converters for distributed generation (DG) resources in Microgrids are introduced and categorized. In the past, some DC/DC converters used to be connected parallel to each other and thus multiple converters were created. But today, many structures have been proposed for this purpose. The main reason behind employing these converters is to get the following features based on their application and topology: fewer semiconductors, less volume, fewer switches' stress, simple control, low power loss, high efficiency, high gain, low manufacturing cost, etc. In this chapter, multi-input converters are categorized into three groups based on their structure: magnetic, electrical, and electromagnetic types. The first group includes topologies that have a multi-winding transformer. The second group consists of topologies that have DC-link and sources connected to the DC-link, and the latter contains topologies that have DC-link and transformer. In this chapter, the topologies of various multi-input DC/DC converters will be reviewed and compared from different aspects such as the battery life, the soft-switching, the source utilization, the isolation between the input and the output. This chapter providing an information source in the selection guide on multi-input DC/DC converters of DER in Microgrids for researchers, designers and application engineers.

12.1 Overview

In recent years, due to the increasing and manufacturing technology of connected converters to distributed energy resources (DER), using DGs has been growing. Using individual DG in the system can cause many problems such as low power density, voltage oscillation, protection issues, etc. [1]. Traditional converters can connect to only one type of DGs with different methods like multi-level converter [2,3], Z-source converter [4,5], Γ-Z-source converter [6–8], and primary DC/DC

[1]Faculty of Electrical Engineering, Shahid Beheshti University, Tehran, Iran
[2]Department of Electrical and Computer Engineering, Ohio State University, Ohio, USA

converter such as buck-boost converter [9–11], but recently, some sources have been able to connect simultaneously to converters and power generation of DGs has transferred to output [12]. These converters are known as multi-input converters, and each one of them has their own structure [13,14]. One additional reason for using multi-input converter is slew-rate limits of DGs such as fuel cell (FC) next to the battery or super-capacitor [15]. For the first time in 1999, a structure was disguised as a multi-input converter, but the practical use of these converters need to be studied. Figure 12.1 shows types of two common multi-input converters. A multi-winding transformer is used in converters with magnetically coupling. Aggregate production fluxes from each of the primary windings induce a voltage in the secondary winding and transfer energy from input to output. In these converters, each of the inputs is controlled by input inductance of dispersion and phase angle [13]. Conversely, converters with electrical coupling do not have electrical isolation between input and output. In this type of structure, non-isolated converters such as buck converters, boost converters are used. Control of converters with electrically coupling is easier than magnetic coupling.

Generally, converters are divided into three categories: converters with (a) magnetically, (b) electrically, and (c) electrically-magnetically couplings as shown in Figure 12.2. These converters could be controlled by sine pulse width modulation (SPWM) or pulse width modulation (PWM) or some control power sources similar to the references [16–18]. The main feature of these converters is their ability to transfer power control from input to output which is only possible with an appropriate control system. Figure 12.3 shows a general form of a multi-input converter.

This chapter introduces the structures of multi-input converters. The multi-input converters with magnetically coupling and with electrically coupling and compound converters are introduced. Finally, a comparison study and conclusion

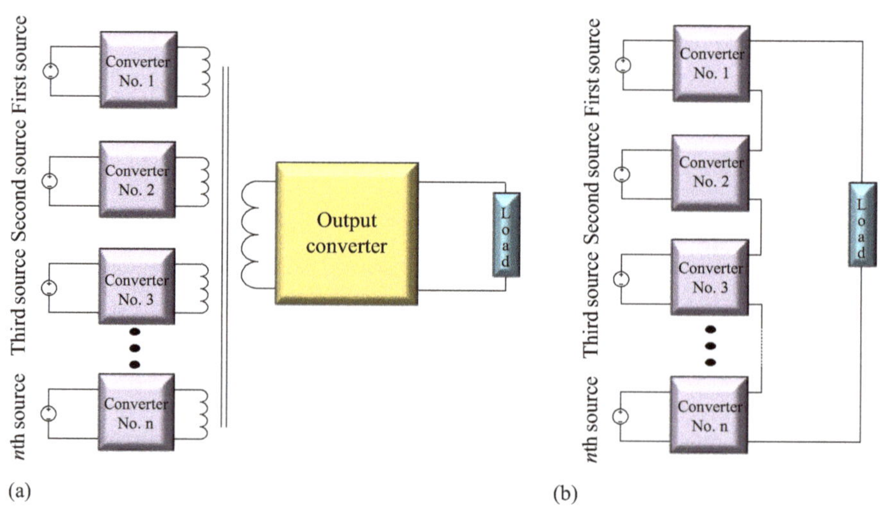

Figure 12.1 Multi-input converters: (a) magnetically coupling and (b) electrically coupling

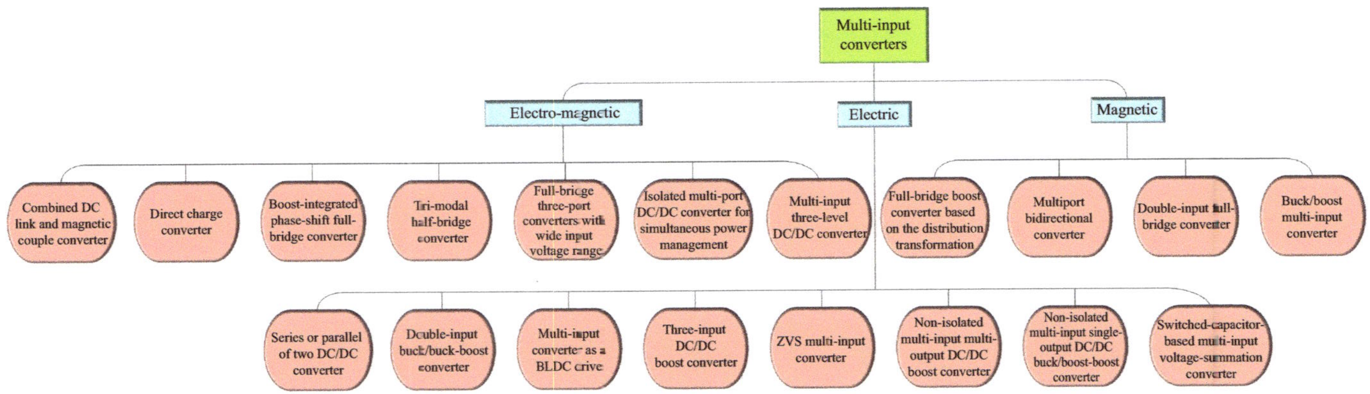

Figure 12.2 Different topologies of the multi-input converters in connection with DGs

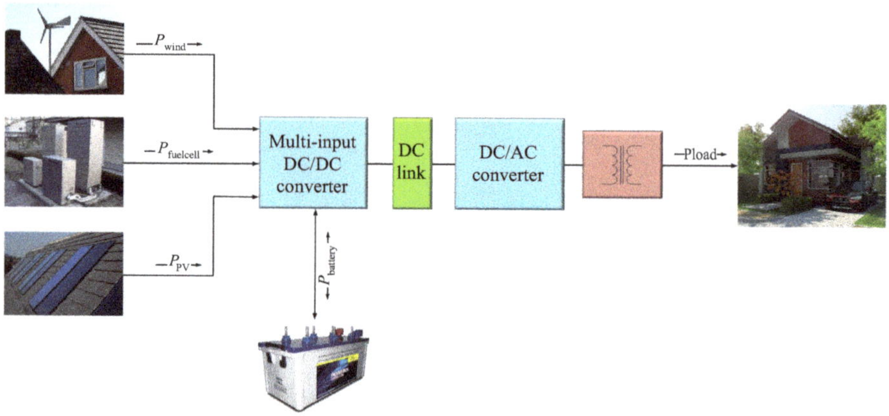

Figure 12.3 The general form of a multi-input converters

will be presented. The comparison helps designers to select an appropriate converter for their applications with different conditions.

12.2 Magnetically coupled multi-input converters

In these types of converters, a multi-winding transformer is used for magnetically coupling. Aggregate production fluxes from each of the primary windings induce a voltage in the secondary winding and transfer energy from input to output. In these converters, each of the inputs is controlled by input inductance of dispersion and phase angle.

12.2.1 Buck-boost multi-input converter

The first topology for magnetically multi-input converter was presented in [19]. This topology is buck-boost multi-input converter that is shown in Figure 12.4.

In this topology, the transformer has two roles. First, an inductor for the buck-boost converter, and second, aggregate production fluxes from each of sources. This converter has three states. First, E_1 charges L and S_1 is on. Second, E_2 charges L and S_2 is on. Third, L supply the load. Output voltage equivalent for this converter is:

$$V_o = \frac{D_1}{N_1(1 - D_1 - D_2)}E_1 + \frac{D_2}{N_2(1 - D_1 - D_2)}E_2 \tag{12.1}$$

D_1 is the duty cycle of S_1, D_2 is the duty cycle of S_2, and N_1 and N_2 are the number of two primary windings. In the first and second state, the output capacitor must be able to supply the load. Because, in these states, D is off. If one of the input could not provide the output (i.e. one of the input is a photovoltaic (PV) panel), the other source could produce the output. In this case, this converter is similar to the buck-boost converter with one source.

In this converter, two sources could not flow power to load with together. It is one of the disadvantages of this converter. Although there are some advantages of

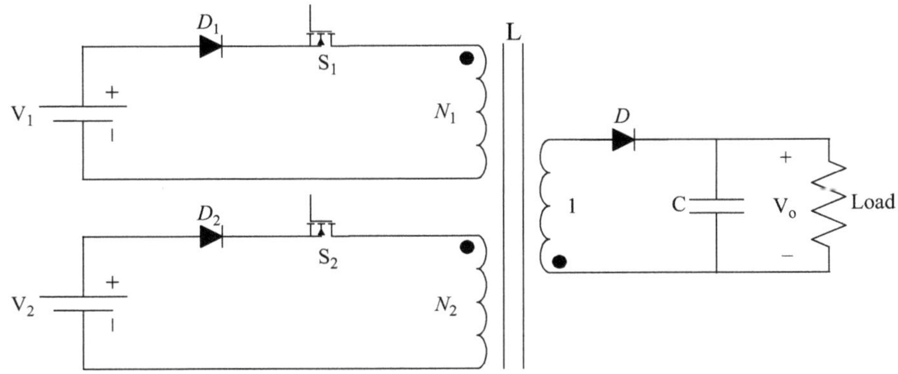

Figure 12.4 Buck-boost multi-input converter [20]

this structure such as no need for complicated control for switching. Moreover, disregarding the battery recharge possibility is another drawback of this structure. If one of the sources connected to the grid, transferring extra power to the grid is impossible. In this topology, soft-switching is not noted.

The PFC can be used to correct power factor when one of the input sources is commercial AC line [21]. This topology can be used with a different number of sources such as a battery, FC, etc. [22].

12.2.2 Double-input full-bridge converter

In a dual-input full-bridge converter, two different converters are utilized in the input and output [23–26]. The input converters operate as the current supplier, and the output converter is a full-bridge diode rectifier. The layout of this converter is shown in Figure 12.5, which is introduced as a multi-input converter based on the flux increase.

This converter works by the flux deviation in the transformer, although, multiple suppliers can be connected in this method. Since the input converters are current types, in the control moments of the converter, both converters have to be short-circuited.

Figure 12.6 shows the operation of the converter. This converter is controlled by phase-shifted PWM control. So, switches operate at 50% duty cycle. Output voltage equivalent for this converter is:

$$V_o = \frac{n_3}{n_1} V_1 + \frac{n_2}{n_1} V_2 \tag{12.2}$$

As some advantages of this structure, the soft switching can be noted. Moreover, the control of the converter is only operated on input, and the output needs no supervision. Some disadvantages of this structure are: disregarding the battery recharge possibility; in addition to this, if one of the inputs is connected to the grid, the return of power in this structure is impossible, so the values of voltage magnitude of both suppliers have to be the same in order to utilize both suppliers, simultaneously.

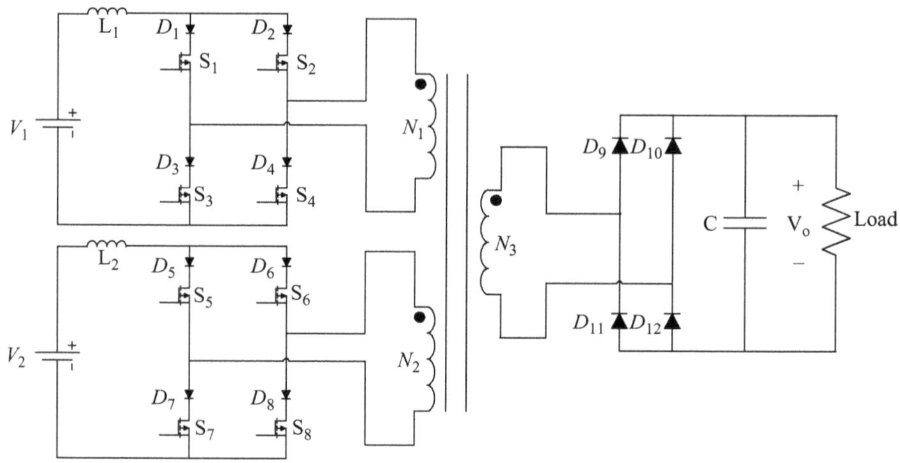

Figure 12.5 Double-input full-bridge converter [23]

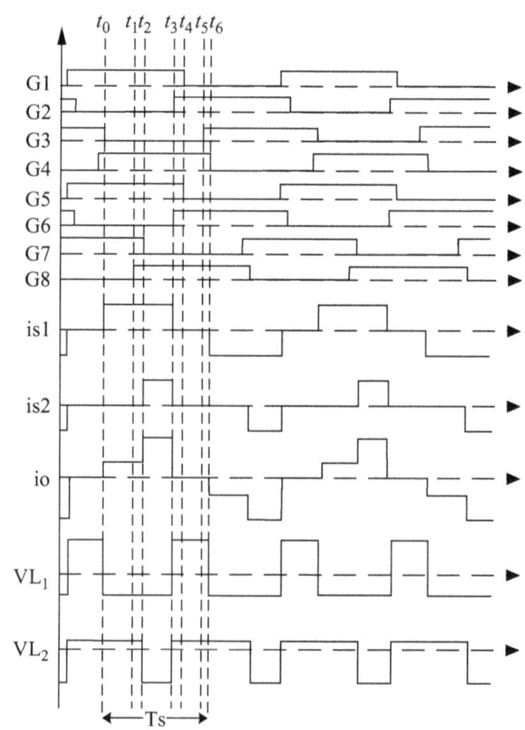

Figure 12.6 Operation modes [23]

In this structure, three voltage suppliers can be utilized instead of the current supplier [27]. Also, the efficiency of the converter could be increased while increasing the number of inputs with the proper design [28]. Furthermore, series resonance tank is utilized on the secondary of the input converters for improve soft-switching. By increasing the number of the windings and the full bridge rectifiers, a 4-input converter can be developed [29].

By combining full-bridge and half-bridge converters, a hybrid structure would be created to supply FC's electrolyzer [30]. This structure can be used to generate power in a FC. In this structure, when the FC is sufficiently charged, the energy from the FC is transferred to the DC link, and when additional power is generated in the system, energy is transferred from the DC link to the electrolyzer. A current-fed full-bridge DC–DC converter can be used as a dual input full-bridge converter [31].

Figure 12.7 illustrates a layout of the 2-input full-bridge converter. Two modes of operation are available for creating a positive voltage in this converter. First, S_1, S_4, and S_5 are turned on at the same time and sum of the voltage of the input $(V_1 + V_2)$ get on the primary of the transformer. Second, the switches S_1 and S_4 are turned on at the same time, and the diode of switch S_6 is turned on, too. In this state, only V_1 supply the primary of the transformer. For creating negative voltage, previous states happens for switches S_2, S_3, S_6 and reverse diode of switch S_5. The major problem of this converter is the impossibility of charging the suppliers. Also, both suppliers have to be distributed generator like PV panels.

12.2.3 Multi-port bidirectional converter

Figure 12.8 shows a schematic which uses half-bridge converters. This converter works by producing a square pulse in suppliers and then rectifying it through the converter. In this converter, three full-bridge converters can be employed instead of

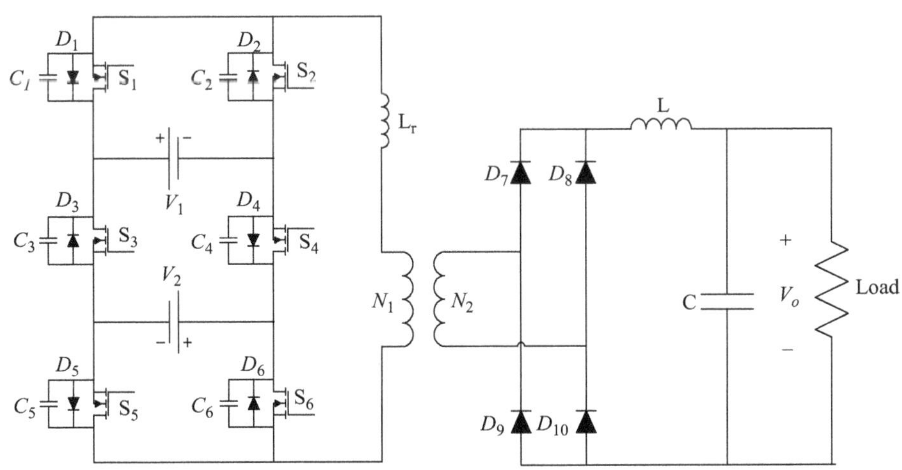

Figure 12.7 Two-input full-bridge converter [32]

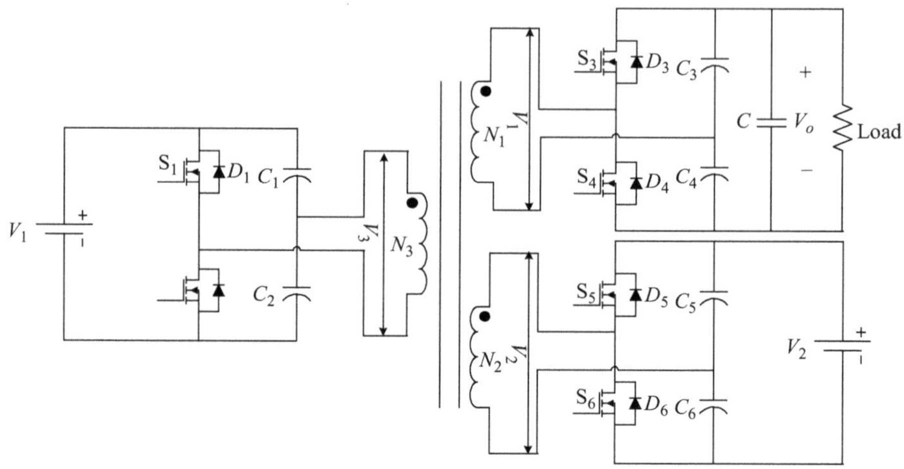

Figure 12.8 Magnetic-coupled half-bridge converter [34]

three half-bridge ones [33]. For this purpose by changing one of the current converters to voltage converter, can be achieved to a high power converter. The benefits of this structure are isolation between input suppliers and output load, different suppliers with different levels of voltage can be connected through transformer winding turn adjustment, battery recharge possibility, soft-switching and the possibility of adding extra suppliers through the increase of the winding turns. Lack of battery recharge control is one of the downsides of this structure. This converter is controlled with phase-shifted PWM control. So, switches operate at 50% duty cycle. The operation of this converter is similar to double-input full-bridge converter. Voltage and current waveforms of the input and output are shown in Figure 12.9.

The standard switching cell converter can be utilized instead of one of the half-bridge converters [35]. The significant benefit of conventional switching is the ripple reduction in the current, taken from the supplier. Also, both half-bridge rectifiers can be changed by boost half-bridge ones to produce the electronic ballasts [36].

In Figure 12.10, three current-fed converters are used. In this structure, the secondary winding of the transformers is connected to a capacitor and capacitors is connected in delta. Capacitors can limit the dynamics of secondary transformer voltage.

12.2.4 Full-bridge boost DC–DC converter based on the distributed transformers

Figure 12.11 shows this type of converter which there are three modes of operation. First, all four switches of both converters are turned on. In this phase, the input inductor is charged and made the boost condition, and the load is supplied through

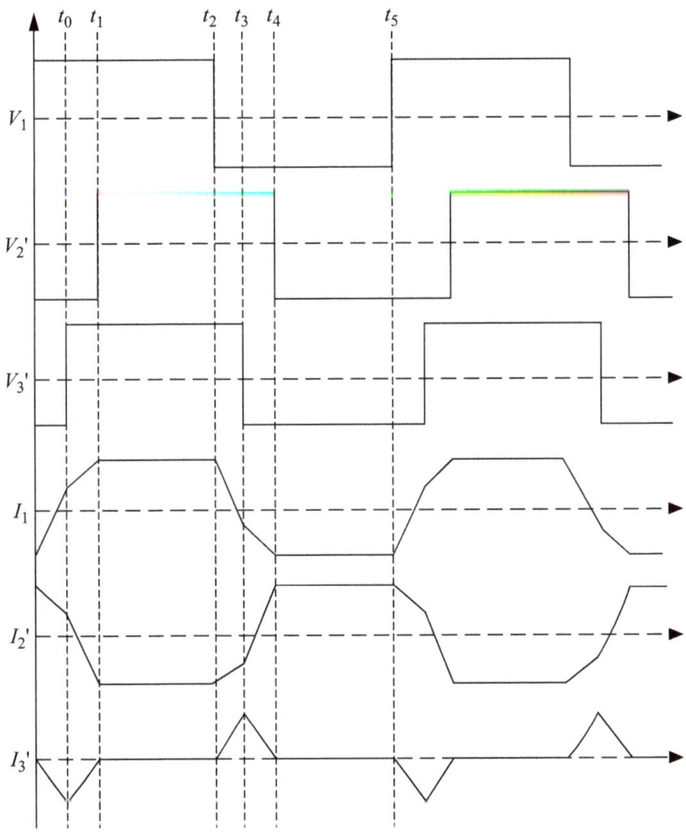

Figure 12.9 Input, output voltage, and current waveforms of the converter [34]

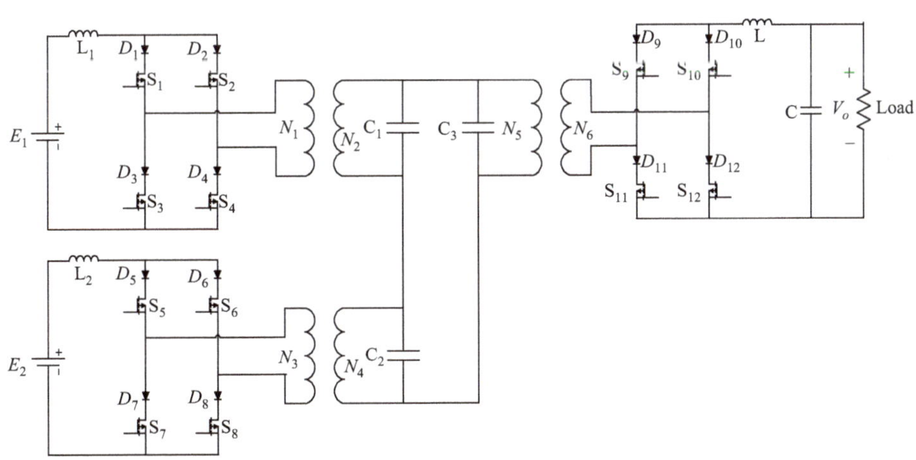

Figure 12.10 Current-fed three-port bi-directional converter [37]

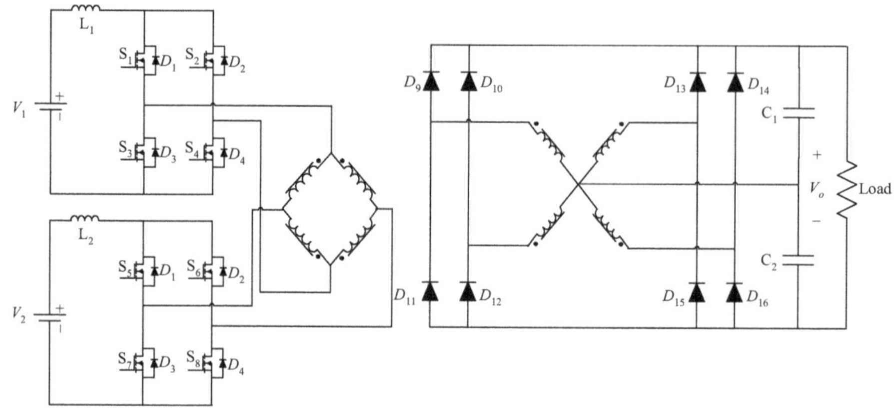

Figure 12.11 Full-bridge boost DC–DC converter based on the distributed transformers [38]

the output capacitors. Second, in the first full-bridge rectifier, S_1 and S_4 or S_2 and S_3 switches are turned on, in this phase the power is transferred through the transformer windings to the load. In this converter, both suppliers cannot supply the power simultaneously, this issue is one of the primary disadvantageous of this structure. The relation between the input voltages and the output voltage is as follows:

$$V_o = \frac{n}{2(1 - D_1)} V_1 \tag{12.3}$$

$$V_o = \frac{n}{2(1 - D_2)} V_2 \tag{12.4}$$

where n is the turning ratio of the transformer, D_1 is the duty cycle of the switches in the first full-bridge converter, and D_2 is the duty cycle of the switches of the second full-bridge converter. The duty cycle of both converters should exceed 50%. The major weakness of this converter is the number of semiconductor elements which may reduce the total efficiency of the circuit. Also, the employed transformer has some complicated points in design and production.

12.2.5 Dual-transformer-based asymmetrical triple-port active bridge isolated DC–DC converter

Dual-transformer-based asymmetrical triple-port active bridge (DT-ATAB) consists of three active power electronic converters and two high-frequency transformers (Figure 12.12). All switches of these converters can be turned-on with zero-voltage-switching (ZVS) to reduce the switching losses. The bidirectional power flow operation is possible between the ports. The DT-ATAB also reduces the circulating powers between the ports for well-matched transformer turns ratios as compared to those in the other existing triple-port active bridge converters

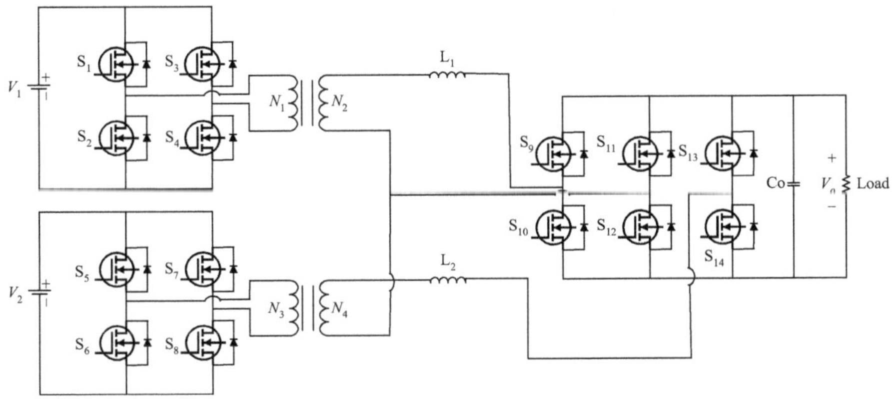

Figure 12.12 Dual-transformer triple-port active bridge DC–DC converter [39]

(TAB). Furthermore, the magnetic short-circuit conditions arising in the three-winding transformer of the TAB are mitigated in DT-ATAB. The illustrated results show that DT-ATAB can be used as a promising multi-port converter (MPC) to interface the multiple sources and load to achieve wide-ranging outputs with the minimal losses.

12.2.6 Summary

The different structures of the magnetically coupled converters had been described in this section. These types of converters have the following benefits: (a) isolation between the input sources and output load, and (b) the ability to create different voltage levels at the output. But they have the following disadvantages: (a) complex circuits due to multiple input wires, (b) the difficulty of controlling energy storage resources, and (c) the need for the same input voltage input in some structures.

12.3 Electrically coupled multi-input converters

The converters with electrical coupling do not have electrical isolation between input and output (such as transformers). In this type of structure, non-isolated converters such as buck converters, boost converters are used. Control of converters with electrically coupling is easier than magnetic coupling.

12.3.1 Series or parallel of two DC/DC converters

The first structure is introduced as the electrical couple with combination of two boost converters is shown in Figure 12.13.

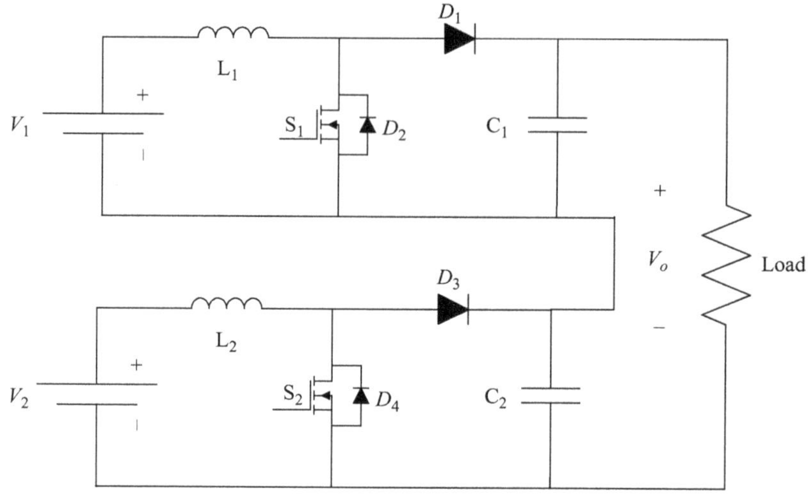

Figure 12.13 The structure of series two DC choppers [40]

This converter is used to combine the energy production through two solar and wind suppliers. In these converters, the output voltage changes as well deviation in switching period of each DC/DC converter the output voltage changes as well which leads to a change in output voltage and subsequently the transferred output power. So, it is used only for battery recharging in this chapter.

The two buck converters can put in parallel (or series) [41]. In the parallel connection, the output voltage of converters have to be the same. Generally connecting the DC rectifiers in series and parallel, have the following pros and cons:

a. In a series of two converters, the current flowing from both get equal, but the output voltage of each are different due to the different power production. Therefore, the voltage regulation in these types of converters is complicated, and the failure of one of the suppliers turns into an unacceptable mode of operation.
b. Paralleling two converters may lead to a spinning current in case there is no appropriate selected control for the converters.

Use of two or more parallel converters is common. Figure 12.14 shows three bi-directional structure are parallel this structure. In this structure, the voltage can be increased or decreased, and the power can be transferred for the battery and super-capacitor recharging in reverse mode. According to Figure 12.14, switch S_2, and diode D_1 make up a boost converter and when the super-capacitor and the battery have to supply the load, the converter functions as a voltage booster.

Switch S_1 and diode D_2 form a buck converter. In case the output voltage exceeds its nominal value, the voltage can be adjusted to the desired value through this converter for the battery and the super-capacitor to be recharged. The benefits of this model are such as; the possibility of charging the battery and the super-capacitor

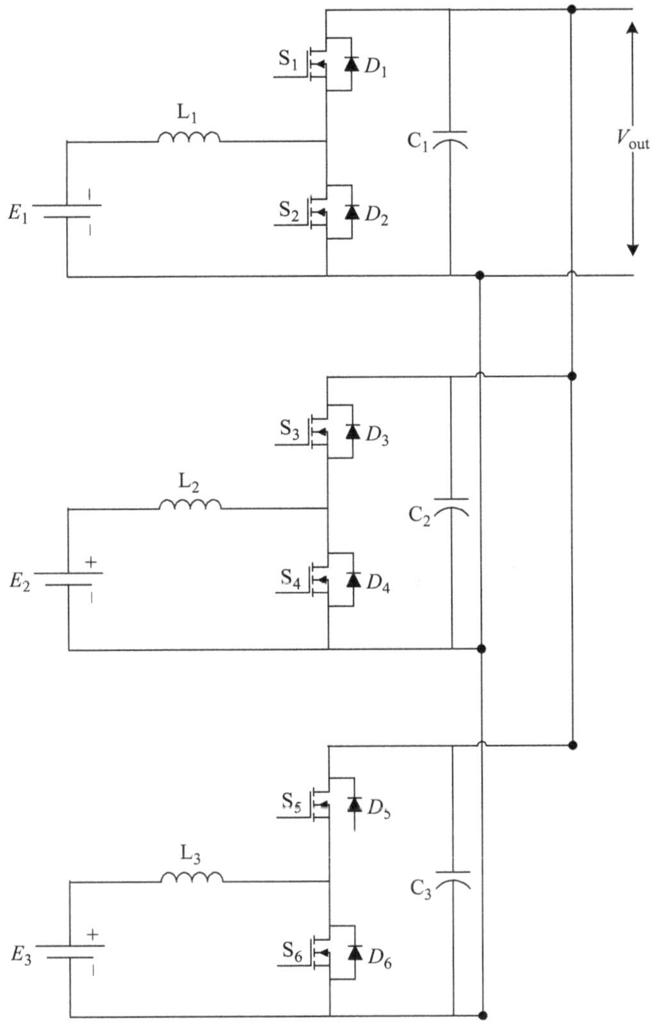

Figure 12.14 Multi-input converter for hybrid vehicles [42]

is available in one structure. Suppliers with different voltage values can work together in this structure. In this layout, the multi-winding transformer is not utilized.

Some disadvantages of this model are as follows; in this converter, three DC/DC converters are paralleled. So the number of the utilized switches do not economize, and that soft-switching is not employed in this model.

In a series converter like Figure 12.13, voltage conversion ratio is:

$$V_o = \frac{V_1}{1 - D_1} + \frac{V_2}{1 - D_2} \tag{12.5}$$

D_1 and D_2 are duty cycle of S_1 and S_2. In a series converter like Figure 12.14, voltage conversion ratio is:

$$V_o = \frac{V_1}{1 - D_2} = \frac{V_2}{1 - D_4} = \frac{V_3}{1 - D_6} \tag{12.6}$$

D_2, D_4, and D_6 are duty cycle of S_2, S_4, and S_6. This structure can be used for the DC microgrid to manage resources [43] with a simple controller [44]. With combing two switched-diode-capacitor voltage accumulator in series or parallel, the output voltage of this converter would be higher than the one of the input [45]. Also, a forward converter can be connected in parallel with a buck-boost converter, that this combination creates a dual-input converter which can charge a battery [46]. In general, different types of buck/boost converter connections (series, parallel and hybrid) could be utilized for creating a multi-input converter [47].

12.3.2 Double input fundamental converter (buck/buck-boost converter)

This type of converter is shown in Figure 12.15 [48]. In this converter, the voltage of E_1 is higher than E_2 and output voltage is the voltage value between E_1 and E_2. When either switch S_1 or S_2 or both are turned on, power is transmitted from the source to the output. But when both switches are off, D_1 and D_2 bypass the inductor. In this structure, E_1, S_1, and D_1 work as a buck converter and, E_2, S_2 and D_2 work as a buck-boost converter. The relation between the input voltages and the output voltage is as follows:

$$V_o = \frac{d_1}{1 - d_2} V_1 + \frac{d_2}{1 - d_2} V_2 \tag{12.7}$$

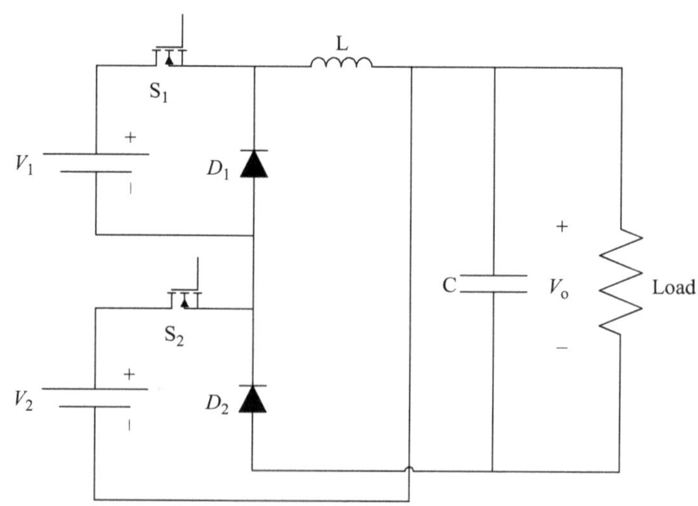

Figure 12.15 Double-input buck/buck-boost [48]

Voltage and current waveforms of the input and output are shown in Figure 12.16. The advantages of this structure are as follows: There is no need to multi-winding transformers, soft- switching occurs without the need for additional equipment, and different voltage sources can be used. Disadvantages of this structure are as follows: adjustable output voltage between the voltage sources, no possibility to recharge the battery using the structure (of course using four quarter switches can solve this issue [49]).

To achieve different value of gain (or output voltage) and also simplify the converter's control, various combination can be implemented. In [50], two structures has been introduced as buck-boost/buck-boost and buck/buck converter. In these structures change the inductor charging time and thus influence on the output voltage. The equation of the output voltage in these converters is given by (12.8) and (12.9).

$$V_o = d_1 V_1 + d_2 V_2 \tag{12.8}$$

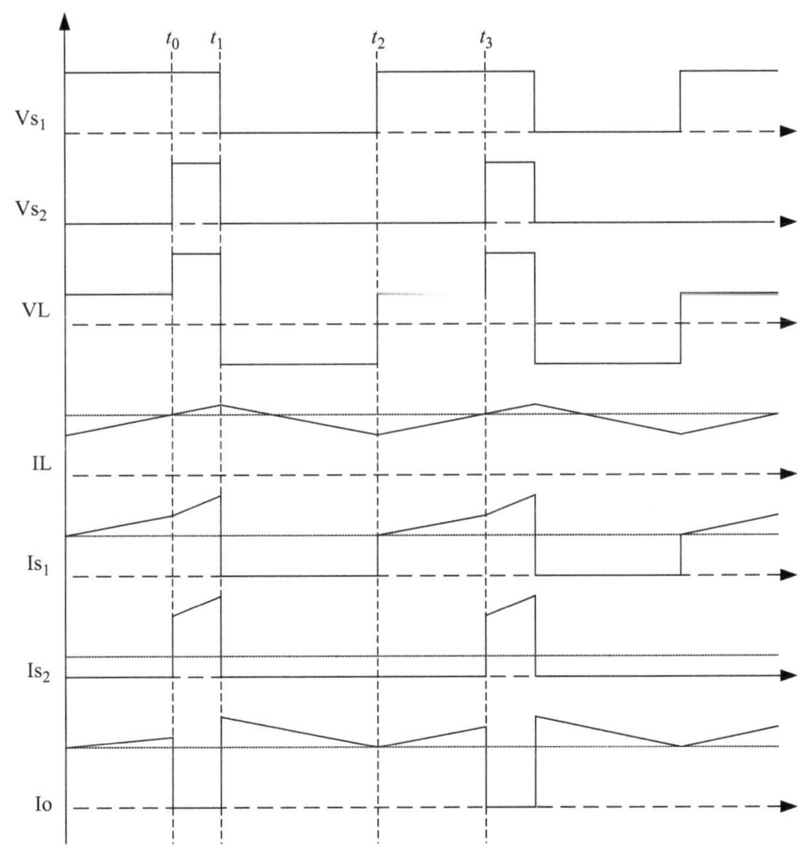

Figure 12.16 Input, output voltage, and current waveforms of the converter [48]

$$V_o = \frac{d_1}{1 - d_1 - d_2} V_1 + \frac{d_2}{1 - d_1 - d_2} V_2 \tag{12.9}$$

In [51], a double-input buck/buck converter has been presented, which is a combination of a charging switch and a series-connected double-input DC/DC converter. This converter robust when it comes to one input intubation or one input short-circuit.

Another buck/buck structure was presented in Figure 12.17. In this structure, an inductive circuit at the tap of the converter and a capacitive circuit at the bottom of the converter is mounted. The use of these circuits creates soft-switching in the converter. The relation between the input voltages and the output voltage is as follows:

$$V_o = \frac{2d_1 - d_2}{2 - d_2} V_1 + \frac{d_2}{2 - d_2} V_2 \quad (d_1 > d_2) \tag{12.10}$$

$$V_o = \frac{d_1}{2 - d_2} V_1 + \frac{d_2}{2 - d_2} V_2 \quad (d_1 \leq d_2) \tag{12.11}$$

In (12.10) and (12.11), d_1 is the duty cycle of switch S_1, and d_2 is the duty cycle of switch S_2. The main advantage of this converter is soft-switching.

Instead of a buck-boost converter, a SEPIC converter can be used (Figure 12.18). In this structure, a buck and a SEPIC converter are utilized. This converter has four operation modes. First, when both switches S_1 and S_2 are off, the power is transferred to the output only through the second supplier. Second, when S_1 is on, and S_2 is off, the energy is transferred to the output through the second

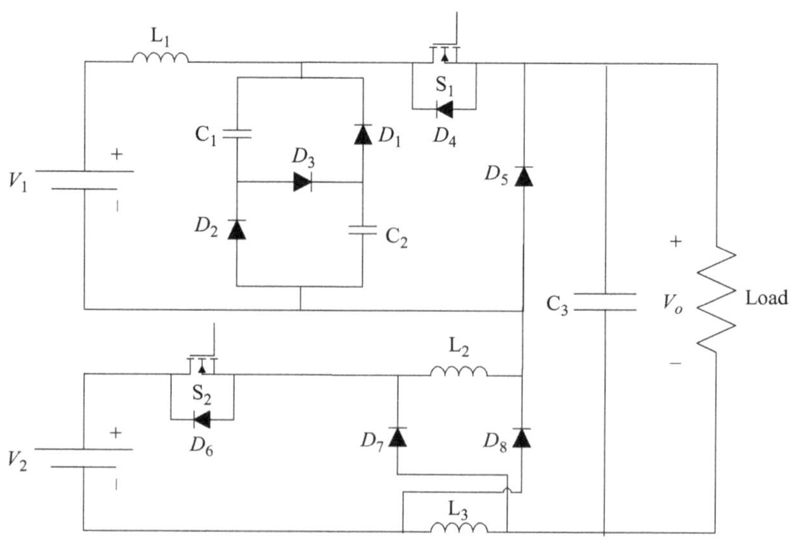

Figure 12.17 Soft-switch buck/buck converter [52,53]

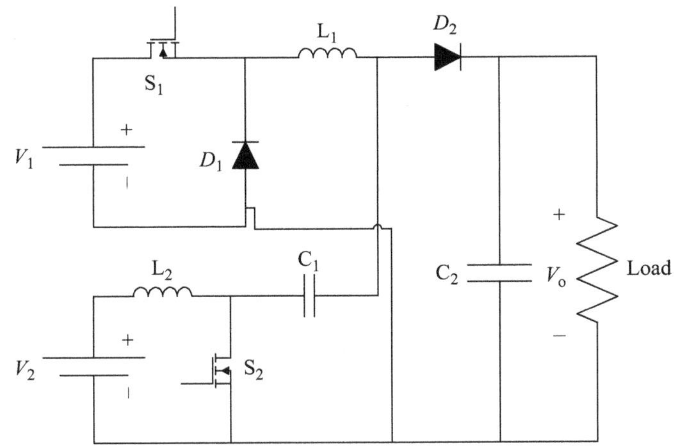

Figure 12.18 Double-input SEPIC/buck converter [56]

supplier. Third, when S_1 is off, and S_2 is on, the inductor L_2 is charged, and the load is supplied through the output capacitor. Fourth, when both switches are on, the inductors L_1 and L_2 are both being charged, and the capacitor supplies the load.

Putting a PFC circuit leads to an enhancement of the ripples in the current taken from the first supplier. In this converter, recharging the rechargeable resources is applicable, and this point of weakness is the significant disadvantage of the proposed layout. Another topology in this category are as follows: dual input SEPIC/Cuk converter (Figure 12.18), SEPIC/SEPIC converter [54,55].

12.3.3 Multi-input DC boost converter supplemented by hybrid PV/FC/battery power system

This topology is illustrated in Figure 12.19. In this structure, FC and solar cell create E_1 and E_2. These two sources are connected to the output using a boost converter, and the battery (E_3) has a full-bridge converter. When S_1 and S_3 are closed, E_1 charges inductor L_1 and when S_1 or S_3 is open, E_1 supplies the load. This process is similar for E_2 and S_2, S_4 switches. When S_1 and S_4 are closed, E_3 is discharged, and when S_2 or S_3 is closed, E_3 is charged. The relation between the input voltages and the output voltage is as follows:

$$V_o = \frac{V_1 + d_4 V_3}{1 - d_1 - d_4} \tag{12.12}$$

$$V_o = \frac{V_2 - d_3 V_3}{1 - d_2 - d_3} \tag{12.13}$$

In (12.12) and (12.13), d_1 is the duty cycle of S_1 and S_3, d_2 is the duty cycle of S_2 and S_4, d_3 is the duty cycle of S_2 and S_3, and d_4 is the duty cycle of S_1 and S_4. The only problem with this converter is no possibility to charge the battery via BLDC generator.

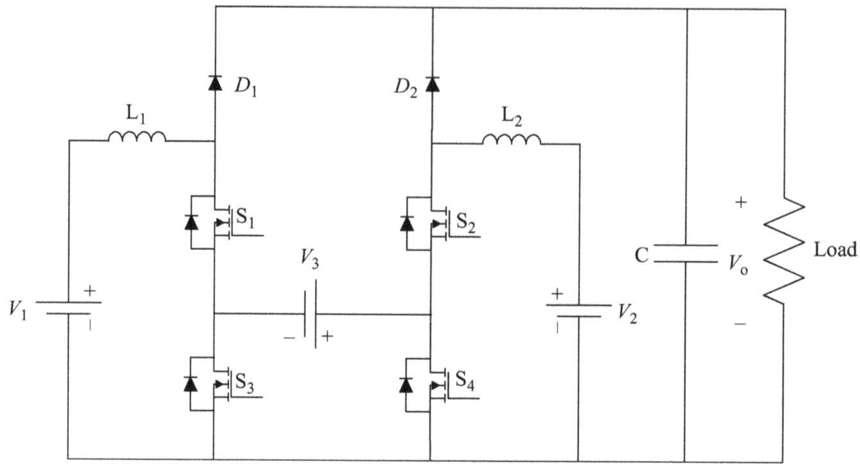

Figure 12.19 Multi-input DC boost converter supplemented by hybrid PV/FC/battery [57]

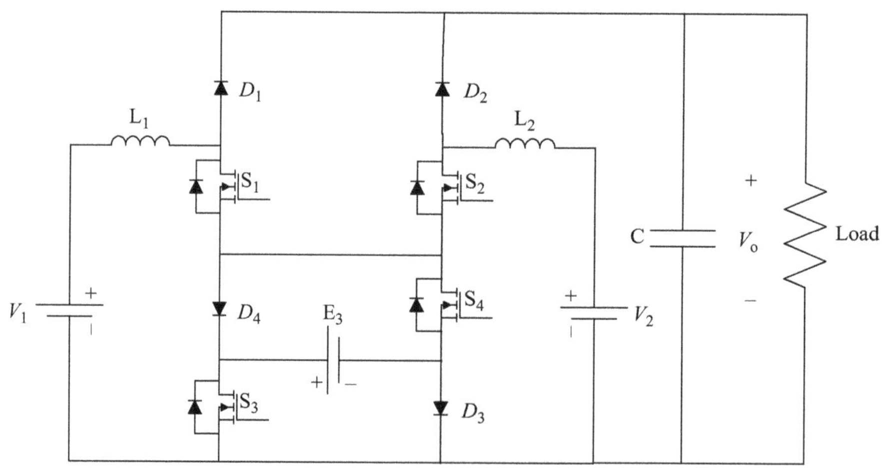

Figure 12.20 Three-input DC–DC boost converter [58]

12.3.4 Three-input DC–DC boost converter

The three-input DC–DC boost converter is presented in Figure 12.20. Table 12.1 shows the operational principle of the converter. In this converter, the battery is discharged, and the reason for this action is unknown which has become the main

Table 12.1 Operational principle of the three-input DC–DC boost converter

S_1	S_2	S_3	S_4	Modes of operation
1	1	0	1	**E_1 and E_2 power control**
0	1	0	1	**E_1 supplies output and E_2 power control**
0	0	0	0	**Zero mode**
1	1	1	1	**E_3 discharge**
1	1	1	0	**E_1 and E_2 power control**
0	1	1	0	**E_1 supplies output and E_2 power control**
1	1	0	0	**E_3 charges by E_1 and E_2**
0	1	0	0	**E_3 charges by E_2**

problem. In this topology, the relation between input and output voltage is as follows:

Mode 1: Battery is charge and no need to charge

$$V_o = \frac{V_1}{1 - d_1 - d_4} = \frac{V_2}{1 - d_2 - d_4} \tag{12.14}$$

Mode 2: Battery is discharging

$$V_o = \frac{V_1}{1 - d_1 - d_3} = \frac{V_2}{1 - d_2 - d_3} \tag{12.15}$$

$$V_3 = \frac{d_1 + d_3 + d_4}{d_2 + d_3 + d_4} V_1 = \frac{d_2 + d_3 + d_4}{d_1 + d_3 + d_4} V_2 \tag{12.16}$$

Mode 3: Battery is charging

$$V_o = \frac{V_1}{1 - d_1 - d_3} = \frac{V_2}{1 - d_2 - d_3} \tag{12.17}$$

$$V_3 = d_1.V_1 = d_2.V_2 \tag{12.18}$$

Another type of DC/DC multi-input boost converter has been introduced in Figure 12.21 that have features like high voltage gain, the possibility of using with a different voltage source. This converter is proper to use in PV applications and can be used as a multi-output converter.

12.3.5 ZVS multi-input converter

A structure presented in Figure 12.22, there are circumstances in which the soft switching is done. Table 12.2 shows the operation principle of the converter. In this table, assumes that the switch 2 is closed. If in this converter, the resource uses a battery instead of E_3, the battery is always charging and discharging, so the battery life is reduced. In Figure 12.23, the voltage and current waveforms of the switches,

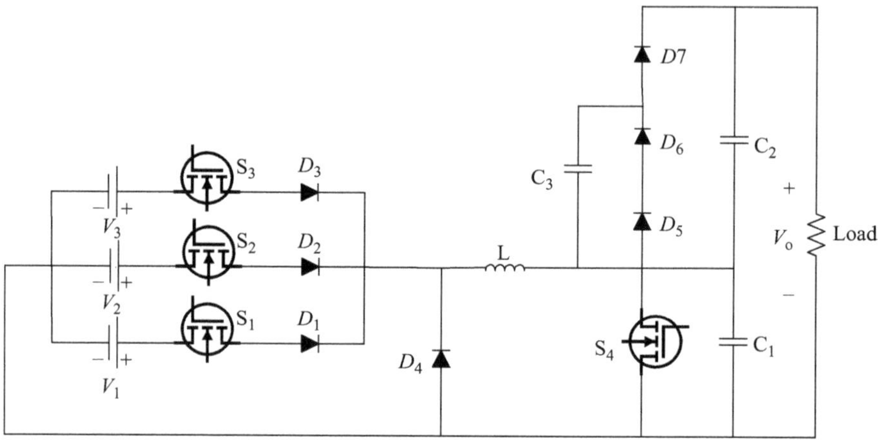

Figure 12.21 Three-input two-output DC–DC boost converter [59]

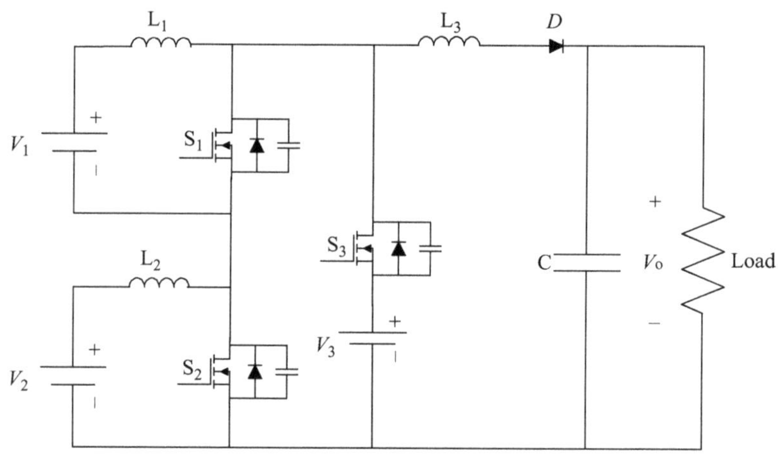

Figure 12.22 ZVS multi-input converter [61]

Table 12.2 Operational principle of the ZVS multi-input converter

Interval	S_1	S_3	Mode of operation
t_0–t_1	1	0	**L_1 is going to charge, and L_3 current is zero**
t_1–t_2	0	0	**S_1 and S_3 capacitor charging**
t_2–t_3	0	0	**S_3 diode on**
t_3–t_4	0	1	**S_3 close in ZVS**
t_4–t_5	0	0	**S_3 opens, S_1 and S_3 capacitor charging**
t_5–t_6	0	0	**S_1 diode on**
t_6–t_7	1	0	**S_1 close in ZVS**

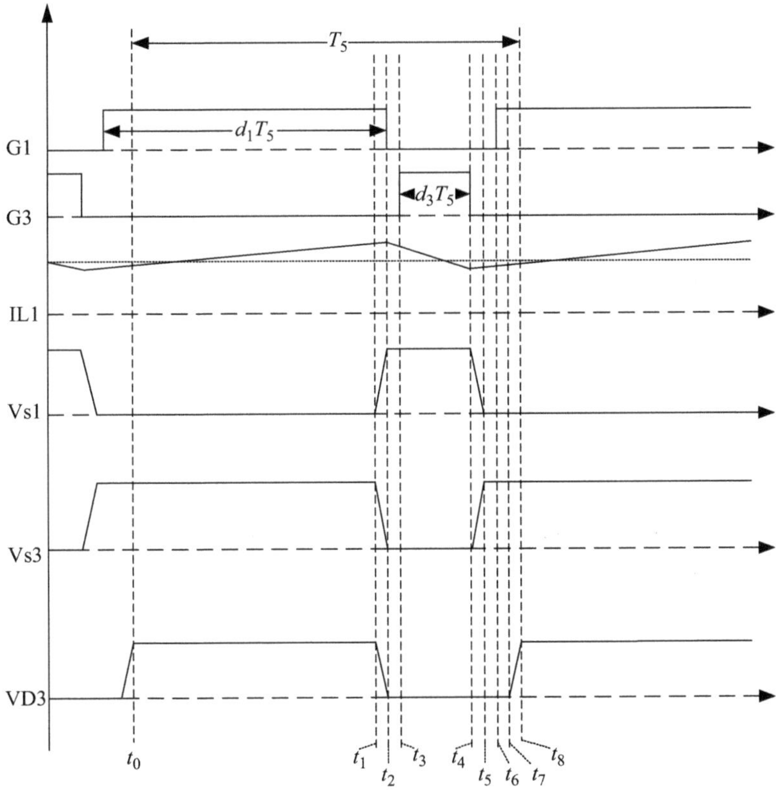

Figure 12.23 Voltage and current waveforms of the converter [61]

are shown. The relation between the input voltages and the output voltage is as follows:

$$V_3 = \frac{1}{1 - d_1} V_1 \tag{12.19}$$

$$V_0 = \frac{(1 - d_1)}{(1 + d_{dcm} - d_1)} V_3 \tag{12.20}$$

In (12.19) and (12.20), d_1 is duty cycle of S_1 and d_{dcm} is dead time duty cycle.

In this structure, instead of an E_3 resource, a capacitor can be used. Using the diode in this structure has caused the charge in the L_3 inductor be transferred to C and E_3 [60].

12.3.6 Non-isolated multi-input multi-output DC/DC boost converter

A non-isolated multi-input multi-output DC/DC boost converter has been shown in Figure 12.24, which can be utilized as an interface for energy sources in electrical vehicles. The outputs of the converter have different voltage levels which can be

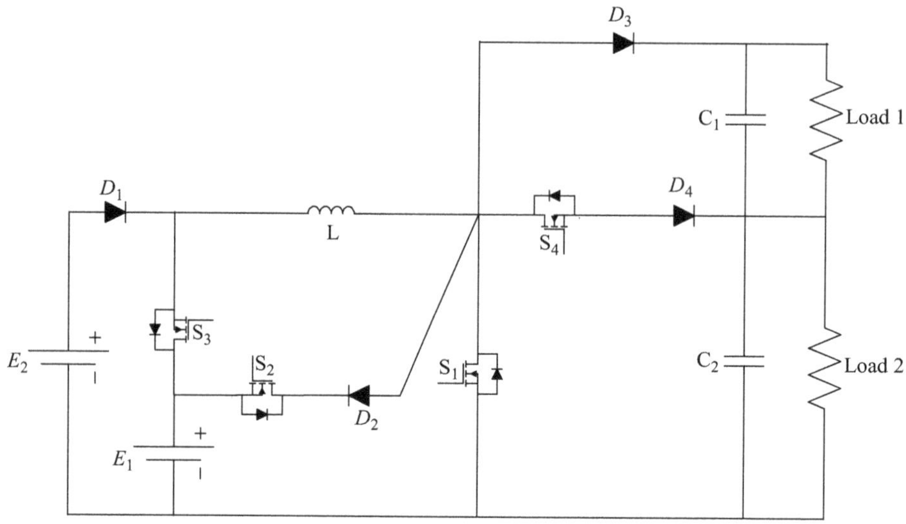

Figure 12.24 Non-isolated multi-input multi-output boost converter [62]

used for interfacing with some of the loads in electrical vehicles. In this topology, E_1 is a storage source like a battery and E_2 is a FC. In this structure, E_1 must be higher than E_2. This converter has four operation modes. First, S_1 and S_3 are turned on and E_1 charges the inductor L. Second, D_1 and S_1 are turned on and E_2 charges the inductor L. Third, D_1 and S_4 are turned on and Load 2 is supplied by E_2. Forth, D_1 and D_3 are turned on and Load 1 is provided through E_2.

12.3.7 Non-isolated multi-input single-output DC/DC buck/boost-boost converter

A multi-input converter has been illustrated in Figure 12.25, which is consisted of the buck/boost and boost converter. In this converter, the delivered power to output from each of the inputs is able to be controlled independently. The relation between inputs and output is as follows:

$$V_{out} = \frac{E_2}{1 - D_2} + \frac{D_1 D_2}{(1 - D_1)(1 - D_2)} E_1 \tag{12.21}$$

In (12.21), D_1 is the duty cycle of S_1 and D_2 is the duty cycle of S_2. This converter has three main operation mode. First, S_1 and S_2 are turned off, so D_1 and D_2 are turned on. In this mode, C_1 is charged by L_1 and C_o is charged by L_2. Second, S_1 and D_2 are turned on. Therefore, S_2 and D_1 are turned off. In this mode, L_1 is charged by E_1 and L_2 has charged C_o like the previous mode of operation. Third, S_1 and S_2 are turned on, so D_1 and D_2 are turned off. In this mode, L_1 is charged by E_1, like the previous mode and L_2 is charged by E_2 while C_1 is discharged.

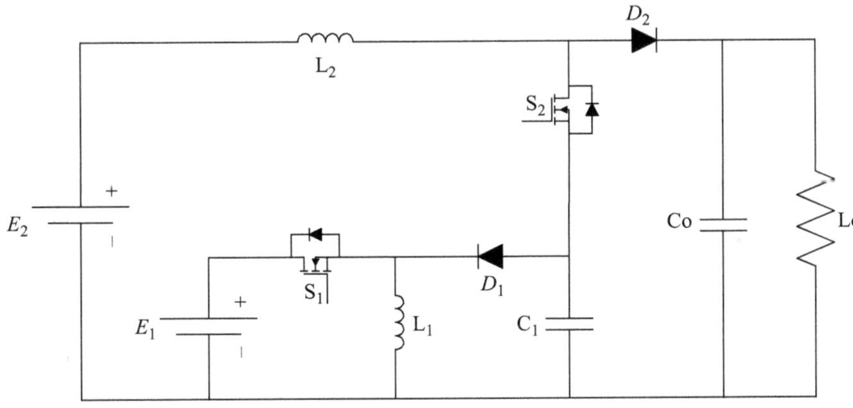

Figure 12.25 Non-isolated multi-input buck/boost-boost converter [63]

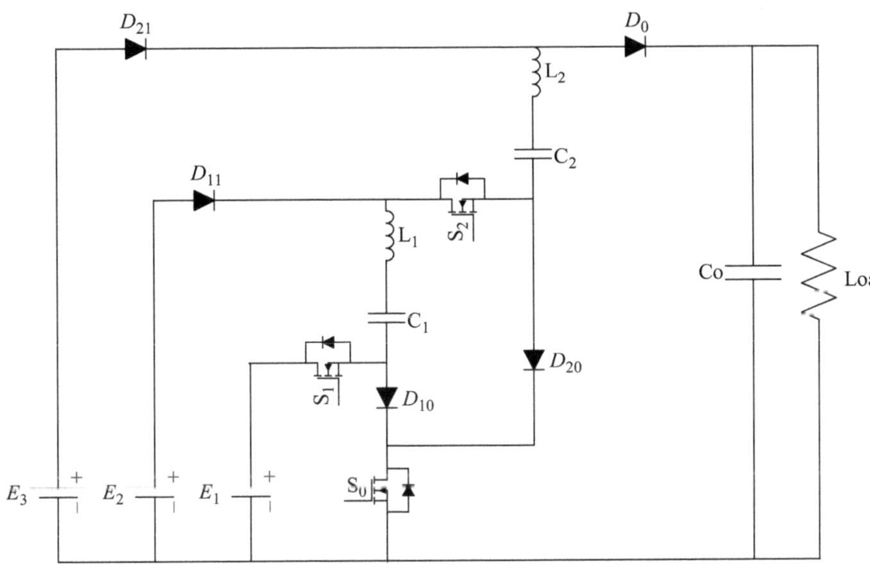

Figure 12.26 Multi-input voltage-summation converter based on switched-capacitor [65]

12.3.8 Switched-capacitor-based multi-input voltage-summation converter

A multi-input converter has been introduced in Figure 12.26, which is designed based on the switched-capacitor instead of magnetic elements. The output voltage of this converter is the summation of input voltages. In this converter, the switches have a 50% duty cycle. The relation between the input voltages and the output voltage is as follows:

$$V_{out} = E_1 + E_2 + E_3 \tag{12.22}$$

In this converter, S_1 and S_2 are turned on simultaneously. This converter has four operation modes. First, S_0 is turned on, and S_1 and S_2 are turned off. In this state, L_1 and C_1 are charged by E_2 and undergo resonance frequency. The same thing also happens for L_2 and C_2. Second, all of the switches are turned off, and because of the resonance current and the reverse biased diodes, the resonance stops. Third, S_1, S_2, and D_0 are turned on, and L_1, C_1, and L_2, C_2 are connected in series with E_1. In this state, capacitors and inductors begin to discharge and supply the load. Forth, S_1 and S_2 are turned on but D_0 is turned off, and the load is provided by the output capacitor C_o.

In [64], a multi-input voltage-subtracting converter based on switched-capacitor has been introduced which is created as the multi-input voltage-summation converter.

12.3.9 SEPIC-based multi-input DC/DC converter

A single-ended primary-inductor converter -based multi-input DC/DC converter is presented in Figure 12.27. The structure inherits all the advantages of the SEPIC converter such as step-up/down capability without inverting the polarity of the regulated output voltage and using a series capacitor to couple energy from the input to the output. The presented converter has a bidirectional input port, in which the energy storage system (ESS) can be connected, and several unidirectional input ports. Hybridizing several alternative energy sources is achievable by the proposed converter. In addition, the load power can be flexibly divided among various power sources. Due to the buck-boost characteristic of the presented converter, it is suitable to charge-discharge the ESS and to extract maximum power from the PV panels. Based on the charging state of the ESS, two operation modes are defined.

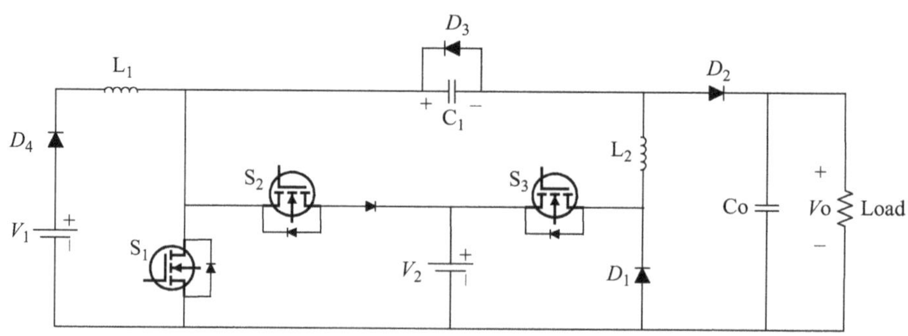

Figure 12.27 SEPIC-based two inputs converter [66]

12.3.10 Bridge-type dual-input DC–DC converters

Two bridge type dual input DC–DC converter topologies are introduced in Figures 12.28 and 12.29, for integrating two input energy sources. Out of the two topologies, the latter topology is an improved version of the former one. Both converters have the capability to simultaneously deliver power to the load, from the input energy sources. The major advantages of the improved converter as compared to the basic topology, are its capability to perform the buck, boost and buck-boost modes of operation using the same structure and the ability to deliver power to the load, even with the failure of any one of the input energy sources. The proposed converters have certain merits like less component count, compact structure and efficient energy utilization.

12.3.11 Three-input DC/DC converter for hybrid PV/FC/battery applications

A three-port DC/DC converter is presented in Figure 12.30 for hybrid PV/FC/battery applications. The proposed structure comprises a conventional buck-boost and a boost converter. Four power switches and four diodes are employed in the

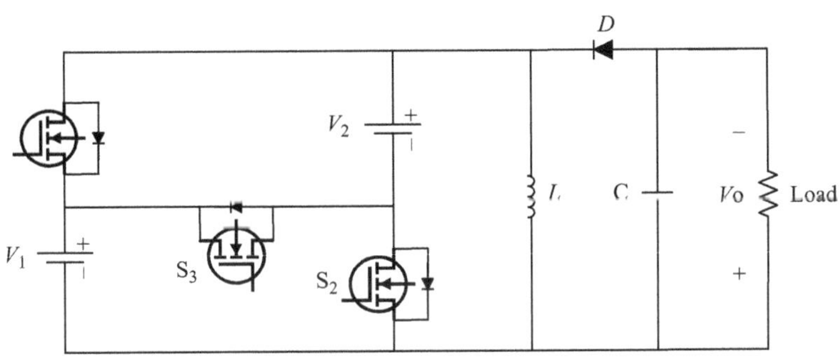

Figure 12.28 Bridge-type dual-input DC–DC converter (BDC) [67]

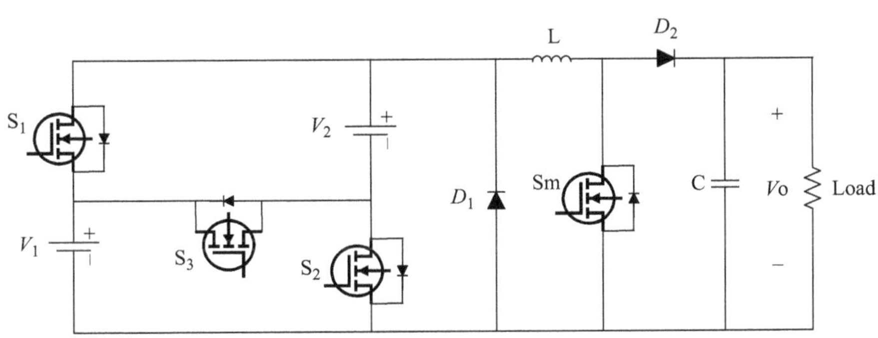

Figure 12.29 Improved bridge-type dual input DC–DC converter (IBDC) [67]

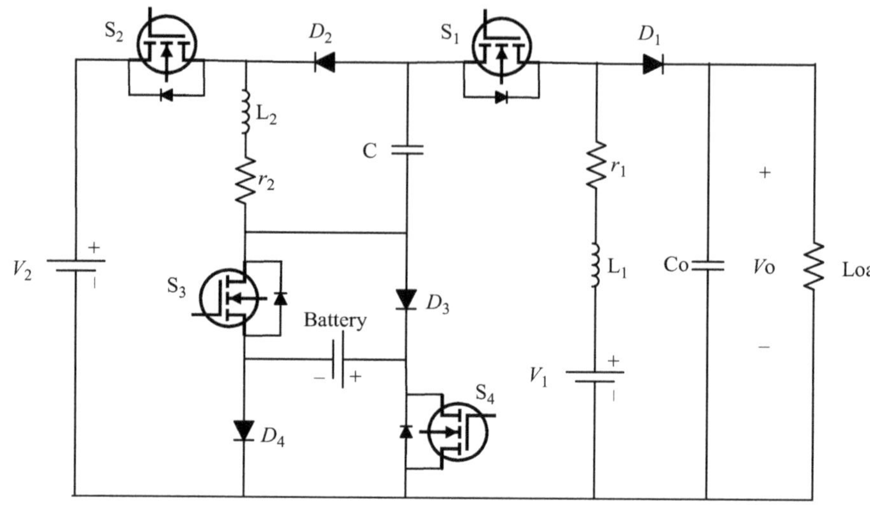

Figure 12.30 Three-input step-up DC–DC boost converter [68]

proposed converter. The voltage gain of the presented converter is more than conventional boost converter. This advantage and having two unidirectional and a bidirectional inputs make the structure a suitable power electronic interface for hybrid generation applications. In addition, there are no limitations in switching modulation. Therefore, tracking the maximum power of the PV source, setting the FC power, controlling the battery power, and calibrating the output voltage can be equipped by controlling duty ratios of the switches. The input power sources can provide power to the load and either charge or discharge the battery individually or simultaneously.

12.3.12 Modular non-isolated multi-input high step-up DC–DC converter with reduced normalized voltage stress and component count

Figure 12.31 proposed a topology can produce higher voltage gains per number of components (switch, diode, capacitor and inductor) than other non-isolated non-coupled inductor-based multi-input (NINCIBMI) topologies. In other words, the proposed topology uses less number of components for achieving the same voltage gain. This property can lead to reduced cost, size, weight, and complexity of topology. Also, proposed topology benefits from continuous input current. Despite the high voltage gain of proposed topology, it has considerably low normalized voltage stress (NVS) on its switches/diodes. Another important advantage of proposed topology is that, as the number of input units increase, the voltage gain increases too, but the NVS on switches/diodes decreases. The proposed topology is suggested for low and medium power applications.

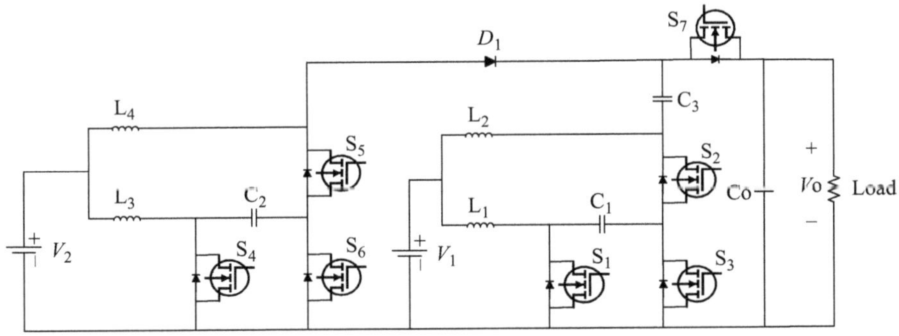

Figure 12.31 NINCIBMI DC–DC converter [69]

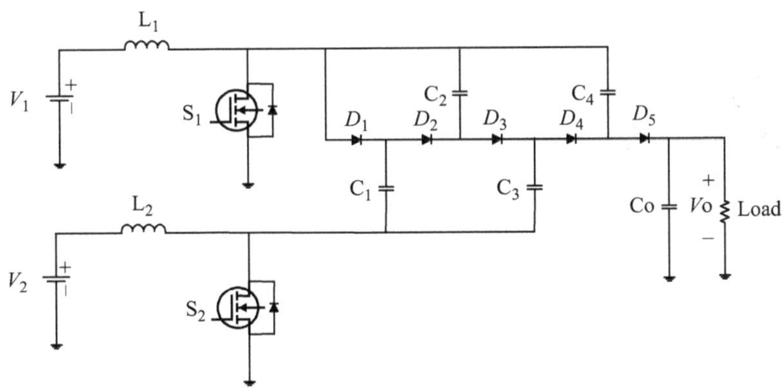

Figure 12.32 Multi-input converter with high-voltage gain and two-input boost stage [70]

12.3.13 Multi-input converter with high-voltage gain and two-input boost stage

Figure 12.32 shows a topology which can be used as multiport converters and draw continuous current from two input sources. They can also draw continuous current from a single source in an interleaved manner. This versatility makes them appealing in renewable applications such as solar farms. The proposed converters can easily achieve a gain of 20 while benefiting from a continuous input current. Such a converter can individually link a PV panel to a 400 V DC bus.

12.3.14 Soft-switched non-isolated, high step-up three-port DC–DC converter for hybrid energy systems

Figure 12.33 illustrates a converter which provides two separated power flow paths from each input sources to the output load. In order to reduce the number of converter components, some components play multiple roles. Accordingly, the energy

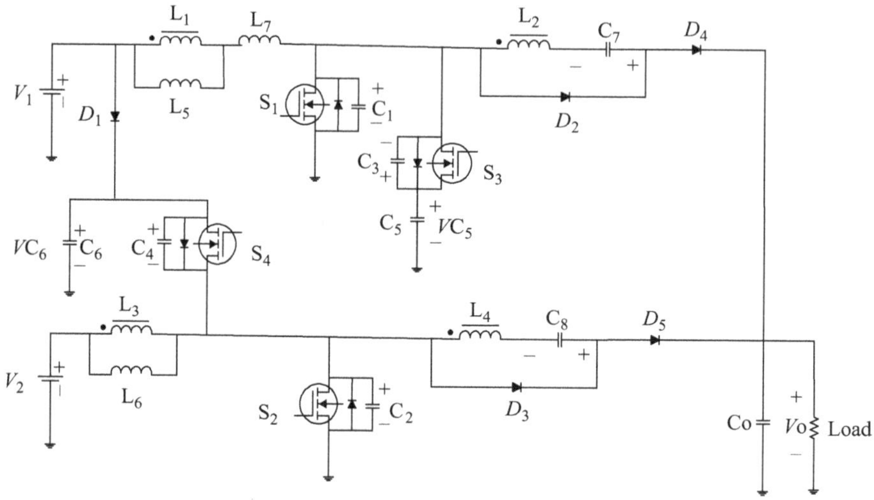

Figure 12.33 Non-isolated high step-up three-port DC–DC converter [71]

storage device is charged with the same components which are used in transferring power to the load. In this converter, coupled inductors technique is used to increase the voltage gain, and to mitigate the leakage inductance effect and to provide soft switching condition, two active clamp circuits are employed. Since the voltages across the switches are clamped, switches with low voltage stress and consequently low conduction loss can be used.

12.3.15 Summary

The different structures of the electrically coupled converters had been described in this section. These types of converters have the following benefits: (a) various assembly non-isolated converters can be created by basic converters (buck, boost, Cuk, sepic, etc.). (b) The control circuit of these converters is simpler than magnetically coupled. (c) These converters can be used as serial, parallel and cascade cells. (d) There are no saturation and magnetic problems, due to the absence the transformer, and (e) The ability to change the control method by changing the switching strategy. But they have the following disadvantages: (a) There is no isolation between input sources and output load, (b) The limitation of output voltage variations, and (c) The increasing cost, switching stress and losses with increasing the number of inputs.

12.4 Electrically-magnetically coupled multi-input converters

The converters with electrically-magnetically coupling try to collect advantages of other two categories; however, facing some of the disadvantages is inevitable. This converters may be named as compound or hybrid converters. In this type of structure, the basic converters and transformers are used for electrically and magnetically coupling, respectively. Also, the control of converters would be harder than other coupling topologies. The different structures of the electrically-magnetically coupled converters are introduced in this section.

12.4.1 Combined DC link and magnetic-coupled converter

A combined structure namely as combined DC link and magnetic couple converter are illustrated in Figure 12.34. Switching method in this converter is similar to Figure 12.35. Two half-bridge converters [HB2 and HB3] are utilized in this model. In multi-input converters, one of the inputs is employed for the super-capacitor to decrease the voltage fluctuation speed of the DC link (due to the high dynamics of

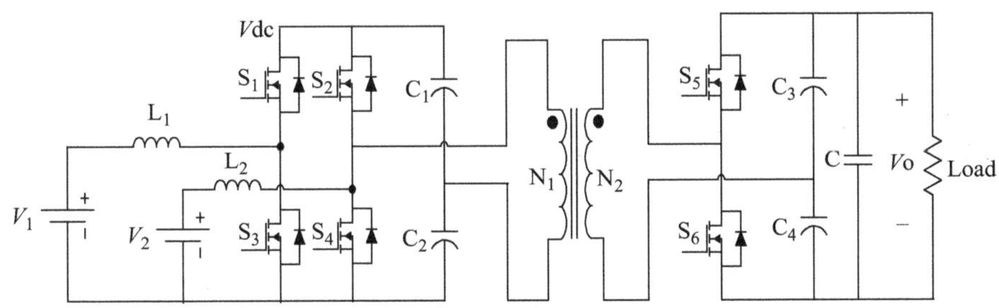

Figure 12.34 Combined DC link and magnetic-coupled converter [72]

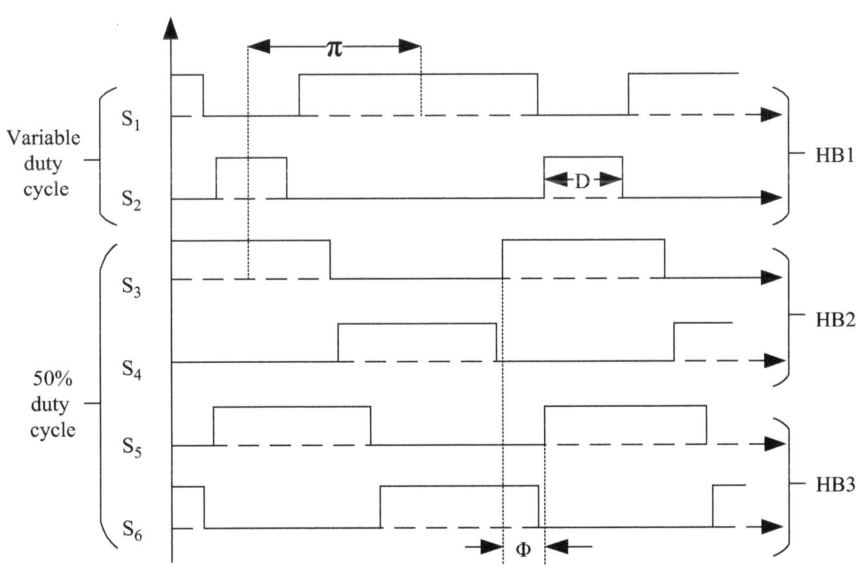

Figure 12.35 Switching strategy [72]

the capacitor). A battery can be utilized in this model, but the slow dynamics of the battery operation is not appropriate for holding the DC link voltage.

In this converter, the HB1 duty cycle is taken from the equation (12.21). HB1 has two operation modes: boost mode for the capacitor discharging and buck mode for the capacitor charging.

$$D = \frac{V_{DC} - V_{SC}}{V_{SC}} \tag{12.23}$$

$$V_{DC} = 2.V_2 \tag{12.24}$$

$$V_{Load} = n.V_{DC} \tag{12.25}$$

In (12.21), V_{DC} is the voltage of the DC link, and V_{SC} is the voltage of E_1. The duty cycle of the HB2 and HB3 is 50%. Controlling the duty cycle is performed through the phase difference of φ between the two converters HB2 and HB3. In this converter, the current passes through the parallel diodes of the switches due to the disconnection of the switches, ZVS phenomenon happens. But, both S_1 and S_2 need snubber circuit for soft switching. The benefits of this structure are as follow; there is isolation between the input and output of the converter. Since only six switches are utilized in the construction, there is a high level of reliability with low price. The return of the power to the HB1 is probable. Therefore, additional circuits are not needed, for soft switching. The weakness of this model is only the unavailability of the soft switching in S_1 and S_2.

This topology can be modified to improve the characteristics. For example: (a) removing the C_1 and C_2, (b) replacing the transformer with a mid-tap transformer, (c) connection one of the sources to the H-bridge converter, and connection the other one to the mid-tap transformer, and (d) connecting three diode legs to the output of the transformer that can rectify the high-frequency voltage and create DC voltage [73].

12.4.2 Direct charge converter

This type of converter presented in Figure 12.36. The reason for this naming is an ESS placing on DC link. Input and output, voltage and current waveforms are

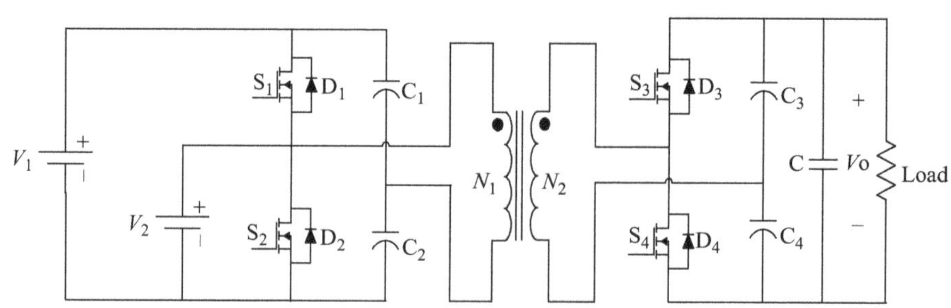

Figure 12.36 Direct charge converter [74]

shown in Figure 12.37. The relation between the input voltages and the output voltage is as follows:

$$V_2 = \frac{d_1}{d_1 + d_2} V_1 \tag{12.26}$$

$$V_o = 2nd_1 V_2 \tag{12.27}$$

In (12.24) and (12.25), d_1 is duty cycle of S_1 and S_4, d_2 is duty cycle of S_2 and S_3, and n is the ratio of the transformer. The advantage of this type against the previous converter is less number of the switch in configuration, but the primary defect is uncontrollable charge and discharge cycle of battery that causes battery lifetime reducing. This converter has more limits on voltage in comparison to another type. This converter can be built of four switches in the primary side of the transformer that will be reviewed in the following part. Output converter can be used as a controlled or uncontrolled. The fundamental problem is that the battery is always charging and discharging, so the battery life will be reduced.

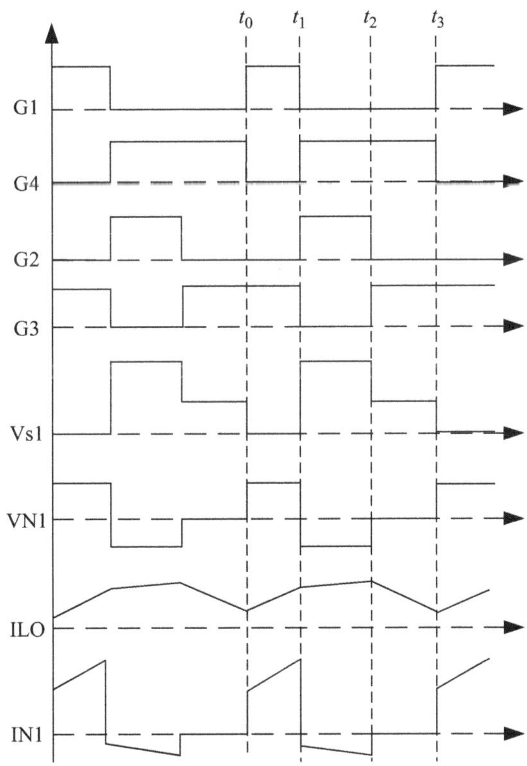

Figure 12.37 Voltage and current waveforms [74]

12.4.3 Boost-integrated phase-shift full-bridge converter

This compound topology is presented in Figure 12.38. Switch on the same leg cannot turn on together (because the short circuit of battery) and the duty cycle of switches is equal. This converter operates like two boost converter that has a phase difference. In Figure 12.39, input and output, voltage and current waveforms, are shown. Some advantages of this structure are: having less cost against another

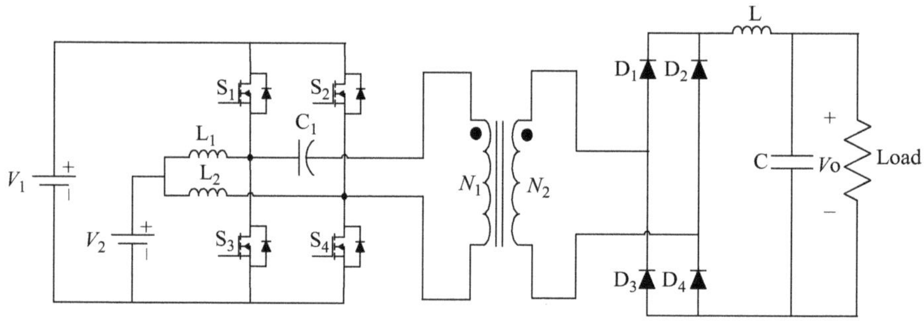

Figure 12.38 Boost-integrated phase-shift full-bridge converter [75]

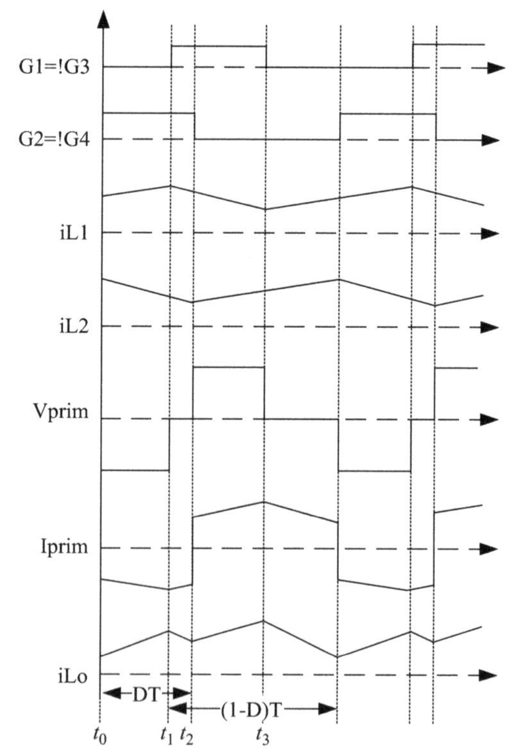

Figure 12.39 Input and output, voltage, and current waveforms [75]

type, soft-switching, the ability of synchronous operation of two sources. On the other hand, disadvantages of this structure are uncontrollable charge and/or discharge cycle in the battery, voltage limit caused by the presence battery on DC link.

In this converter, the relation between the input voltages and the output voltage is as follows:

$$V_2 = D.V_1 \tag{12.28}$$

$$V_o = 2.\varphi_{eff}.n.V_1 \tag{12.29}$$

$$\varphi_{eff} = \min(\varphi, D, 1 - D) \tag{12.30}$$

In (12.26), (12.27), and (12.28), D is the duty cycle of each phase-leg and n is the transformer turns' ratio, and φ_{eff} is the effective value of the phase-shift between the two switching phase-legs.

This topology can be modified with a partial difference. This difference is the elimination of the one inductance in the input, to allows converter operate with the different duty cycle in the switches. The voltage gain in this topology is more than the other [75,76]. Also, other configuration could be created, that is a compound of half-bridge and two parallel conical switching cell. In this topology, two conical switching cell create DC voltage on DC link, and half-bridge converter creates a square wave on the primary winding of the transformer, and in the secondary winding of the transformer, a full-bridge converter create a DC voltage [77].

12.4.4 Tri-modal half-bridge converter

Tri-modal half-bridge converter is presented in Figure 12.40. In this type, neither of the switches turn on each other, and their duty cycle is not equal. This configuration has three operation mode. First, S_1 is on and positive voltage delivered to the transformer. Second, S_2 is on and negative voltage provided to the transformer (charge or discharge of the second source occurred in this duration). Third, S_3 is on, and the primary of the transformer is shorted. Input and output, voltage and current waveforms are shown in Figure 12.41. The relation between the input voltages and the output voltage is as follows:

$$V_2 = \frac{D_2}{D_1 + D_2} V_1 \tag{12.31}$$

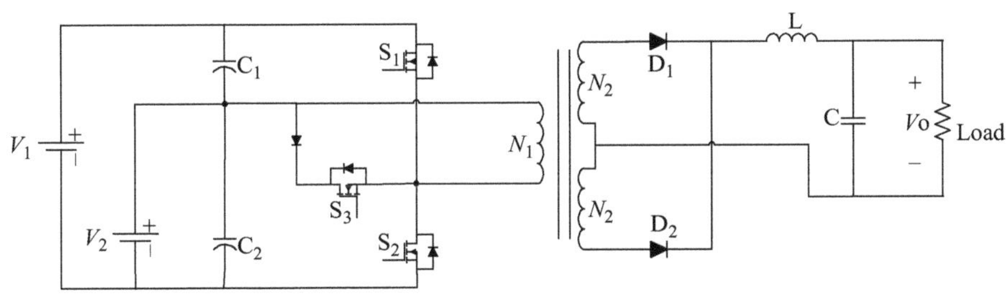

Figure 12.40 Tri-modal half-bridge converter [78]

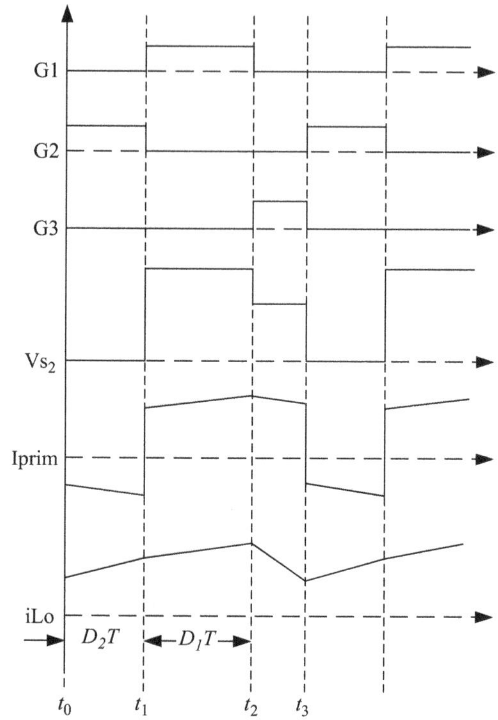

Figure 12.41 Input and output, voltage, and current waveforms [78]

$$V_o = \frac{2D_1D_2}{D_1 + D_2} nV_1 \tag{12.32}$$

In (12.29) and (12.30), D_1 is the duty cycle of S_1, D_2 is the duty cycle of S_2, and n is the transformer ratio. Advantages of this configuration are less stress on switches, possibility to charge the battery, soft-switching (if there is common on time between the duration of switches S_2 and S_3). The main disadvantage of this topology is unable to control on charge or discharge of the battery.

In this topology, the capacitor C_1 may be connected to the ground instead of the terminal of the battery which makes no difference in performance, but there is more control over charging or discharging of the battery. Also, by adding a diode and a switch, other sources can be added to the converter [79].

12.4.5 Full-bridge three-port converters with wide input voltage range

This topology which is presented in Figure 12.42 is like a full-bridge converter. There is four operation mode. First, S_1 and S_4 are on and power transfer from E_1 to output. Second, S_2 and S_3 are on and power transfer from E_2 to output. Third, S_1 and S_2 are on, and E_2 is charged by E_1 and power transfer from E_1 to output. Forth,

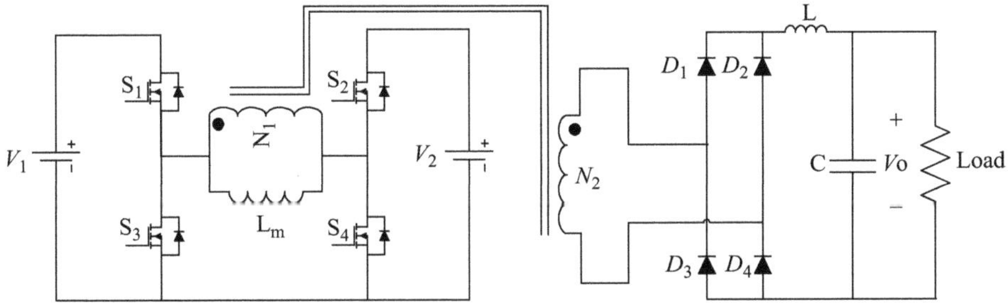

Figure 12.42 Full-bridge three-port converters with wide input voltage range [80]

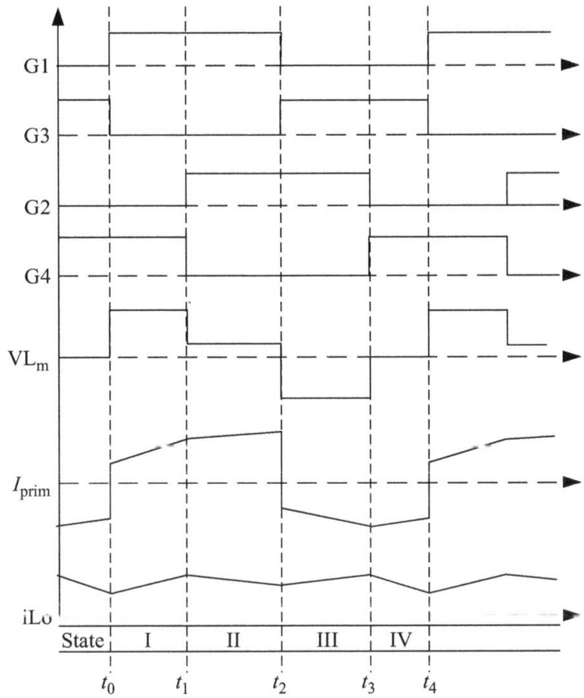

Figure 12.43 Key waveforms of the topology [80]

S_3 and S_4 are on, and L_m supplies the load. In this configuration, soft-switching does not happen that it is a disadvantage of this converter. In Figure 12.43, input and output, voltage and current waveforms, are shown. The relation between the input voltages and the output voltage is as follows:

$$V_1 = \frac{D_2}{D_1} V_2 \tag{12.33}$$

$$V_o = 2nD_3V_2 \tag{12.34}$$

where D_1 and D_2 are the duty cycles of S_1 and S_2 in the steady state, respectively. D_3 is the overlapped section of the driven signal for S_1 and S_4.

12.4.6 Isolated multi-port DC–DC converter for simultaneous power management

This converter can boost input voltages and control the converter under the MPPT condition for renewable energy resources. This structure is shown in Figure 12.44. This topology operates as a three-boost converter that has three operation modes: (1) all switches are on, and the inductors of the sources are charged. In this condition, the primary winding of the transformer has a voltage of C_1. (2) One of the switches is on, but other switches are off. In this condition, power is transferred to the output. (3) All of the switches are off, and all sources transfer energy to the output.

12.4.7 Multi-input three-level DC/DC converter

Figure 12.45 shows a multi-input three-level DC/DC converter has been studied which is consisted of two pulsating current-source cells. The secondary side of the transformer includes the full-bridge converter. The secondary side of the rectifier can be a half-bridge rectifier as well. L_1 and L_2 are charged when S_1 and S_2 are turned on. When S_1 or S_2 are turned off the energy stored in the inductor is transferred to the load. The duty cycle of these switches should be higher than 50%. S_3 is turned on, then S_2 is turned off, and S_4 is turned on, and then subsequently S_1 is turned off. Advantages of this configuration include isolation between the input voltage and output voltage. This converter can control the output power of sources. The main disadvantage of this converter is the lack of control on the charging or discharging of the battery.

12.4.8 Summary

The different structures of the electrically-magnetically coupled converters had been described in this section. These type of converters have the following features: (a) There is an isolation between input sources and output load. (b) Degrees of freedom is high, it

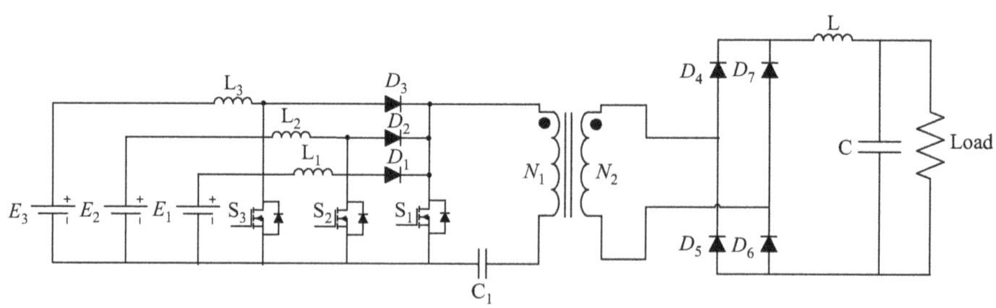

Figure 12.44 Isolated multiport DC–DC converter for simultaneous power management [81]

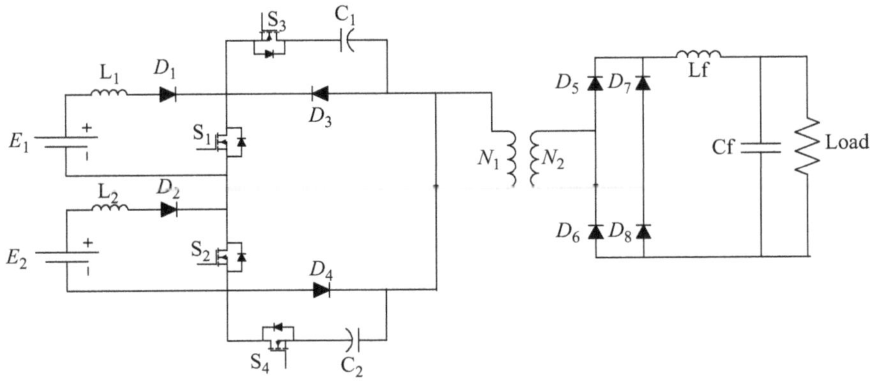

Figure 12.45 Multi-input three-level DC/DC converter [82]

means that the various structure can be created by combination of conventional converters (electrically) and transformers (magnetically). (c) The different voltage levels could be achieved in the output. (d) It would be more suitable for multi-output converters. (e) The control circuit of these converters is complex than others, but more flexible methods can be used. (f) These converters can be used as serial, parallel, cascade, and hybrid cells. (g) The saturation phenomena and magnetic problems may be occurred. (f) Ability to support different energy sources with different condition.

12.5 Comparative study and assessment

In Table 12.3, the results are summarized. In this table, voltage conversion ratio, number of semiconductors, number of inductors and capacitors, and number of transformers for all of the converters are described. According to this table, voltage gain in magnetically coupling converters have relation with transformers ratio, and transformer determined voltage gain, but in electrically coupling converters, the duty cycle of switches established voltage gain. Some of the converters have several voltage conversion ratios. This issue matters because of the difference between the input voltages or the condition changing of the battery from charging to discharging. The number of semiconductors and passive components in electrically coupling converters is less than the one of the other types of the multi-input converters. This means that the cost of the converter is low, but there is no isolation between input suppliers and the output load. In electrically coupled converters, the number of capacitors is less than the one of the other types of the multi-input converters, because DC-link needs only one capacitor in these converters. The multi-winding transformer is an inseparable component in the magnetically coupled converter, but in electrically-magnetically coupled converters, two-winding transformers are commonly utilized. This table shows electrically-magnetically coupling converters have features of the magnetically coupling converter such as isolation and electrically coupling converter such as less number of elements.

Table 12.3 Gain and number of elements evaluation in different converters

Converter	Voltage conversion ratio	No. of semi-conductors (D: Diodes) (S: Switches)	No. of capacitors	No. of inductors	No. of trans-formers (3W: three-winding) (2W: two-winding)
Buck/Boost multi-input converter [20]	$V_o = \frac{d_1}{N_1(1-d_1-d_2)}V_1 + \frac{d_2}{N_2(1-d_1-d_2)}V_2$	3 D 2 S	1	0	**1 (3W)**
Double input full-bridge converter [23]	$V_o = \frac{n_3}{n_1}V_1 + \frac{n_2}{n_1}V_2$	12 D 8 S	1	2	**1 (3W)**
Multiport bidirectional converter [34]	$V_o = \frac{n_3}{n_1}V_{S1} + \frac{n_2}{n_1}V_{S2}$	6 S	7	0	**1 (3W)**
Full-bridge boost converter-based distribution transformer [83]	$V_o = \frac{n}{2(1-d_1)}V_1$ $V_o = \frac{n}{2(1-d_2)}V_2$	8 D 8 S	2	2	**1 (8W)**
Dual-transformer based asymmetrical triple-port active bridge DC–DC converter [84]	-	14 S 0 D	1	2	**2 (2W)**
Series or parallel of two DC/DC converters [42]	$V_o = \frac{V_1}{1-d_1} + \frac{V_2}{1-d_2}$ $V_o = \frac{V_1}{1-d_2} = \frac{V_2}{1-d_4}$	4 S or 2 D 2 S	2	2	**0**
Double-input buck/buck-boost converter [85]	$V_o = \frac{d_1}{1-d_2}V_1 + \frac{d_2}{1-d_2}V_2$	2 D 2 S	1	1	**0**

A multi-input converter as a BLDC drive [57]	$V_o = \frac{V_1 + d_4 V_3}{1 - d_1 - d_4}$ $V_o = \frac{V_2 - d_3 V_3}{1 - d_2 - d_3}$	2 D 4 S	1	2	**0**
Three-input DC/DC boost converter [58]	Mode 1: $V_o = \frac{V_1}{1-d_1-d_4} = \frac{V_2}{1-d_2-d_4}$ Mode 2: $V_o = \frac{V_1}{1 - d_1 - d_3} = \frac{V_2}{1 - d_2 - d_3}$ $V_3 = \frac{d_1 + d_3 + d_4}{d_2 + d_3 + d_4} V_1 = \frac{d_2 + d_3 + d_4}{d_1 + d_3 + d_4} V_2$ Mode 3: $V_o = \frac{V_1}{1 - d_1 - d_3} = \frac{V_2}{1 - d_2 - d_3}$ $V_3 = d_1.V_1 = d_2.V_2$	4 D 4 S	1	2	**0**
ZVS multi-input converter [61]	$V_3 = \frac{1}{1 - d_1} V_1$ $V_o = \frac{(1 - d_1)}{(1 + d_{dcm} - d_1)} V_3$	1 D 3 S	1	3	**0**
Non-isolated multi-input multi-output DC/DC boost converter [62]	-	4 D 4 S	2	1	**0**
Non-isolated multi-input single-output DC/DC buck/boost-Boost converter [63]	$V_{out} = \frac{E_2}{1-D_2} + \frac{D_1 D_2}{(1-D_1)(1-D_2)} E_1$	2 D 2 S	2	2	**0**

(Continues)

Table 12.3 (Continued)

Converter	Voltage conversion ratio	No. of semi-conductors (D: Diodes) (S: Switches)	No. of capacitors	No. of inductors	No. of trans-formers (3W: three-winding) (2W: two-winding)
Switched-capacitor based multi-input voltage-summation converter [65]	$V_{out} = E_1 + E_2 + E_3$	5 D 3 S	3	2	**0**
SEPIC-based two inputs converter [86]	Mode I: $V_0 = \frac{d_1 V_1}{(1-d_1)(1-d_3)} + \frac{d_3}{1-d_3} V_2$ Mode II: $V_0 = \frac{(d_1+d_2)V_1 - d_2 V_2}{1-d_1-d_2}$	3 S 4 D	1	2	**0**
Improved bridge type dual input DC–DC converter (IBDC) [87]	Mode I: $V_0 = d_1 V_1 + d_2 V_2 + (V_1 + V_2)d_3$ Mode II: $V_0 = \frac{d_1 V_1 + d_2 V_2 + (V_1+V_2)d_3}{(1-d_1-d_2-d_3)}$ Mode III: $V_0 = \frac{V_1(d_1+d_3)+V_2(1-d_1)}{(1-d_1-d_3)}$	4 S 2 D	1	1	**0**
Three-input step-up DC–DC boost converter [88]	Mode I: $V_0 = \frac{V_1 + d_1\left(\frac{d_2 V_2}{1-d_1}\right)}{1-d_1}$ Mode II: $V_0 = \frac{V_1 + d_1\left(\frac{d_2 V_2 + d_3 V_{batt}}{1-d_2}\right) + d_3 V_{batt}}{1-d_1}$	4 S 4 D	2	2	**0**
Non-isolated non-coupled inductor based multi-input (NIN-CIBMI) DC–DC converter [89]	$V_0 = \frac{2-d_1}{(1-d_1)^2} V_1 + \frac{1}{(1-d_2)^2} V_2$	7 S 1 D (bidirectional) or 8 D (unidirectional)	3	4	**0**

Multi-input converter with high-voltage gain and two input boost stage [90]	$N \text{ is odd}: V_0 = \frac{N+1}{2}\left[\frac{V_1}{1-d_1} + \frac{V_2}{1-d_2}\right]$ $N \text{ is even}: V_0 = \frac{N+2}{2}\left(\frac{V_1}{1-d_1}\right) + \frac{N}{2}\left(\frac{V_2}{1-d_2}\right)$	2 S 5 D	5	2	**0**
Non-isolated high step-up three-port DC–DC converter [91]	$V_0 = \frac{1+n}{1-D} V_{in}$	4 S 5 D	9	7	**0**
Combined DC link and magnetic couple converter [72]	$D = \frac{V_{DC} - V_{SC}}{V_{SC}}$ $V_{DC} = 2.V_2$ $V_{Load} = n.V_{DC}$	6 S	5	2	**1 (2W)**
Direct charge converter [74]	$V_2 = \frac{d_1}{d_1 + d_2} V_1$ $V_o = 2nd_1V_2$	4 S	5	0	**1 (2W)**
Boost-integrated phase-shift full-bridge converter [75]	$V_2 = D.V_1$ $V_o = 2.\varphi_{eff}.n.V_1$ $\varphi_{eff} = \min(\varphi, D, 1-D)$	4 D4 S	2	3	**1 (2W)**
Tri-modal half-bridge converter [78]	$V_2 = \frac{d_2}{d_1 + d_2} V_1$ $V_o = \frac{2d_1d_2}{d_1 + d_2} nV_1$	3 D 3 S	3	1	**1 (3W)**
Full-bridge three-port converters with Wide Input Voltage Range [80]	$V_1 = \frac{d_2}{d_1} V_2$ $V_o = 2nd_3V_2$	4 D 4 S	1	1	**1 (2W)**
Isolated multi-port DC–DC converter for simultaneous power management [81]	-	7 D 3 S	2	4	**1 (2W)**
Multi-input three-level DC/DC converter [82]	$V_0 = V_{dc,cri} = \frac{2V_{input}}{D_2 - D_1 + 1}$	8 D 4 S	3	3	**1 (2W)**

Converters mentioned in the previous sections have different features, which are summarized in Table 12.4 and the characteristics of this converters are presented. According to this table, the best choice for the multi-input converter is based on the needed user requirement such as; voltage level, gain level, load variation, source variation, bidirectional power flow, efficiency, losses, cost, control complexity, number of active elements, type of sources, source utility, isolation, battery charging capability, soft-switching, environment, etc. Characteristics summarized in Table 12.4 are categorized in Figure 12.46. This figure can be used as a guide for the design of multi-input converters. For example: Combined DC link and magnetic couple converter and boost-integrated phase-shift full-bridge converter are one of the good candidate for this applications.

According to Table 12.4 and Figure 12.46, the converters can be compared regarding the isolation, soft switching, source utilization, and battery charging. Source utilization is one of the critical features in a multi-input converter that is noticed in the most converters. Battery charging means that in the converter, the battery can be used as a source in the converter and the battery can sometimes be charged. The term “Isolation” is used since an isolation transformer is included use in the converter. Soft switching means that the converter can operate in high frequency and the switching loss in the converter is extremely low.

Among the introduced structures, the soft switching feature is only available in second input converter using a full-bridge converter and other structures are not applicable for such a feature. If the soft switching feature is ignored in such converters, the volume of the proposed converter may exceed the ones of the existing converters. The higher volume which is the necessity for the cooling process of the switches leads to employing the bigger coolers.

The isolation of the suppliers from each other (in multi-input magnetically converters) in addition to loading from the suppliers which makes it possible to utilize different suppliers with different voltages are two advantages of this converter. On the other hand, the disadvantages of the isolation are the difficulties of designing multiple winding transformers and also increasing in circuit volume. Presence and absence of the isolation depend tightly upon the designing of the system and is not solely a unique criterion for the converter’s appropriateness or inappropriateness. In this section, only four structures have employed the isolation feature, which is the magnetically and electromagnetically coupled converters.

Battery charging issue that must be considered in the design of multi-input converters. Converters in which there is no possibility of charging and discharging batteries are not suitable for operation disconnected of the grid because some DGs could not generate power at all times and must become a battery for energy storage and provide energy to the load.

12.6 Conclusion

The primary purpose of this chapter is to provide state of the art account for multi-input DC/DC converters. A large part of the following application uses multi-input DC/DC converters: DC micro-grid, battery charger, electric vehicle charging

Table 12.4 Feature comparison of the converters

Converter types	Soft-switching	Battery charging	Isolation	Source utility
Buck/boost multi-input converter [20]			*	
Double-input full-bridge converter [23]			*	*
Multiport bidirectional converter [34]	*		*	*
Full-bridge boost converter-based distribution transformer [83]			*	
Dual-transformer-based asymmetrical triple-port active bridge DC–DC converter [84]	*	*	*	*
Series or parallel of two DC/DC converters [42]				*
Double-input buck/buck-boost converter [85]	*			*
A multi-input converter as a BLDC drive [57]		*		*
Three-input DC/DC boost converter [58]				*
ZVS multi-input converter [61]	*			*
Non-isolated multi-input multi-output DC/DC boost converter [62]		*		*
Non-isolated multi-input single-output DC/DC buck/boost-boost converter [63]				*
Switched-capacitor-based multi-input voltage-summation converter [65]	*			
SEPIC-based two inputs converter [86]		*		*
Improved bridge type dual input DC–DC converter (IBDC) [87]		*		*
Three-input step-up DC–DC boost converter [88]		*		*
Non-isolated non-coupled inductor-based multi-input (NINCIBMI) DC–DC converter [89]		*		*
Multi-input converter with high-voltage gain and two input boost stage [90]				*
Non-isolated high step-up three-port DC–DC converter [91]	*	*		*
Combined DC link and magnetic couple converter [72]	*	*	*	*
Direct charge converter [74]		*	*	*
Boost-integrated phase-shift full-bridge converter [75]	*	*	*	*
Tri-modal half-bridge converter [78]		*	*	
Full-bridge three-port converters with wide input voltage range [80]		*	*	
Isolated multi-port DC–DC converter for simultaneous power management [81]			*	*
Multi-input three-level DC/DC converter [82]			*	*

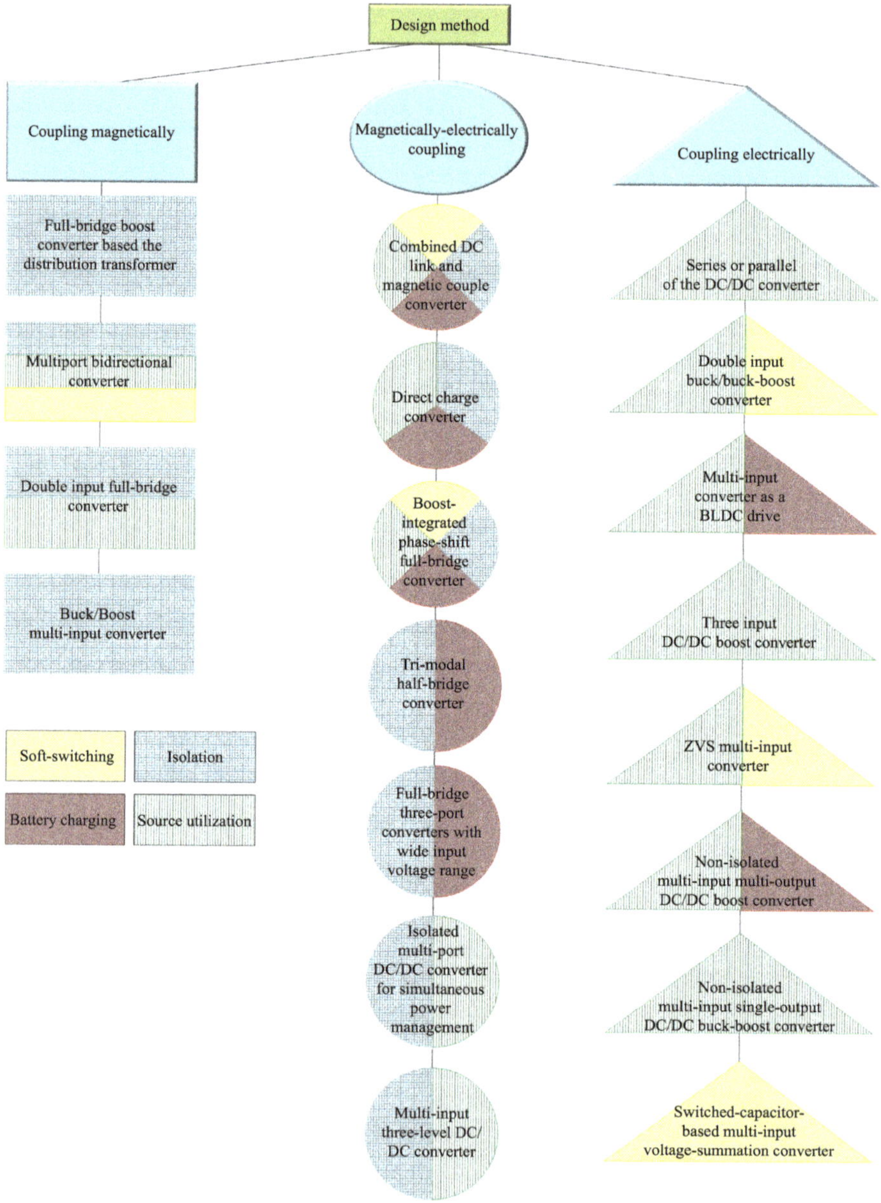

Figure 12.46 A design guide for selecting proper converter

stations and DC-link of FACTS devices, etc. These converters are able to connect to DG units like wind turbines, solar panels, batteries, FCs, and microturbines.

The multi-input DC/DC converters described in this chapter have been divided into three main groups of magnetic, electric and electromagnetic types that are compared with one another. The general comparison among the configurations includes battery life, soft-switching, source utilization, the input and output isolation, and some elements employed in each topology. According to the explained issues, and comparisons one can say that the multi-input DC/DC converters are a desirable option when more than one DG unit are connected with each other because of the inherent characteristics such as less semiconductor utilization and occupying less volume. In general, these converters have the advantages and disadvantages such as:

a. Simultaneous usage of two or more suppliers is possible,
b. Lower cost due to the utilization of the fewer passive and semiconductor elements,
c. The smaller size of the converter,
d. Higher power production due to the utilization of some suppliers,
e. Better monitoring of the energy management resources.

An increasing number of potential applications for multi-input DC/DC converter are presented in this literature. These applications also include DC-link of flexible alternating current transmission system like UPQC for creating DG-UPQC. This application causes FACTS devices to be able to supply the load when the grid is disconnected. Also, the DG application for creating AC or DC voltage and providing connected grid loads or stand-alone loads. Exciting fields of application also include spacecraft and aircraft.

The optimum options for the creation of the multi-input converters are the combined electromagnetic converters since the investigated features inside of them are more appropriate than the other two structures. It is better to utilize this type of the multi-input DC/DC converter for academic researches and industrial applications.

In the future, the use of DG sources in different type will increase and interest in the use of them together in one area increases. This matter will cause, new topologies are introduced that can connect to more than two DG sources like multi-input converters. The use of these topologies will decrease converters size strongly. In the future, research will focus on management systems for optimal utilization of DG sources to maximize system efficiency. The researcher and suppliers will try to design and manufacture converters that can supply DC and AC loads simultaneously, without complex control system are used in the converter.

According to studies, multi-input converters can have different attributes. The features will be considered by manufacturers and consumers is the number of inputs, the number of outputs, soft switching, battery charging control, bi-directional power flow, less noise, hybrid operation, optimum and straight forward control, isolation between input and output, smaller volume, maximum use of small and large sources simultaneously, etc. Some of these features investigate in different topologies.

References

[1] A. Eid, "Utility integration of PV-wind-fuel cell hybrid distributed generation systems under variable load demands," *International Journal of Electrical Power & Energy Systems*, vol. 62, no. 0, pp. 689–99, 2014.

[2] M. Barghi Latran and A. Teke, "Investigation of multilevel multifunctional grid connected inverter topologies and control strategies used in photovoltaic systems," *Renewable and Sustainable Energy Reviews*, vol. 42, no. 0, pp. 361–76, 2015.

[3] H. Torkaman, M. Fakhari, H. Karimi, and B. Taheri, "New frequency modulation strategy with SHE for H-bridge multilevel inverters," in *2018 4th International Conference on Electrical Energy Systems (ICEES)*, pp. 157–61: IEEE, 2018.

[4] D. Sun, B. Ge, D. Bi, and F. Z. Peng, "Analysis and control of quasi-Z source inverter with battery for grid-connected PV system," *International Journal of Electrical Power & Energy Systems*, vol. 46, no. 0, pp. 234–40, 2013.

[5] H. Torkaman, and T. Hemmati, "Hybrid Z-source DC–DC converter with ZVZCS and power transformer resetting: design, modeling, and fabrication," *Iranian Journal of Electrical & Electronic Engineering*, vol. 14, no. 1, p. 49, 2018.

[6] H. Torkaman, E. Afjei, A. Keyhani, and M. Poursmaeil, "Control and management of a hybrid AC/DC microgrid based on Gamma-Z-Source interlinking converter and multi-port converter for DERs integration," *IET Generation, Transmission & Distribution*, 2019.

[7] M. Poursmaeil, S. M. Dizgah, H. Torkaman, and E. Afjei, "Autonomous control and operation of an interconnected AC/DC microgrid with Gamma-Z-Source interlinking converter," in *Smart Grid Conference (SGC)*, pp. 1–6, 2017.

[8] M. Poursmaeil, S. M. Dizgah, H. Torkaman, and E. Afjei, "Small signal modeling, analysis and control of Gamma-Z-Source Inverter," in *Iranian Conference on Electrical Engineering (ICEE)*, pp. 1216–22, 2017.

[9] M. H. Taghvaee, M. A. M. Radzi, S. M. Moosavain, H. Hizam, and M. Hamiruce Marhaban, "A current and future study on non-isolated DC–DC converters for photovoltaic applications," *Renewable and Sustainable Energy Reviews*, vol. 17, no. 0, pp. 216–27, 2013.

[10] A. Sarikhani, B. Allahverdinejad, and H. Torkaman, "A non-isolated buck-boost DC–DC converter with single switch," in *Power Electronics, Drives Systems and Technologies Conference (PEDSTC), 2018 9th Annual*, pp. 369–73: IEEE, 2018.

[11] H. Torkaman, N. Karami, and M. Nezamabadi, "Design, simulation, validation and comparison of new high step-up soft switched converter for fuel cell energy system," *Journal of Energy Management and Technology*, vol. 1, no. 1, pp. 53–60, 2017.

[12] L. Kumar, and S. Jain, "A multiple source DC/DC converter topology," *International Journal of Electrical Power & Energy Systems*, vol. 51, no. 0, pp. 278–291, 2013.

[13] J. Wei, and B. Fahimi, "Multi-port power electric interface for renewable energy sources," in *24th Annual IEEE Applied Power Electronics Conference and Exposition*, Washington, February 2009, pp. 347–52.

[14] S. Khosrogorji, H. Torkaman, and F. Karimi, "A short review on multi-input DC/DC converters topologies," in *Power Electronics, Drives Systems & Technologies Conference (PEDSTC), 2015 6th*, pp. 650–54: IEEE, 2015.

[15] M. Einan, H. Torkaman, and M. Pourgholi, "Optimized fuzzy-cuckoo controller for active power control of battery energy storage system, photovoltaic, fuel cell and wind turbine in an isolated micro-grid," *Batteries*, vol. 3, no. 3, p. 23, 2017.

[16] P. G. Arul, V. K. Ramachandaramurthy, and R. K. Rajkumar, "Control strategies for a hybrid renewable energy system: a review," *Renewable and Sustainable Energy Reviews*, vol. 42, no. 0, pp. 597–08, 2015.

[17] A. Mohammed, J. Pasupuleti, T. Khatib, and W. Elmenreich, "A review of process and operational system control of hybrid photovoltaic/diesel generator systems," *Renewable and Sustainable Energy Reviews*, vol. 44, no. 0, pp. 436–46, 2015.

[18] A. H. Fathima, and K. Palanisamy, "Optimization in microgrids with hybrid energy systems – a review," *Renewable and Sustainable Energy Reviews*, vol. 45, no. 0, pp. 431–46, 2015.

[19] H. Matsuo, T. Shigemizu, F. Kurokawa, and N. Watanabe, "Characteristics of the multiple-input DC-DC converter," in *24th Annual IEEE Power Electronics Specialists Conference*, Seattle, June 1993, pp. 115–20.

[20] H. Matsuo, L. Wenzhong, F. Kurokawa, T. Shigemizu, and N. Watanabe, "Characteristics of the multiple-input DC-DC converter," *IEEE Transactions on Industrial Electronics*, vol. 51, no. 3, pp. 625–31, 2004.

[21] H. Matsuo, K. Kobayashi, Y. Sekine, M. Asano, and W. Lin, "Novel solar cell power supply system using the multiple-input DC-DC converter," in *20th International Telecommunications Energy Conference*, San Francisco, October 1998, pp. 797–802.

[22] C. Yaow-Ming, and L. Yuan-Chuan, "Development of multi-port converters for hybrid wind-photovoltaic power system," in *Proceedings of IEEE Region 10 International Conference on Electrical and Electronic Technology*, Singapore, August 2001, vol. 2, pp. 804–8.

[23] Y. M. Chen, Y. C. Liu, F. Y. Wu, and T. F. Wu, "Multi-input DC/DC converter based on the flux additivity," in *36th IAS Annual Meeting Industry Applications Conference*, Chicago, vol. 3, pp. 1866–73, 2001.

[24] C. Yaow-Ming, L. Yuan-Chuan, and W. Feng-Yu, "Multi-input DC/DC converter based on the multiwinding transformer for renewable energy applications," *IEEE Transactions on Industrial Electronics*, vol. 38, no. 4, pp. 1096–104, 2002.

[25] K. P. Yalamanchili, and M. Ferdowsi, "Review of multiple input DC-DC converters for electric and hybrid vehicles," in *IEEE Vehicle Power and Propulsion Conference*, Chicago, September 2005, pp. 160–63.

[26] M. Gavris, O. Cornea, and N. Muntean, "Multiple input DC-DC topologies in renewable energy systems - a general review," in *IEEE 3rd International Symposium on Exploitation of Renewable Energy Sources (EXPRES)*, Subotica, March 2011, pp. 123–28.

[27] S. Xiaofeng, P. Guangming, Y. Shuai, and C. Zhe, "A novel multi-port dc/dc converter with bi-directional storage unit," in *7th International Power Electronics and Motion Control Conference (IPEMC)*, Harbin, China, June 2012, vol. 3, pp. 1771–75.

[28] M. Suetomi, H. Matsuo, K. Hirakida, *et al.* "A novel multi-input DC-DC converter with high power efficiency," in *IEEE 33rd International Telecommunications Energy Conference (INTELEC)* Amsterdam, October 2011, pp. 1–5.

[29] S. Falcones, R. Ayyanar, and M. Xiaolin, "A DC/DC multiport-converter-based solid-state transformer integrating distributed generation and storage," *IEEE Transactions on Power Electronics*, vol. 28, no. 5, pp. 2192–203, 2013.

[30] A. Andrijanovits, A. Blinov, O. Husev, and D. Vinnikov, "Multiport converter with integrated energy storage for hydrogen buffer interfacing with renewable energy systems," in *IEEE International Conference on Industrial Technology (ICIT)*, Athens, 2012, pp. 230–35.

[31] P. Veena, V. Indragandhi, R. Jeyabharath, and V. Subramaniyaswamy, "Review of grid integration schemes for renewable power generation system," *Renewable and Sustainable Energy Reviews*, vol. 34, no. 0, pp. 628–41, 2014.

[32] L. Yan, Z. Chuang, C. Jia Yao, D. Rui, and Z. Yujin, "Optimizing design of soft-switching dual-input full-bridge DC/DC converter," in *IEEE Vehicle Power and Propulsion Conference (VPPC)*, Chicago, September 2011, pp. 1–6.

[33] N. M. B. Eloundou, F. Gustin, and A. Berthon, "Multi-source high frequency link DC-DC converter for EV or HEV applications," in *6th International Power Electronics and Motion Control Conference*, Wuhan, May 2009, pp. 1282–87.

[34] H. Tao, A. Kotsopoulos, J. L. Duarte, and M. A. M. Hendrix, "Family of multiport bidirectional DC-DC converters," *IEE Proceedings - Electric Power Applications*, vol. 153, no. 3, pp. 451–58, 2006.

[35] T. Haimin, J. L. Duarte, and M. A. Hendrix, "Three-port triple-half-bridge bidirectional converter with zero-voltage switching," *IEEE Transactions on Power Electronics*, vol. 23, no. 2, pp. 782–92, 2008.

[36] L. ShengYong, Z. Xing, G. Haibin, and X. Jun, "Multiport DC/DC converter for stand-alone photovoltaic lighting system with battery storage," in *International Conference on Electrical and Control Engineering (ICECE)*, Wuhan, June 2010, pp. 3894–97.

[37] H. Krishnaswami, and N. Mohan, "A current-fed three-port bi-directional DC-DC converter," in *29th International Telecommunications Energy Conference* Rome, 2007, pp. 523–26.

[38] Z. Zhang, O. C. Thomsen, M. A. Andersen, and H. R. Nielsen, "Dual-input isolated full-bridge boost DC–DC converter based on the distributed transformers," *IET Power Electronics*, vol. 5, no. 7, pp. 1074–83, 2012.

[39] V. N. S. R. Jakka, A. Shukla, and G. D. Demetriades, "Dual-transformer-based asymmetrical triple-port active bridge (DT-ATAB) isolated DC–DC converter," *IEEE Transactions on Industrial Electronics*, vol. 64, no. 6, pp. 4549–60, 2017.

[40] L. Solero, F. Caricchi, F. Crescimbini, O. Honorati, and F. Mezzetti, "Performance of a 10 kW power electronic interface for combined wind/PV isolated generating systems," in *27th Annual IEEE Power Electronics Specialists Conference*, Baveno, June 1996, vol. 2, pp. 1027–32.

[41] F. D. Rodriguez, and W. G. Imes, "Analysis and modeling of a two-input DC/DC converter with two controlled variables and four switched networks," in *Proceedings of the 31st Intersociety Energy Conversion Engineering Conference*, Washington, August 1996, vol. 1, pp. 322–27.

[42] A. Di Napoli, F. Crescimbini, S. Rodo, and L. Solero, "Multiple input DC-DC power converter for fuel-cell powered hybrid vehicles," in *IEEE 33rd Annual Power Electronics Specialists Conference*, Cairns, 2002, vol. 4, pp. 1685–90.

[43] Z. Junjun, W. Hongfei, H. Jun, X. Yan, and M. Xudong, "A novel multi-port bidirectional converter for interfacing distributed DC micro-grid," in *IEEE 23rd International Symposium on Industrial Electronics (ISIE)*, Istanbul, June 2014, pp. 2344–48.

[44] R. Majumder, and G. Bag, "Parallel operation of converter interfaced multiple microgrids," *International Journal of Electrical Power & Energy Systems*, vol. 55, no. 0, pp. 486–96, 2014.

[45] H. Shiying, C. Jianfei, S. Tao, and B. Xiaohui, "Multi-input step-up converters based on the switched-diode-capacitor voltage accumulator," *IEEE Transactions on Power Electronics*, vol. 31, no. 1, pp. 381–93, 2016.

[46] C. Cheng-Wei, L. Chien-Yao, C. Kun-Hung, and C. Yaow-Ming, "Modeling and controller design of a semiisolated multiinput converter for a hybrid PV/wind power charger system," *IEEE Transactions on Power Electronics*, vol. 30, no. 9, pp. 4843–53, 2015.

[47] W. Hongfei, Z. Junjun, and X. Yan, "A family of multiport buck/boost converters based on DC-link-inductors (DLIs)," *IEEE Transactions on Power Electronics*, vol. 30, no. 2, pp. 735–46, 2015.

[48] C. Yaow-Ming, L. Yuan-Chuan, and L. Sheng-Hsien, "Double-input PWM DC/DC converter for high/low voltage sources," in *25th International Telecommunications Energy Conference*, Yokohama, Japan, October 2003, pp. 27–32.

[49] K. Gummi, and M. Ferdowsi, "Derivation of new double-input DC-DC converters using H-Bridge cells as building blocks," in *34th Annual Conference of IEEE Industrial Electronics*, Orlando, 2008, pp. 2806–11.

[50] V. A. Prabhala, D. Somayajula, and M. Ferdowsi, "Power sharing in a double-input buck converter using dead-time control," in *IEEE Energy Conversion Congress and Exposition*, San Jose, September 2009, pp. 2621–26.

[51] S. Xiaofeng, Z. Yue, W. Wei, W. Baocheng, and Z. Zhe, "Alternative source-port-tolerant series-connected double-input DC/DC converter," *IEEE Transactions on Power Electronics*, vol. 30, no. 5, pp. 2733–42, 2015.

[52] M. Gavris, N. Muntean, and O. Cornea, "A new dual-input hybrid buck DC-DC converter," in *International Aegean Conference on Electrical Machines and Power Electronics and 2011 Electromotion Joint Conference (ACEMP)*, Istanbul, 2011, pp. 109–14.

[53] M. Gavris, L. Cadariu, B. Caruntu, O. Cornea, and N. Muntean, “Analysis of the dual input hybrid buck DC-DC converter in boundary conduction mode,” in *International Conference on Optimization of Electrical and Electronic Equipment (OPTIM)*, Bran, May 2014, pp. 601–6.

[54] Y. Sheng-Yang, Z. Ruichen, and A. Kwasinski, “Design considerations of a multiple-input isolated single ended primary inductor converter (SEPIC) for distributed generation sources,” in *IEEE Energy Conversion Congress and Exposition (ECCE)*, Phoenix, September 2011, pp. 3960–67.

[55] S. Y. Yu, and A. Kwasinski, “Analysis of soft-switching isolated time-sharing multiple-input converters for DC distribution systems,” *IEEE Transactions on Power Electronics*, vol. 28, no. 4, pp. 1783–94, 2013.

[56] M. Veerachary, “Buck-integrated SEPIC converter for photovoltaic power conversion,” in *31st International Telecommunications Energy Conference*, Incheon, October 2009, pp. 1–5.

[57] M. R. Feyzi, S. A. Mozaffari Niapour, F. Nejabatkhah, S. Danyali, and A. Feizi, “Brushless DC motor drive based on multi-input DC boost converter supplemented by hybrid PV/FC/battery power system,” in *24th Canadian Conference on Electrical and Computer Engineering (CCECE)*, Niagara Falls, May 2011, pp. 442–46.

[58] F. Nejabatkhah, S. Danyali, S. H. Hosseini, M. Sabahi, and S. M. Niapour, “Modeling and control of a new three-input DC–DC boost converter for hybrid PV/FC/battery power system,” *IEEE Transactions on Power Electronics*, vol. 27, no. 5, pp. 2309–24, 2012.

[59] E. Babaei, and O. Abbasi, “Structure for multi-input multi-output DC/DC boost converter,” *IET Power Electronics*, vol. 9, no. 1, pp. 9–19, 2016.

[60] W. Rong-Jong, L. Chung-You, and C. Bo-Han, “High-efficiency DC/DC converter with two input power sources,” *IEEE Transactions on Power Electronics*, vol. 27, no. 4, pp. 1862–75, 2012.

[61] W. Rong-Jong, L. Chung-You, L. Jun-Jie, and C. Yung-Ruei, “Newly designed ZVS multi-input converter,” *IEEE Transactions on Industrial Electronics*, vol. 58, no. 2, pp. 555–66, 2011.

[62] A. Nahavandi, M. T. Hagh, M. B. B. Sharifian, and S. Danyali, “A non-isolated multi input multioutput DC/DC boost converter for electric vehicle applications,” *IEEE Transactions on Power Electronics*, vol. 30, no. 4, pp. 1818–35, 2015.

[63] M. R. Banaei, H. Ardi, R. Alizadeh, and A. Farakhor, “Non-isolated multi-input-single-output DC/DC converter for photovoltaic power generation systems,” *IET Power Electronics*, vol. 7, no. 11, pp. 2806–16, 2014.

[64] Y. Ye, and K. W. E. Cheng, “Multi-port voltage-subtracting circuit based on resonant switched-capacitor,” *IET Power Electronics*, vol. 5, no. 6, pp. 693–701, 2012.

[65] Y. Yuan-mao, and K. W. E. Cheng, “Multi-input voltage-summation converter based on switched-capacitor,” *IET Power Electronics*, vol. 6, no. 9, pp. 1909–16, 2013.

[66] S. K. Haghighian, S. Tohidi, M. R. Feyzi, and M. Sabahi, "Design and analysis of a novel SEPIC-based multi-input DC/DC converter," *IET Power Electronics*, vol. 10, no. 12, pp. 1393–402, 2017.

[67] S. Athikkal, G. G. Kumar, K. Sundaramoorthy, and A. Sankar, "Performance analysis of novel bridge type dual input DC-DC converters," *IEEE Access*, vol. 5, pp. 15340–53, 2017.

[68] F. Kardan, R. Alizadeh, and M. R. Banaei, "A new three input DC/DC converter for hybrid PV/FC/battery applications," *IEEE Journal of Emerging and Selected Topics in Power Electronics*, vol. 5, no. 4, pp. 1771–78, 2017.

[69] K. Varesi, S. H. Hosseini, M. Sabahi, and E. Babaei, "Modular non-isolated multi-input high step-up DC–DC converter with reduced normalised voltage stress and component count," *IET Power Electronics*, vol. 11, no. 6, pp. 1092–100, 2018.

[70] V. A. K. Prabhala, P. Fajri, V. S. P. Gouribhatla, B. P. Baddipadiga, and M. Ferdowsi, "A DC–DC converter with high voltage gain and two input boost stages," *IEEE Transactions on Power Electronics*, vol. 31, no. 6, pp. 4206–15, 2016.

[71] R. Faraji, and H. Farzanehfard, "Soft-switched non-isolated high step-up three-port DC-DC converter for hybrid energy systems," *IEEE Transactions on Power Electronics*, vol. 33, no. 12, pp. 10101–10111, 2018.

[72] H. Tao, A. Kotsopoulos, J. L. Duarte, and M. A. M. Hendrix, "Multi-input bidirectional DC-DC converter combining DC-link and magnetic-coupling for fuel cell systems," in *40th IAS Annual Meeting Industry Applications Conference*, 2005, vol. 3, pp. 2021–28.

[73] H. Wu, K. Sun, L. Zhu, and Y. Xing, "An interleaved half-bridge three-port converter with enhanced power transfer capability using three-leg rectifier for renewable energy applications," *IEEE Journal of Emerging and Selected Topics in Power Electronics*, vol. 4, no. 2, pp. 606–616, 2016.

[74] W. Hongfei, X. Yan, C. Runruo, Z. Junjun, S. Kai, and G. Hongjuan, "A three-port half-bridge converter with synchronous rectification for renewable energy application," in *IEEE Energy Conversion Congress and Exposition*, Phoenix, 2011, pp. 3343–49.

[75] H. Al-Atrash, and I. Batarseh, "Boost-integrated phase-shift full-bridge converter for three-port interface," in *IEEE Power Electronics Specialists Conference*, Orlando, 2007, pp. 2313–21.

[76] H. Al-Atrash, "Integrated Topologies and Digital Control for Satellite Power Management and Distribution Systems," PhD Thesis, University of Central Florida, 2007.

[77] H. M. Chou, "Multi-Port DC-DC Power Converter for Renewable Energy Application," M. S., Texas A&M University, 2009.

[78] H. Al-Atrash, T. Feng, and I. Batarseh, "Tri-modal half-bridge converter topology for three-port interface," *IEEE Transactions on Power Electronics*, vol. 22, no. 1, pp. 341–45, 2007.

[79] Q. Zhijun, O. Abdel-Rahman, and I. Batarseh, "An integrated four-port DC/DC converter for renewable energy applications," *IEEE Transactions on Power Electronics* vol. 25, no. 7, pp. 1877–87, 2010.

[80] W. Hongfei, S. Kai, C. Runruo, H. Haibing, and X. Yan, "Full-bridge three-port converters with wide input voltage range for renewable power systems," *IEEE Transactions on Power Electronics*, vol. 27, no. 9, pp. 3965–74, 2012.

[81] Z. Jianwu, Q. Wei, Q. Liyan, and J. Yanping, "An isolated multiport DC/DC converter for simultaneous power management of multiple different renewable energy sources," *IEEE Journal of Emerging and Selected Topics in Power Electronics*, vol. 2, no. 1, pp. 70–78, 2014.

[82] S. Dusmez, L. Xiong, and B. Akin, "A new multiinput three-level DC/DC converter," *IEEE Transactions on Power Electronics*, vol. 31, no. 2, pp. 1230–40, 2016.

[83] Z. Zhang, O. C. Thomsen, M. A. E. Andersen, and H. R. Nielsen, "A novel dual-input isolated current-fed DC-DC converter for renewable energy system," in *26th Annual IEEE Applied Power Electronics Conference and Exposition (APEC)*, Fort Worth, March 2011, pp. 1494–501.

[84] V. N. S. R. Jakka, A. Shukla, and G. D. Demetriades, "Dual-transformer-based asymmetrical triple-port active bridge (DT-ATAB) isolated DC–DC converter," *IEEE Transactions on Industrial Electronics*, vol. 64, no. 6, pp. 4549–60, 2017.

[85] C. Yaow-Ming, L. Yuan-Chuan, and L. Sheng-Hsien, "Double-input PWM DC/DC converter for high-/low-voltage sources," *IEEE Transactions on Industrial Electronics*, vol. 53, no. 5, pp. 1538–45, 2006.

[86] S. K. Haghighian, S. Tohidi, M. R. Feyzi, and M. Sabahi, "Design and analysis of a novel SEPIC-based multi-input DC/DC converter," *IET Power Electronics*, vol. 10, no. 12, pp. 1393–402, 2017.

[87] S. Athikkal, G. G. Kumar, K. Sundaramoorthy, and A. Sankar, "Performance analysis of novel bridge type dual input DC-DC converters," *IEEE Access*, vol. 5, pp. 15340–53, 2017.

[88] F. Kardan, R. Alizadeh, and M. R. Banaei, "A new three input DC/DC converter for hybrid PV/FC/battery applications," *IEEE Journal of Emerging and Selected Topics in Power Electronics*, vol. 5, no. 4, pp. 1771–78, 2017.

[89] K. Varesi, S. H. Hosseini, M. Sabahi, and E. Babaei, "Modular non-isolated multi-input high step-up DC–DC converter with reduced normalised voltage stress and component count," *IET Power Electronics*, vol. 11, no. 6, pp. 1092–100, 2018.

[90] V. A. K. Prabhala, P. Fajri, V. S. P. Gouribhatla, B. P. Baddipadiga, and M. Ferdowsi, "A DC–DC converter with high voltage gain and two input boost stages," *IEEE Transactions on Power Electronics*, vol. 31, no. 6, pp. 4206–15, 2016.

[91] R. Faraji, and H. Farzanehfard, "Soft-switched non-isolated high step-up three-port DC–DC converter for hybrid energy systems," *IEEE Transactions on Power Electronics*, 2018.

Chapter 13

Modeling and control of DC–DC converters for DC microgrid application

Man Mohan Garg[1], Mukesh Kumar Pathak[2], and Rajeev Kumar Chauhan[3]

In a DC microgrid system connected to photovoltaic distributed generation system, DC–DC converters play an important role to perform various functions. In this chapter, a DC microgrid system is presented in which DC–DC converter is utilized to regulate the DC bus voltage under different operating conditions. This chapter focuses on design, modeling, and control of a DC–DC converter. Considering DC–DC buck converter as an example, a systematic procedure is explained to obtain its mathematical model using the state-space averaging (SSA) approach. The steady-state and dynamic models of the buck converter connected to PV systems are derived. Finally, a control design algorithm is discussed to regulate the DC bus voltage in the presence of variations in PV voltage, DC busload, and reference DC bus voltage. The simulation and experimental results are demonstrated in order to validate the controller performance under various aforementioned operating conditions.

13.1 Introduction

The use of renewable energy sources to feed a distributed direct current microgrid is becoming prominent over the traditional centralized alternative current (AC) power generating stations [1–3]. DC microgrid is becoming more popular in low-voltage applications because of its advantages over AC microgrid. DC microgrid needs less number of power processing stages for interfacing with DC distributed energy sources such as solar photovoltaic (PV) and storage system. This also helps in achieving high-efficiency and cost-effective system [4,5]. DC microgrid is a suitable source of power supply for households, parks, LED lightings, variable speed motor drives, commercial buildings in terms of high reliability and improved

[1]Department of Electrical Engineering, National Institute of Technology Rourkela, Rourkela, Odisha, India
[2]Department of Electrical Engineering, Indian Institute of Technology Roorkee, Roorkee, Uttrakhand, India
[3]Department of Electrical Engineering, Dayalbagh Educational Institute, Agra, Uttar Pradesh, India

power quality [6,7]. DC microgrid is also free from issues related to frequency stability, reactive power, skin effect in comparison to it AC counterpart. DC–DC converters are an integral part of DC microgrid because they are used as main power conditioning units. Therefore, the selection of a suitable DC–DC converter becomes important as the overall performance and reliability of the complete system depend on it [8], [9]. The different DC–DC converter configuration can be used in DC microgrid system. The various challenges faced for regulating DC voltage have been described in the literature [10–14].

In this chapter, a DC microgrid system as shown in Figure 13.1 is considered. This system describes that a DC microgrid is having two DC buses with different DC bus voltages. Each DC bus may be connected to various distributed energy resources. However, for the sack of simplicity only wind sources are shown, which are connected to respective DC buses with a suitable AC–DC converter. The various DC loads are connected on both buses. In this chapter, the aim is to interface an incoming solar PV distributed energy source with either of the DC buses. The solar PV distributed energy source consists of a bidirectional DC–DC converter interfaced with a battery and a DC–DC converter whose output is to be connected at either of the DC buses. In this architecture, the beauty is that the battery storage is directly connected to solar PV. This configuration requires a low-voltage battery which results in a comparatively low step-up gain for bidirectional DC–DC converter [15]. If the storage system is connected to high DC-link voltage, it requires a stack of multiple batteries or high step-up gain of bidirectional converter. The stacking up of batteries requires a charge balancing circuit making the system architecture more complex, bulky and costly. The highly efficient converter topology can be used for DC–DC bidirectional power flow [16,17]. The DC–DC bidirectional converter also serves a purpose to perform the maximum power point tracking (MPPT) of the solar PV system. The semiconductor switches of the

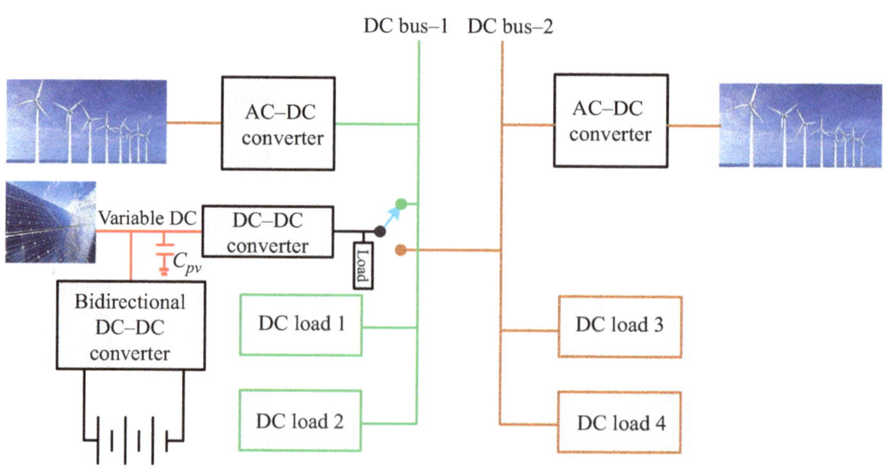

Figure 13.1 System architecture of a DC microgrid

bidirectional DC–DC converter are modulated to ensure that the maximum power is extracted from solar PV. The PV voltage will be decided by MPPT algorithm [18–24]. Depending upon the atmospheric conditions (irradiance and cell temperature), the PV voltage will vary to harvest maximum power. Therefore, the PV voltage of the system under consideration is of floating/variable nature. The function of DC–DC converter is to connect this PV power to the DC Grid. The output voltage of the DC–DC converter must be regulated to DC bus voltage irrespective of the variation in PV voltage (i.e. input to DC–DC converter) or load at the DC bus. The DC–DC converter can be of buck, boost or buck-boost type depending upon the rating of PV array voltage and DC bus voltage. In this analysis, it has been considered that the PV voltage is always higher than the DC bus voltage, therefore a DC–DC buck converter would serve the purpose.

13.2 Interfacing of PV source to DC bus

As discussed in the previous section, the aim of this chapter is to interface the PV distributed energy source to DC microgrid system. For harvesting maximum power, the PV voltage will be of variable nature. Similarly, if there is any variation in the load at DC bus, then also DC–DC converter should be able to maintain DC bus voltage. It is also possible that this PV distributed energy source may be connected to DC bus-2 which is having different bus voltage but lesser than the PV voltage. Therefore, now our problem is formulated to design a control scheme for modulating the DC–DC buck converter switch in such a way that its output voltage is regulated under variations in input voltage, load current or reference output voltage. The complete analysis, modeling, and control of the buck converter are described in the following sections. The input voltage to buck converter is the same as PV voltage V_{pv} and output voltage V_o of buck converter is the same as DC bus voltage V_{dc_bus}. The load resistance R represents the local load connected to the PV source.

13.3 Analysis of non-ideal DC–DC buck converter

The fundamental circuit diagram of a non-ideal DC–DC buck converter is shown in Figure 13.2. It is made up of a power semiconductor switch S with on-resistance r_{sw} (which may be a MOSFET or IGBT), a fast recovery diode D_d (having on-resistance r_d and forward voltage drop V_F), an inductor L in series with an equivalent series resistance r_L, capacitor C in series with an equivalent series resistance r_C and load resistance R [25,26]. For proper design and modeling of the converter dynamics, these different non-idealities have been considered. These parasitic resistances are very small in comparison to load resistance R. Depending upon the application, the buck converter circuit may operate either in continuous-current conduction mode (CCM) or discontinuous-current conduction mode. However, in the present analysis, the buck converter is assumed to operate in CCM with steady-state duty cycle D and switching period T. Duty cycle D is defined as the ratio of time duration for which switch is on (t_{on}) to the total switching period (T).

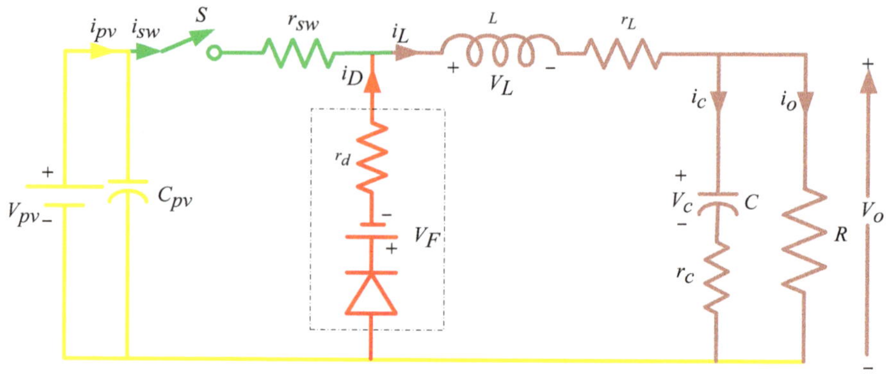

Figure 13.2 PV-source fed non-ideal DC–DC buck converter

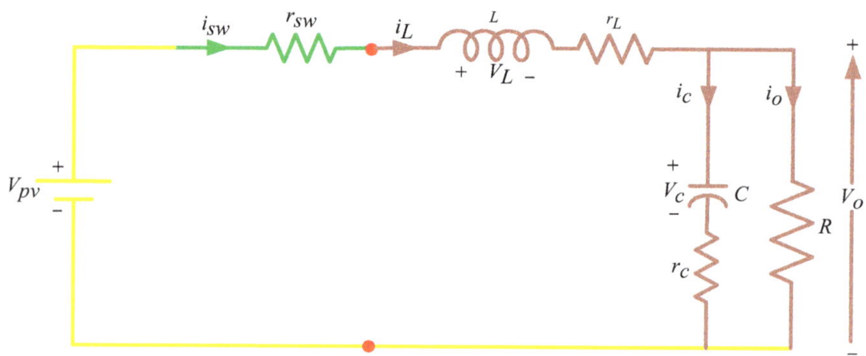

Figure 13.3 Equivalent circuit of PV system when the switch is on

In CCM, the buck converter has two circuit states (a) when semiconductor switch conducts and diode is reverse biased (b) when switch does not conduct and diode is forward biased. The analysis of non-ideal buck converter in these two states is discussed in detail as follows:

13.3.1 Semiconductor switch is on (time interval $0 < t \leq DT$)

The equivalent circuit of buck converter during switch-on is shown in Figure 13.3 The switch is represented by its on-resistance r_{sw} and diode is represented as an open circuit. The switch current is same as inductor current (i_L) and the current through the diode is zero. The PV source provides the power for inductor energy storage and load resistance. In this mode, the capacitor is first discharged by the load and then charged by the PV current.

The inductor voltage for this mode is given as

$$L\frac{di_L(t)}{dt} = v_{pv}(t) - (r_L + r_{sw})i_L(t) - v_o(t) \tag{13.1}$$

The output capacitor current is

$$i_c(t) = i_L(t) - \frac{v_o(t)}{R} \tag{13.2}$$

The output voltage is

$$v_o(t) = v_c(t) + r_c i_c(t) \tag{13.3}$$

13.3.2 *Semiconductor switch is off (time interval $DT < t \leq T$)*

The equivalent circuit of the PV system during switch-off is shown in Figure 13.4 The switch is represented by open circuit and diode is represented by its on-resistance r_d and forward voltage drop V_F. The diode current and inductor current (i_L) are same. The source is supplying no current and therefore, the current through switch is zero. The energy stored in the inductor is used to feed the power to load resistance. In this mode, the capacitor is charged initially by inductor energy and then discharged by load current.

The inductor voltage for this mode is given as

$$L\frac{di_L(t)}{dt} = -(r_L + r_d)i_L(t) - V_F - v_o(t) \tag{13.4}$$

The output capacitor current is

$$i_c(t) = i_L(t) - \frac{v_o(t)}{R} \tag{13.5}$$

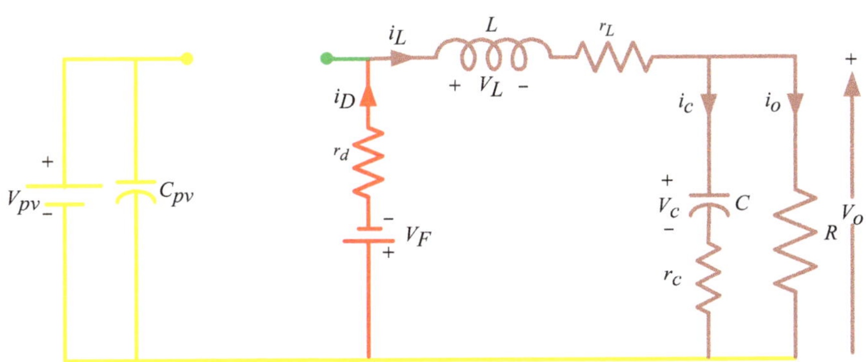

Figure 13.4. Equivalent circuit of PV system when the switch is off

The output voltage is

$$v_o(t) = v_c(t) + r_c i_c(t) \tag{13.6}$$

13.3.3 Steady-state analysis

In the steady state, the voltage across inductor and current through capacitor should be zero. Therefore, using (13.1),(13.4) and (13.2)–(13.5), we get

$$\begin{aligned} V_L &= \frac{1}{T}\int_0^T v_L(t)\,dt \\ &= \big(V_{pv} - (r_L + r_{sw})I_L - V_o\big)D + \big(-(r_L + r_d)I_L - V_F - V_o\big)(1 - D) = 0 \end{aligned} \tag{13.7}$$

$$\Rightarrow DV_{pv} = D'V_F + V_o + (r_L + Dr_{sw} + D'r_d)I_L \tag{13.8}$$

$$I_c = \frac{1}{T}\int_0^T i_c(t)\,dt = \left(I_L - \frac{V_o}{R}\right)D + \left(I_L - \frac{V_o}{R}\right)(1 - D) = 0 \tag{13.9}$$

$$\Rightarrow I_L = \frac{V_o}{R} = I_o \tag{13.10}$$

Simplifying (13.3) and (13.6), we get the average output voltage $V_o = V_c$

Substituting (13.10) into (13.8), the final expression for the output voltage is calculated as

$$V_o = \frac{DV_{pv} - (1 - D)V_F}{1 + \frac{(r_L + Dr_{sw} + (1-D)r_d)}{R}} \tag{13.11}$$

From the output voltage expression in (13.11), it is evident that the output voltage is lesser in the presence of non-idealities for a particular duty cycle.

13.4 Mathematical modeling of PV-source fed non-ideal DC–DC buck converter system

In this section, the mathematical models of PV-source fed non-ideal DC–DC buck converter system are obtained. The modeling of the PV system is carried out by considering the buck converter operation in CCM. In CCM operation, the converter operates in two modes (a) when MOSFET switch is on and diode remains off (b) when MOSFET switch is off and diode conducts. The circuit description and operation remain the same as discussed in the previous section. There are numerous modeling techniques to obtain the small-signal transfer function of the DC–DC converters. However, the SSA technique is used to obtain the different transfer function models of the converter due to its ease of implementation [27,28]. To obtain the small-signal model and different transfer functions using SSA technique, the steps given below are followed:

Step 1: Obtaining the state equations for each circuit state

Switch on (time interval $0 < t \le DT$)

For this mode, the circuit operation is not discussed again. It is the same as discussed in the previous section. Equations (13.1)–(13.3) are rewritten:

$$L\frac{di_L(t)}{dt} = v_{pv}(t) - (r_{sw} + r_L)i_L(t) - v_o(t) \tag{13.12}$$

$$C\frac{dv_c(t)}{dt} = i_L(t) - \frac{v_o(t)}{R} \tag{13.13}$$

$$v_o(t) = v_c(t) + r_c i_c(t) \tag{13.14}$$

Substituting (13.13) into (13.14),

$$v_o(t) = v_c(t) + r_c\left(i_L(t) - \frac{v_o(t)}{R}\right) \Rightarrow \quad v_o(t) = \frac{r_c R}{R + r_c} i_L(t) + \frac{R}{R + r_c} v_c(t) \tag{13.15}$$

Substituting (13.15) into (13.12), we get

$$L\frac{di_L(t)}{dt} = -\left(r_{sw} + r_L + \frac{r_c R}{R + r_c}\right) i_L(t) - \frac{R}{R + r_c} v_c(t) + v_{pv}(t) \tag{13.16}$$

Substituting (13.15) into (13.13),

$$C\frac{dv_c(t)}{dt} = \frac{R}{R + r_c} i_L(t) - \frac{1}{R + r_c} v_c(t) \tag{13.17}$$

Equations (13.16),(13.17), and (13.15) can be written in state-space form as

$$\begin{aligned} \dot{x} &= A_1 x + B_1 u \\ y &= C_1 x + E_1 u \end{aligned} \tag{13.18}$$

where

$$\begin{aligned} &x = \begin{bmatrix} i_L & v_c \end{bmatrix}^T,\ u = v_{pv},\ y = v_o \\ &A_1 = \begin{bmatrix} -\dfrac{1}{L}\left(r_{sw} + r_L + \dfrac{r_c R}{R + r_c}\right) & -\dfrac{1}{L}\left(\dfrac{R}{R + r_c}\right) \\ \dfrac{1}{C}\left(\dfrac{R}{R + r_c}\right) & -\dfrac{1}{C}\left(\dfrac{1}{R + r_c}\right) \end{bmatrix}, \\ &B_1 = \begin{bmatrix} \dfrac{1}{L} \\ 0 \end{bmatrix} \\ &C_1 = \begin{bmatrix} \dfrac{r_c R}{R + r_c} & \dfrac{R}{R + r_c} \end{bmatrix}, \\ &E_1 = [0] \end{aligned} \tag{13.19}$$

Switch off (time interval $DT < t \leq T$)

For this mode also, the circuit operation is not discussed as it has been already discussed in the previous section. Equations (13.4)–(13.6) are rewritten as

$$v_L(t) = L\frac{di_L(t)}{dt} = -V_F - (r_d + r_L)i_L(t) - v_o(t) \tag{13.20}$$

$$i_c(t) = C\frac{dv_c(t)}{dt} = i_L(t) - \frac{v_o(t)}{R} \tag{13.21}$$

$$v_o(t) = v_c(t) + r_c i_c(t) \tag{13.22}$$

Substituting (13.21) into (13.22), we get

$$v_o(t) = v_c(t) + r_c\left(i_L(t) - \frac{v_o(t)}{R}\right) \Rightarrow \quad v_o(t) = \frac{r_c R}{R + r_c} i_L(t) + \frac{R}{R + r_c} v_c(t) \tag{13.23}$$

Substituting (13.23) into (13.20), we get

$$L\frac{di_L(t)}{dt} = -\left(r_d + r_L + \frac{r_c R}{R + r_c}\right)i_L(t) - \frac{R}{R + r_c}v_c(t) - V_F \tag{13.24}$$

Substituting (13.23) into (13.21), we have

$$C\frac{dv_c(t)}{dt} = \frac{R}{R + r_c}i_L(t) - \frac{1}{R + r_c}v_c(t) \tag{13.25}$$

Equations (13.24), (13.25), and (13.23) can be written in state-space form as

$$\begin{aligned} \dot{x} &= A_2 x + B_2 u \\ y &= C_2 x + E_2 u \end{aligned} \tag{13.26}$$

where

$$\begin{aligned}
A_2 &= \begin{bmatrix} -\frac{1}{L}\left(r_d + r_L + \frac{r_c R}{R + r_c}\right) & -\frac{1}{L}\left(\frac{R}{R + r_c}\right) \\ \frac{1}{C}\left(\frac{R}{R + r_c}\right) & -\frac{1}{C}\left(\frac{1}{R + r_c}\right) \end{bmatrix}, \\
B_2 &= \begin{bmatrix} 0 \\ 0 \end{bmatrix}, \\
C_2 &= \begin{bmatrix} \frac{r_c R}{R + r_c} & \frac{R}{R + r_c} \end{bmatrix}, \\
E_2 &= [0]
\end{aligned} \tag{13.27}$$

Step 2: Finding the averaged state-space model

Having obtained the state-space equation in individual mode, in this step, these equations are averaged over one switching period by taking their respective duration as weights [28]. The corresponding averaged state-space model of non-ideal buck converter can be given as

$$\begin{bmatrix} \dfrac{d\bar{i}_L(t)}{dt} \\ \dfrac{d\bar{v}_c(t)}{dt} \end{bmatrix} = A'(t)\begin{bmatrix} \bar{i}_L(t) \\ \bar{v}_c(t) \end{bmatrix} + B'(t)\bar{v}_{pv}(t) \tag{13.28}$$

$$\bar{v}_o(t) = C'(t)\begin{bmatrix} \bar{i}_L(t) \\ \bar{v}_c(t) \end{bmatrix} + E'(t)\bar{v}_{pv}(t) \tag{13.29}$$

where

$$\begin{aligned} A'(t) &= A_1 d(t) + A_2(1 - d(t)), B'(t) = B_1 d(t) + B_2(1 - d(t)), \\ C'(t) &= C_1 d(t) + C_2(1 - d(t)), E'(t) = E_1 d(t) + E_2(1 - d(t)) \end{aligned} \tag{13.30}$$

Step 3: Linearization by introducing small AC perturbations in state variables

In order to obtain a linearized model around a DC operating point, the small AC perturbations are introduced as mentioned below.

$$\begin{aligned} &\bar{i}_L(t) = I_L + \tilde{i}_L(t), d(t) = D + \tilde{d}(t), \bar{v}_{pv}(t) = V_{pv} + \tilde{v}_{pv}(t) \\ &\bar{v}_c(t) = V_c + \tilde{v}_c(t), \bar{v}_o(t) = V_o + \tilde{v}_o(t) \end{aligned}$$

Substituting these values into (13.28)–(13.30), the steady-state (DC) and small-signal (AC) models of the PV system are given as

Steady-state (DC) model:

$$\begin{bmatrix} I_L \\ V_c \end{bmatrix} = -A^{-1}BV_{pv} \tag{13.31}$$

$$V_o = C\begin{bmatrix} I_L \\ V_c \end{bmatrix} + EV_{pv} \tag{13.32}$$

Small-signal (AC) model:

$$\begin{bmatrix} \dfrac{d\tilde{i}_L(t)}{dt} \\ \dfrac{d\tilde{v}_c(t)}{dt} \end{bmatrix} = A\begin{bmatrix} \tilde{i}_L(t) \\ \tilde{v}_c(t) \end{bmatrix} + B\tilde{v}_{pv}(t) + B_d\tilde{d}(t) \tag{13.33}$$

$$\tilde{v}_o(t) = C\begin{bmatrix} \tilde{i}_L(t) \\ \tilde{v}_c(t) \end{bmatrix} + E\tilde{v}_{pv}(t) + E_d\tilde{d}(t) \tag{13.34}$$

where

$$\begin{aligned}
A &= A_1D + A_2(1 - D),\\
B &= B_1D + B_2(1 - D),\\
B_d &= (A_1 - A_2)X + (B_1 - B_2)U\\
C &= C_1D + C_2(1 - D),\\
E &= E_1D + E_2(1 - D),\\
E_d &= (C_1 - C_2)X + (E_1 - E_2)U
\end{aligned} \tag{13.35}$$

Substituting corresponding values from (13.19), (13.27), (13.31), and (13.32) into (13.35):

$$A = \begin{bmatrix} -\frac{1}{L}\left(r_x + r_L + \frac{r_cR}{R+r_c}\right) & -\frac{1}{L}\left(\frac{R}{R+r_c}\right) \\ \frac{1}{C}\left(\frac{R}{R+r_c}\right) & -\frac{1}{C}\left(\frac{1}{R+r_c}\right) \end{bmatrix}, B = \begin{bmatrix} \frac{D}{L} \\ 0 \end{bmatrix}, r_x = Dr_{sw} + D'r_d$$

$$C = \begin{bmatrix} \frac{r_cR}{R+r_c} & \frac{R}{R+r_c} \end{bmatrix}, E = [0], B_d = \begin{bmatrix} \frac{V_{pv} + V_F - (r_{sw} - r_d)I_L}{L} \\ 0 \end{bmatrix}, E_d = 0 \tag{13.36}$$

The steady-state values can also be obtained by putting these matrices and vector values in (13.31)–(13.32). We get

$$I_L = I_o = \frac{V_o}{R} \tag{13.37}$$

$$V_o = V_c = \frac{DV_{pv} - D'V_F}{1 + \frac{1}{R}(r_L + r_x)} \tag{13.38}$$

It is worthwhile to note that the DC values obtained in (13.37)–(13.38) are exactly same as obtained in (13.10) and (13.11).

Step 4: Deriving the important transfer functions

In this step, the transfer functions of the PV system are derived which will be used for controller design.

Taking Laplace transfer function of (13.33) and (13.34), we have

$$s\begin{bmatrix} \tilde{i}_L(s) \\ \tilde{v}_c(s) \end{bmatrix} = A\begin{bmatrix} \tilde{i}_{Ls} \\ \tilde{v}_c(s) \end{bmatrix} + B\tilde{v}_{pv}(s) + B_d\tilde{d}(s) \tag{13.39}$$

$$\tilde{v}_o(s) = C\begin{bmatrix} \tilde{i}_L(s) \\ \tilde{v}_c(s) \end{bmatrix} + E\tilde{v}_{pv}(s) + E_d\tilde{d}(s) \tag{13.40}$$

Simplifying (13.39),

$$\begin{bmatrix} \tilde{i}_L(s) \\ \tilde{v}_c(s) \end{bmatrix} = (sI - A)^{-1}B\tilde{v}_{pv}(s) + (sI - A)^{-1}B_d\tilde{d}(s) \tag{13.41}$$

Substituting (13.41) into (13.40), we get

$$\tilde{v}_o(s) = \left(C(sI - A)^{-1}B + E\right)\tilde{v}_g(s) + \left(C(sI - A)^{-1}B_d + E_d\right)\tilde{d}(s) \quad (13.42)$$

Using (13.41), four transfer functions $\frac{\tilde{i}_L(s)}{\tilde{v}_{pv}(s)}, \frac{\tilde{i}_L(s)}{\tilde{d}(s)}, \frac{\tilde{v}_C(s)}{\tilde{v}_{pv}(s)}, \frac{\tilde{v}_C(s)}{\tilde{d}(s)}$ and using (13.42), two transfer functions $\frac{\tilde{v}_o(s)}{\tilde{v}_{pv}(s)}, \frac{\tilde{v}_o(s)}{\tilde{d}(s)}$ can be obtained.

In this chapter, we shall derive two transfer functions. Other transfer functions can be derived in the same way.

PV voltage to output voltage transfer function $\frac{\tilde{v}_o(s)}{\tilde{v}_{pv}(s)}$

This transfer function describes how the variations in PV voltage may affect the output voltage. From (13.42), the PV voltage to output voltage transfer function is

$$\frac{\tilde{v}_o(s)}{\tilde{v}_{pv}(s)} = C(sI - A)^{-1}B + E \quad (13.43)$$

Substituting respective matrices A, B, C, and E, we get

$$\frac{\tilde{v}_o(s)}{\tilde{v}_{pv}(s)} = \frac{RD(r_cCs + 1)}{LC(R + r_c)s^2 + (C(R + r_c)(r_x + r_L) + r_cRC + L)s + (r_x + r_L + R)} \quad (13.44)$$

Duty cycle to output voltage transfer function $\frac{v_o(s)}{\tilde{d}(s)}$

This transfer function describes how the variations in duty cycle reflect on the DC bus voltage. For this, the effect of disturbances in PV voltage is considered zero. From (13.42), the duty cycle to output voltage transfer function is

$$\frac{\tilde{v}_o(s)}{\tilde{d}(s)} = C(sI - A)^{-1}B_d + E_d \quad (13.45)$$

Substituting respective matrices A, B_d, C, and E_d, we get

$$\frac{\tilde{v}_o(s)}{\tilde{d}(s)} = \frac{R\left(V_{pv} + V_F - (r_{sw} - r_d)I_L\right)(r_cCs + 1)}{LC(R + r_c)s^2 + (C(R + r_c)(r_x + r_L) + r_cRC + L)s + (r_x + r_L + R)} \quad (13.46)$$

As the non-idealities of buck converter elements have been included in modeling therefore, the more accurate model of the PV system is obtained in comparison to its ideal counterpart. This model will be used for controller design in the next section.

13.5 Closed-loop control of the PV system

The closed-loop control configuration of a PV-source fed DC–DC buck converter using PI controller is shown in Figure 13.5 and its block diagram representation in Figure 13.6. The converter output voltage (v_o) is compared with the reference DC bus voltage ($v_{dc,ref}$). The difference of both is fed to PI controller $G_c(s)$ which provides a processed control voltage v_c. This control voltage is compared with a fixed frequency sawtooth waveform having fixed amplitude. In this chapter, sawtooth peak voltage $V_{sw} = 1$ is taken for simplicity. The output of PWM modulator is switching pulses which are applied to semiconductor switch after proper amplification and isolation.

In order to regulate the DC bus voltage in the presence of disturbances in PV voltage, DC load or reference DC bus voltage itself, a controller must be designed which would modulate the duty cycle of converter switch appropriately. In this chapter, a PI controller is used for this purpose. To tune the parameters of this PI controller, an algorithm based on stability boundary locus is used.

13.5.1 Tuning of PI controller

The stability boundary locus approach [29] is used to properly tune the parameters of the PI controller for the PV system under consideration. The PI controller parameters are graphically obtained based on the desired gain margin and phase

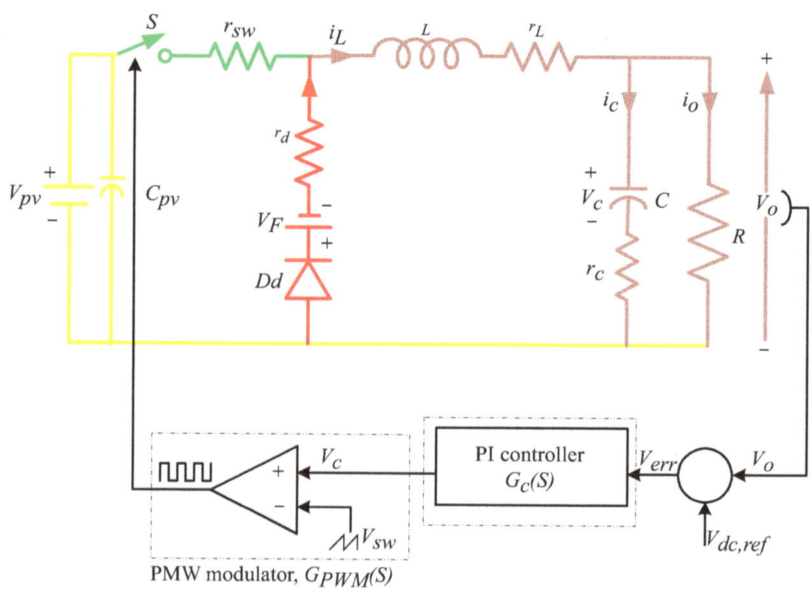

Figure 13.5 Closed-loop control of PI controlled PV system

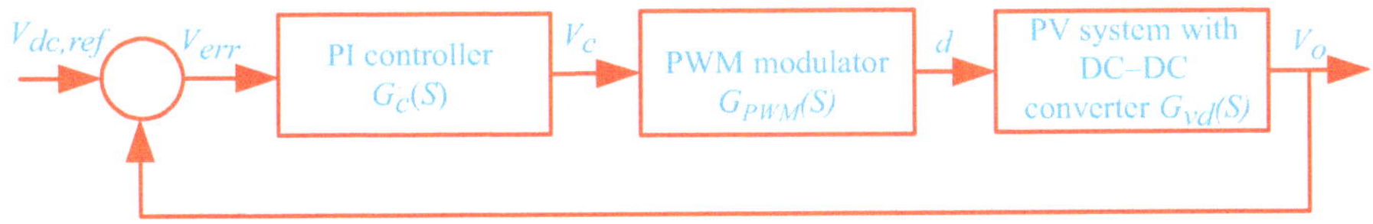

Figure 13.6 Block diagram of closed-loop PI control scheme for PV system

margin. The system under consideration is of second order and has infinite gain margin, therefore only phase margin is specified to tune the PI controller parameters. The procedure is explained in the following steps:

Step 1: Let the duty cycle to output voltage transfer function of the PV system as

$$T_1(s) = \frac{\tilde{v}_o(s)}{\tilde{d}(s)} = \frac{b_1 s + b_0}{a_2 s^2 + a_1 s + a_0} \tag{13.47}$$

The transfer function of PI controller is defined as

$$G_c(s) = K_p + \frac{K_i}{s} \tag{13.48}$$

where K_p and K_i are parameters of PI controller.

Step 2: Consider that the compensated PV system must have at least φ (in degree) phase margin. To satisfy this phase margin requirement, the characteristic equation of closed-loop system should be

$$1 + e^{-j\varphi} T_1(s) G_c(s) = 0 \tag{13.49}$$

Substituting (13.47) and (13.48) into the above equation, we get

$$1 + (\cos\varphi - j\sin\varphi)\left(\frac{b_1 s + b_0}{a_2 s^2 + a_1 s + a_0}\right)\left(K_p + \frac{K_i}{s}\right) = 0 \tag{13.50}$$

where

$$e^{-j\varphi} = (\cos\varphi - j\sin\varphi) \tag{13.51}$$

Replacing $s = j\omega$ into (13.50) and simplifying, we get

$$(K_p X_1 + K_i X_2) + j(K_p X_4 + K_i X_5) = X_3 + jX_6 \tag{13.52}$$

where

$$X_1 = [-\omega^2 b_1 \cos\varphi + \omega b_0 \sin\varphi], X_2 = [b_0 \cos\varphi + \omega b_1 \sin\varphi], X_3 = \omega^2 a_1$$
$$X_4 = [\omega^2 b_1 \sin\varphi + \omega b_0 \cos\varphi], X_5 = [-b_0 \sin\varphi + \omega b_1 \cos\varphi], X_6 = \omega^3 a_2 - \omega a_0$$

Separating real and imaginary parts in the above equation and then simplifying the expression for K_p and K_i are obtained as follows:

$$K_p(\omega,\varphi)=\frac{(-a_2b_1\sin\varphi\omega^3+(-a_2b_0+a_1b_1)\cos\varphi\omega^2+(-a_1b_0+a_0b_1)\sin\varphi\omega+a_0b_0\cos\varphi)}{-(b_1^2\omega^2+b_0^2)} \tag{13.53}$$

$$K_i(\omega,\varphi)=\frac{\omega(-a_2b_1\cos\varphi\omega^3+(a_2b_0-a_1b_1)\sin\varphi\omega^2+(a_0b_1-a_1b_0)\cos\varphi\omega-a_0b_0\sin\varphi)}{-(b_1^2\omega^2+b_0^2)} \tag{13.54}$$

For a specified phase margin φ and at varying frequency ω, the above two equations provide the stability boundary locus for the desired phase margin φ. Any pair of K_p and K_i chosen within this stability region guarantees to meet the specified phase margin.

13.6 Design example

In this section, a design example of PV system is exercised in order to show the efficacy of the proposed PI controller. The different parameters of the system are shown in Table 13.1. It is considered that due to variations in temperature and irradiance, to achieve maximum power point, PV voltage may deviate within a range of 16–24 V. Similarly, the load on PV system may vary from 11 to 22 Ω.

By substituting the corresponding values in the mathematical model derived in (13.46), the duty-cycle to output-voltage transfer function is

$$G_{vd}(s)=\frac{\tilde{v}_o(s)}{\tilde{d}(s)}=\frac{4428s+1.757\times10^8}{s^2+1518s+1.074\times10^7} \tag{13.55}$$

Table 13.1 DC–DC buck converter specifications

Parameters	**Values**
PV voltage, V_{pv}	16–24 V
DC bus voltage, V_o	12 V
Load, R	11–22 Ω
Inductance, L	1.1 mH
Inductance ESR	0.2 Ω
DC link capacitance	84 μF
Capacitance ESR	0.25 Ω
Switch-on resistance, r_{sw}	0.05 Ω
Diode forward resistance, r_d	0.03 Ω
Diode forward voltage, V_F	0.8 V
Switching frequency, f	20 kHz

The magnitude of fixed-frequency sawtooth waveform is unity, therefore the uncompensated loop transfer function of the system is same as duty-cycle to output-voltage transfer function and is described by

$$T(s) = G_{PWM}(s)G_{vd}(s) = \frac{1}{V_{sw}}G_{vd}(s) = \frac{4428s + 1.757 \times 10^8}{s^2 + 1518s + 1.074 \times 10^7} \quad (13.56)$$

The phase margin of the system is 26° at a gain crossover frequency of 2.23 kHz and gain margin is infinite as obtained from Bode plot of the uncompensated system in Figure 13.7. As the phase margin is not good enough, the transient response of the system would be poor with large oscillations. Further, the Bode plot informs that DC gain of the system is constant which results in steady-state error. This is obvious because the system in (13.56) is of type zero.

In order to improve the steady-state tracking accuracy and reduce oscillations in the transient response, a PI controller is designed for this system. The PI controller is tuned using the same approach as explained in detail in the previous section.

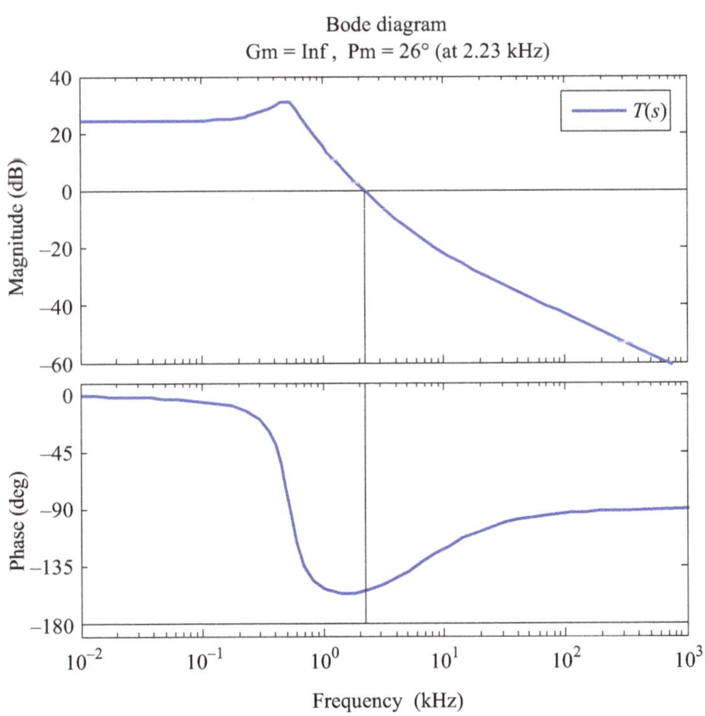

Figure 13.7 Bode plot of uncompensated PV system

By substituting the corresponding coefficients from (13.56) into (13.53)–(13.54), we get

$$K_p(\omega,\varphi)=\frac{\left(-4428\sin\varphi\omega^3-1.69\times10^8\cos\varphi\omega^2-2.19\times10^{11}\sin\varphi\omega+1.88\times10^{15}\cos\varphi\right)}{-\left(1.96\times10^7\omega^2+3.08\times10^{16}\right)} \tag{13.57}$$

$$K_i(\omega,\varphi)=\frac{\omega\left(-4428\cos\varphi\omega^3+1.69\times10^8\sin\varphi\omega^2-2.19\times10^{11}\cos\varphi\omega-1.88\times10^{15}\sin\varphi\right)}{-\left(1.96\times10^7\omega^2+3.08\times10^{16}\right)} \tag{13.58}$$

These two equations give various combinations of K_p and K_i for a specified phase margin. In order to find the pair of K_p and K_i which gives a stable and guaranteed phase margin, a stability boundary locus is obtained in K_p-K_i plane by varying ω. Let the desired phase margin of the overall system be 75°. The corresponding stability boundary locus in K_p-K_i plane is shown in Figure 13.8. Every pair of K_p and K_i within this stability area results in a phase margin which is at least 75°. Choosing $K_p = 0.02$ and $K_i = 20$, PI controller transfer function is

$$G_c(s) = \frac{0.02s + 20}{s} \tag{13.59}$$

With the implementation of this PI controller in cascade with uncompensated system, the corresponding Bode plot is shown in Figure 13.9. The phase margin of

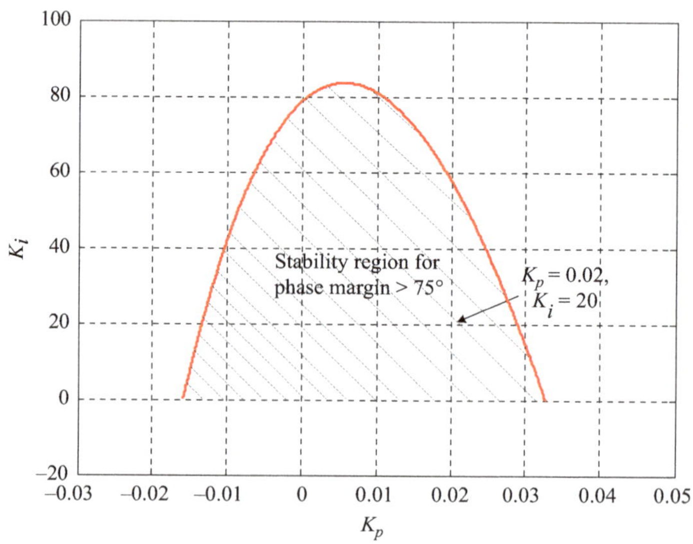

Figure 13.8 Stability boundary locus

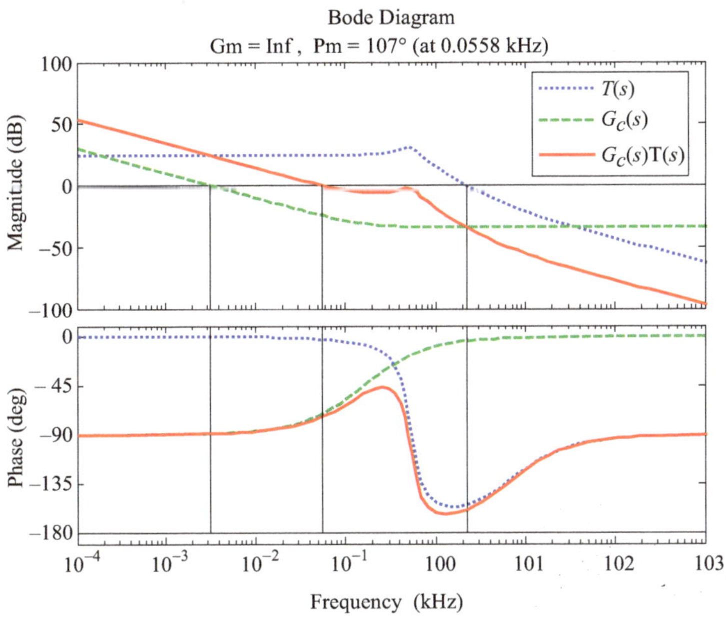

Figure 13.9 Bode plot of the system with PI controller

the closed-loop system has now increased from 26° to 107°, which helps in reducing transient oscillations. Furthermore, the gain of the system in low-frequency region has slope of −20 dB per decade which eliminates the steady state error. The increased gain in high-frequency region would better attenuate the high-frequency disturbances.

13.7 Case studies with simulation and experimental results

The aim of this chapter is to regulate the converter output voltage to reference DC bus voltage in the presence of different disturbances in the system. For this purpose, a PI controller has been designed in the previous section. In order to investigate the performance of the designed PI controller, three cases would be discussed as follows:

Case 1 – Change in PV Voltage: It is evident that to harvest maximum power under the varying atmospheric conditions (irradiance and cell temperature), the solar PV voltage would fluctuate which is directly input to the DC–DC buck converter.

Case 2 – Change in DC load: As the load connected to PV system may change resulting in deviation of its output voltage. PI controller should be able to regulate the output voltage to reference DC bus voltage.

Case 3 – Change in reference DC bus voltage: In this case, it is considered that the PV distributed energy source is shifted to another DC bus. It is expected that the output of DC–DC buck converter should follow the new DC bus reference voltage.

In order to study the above three cases, extensive simulation and experimental results are obtained which is discussed as follows:

13.7.1 Change in PV voltage

It is considered that the solar PV voltage may fluctuate between 16 and 24 V for getting maximum power point under the varying atmospheric conditions (irradiance and cell temperature) with constant load. The reference voltage of DC bus is 12 V.

Both simulation and experimental results are shown in Figure 13.10. The PV voltage changes from 16 to 24 V and converter is connected to a load resistance of 22 Ω. It can be seen that the PI controller is able to regulate the DC bus voltage to 12 V during this transition with a settling time of 20 ms. The simulation and experimental results for the variation of PV voltage from 24 to 16 V are shown in Figure 13.11. it is observed that the DC bus voltage is regulated to 12 V during this transition with settling time of 12 ms. The change in PV voltage from 16 to 24 V and 24 to 16 V with a load of 11 Ω is shown in Figures 13.12 and 13.13, respectively. These results depict that DC bus voltage settled back with small oscillations and settling time with variation in PV voltage.

13.7.2 Change in DC load

In this case, the performance of PI controller is investigated when load connected to PV system changes. The reference voltage of DC bus is 12 V. It is considered that the load resistance fluctuates between 11 and 22 Ω and vice versa with constant PV voltage.

In Figure 13.14, the response of the system is shown when load changes from 22 to 11 Ω and vice versa with constant PV voltage of 24 V. In this transition, the DC bus voltage is regulated to 12 V with settling time of 20 ms. For PV voltage constant at 16 V and load variation from 22 to 11 Ω and vice versa, the simulation and experimental results are shown in Figure 13.15. In this case also, DC bus voltage is regulated with similar kind of performance.

13.7.3 Change in reference DC bus voltage

In this case, the performance of PI controller is observed, when there is step change in the reference DC bus voltage i.e. the PV distributed energy source is shifted to from one DC bus to other DC bus. As shown in Figure 13.16, the reference DC bus voltage changes from 12 to 15 V and vice versa. For this case, PV voltage is 16 V and load resistance is 11 Ω. It is found that in both the cases, the output voltage of buck converter attains the new reference voltage within 15–20 ms without any oscillations. The simulation and experimental results are in close proximity.

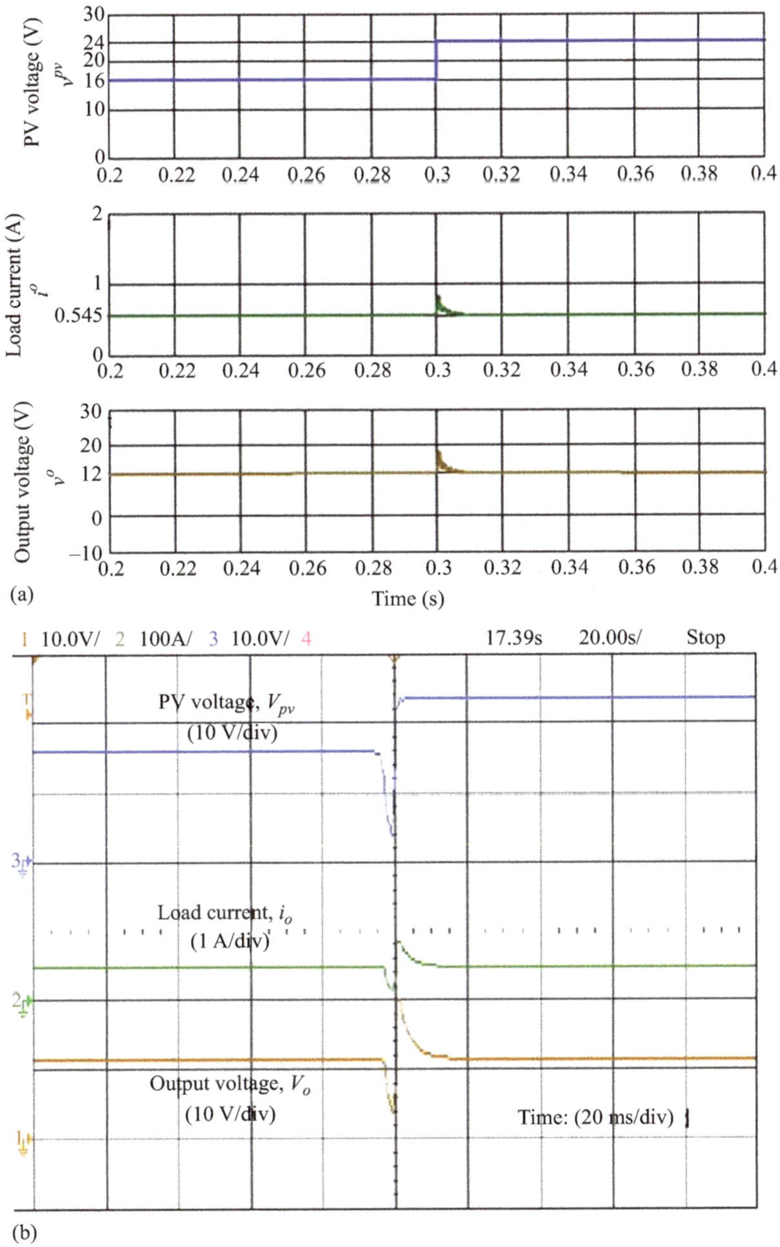

Figure 13.10 Variation in PV voltage (16–24 V with load resistance 22 Ω): (a) Simulation (b) Experimental

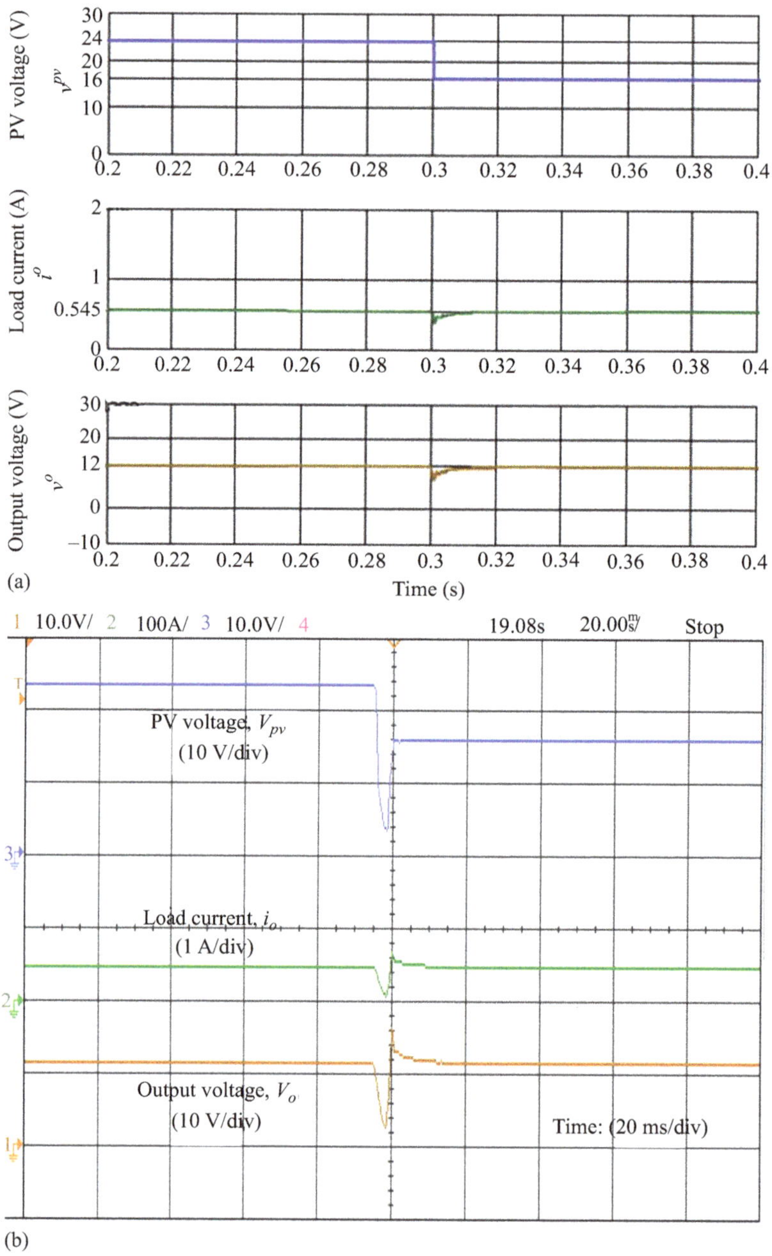

Figure 13.11 Variation in PV voltage (24–16 V with load resistance 22 Ω): (a) Simulation (b) Experimental

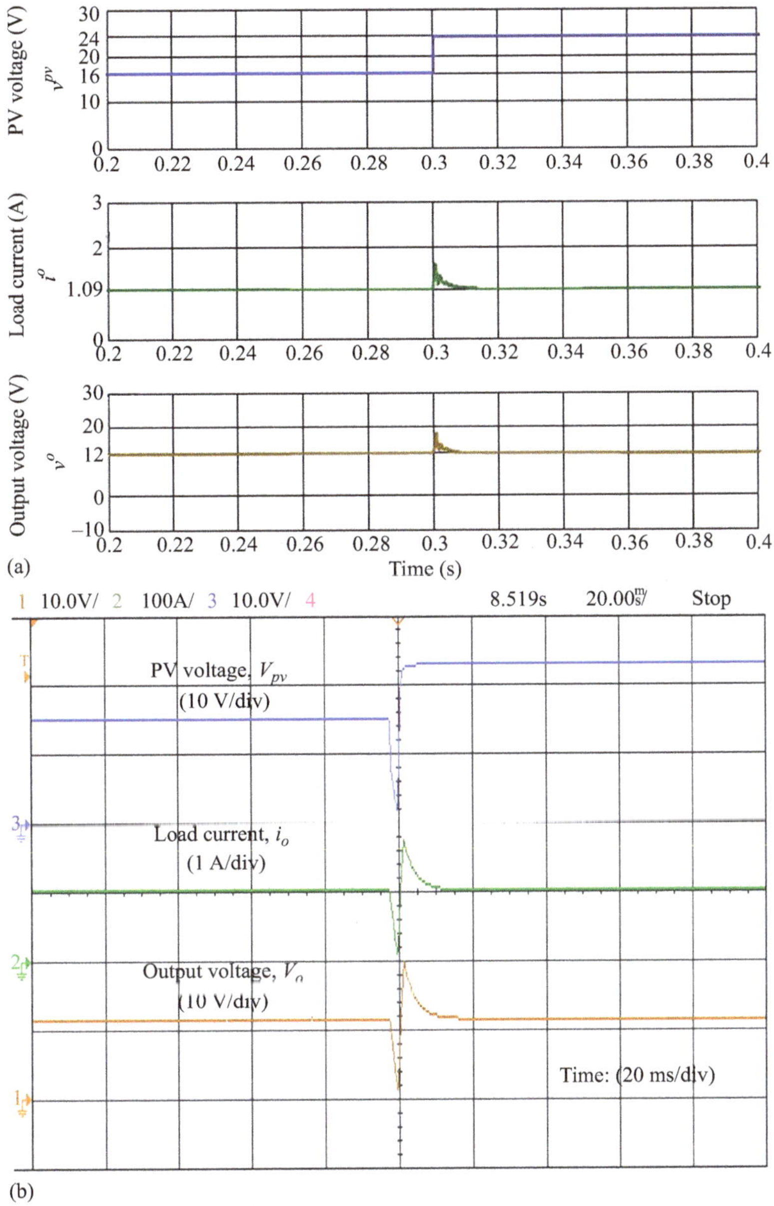

Figure 13.12 Variation in PV voltage (16–24 V with load resistance 11 Ω): (a) Simulation (b) Experimental

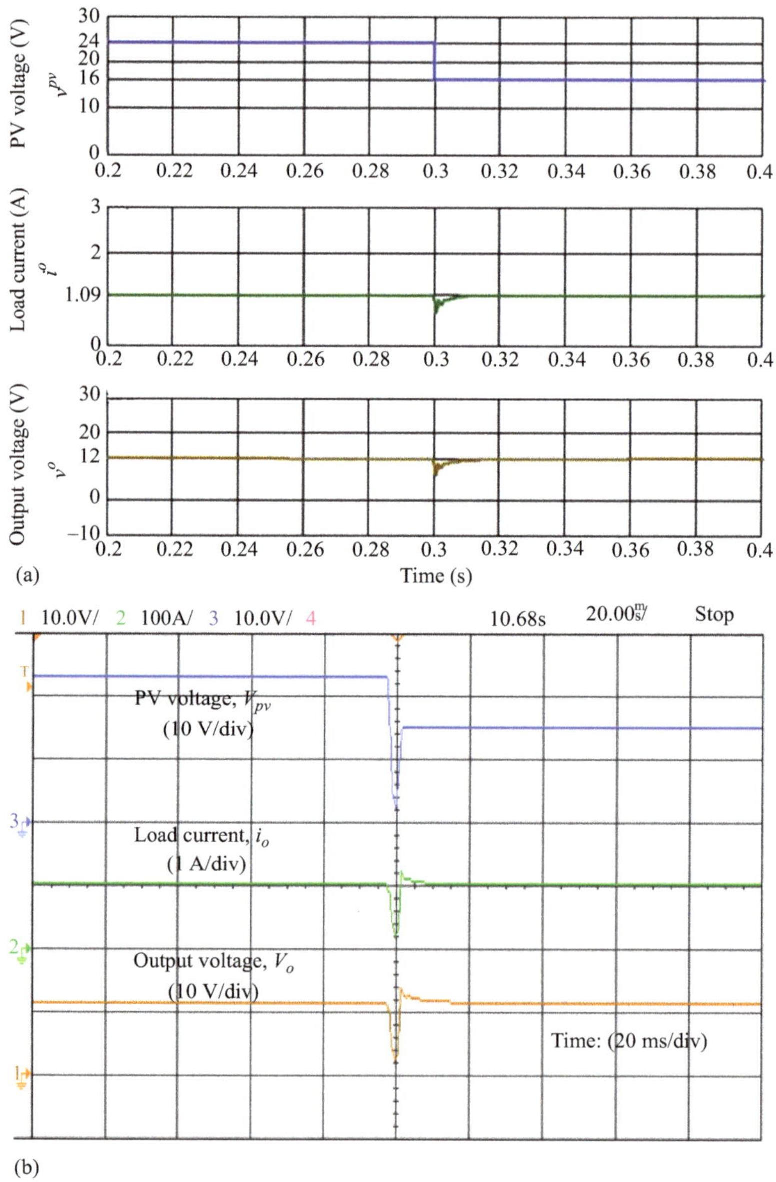

Figure 13.13 Variation in PV voltage (24–16 V with load resistance 11 Ω): (a) Simulation (b) Experimental

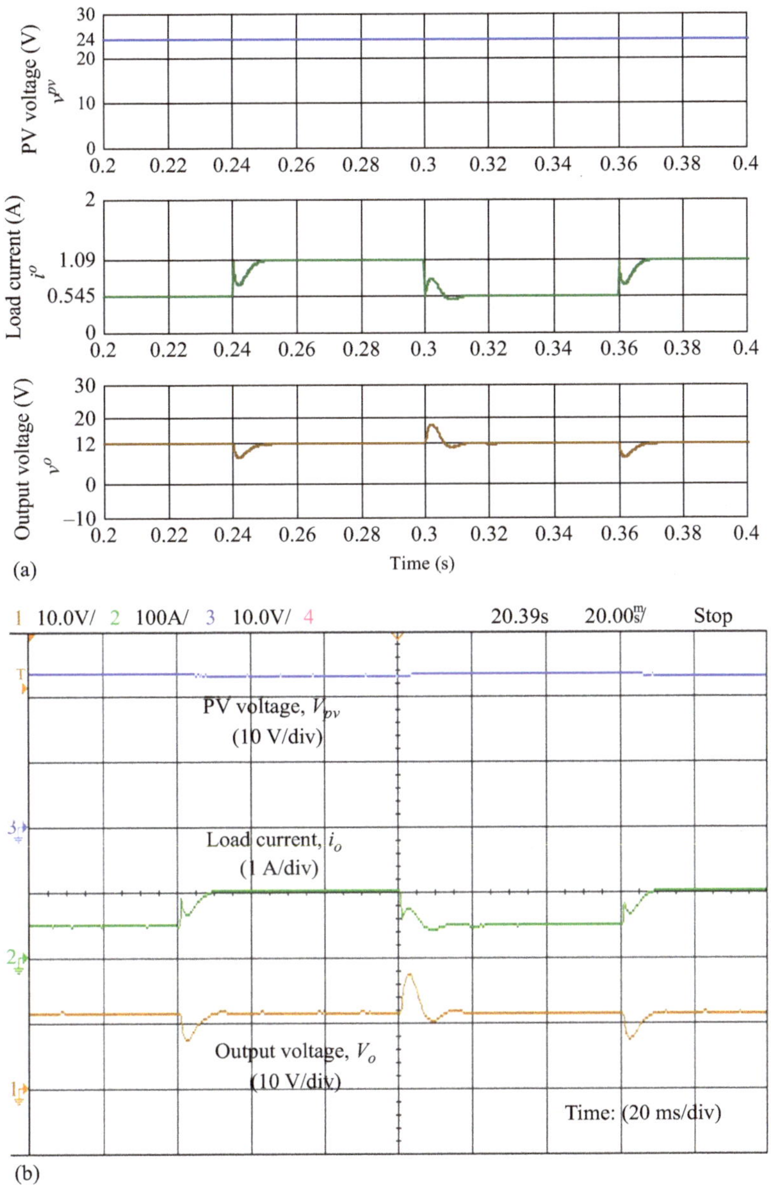

Figure 13.14 Variation in load (22–11 Ω and vice versa with PV voltage 24 V): (a) Simulation (b) Experimental

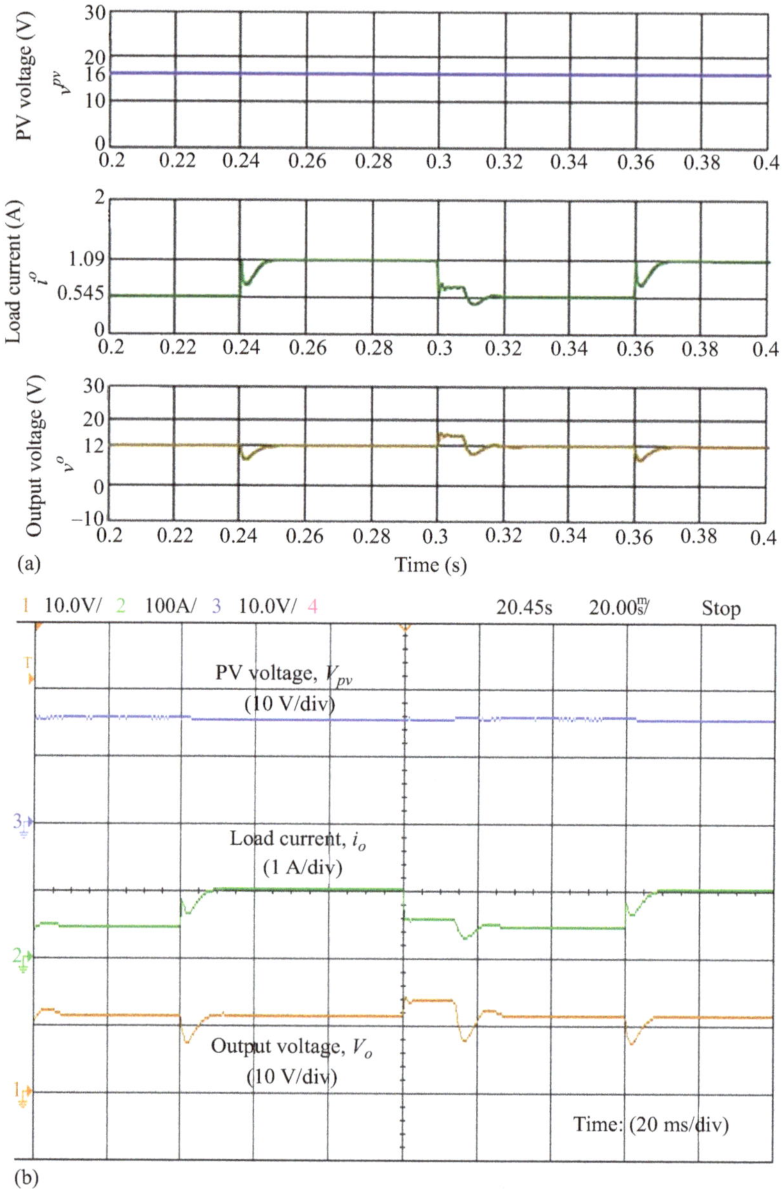

Figure 13.15 Variation in load (22–11 Ω and vice versa with PV voltage 16 V): (a) Simulation (b) Experimental

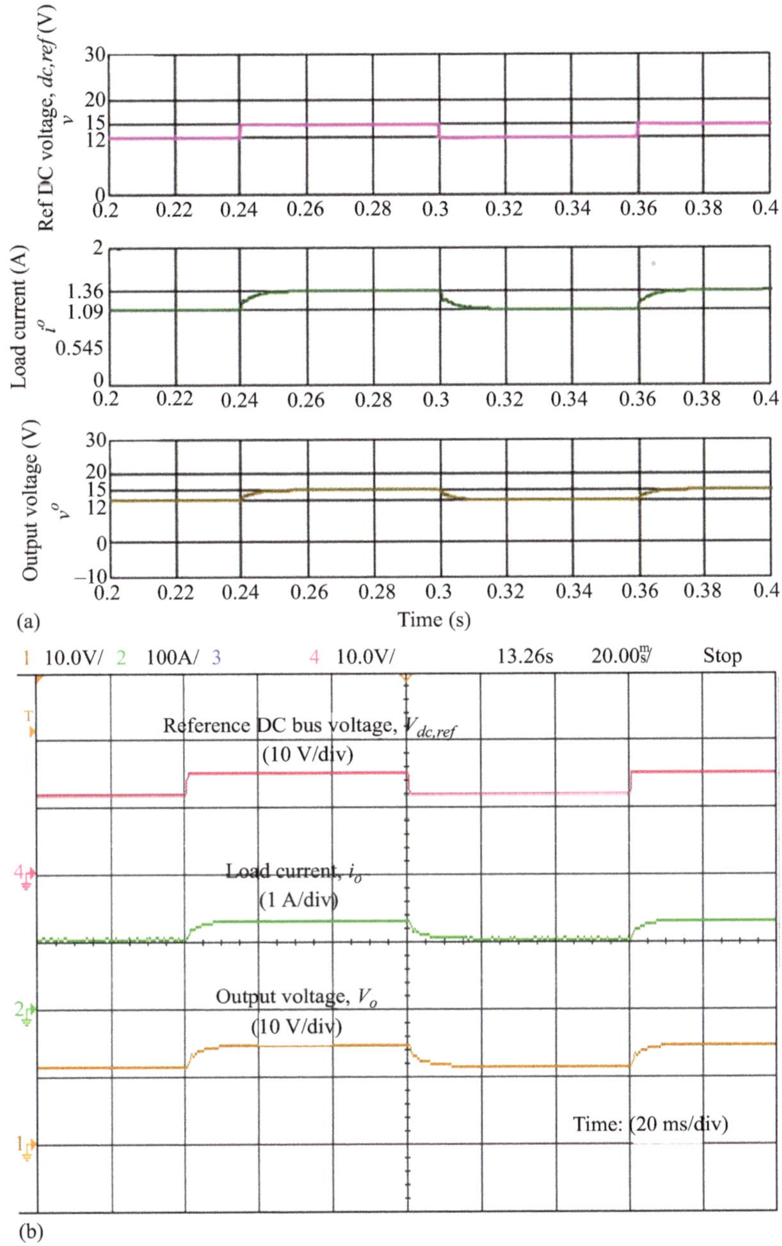

Figure 13.16 Variation in reference DC bus voltage (12–15 V and vice versa): (a) Simulation (b) Experimental

13.8 Conclusion

In this chapter, a complete analysis of DC–DC converter interfaced between PV distributed energy source and DC bus has been presented. The non-ideality present in the converter elements has been considered for obtaining accurate mathematical model. Then, a simple graphical approach has been used to determine the parameters of PI Controller for specified phase margin. A family of stabilizing PI controllers is obtained for a particular value of phase margin. Any combination of K_p–K_i within this family guarantees the minimum specified phase margin. The gain in low-frequency region and phase margin of the closed-loop system has significantly been increased. The performance of the controller was tested under three different cases and it was observed that the DC bus voltage is regulated to desired value on all cases. The experimental results match with the simulation results in close proximity proving the accuracy of mathematical model and controller designed. This chapter has analyzed the microgrid system for a PV-source fed DC–DC buck converter prototype of low voltage and low power. However, the same procedure can be followed up for the high power level and other DC–DC converter configurations.

References

[1] L. Che, M. Shahidehpour, A. Alabdulwahab, and Y. Al-Turki, “Hierarchical Coordination of a Community Microgrid With AC and DC Microgrids,” *IEEE Trans. Smart Grid*, vol. 6, no. 6, pp. 3042–3051, 2015.

[2] F. Nejabatkhah and Y. W. Li, “Overview of Power Management Strategies of Hybrid AC/DC Microgrid,” *IEEE Trans. Power Electron.*, vol. 30, no. 12, pp. 7072–7089, 2015.

[3] Y. Xia, W. Wei, M. Yu, X. Wang, and Y. Peng, “Power Management for a Hybrid AC/DC Microgrid With Multiple Subgrids,” *IEEE Trans. Power Electron.*, vol. 33, no. 4, pp. 3520–3533, 2018.

[4] K. M. Rogers, R. Klump, H. Khurana, A. A. Aquino-Lugo, and T. J. Overbye, “An Authenticated Control Framework for Distributed Voltage Support on the Smart Grid,” *IEEE Trans. Smart Grid*, vol. 1, no. 1, pp. 40–47, 2010.

[5] T. Dragicevic, X. Lu, J. C. Vasquez, and J. M. Guerrero, “DC Microgrids—Part II: A Review of Power Architectures, Applications, and Standardization Issues,” *IEEE Trans. Power Electron.*, vol. 31, no. 5, pp. 3528–3549, 2016.

[6] P. Wang, C. Jin, D. Zhu, Y. Tang, P. C. Loh, and F. H. Choo, “Distributed Control for Autonomous Operation of a Three-Port AC/DC/DS Hybrid Microgrid,” *IEEE Trans. Ind. Electron.*, vol. 62, no. 2, pp. 1279–1290, 2015.

[7] H-J. Chiu and S-J. Cheng, “LED Backlight Driving System for Large-Scale LCD Panels,” *IEEE Trans. Ind. Electron.*, vol. 54, no. 5, pp. 2751–2760, 2007.

[8] J. Mirez, “A modeling and simulation of optimized interconnection between DC microgrids with novel strategies of voltage, power and control,” in *2017 IEEE Second International Conference on DC Microgrids (ICDCM)*, 2017, pp. 536–541.

[9] D. Habumugisha, S. Chowdhury, and S. P. Chowdhury, "A DC–DC interleaved forward converter to step-up DC voltage for DC Microgrid applications," in *2013 IEEE Power & Energy Society General Meeting*, 2013, pp. 1–5.

[10] Q. Ye, R. Mo, and H. Li, "Low-Frequency Resonance Suppression of a Dual-Active-Bridge DC/DC converter Enabled DC Microgrid," *IEEE J. Emerg. Sel. Top. Power Electron.*, vol. 5, no. 3, pp. 982–994, 2017.

[11] X. Li, Li Guo, Yunwei Li, *et al.* "Flexible Interlinking and Coordinated Power Control of Multiple DC Microgrids Clusters," *IEEE Trans. Sustain. Energy*, vol. 9, no. 2, pp. 904–915, 2018.

[12] V. Nasirian, S. Moayedi, A. Davoudi, and F. L. Lewis, "Distributed Cooperative Control of DC Microgrids," *IEEE Trans. Power Electron.*, vol. 30, no. 4, pp. 2288–2303, 2015.

[13] S. Lu, L. Wang, T.-M. Lo, and A. V. Prokhorov, "Integration of Wind Power and Wave Power Generation Systems Using a DC Microgrid," *IEEE Trans. Ind. Appl.*, vol. 51, no. 4, pp. 2753–2761, 2015.

[14] M. Kwon and S. Choi, "Control Scheme for Autonomous and Smooth Mode Switching of Bidirectional DC–DC Converters in a DC Microgrid," *IEEE Trans. Power Electron.*, vol. 33, no. 8, pp. 7094–7104, 2018.

[15] L. H. S. C. Barreto, P. P. Praca, D. S. Oliveira, and R. N. A. L. Silva, "High-Voltage Gain Boost Converter Based on Three-State Commutation Cell for Battery Charging Using PV Panels in a Single Conversion Stage," *IEEE Trans. Power Electron.*, vol. 29, no. 1, pp. 150–158, 2014.

[16] K. Filsoof and P. W. Lehn, "A Bidirectional Multiple-Input Multiple-Output Modular Multilevel DC–DC Converter and its Control Design," *IEEE Trans. Power Electron.*, vol. 31, no. 4, pp. 2767–2779, 2016.

[17] Y.-F. Wang, L.-K. Xue, C.-S. Wang, P. Wang, and W. Li, "Interleaved High-Conversion-Ratio Bidirectional DC–DC Converter for Distributed Energy-Storage Systems—Circuit Generation, Analysis, and Design," *IEEE Trans. Power Electron.*, vol. 31, no. 8, pp. 5547–5561, 2016.

[18] E. I. Batzelis, G. E. Kampitsis, S. A. Papathanassiou, and S. N. Manias, "Direct MPP Calculation in Terms of the Single-Diode PV Model Parameters," *IEEE Trans. Energy Convers.*, vol. 30, no. 1, pp. 226–236, 2015.

[19] L. M. Elobaid, A. K. Abdelsalam, and E. E. Zakzouk, "Artificial Neural Network-Based Photovoltaic Maximum Power Point Tracking Techniques: a survey," *IET Renew. Power Gener.*, vol. 9, no. 8, pp. 1043–1063, 2015.

[20] B. W. Williams, A. A. Helal, M. A. Elsaharty, A. K. Abdelsalam, and N. E. Zakzouk, "Improved Performance Low-Cost Incremental Conductance PV MPPT Technique," *IET Renew. Power Gener.*, vol. 10, no. 4, pp. 561–574, 2016.

[21] N. A. Rahim, A. Che Soh, M. A. M. Radzi, and M. A. A. M. Zainuri, "Development of Adaptive Perturb and Observe-Fuzzy Control Maximum Power Point Tracking for Photovoltaic Boost DC–DC Converter," *IET Renew. Power Gener.*, vol. 8, no. 2, pp. 183–194, 2014.

[22] J. D. Bastidas-Rodriguez, E. Franco, G. Petrone, C. A. Ramos-Paja, and G. Spagnuolo, "Maximum Power Point Tracking Architectures for Photovoltaic Systems in Mismatching Conditions: a review," *IET Power Electron.*, vol. 7, no. 6, pp. 1396–1413, 2014.

[23] A. Thangavelu, S. Vairakannu, and D. Parvathyshankar, "Linear Open Circuit Voltage-Variable Step-Size-Incremental Conductance Strategy-Based Hybrid MPPT Controller for Remote Power Applications," *IET Power Electron.*, vol. 10, no. 11, pp. 1363–1376, 2017.

[24] A. Hussain, M. M. Garg, M. P. Korukonda, S. Hasan, and L. Behera, "A Parameter Estimation Based MPPT Method for a PV System using Lyapunov Control Scheme," *IEEE Trans. Sustain. Energy*, vol. 10, no. 4, pp. 2123–2132, 2019.

[25] R. W. Erickson and D. Maksimovic, *Fundamentals of Power Electronics*, 2nd ed. Kluwer Academic Publishers, Norwell, 2001.

[26] N. Mohan, T. M. Undeland, and W. P. Robbins, Power Electronics-Converters, *Applications, and Design*, 3rd ed. (John Wiley), New York, 2003.

[27] R. D. Middlebrook and S. Cuk, "A general unified approach to modelling switching converter power stages," in *IEEE Power Electronics Specialists Conference*, 1976, pp. 73–86.

[28] M. M. Garg, Y. V. Hote, and M. K. Pathak, "Leverrier's algorithm based modeling of higher-order DC–DC converters," in *IEEE India International Conference on Power Electronics*, 2012, pp. 1–6.

[29] N. Tan, I. Kaya, C. Yeroglu, and D. P. Atherton, "Computation of Stabilizing PI and PID Controllers using the Stability Boundary Locus," *Energy Convers. Manag.*, vol. 47, no. 18–19, pp. 3045–3058, 2006.

Part IV

Case studies

Chapter 14
Case studies of microgrids systems

Emilio Ghiani[1], *Fabrizio Pilo*[1], *Gian Giuseppe Soma*[1], *Enrico Elio De Tuglie*[2], *Alessia Cagnano*[2], *and Stefania Conti*[3]

Microgrids (MGs) are becoming more attractive due to their ability to reduce energy costs for customers, allow the full exploitation of renewable energy sources and increase security, reliability and resiliency of the power distribution system. Moreover, the integration of MGs into power systems will alleviate the consequences of sudden grid outages on the end users and provide higher power security to critical loads. In addition, thanks to their ability to produce energy close to the consumption points, MGs offer a possible solution for reliable energy supply into the areas where the extension of power grids is considered technically and economically unfeasible. Nonetheless, their practical implementation into actual distribution networks is still hindered by several technical and economic issues.

This chapter aims to show, by using several Case Studies of MGs, the value of each MG system in providing energy efficiency, ancillary services, demand response and electricity bill reduction. In particular, the possibility offered by such systems to satisfy the energy load requirements of several classes of electric users, ranging from buildings and business facilities to rural communities, will be investigated. Subsequently, the techno economic feasibility and the optimal sizing of components, as well as the economic performances of each MG system, will be compared with those achievable with traditional load feeding by the public distribution network.

14.1 Introduction to energy optimization with microgrids

The persisting growth of variable renewable generation – predominantly at the distribution level – is gradually altering the operating conditions of both electricity transmission and distribution networks.

[1]Department of Electrical and Electronic Engineering, University of Cagliari, Cagliari, Italy
[2]Department of Electrical and Information Engineering, Polytechnic University of Bari, Bari, Italy
[3]Department of Civil Engineering and Architecture, University of Catania, Catania, Italy

The power distribution sector is foreseen to be the more involved in the evolution of the power system towards future smart grid scenarios, in which active networks will be more intelligently capable to integrate distributed generation (DG), renewable energy sources (RES), demand side management (DSM) and energy storage systems (ESS). The transition towards active distribution networks (ADN) requires the implementation of new concepts with novel electrical system structure, as well as the addition of novel types of control systems and equipment able to manage bidirectional power flows. A possible evolution of the traditional power distribution configuration is then to move towards the adoption of small-scale distribution networks equipped with DG and ESS, which can be operated autonomously and disconnected, when necessary, from the main power grid referred to as microgrids (MGs) [1].

In the present changing context, the MGs are becoming more and more attractive due to their ability to reduce energy costs for customers, allow the full exploitation of RES and increase security, reliability and resiliency of the power distribution system. Moreover, the integration of MGs into power systems will alleviate the consequences of sudden grid outages on the end users and provide higher power security to critical loads. In addition, thanks to their ability to produce energy close to the consumption points, MGs offer a possible solution for reliable energy supply into the areas where the extension of power grids is considered technically and economically unfeasible. Nonetheless, their practical implementation into actual distribution networks is still hindered by several technical and economic issues.

In this section, some fundamental concepts related to MG optimization procedures are introduced; the MG optimization needs to find out technically feasible configurations that are able to serve the electric load and also to satisfy any other constraint imposed by the designer, principally the economic issues, sorting the technically feasible configurations by their economic performances.

In this chapter, different MG case studies will be discussed. Before discussion, in the following sections, preliminary definitions of basic concepts subsequently used will be given.

14.2 Net present cost

In MG optimization, the main economic parameter considered to identify the most cost-effective solution is the net present cost (NPC), which is defined as the sum of the discounted cash flows calculated over the system lifetime. Cash flows are obtained from the difference between the annual costs and revenues of the system, which are discounted to the present by the use of an interest rate. NPC is evaluated with (14.1):

$$\mathrm{NPC}(i, N) = \sum_{k=1}^{N} \frac{\left(C_{K,TOT} - R_{K,TOT}\right)}{(1+i)^{N}} \tag{14.1}$$

where $C_{K,TOT}$ is total annualized cost of the system including CAPEX and OPEX (represented by the sum of the costs for each system component) [€/year]; $R_{K,TOT}$ is total annual revenues of the system [€/year]; i is interest rate [%] or discount factor; N is project lifetime [year].

The NPC (or life-cycle cost) of a component is the present value of all the installation and operation costs of the component over the project lifetime minus the present value of all the revenues which it earns over the project lifetime. The capital cost occurs at the start of the project, or 'year zero'. The annual operation and maintenance (O&M) and fuel costs occur at the end of each year. The replacement costs occur every defined number of years.

This equation discounts each year's cash flow back to the present, then deducts the initial investment, which gives a net value of the investment in today's Euros. The discount factor is the key input for the NPC calculation, which indicates the project's risk and investors return expectations on the investment., The cost of capital for an investor is the minimum rate that it must earn from investment projects in order to satisfy the required rates of its suppliers of finance, so it is used as the discount factor when calculating a NPC. In general, when evaluating different investment alternatives, a positive NPC during the service lifetime indicates that the investment has a positive effective.

14.3 Internal rate of return

The internal rate of return (IRR) expresses the time value of the annual savings of the equipment investment so that it is comparable with the interest rate one might earn by investing cash in a bank account or in securities. It is a discount rate, the one that makes the present value of the future cash flows equal to the cost of the investment.

14.4 Discounted payback time

The discounted payback time (DPT) measure indicates the number of years required to recover the initial investment and represents an alternative means of evaluating the cost–benefit of a project. It is well known that a more effective investment is represented by a smaller DPT value. Basically, the DPT represents the smallest value of variable k so that the value of the NPC in (14.1) is equal to zero.

14.5 Levelized cost of electricity

The levelized cost of energy (LCOE) is another economic parameter commonly used to compare alternative power generating technologies which represents the average cost for a kWh of energy produced by a system. It is calculated for a reference year by dividing the cumulative system costs by the total amount of energy produced during the same time period.

The LCOE is evaluated as the average cost per kWh of electrical energy produced by the system. The LCOE is evaluated as the rate by the annualized cost of producing electricity and the total electric energy production as in (14.2):

$$\mathrm{LCOE} = \frac{C_{Y,TOT}}{E_L + E_{G,S}} \tag{14.2}$$

where $C_{Y,TOT}$ is total annualized cost of the system [€/year]; E_L is load energy served [kWh/year] and $E_{G,S}$ is grid energy sales [kWh/year].

14.6 Battery wear cost

In MG, when ESS are involved, the battery wear cost (BWC), that is, the cost of cycling energy through the storage bank, can be evaluated as in (14.3):

$$\mathrm{BWC} = \frac{C_R}{N \cdot Q \cdot \sqrt{\eta_{RT}}} \tag{14.3}$$

where C_R is replacement cost of battery bank [€]; N is the number of batteries in the battery bank; Q is lifetime throughput of a single battery (amount of energy that can be cycled through a battery before it requires replacement) [kWh]; η_{RT} is round-trip efficiency.

The BWC parameter is used during optimization when the production from renewables is not sufficient to meet the load in order to decide whether to buy energy from the grid or use the energy stored in the battery.

14.7 Load and generation modelling in MGs

14.7.1 MG's typical radial architecture

A general architecture of an MG is represented in Figure 14.1.

The operation of an MG is based on the control capabilities of DG including micro-generators, such as micro-turbines, small wind turbines (WT) and photovoltaic (PV) arrays, together with storage devices, such as flywheels, energy capacitors and batteries and controllable (flexible) loads (e.g. represented by electric vehicles or active demand) at the distribution level. These control capabilities allow the MG to operate when isolated from the upstream interconnected distribution network in case of grid events (i.e. faults, blackouts and voltage collapses) or other external disturbances, during grid maintenance and when the grid power quality level falls below standard levels thus increasing the quality of supply [1].

14.7.2 Modelling of loads in MGs

The determination of the customer load characteristics is very important for distribution planners and operators to support various functions, such as MG design and system expansion planning, DSM, system operation efficiency and service quality. As for planning and operation of MGs it is fundamental to define the load behaviour, at least on an hourly basis or even with higher granularity, e.g. on a 5–10–15 min basis.

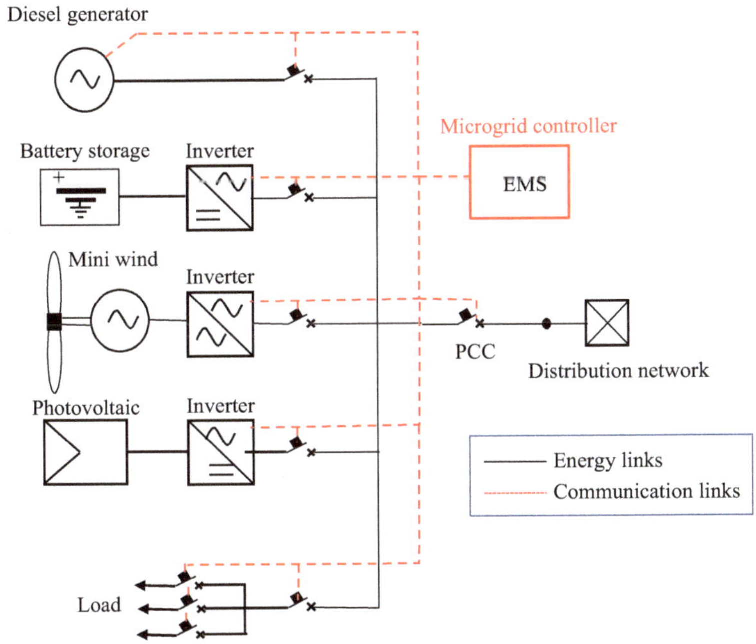

Figure 14.1 General MG architecture

The load characteristics of the energy consumption in an electric distribution system vary with the category of the customers. In Figure 14.2, the demand patterns for typical MV loads have been reported, in particular for industrial, commercial, residential and public lighting facilities.

The power absorption of industrial loads depends on the crew working periods and it has a reduction during lunch breaks (Figure 14.2(a)).

Commercial load demand is rather flat because of a quasi-constant consumption during daytimes; during night times the request of energy is due to uninterruptible loads such as refrigeration, lighting, alarm systems, information technology devices (Figure 14.2(b)).

The public lighting shows a pattern variable with seasons, and it depends on the duration of the day (e.g. longer during summer, shorter during winter). A typical daily load curve related to the public lighting is depicted in Figure 14.2(c).

Finally, residential customers require energy mainly for air conditioners and other electric appliances. The load pattern varies by days (working days or holidays), months, seasons and it depends on the external temperature and meteorological events (Figure 14.2(d)).

The daily load (generation) curves need to be discretized at least into 24 intervals (each one of the duration of 1 h) as shown in Figure 14.3. For each time interval, the uncertainty on the loads (generators) power has been considered by using suitable normal pdf.

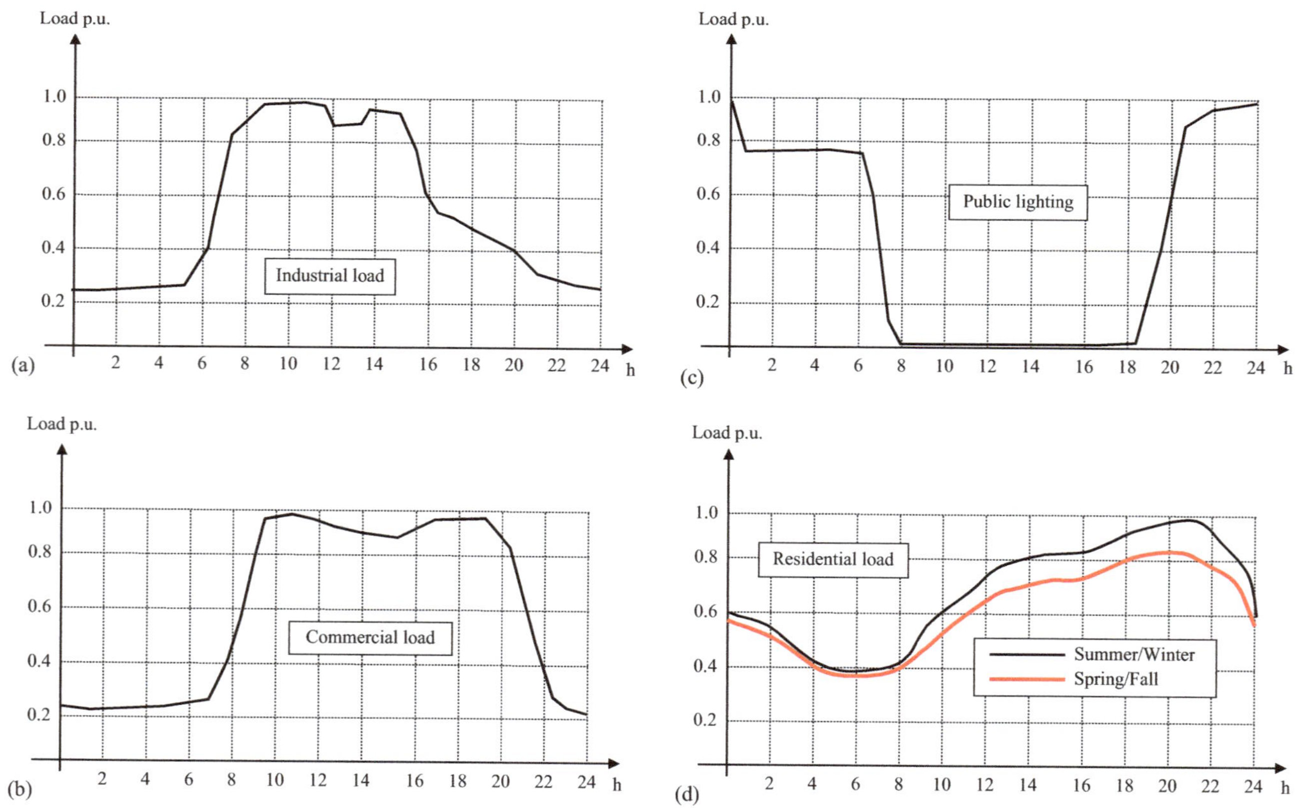

Figure 14.2 Typical MV load to be feed in MGs

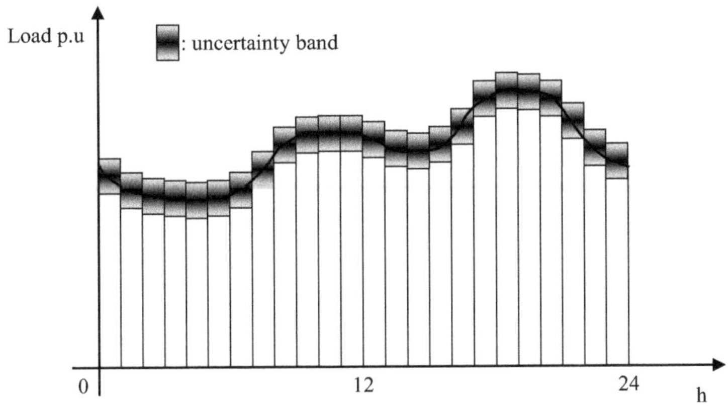

Figure 14.3 Daily load curve with uncertainties

14.7.3 Modelling of renewable output power in MGs

In the power system, MGs are deeply involved in the challenge of integrating non-despatchable energy generators based on renewable sources (e.g. solar, wind), whose electricity production depends on meteorological conditions and/or the time of the day. Among the various renewable generation plants, PV and wind generators can be deployed more flexibly, in terms of location, than other renewable resources. In particular, small-scale installation of PV panels can be roof-mounted on commercial or residential buildings or ground-mounted close to the load in the proximity of commercial and industrial areas, while small wind generators are quite common in rural areas.

Typical models for representing the most frequently installed PV and wind generation systems in distribution networks are described in the following sections.

14.7.3.1 PV generators modelling

The amount of energy that can be produced by a PV system is directly dependent on the irradiance and the angle at which PV cells are radiated. The variability of the irradiance, and then of the power production, depends on meteorological conditions (Figure 14.4).

Then, the power output of PV generators can be assumed linearly dependent on the global solar irradiance. The proportionality has to consider:

1. a parameter related to a superimposition of a cloud distribution model to the global solar radiation time series during clear atmospheric conditions;
2. a parameter that accounts for the environmental and electrical factors that can influence the production;
3. the performances of the PV system, according to (14.4):

$$P_{PV_{hm}} = k_{cloud} \cdot k_{BOS} \cdot \frac{G_{hm}}{G_{stc}} \cdot P_{PV_r} \tag{14.4}$$

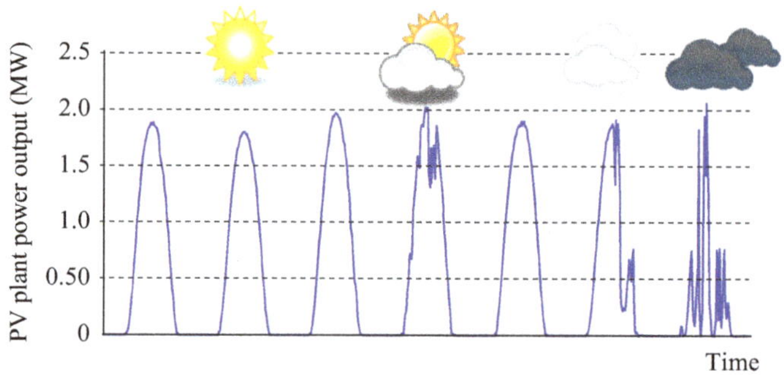

Figure 14.4 Typical variability of the irradiance and production of PV systems

where G_{stc} is the global solar irradiance in the standard test condition (1,000 W/m^2); G_{hm} is the global solar irradiance for clear atmospheric conditions in the *h*th hour of a generic day of the *m*th month of the year; P_{PV_r} is the rated power of the PV plant; k_{cloud} is a reduction factor due to the presence of clouds and k_{BOS} is a reduction factor that takes into account the performance of the whole balance of system (BOS) that includes all the components of a PV system other than the PV panels (wiring, switches, solar inverter) in real operating conditions.

Many environmental and electrical factors can influence the k_{BOS} value [2]:

- variable solar irradiation
- temperature of the PV module
- PV module power dissipation
- PV module shaded or soiled
- cabling, switchboards and circuit losses
- efficiency of the DC/AC inverter
- efficiency of the transformer (if needed).

The solar in-plane irradiation can be directly measured with a solarimeter/pyranometer or by using different databases available in the literature [3], and in the specific web resources [4] of each installation site.

In order to properly model the power output of a PV system, it would be necessary to know the relative geometry of the cloud distribution at the site of the PV plant. In practice, some assumption can be made to represent various conditions when the PV plant is shaded: in partially cloudy days the global solar irradiance can be reduced to 50%, while for overcast days a global irradiance reduction of 90% can be assumed instead ($k_{cloud} = 0.1$).

14.7.3.2 Wind generators modelling

In the wind generation plant, the output production depends on the availability of the primary source (wind) that can fluctuate at various timescales; wind is subjected to seasonal variations, daily and hourly changes, as well as to very short-term fluctuations in an intra-minute and inter-minute time frame, which can be difficult to manage in electrically weak distribution networks (Figure 14.5).

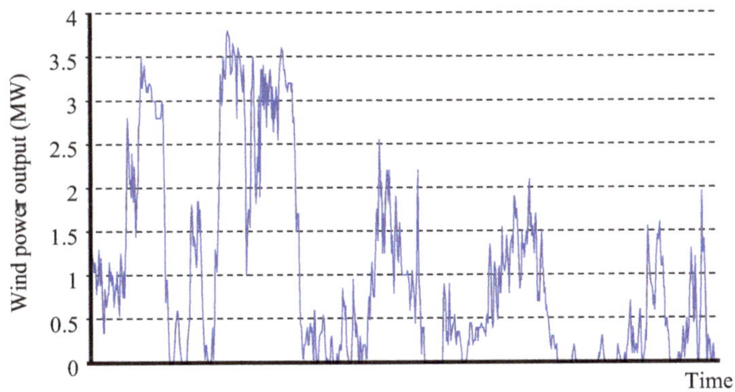

Figure 14.5 Typical variability of the production of wind systems

The output power of wind production systems can be modelled by considering time series of the primary source and applying them to the generator model in order to calculate the hourly power output; otherwise, suitable time series may be randomly formed in accordance with the probabilistic nature of the primary source.

Starting from time series, the output of the turbine can be evaluated using the following non-linear relationship (14.5):

$$P_{WF_h} = \begin{cases} 0 & 0 \le v_h \le v_{ci} \\ P_{Wt_r} \cdot \dfrac{v_h - v_{ci}}{v_r - v_{ci}} & v_{ci} < v_h < v_r \\ P_{Wt_r} & v_r \le v_h \le v_{co} \\ 0 & v_h > v_{co} \end{cases} \tag{14.5}$$

where v_{ci}, v_r and v_{co} are, respectively, the cut-in, rated and cut-off wind speed of the generic WT of a specific wind farm; v_h is the actual wind speed at hour h and P_{Wt_r} is the rated power of the WT.

14.8 Case study 1: MG for energy efficiency

14.8.1 Case study description

In this case study, a low voltage (LV) MG, containing renewable generation, loads, storage devices and equipped with a communication and control system, capable of coordinating the energy resources, is considered. This LV MG can operate both in grid-connected and islanded mode and exchange power with the distribution network. The MG energy management system (EMS) maximizes the system efficiency and improves the quality of supply to customers. Suitable communication systems allow the EMS to send set point and control signals to generators, loads and storage devices and to receive the necessary inputs from measurement devices and other external sources (e.g. weather forecasts, energy prices).

The optimal size of the MG components has been designed with the support of HOMER, a simulation tool developed by the National Renewable Energy

Laboratory (NREL) [5]. HOMER simulates the operation of a system by making hourly energy balance calculations for determining the flow of energy to/from each component of the system (load/generation/storage). Technology options, component costs and, in case of renewable generation, data about resource availability can be easily chosen by means of a graphical user interface (GUI) [3,4]. The modeller can compare many different design options based on their technical and economic merits [5,6]. The capital and operational expenditures and the environmental impact of different designs are used to find the final optimal solution.

The load considered in this case study is a typical industrial load with an electrical demand of 220 kWh/day for weekdays. The load profile is shown in Figure 14.6 and corresponds to an existing manufacturing site with working hours from 8:00 to 18:00, Monday to Friday, with a slightly fluctuating load of about 15 kW during the day and a load of 5 kW due to the operation of an industrial furnace during night hours. During weekends, a baseload of 2 kW is assumed in order to consider some lighting and secondary loads. HOMER allows to introduce randomness in simulations by considering hourly and daily perturbation factors; in this case 20% hour and 10% daily variations of the load have been assumed. According to the considered daily profile, the energy consumed is 62,050 kWh/year.

The MG architecture corresponds to the one depicted in Figure 14.1. The various components are described as follows.

14.8.1.1 PV plant

Figure 14.7 shows the typical daily radiation for some months of the year. A more precise assessment of the energy produced is obtained considering the performance ratio and solar resource data of [3].

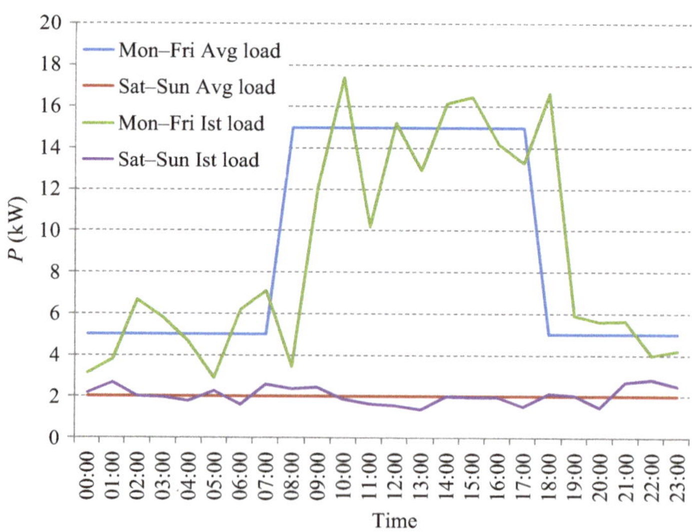

Figure 14.6 Daily LV load profile (average and instantaneous values)

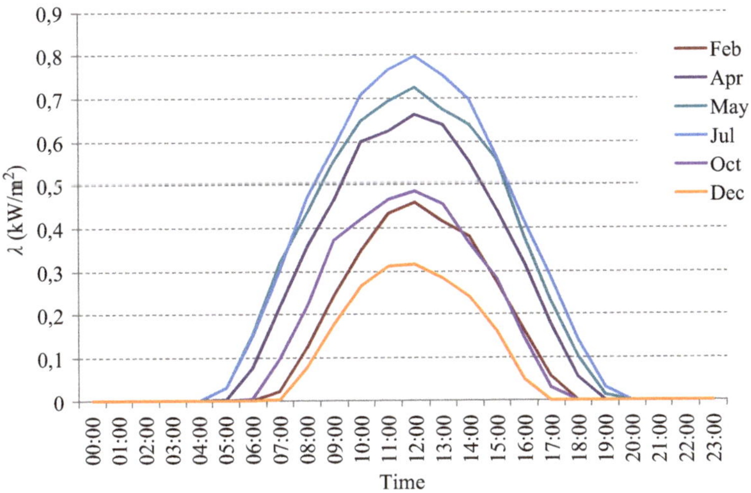

Figure 14.7 Daily radiation

14.8.1.2 Mini WT

The possibility to adopt a small WT of different sizes is considered during the optimization stage.

The wind resource data related to the installation site are assumed in the optimization software. Real values of the hourly wind speed data for the site at the height of the nacelle (e.g. 25 m) have been used. The simulated average wind speed profiles in the site for some months of the year are depicted in Figure 14.8.

14.8.1.3 Diesel generator

A conventional 20 kW diesel generator is considered to be used during off-grid operation as a backup source to cope with failures or RES unavailability to generate power. The physical properties of the fuel (diesel) are provided by HOMER's library; the fuel cost can be provided by the user, in particular a cost of 1.30 €/l has been assumed during optimization.

14.8.1.4 Storage

A battery bank connected to other components through a DC/AC converter is part of the options examined for a proper MG design. In this case study, a lead–acid GEL VRLA battery bank is used. Different sizes of batteries are used in simulations in order to identify the optimum capacity by considering both technical and economic aspects. Limits are defined in terms of state of charge (SoC) and speed for charge and discharge. The planning software determines how much energy can cycle through the battery before its replacement is needed. The despatch strategy associated with the battery operation is defined as the 'load following' strategy, which means that the batteries are charged by the surplus electricity produced only by renewable sources.

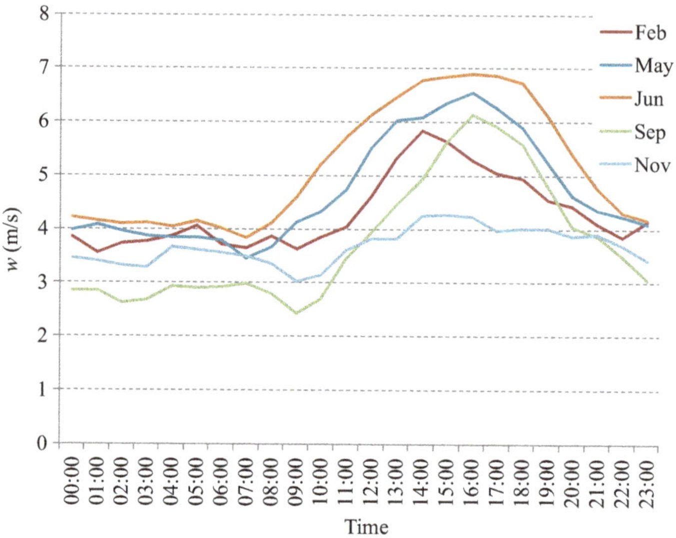

Figure 14.8 Monthly average wind speed profile at 25 m height

Table 14.1 Homer MG sizing options

MG DER	Capital cost		O&M cost		Size options	
Photovoltaic	1,000	[€/kW]	30	[€/kW year]	9–18–27	[kW]
Mini wind	4,500	[€/kW]	135	[€/kW year]	10–20–30	[kW]
Generation set	500	[€/kW]	0.260	[€/h]	20	[kW]
Storage	187	[€/kWh]	100	[€/year]	860–1,440–1,720	[Ah]
AC/DC converter	270	[€/kW]	100	[€/year]	18	[kW]

14.8.1.5 MG components sizing options

HOMER assists the designer in the sizing of the different MG components (wind/PV generators and battery bank) considering different economic factors. The sizing options represented in Table 14.1 have been considered in the optimization process for the components of the MG.

14.8.1.6 Cost of energy from the public network

The annual cost of energy for the user depends on the contract signed with the provider, and usually the retail electricity prices for an electricity consumer can be complex, as they combine various charges such as energy and capacity costs. For instance, according to the Italian context, the two typical contracts for non-domestic users have been assumed in the simulation.

Table 14.2 Values of regulated TOU tariff

Period of consumption	F1	F2	F3
	08:00–19:00	07:00–08:00 19:00–23:00 Sat – 07:00–23:00	23:00–07:00 Weekends
Cost of energy (€/kWh)	0.170	0.171	0.156

In the first contract, a time of use (TOU) tariff is considered: the price values are updated by the Italian Authority ARERA every 3 months and they are influenced by the oil & gas market prices. The average prices assumed in the simulation are shown in Table 14.2.

In the second contract, an hourly variable pricing is applied considering the cost of the energy in the Italian Power Exchange Market (IPEX), where a single national price [Prezzo Unico Nazionale (PUN)] is set for each hour of the day. The PUN varies hourly, daily and monthly, and in Italy, it is substantially influenced by the renewable production that is mainly concentrated in the middle hours of the day (e.g. PV production).

The values of the PUN for 2014 have been considered in building the pricing profile used in the simulation. A profit margin of 10 €/MWh has been added to the reference PUN index (it corresponds to a profit margin of about 15%).

In Figure 14.9 a typical behaviour of the pricing based on IPEX is showed. Peak prices occur in the hours 18:00–21:00, whereas the energy price is lower in the middle hours of the day, generally when renewable production is high.

The revenues for selling the energy produced by the MG are calculated considering the zonal price of the IPEX. This price varies during the day and it is lower during the middle hours when renewable production is high.

14.9 Results

Simulations results show the behaviour of the MG in grid-connected state and its capability of achieving economic benefits if compared with the traditional supply of passive load from the grid [assumed as the reference case (RC)]. The software HOMER finds out several MG configurations able to serve the load and obtain economic benefits.

Table 14.3 summarizes the 10 best possible MG configuration obtained, sorted by the NPC calculated considering the IPEX tariff. An additional constraint is assumed to discard all the MG configurations with a RES self-consumption below 70%.

In the RC, the NPC for purchasing the yearly consumed energy is 160 k€ when the TOU tariff is applied, whereas it is 149 k€ if the energy is bought at PUN-based prices. The LCOE is 0.207 €/kWh if the TOU tariff is applied, whereas it is 0.192 €/kWh if the energy is bought at PUN-based prices. Definitely, the PUN prices are more convenient than the TOU prices.

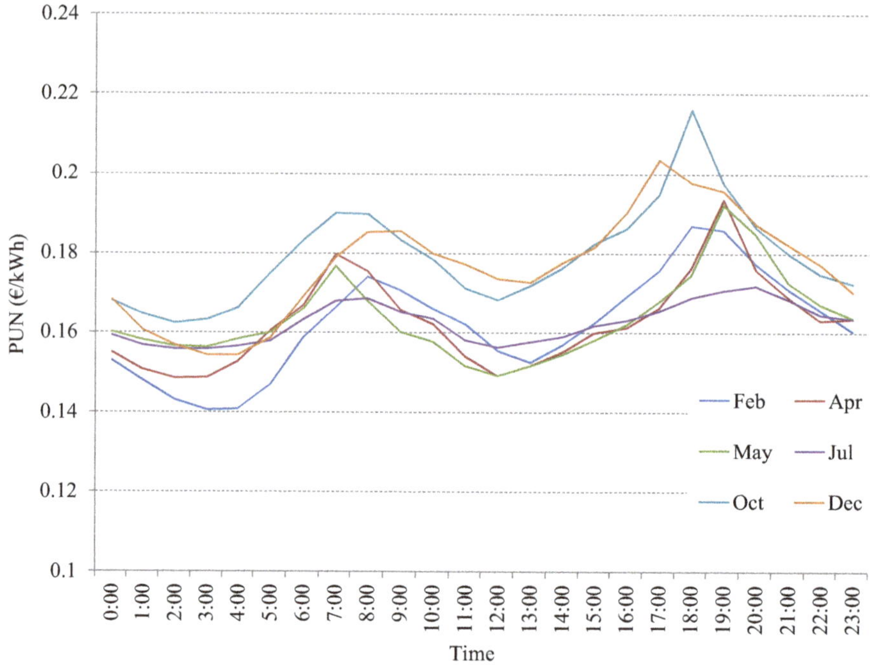

Figure 14.9 Monthly average Italian PUN index in 2014

According to the results given in Table 14.3, the following considerations and remarks can be done:

- the largest sizes of WTs are not considered as a possible candidate for the MG. The optimization privileges the smallest sizes because the installation cost is still quite high and also because the average wind speed in the site is of about 4.36 m/s, so the yearly production is not as high as it could be with better wind profiles;
- for the PV systems, the 18 and 27 kWp sizes are often chosen. This is due to the fact that the consumption profile of the customer closely matches with the PV production profile.

In order to achieve in the real world, the results obtained and an EMS able to control the generation and storage has to be used in the MG. The EMS will control the energy sources by determining when the storage device has to be discharged/recharged with the energy produced by renewable sources and when the exceeding energy is preferable to be sold to the market or self-consumed.

In Figure 14.10, the typical power flow and battery charge/discharge cycle for two consecutive days are shown as determined by the EMS. In particular, the diagram shows that the utilization of batteries is for limited periods and the discharge occurs up to the SOC reaches the 50%, whereas the charging is possible only when exceeding renewable energy is available (see Figure 14.10, day 2).

Table 14.3 Case study 1: optimization results

Case#	MG configuration			NPC (€€)		Self-consumption (kWh/year)		Self-consumption rate (%)		Exceeding energy Grid sale (kWh/year)		LCOE	
	Wind	PV	Storage	TOU	IPEX	TOU	IPEX	TOU	IPEX	TOU	IPEX	TOU	IPEX
1	–	18	–	133,085	126,571	21,280	21,280	76.51	76.51	6,533	6,533	0.172	0.164
2	–	9	–	144,646	135,916	11,800	11,800	84.85	84.85	2,107	2,107	0.187	0.176
3	–	27	1,720	143,894	138,149	29,508	29,941	70.73	71.77	12,212	11,779	0.186	0.179
4	–	18	860	146,175	139,110	21,297	21,628	76.57	77.76	6,516	6,185	0.189	0.180
5	–	18	1,720	151,091	143,473	22,440	23,141	80.68	83.20	5,373	4,672	0.195	0.186
6	–	18	1,440	151,426	143,974	21,301	21,872	76.59	78.64	6,512	5,941	0.196	0.186
7	–	9	860	157,775	148,545	11,809	12,092	84.91	86.95	2,098	1,815	0.204	0.192
RC	–	–	–	159,987	148,732	0	0	0	0	0	0	0.207	0.192
8	10	9	–	156,451	150,661	31,193	31,193	71.71	71.71	12,308	12,308	0.202	0.195
9	–	9	1,440	163,005	153,476	11,820	12,266	84.99	88.20	2,087	1,641	0.211	0.198
10	–	9	1,720	163,243	153,649	12,526	13,069	90.07	93.97	1,381	838	0.211	0.199

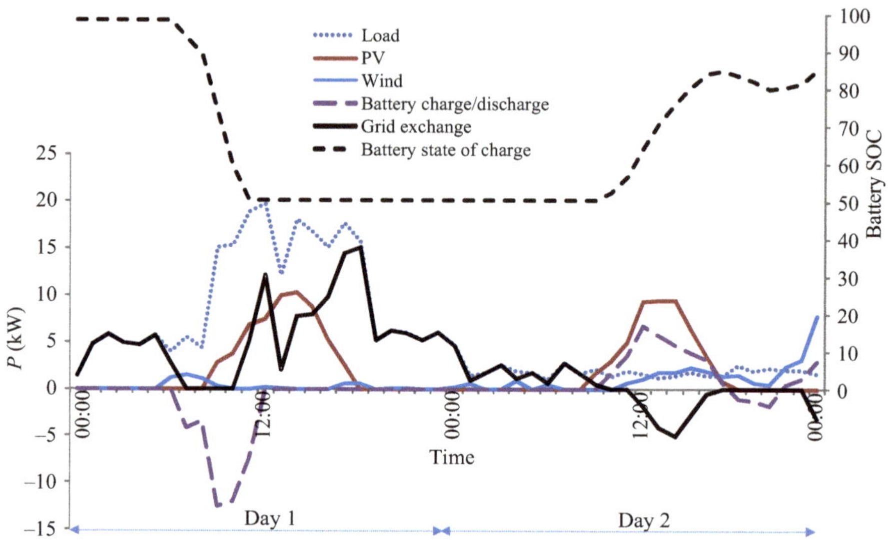

Figure 14.10 MG power flow and battery charge/discharge cycle

Results presented in this case study allow to remark the following points:

- the optimal MG configuration does not correspond to the option with the highest self-consumption rate. This is due to the fact that with a larger PV system there will be higher grid sales and lower grid purchases;
- several configurations with PV allow to reach an LCOE lower than the one of the reference case (RC), testifying that at the current price of electricity in Italy the grid parity has been already reached;
- lower PV sizes, with higher self-consumption rates, have faster payback periods;
- the presence of a storage system permits to increase the self-consumption level, but the current cost of batteries is still high, so that the economic performances are lower if compared with those achievable without batteries.

14.10 Case study 2: MGs for rural electrification

Around 18% of the world population lives without electricity access, most of them are concentrated in sub-Saharan rural areas [7]. Traditional approaches to rural areas electrification imply capital intensive infrastructures and large investments, while MGs based on renewable sources and storage systems can be easily implementable and lead to cost-effective solutions. At present, MGs are technologically and operationally ready to provide communities with electricity services, particularly in rural and suburban areas of less developed countries and also as an alternative to rural electrification [8].

14.10.1 Case study description

This study aims in identifying the optimal MG configuration that could be used to electrify a typical rural village; the village has been assumed, without loss of generality, to be located in Morocco (Africa). Morocco does not have natural oil resources and the energy production mostly depends on non-renewable resources imported by other countries. Nevertheless, the Moroccan renewable resources potential is significant [8]. The exploitation of RES in Morocco has been experienced already for the electrification of remote areas by the Programme d'Electrification Rurale Globale (PERG) programme, launched in 1996, which permitted to increase the electrification rates from 18% in 1995 to 98% in 2011. The programme mainly consisted of the realization of stand-alone diesel generation systems or PV kits used for the electrification of villages [9].

In this study, the technical and economic aspects of assembling a more complex MG [10] that includes PV, WT, as well as ESS and a diesel generator are investigated and compared with those of the RC, represented by the traditional diesel-only supply system. In addition, the possibility of extending the utility grid is also explored by the evaluation of the break-even grid extension, defined as the distance from the grid at which the NPC of the grid extension is equal to the NPC of the stand-alone system.

A detailed model of the system is provided as input to the optimization tool, including a reference load profile, technology alternatives, components costs and resource availability. A project timescale of 20 years and an interest rate of 5% are considered for simulations. A 20-year lifespan is also assumed for the PV and WT systems.

14.10.2 Loads

A village of about 1,000 inhabitants (200 houses) has been considered, with street lights and community services (i.e. 1 primary school, 1 dispensary, 20 small shops, 20 cell phone charging centres, 2 water pumps and 2 milling machines) [11]. A typical rural community load profile is used in simulations (Figure 14.11).

In order to make the load profile more realistic, with a unique pattern for each day of the year, a 64% hourly and a 64.5% daily perturbation factors are introduced. The global village load profile assumed is shown in Figure 14.11; according to the daily profile considered, the consumed energy is 224,839 kWh/year.

14.10.3 PV and wind generation

The database Photovoltaic Geographical Information System (PVGIS) has been used to obtain solar radiation data for the Imlil region in Morocco [4].

Average wind speed data are obtained from meteorological sources [12]. From average monthly wind speed data, the optimization tool is able to synthesize 8,760 speed values for each hour of the reference year.

The wind potential in Morocco has a significant variation between different sites on the territory, with average speed values that vary from less than 3 m/s to more than 8 m/s [8]; in order to evaluate how the convenience of WTs employment

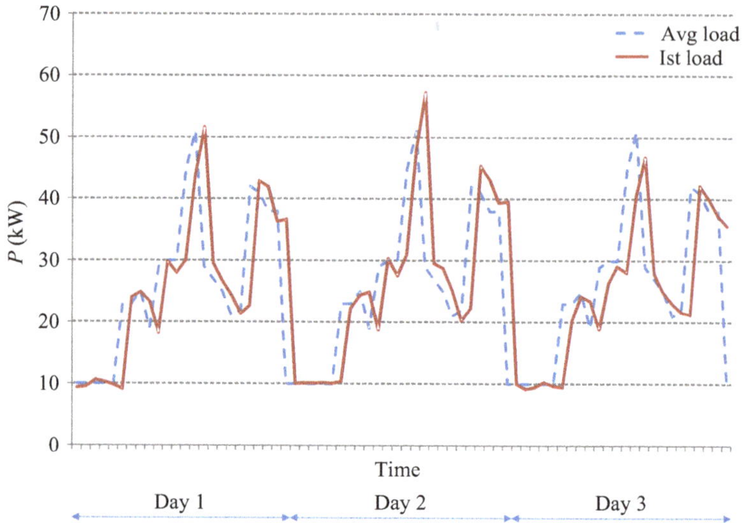

Figure 14.11 Daily load profile (average and instantaneous values), rural case study

is affected by the variations in wind resource conditions, a sensitivity analysis is carried out, so that hourly wind speed data have been scaled considering different yearly wind average values ranging from 2.5 to 7.5 m/s.

14.10.4 ESS

The MG includes an ESS constituted by lead–acid VRLA batteries. Different sizes of batteries are used in simulations in order to identify the optimum capacity by considering both technical and economic aspects. Twenty-four batteries are connected in series to form a 48 V string, and then three strings are connected in parallel to form the battery bank. Batteries are connected to other MG components via a DC/AC converter. On the basis of the daily load average consumption, the ESS sizes that permit to obtain a minimum operation time ranging from about 5 to 11 h have been considered.

14.10.5 Diesel generator

A conventional 60 kW diesel generator, capable to cover the village peak load, is included as a backup source. The lifetime of a conventional generator is evaluated in terms of its hours of operation; in this case, a lifespan of 50,000 h is assumed. The physical properties of fuel (diesel) are provided by HOMER's library; a diesel cost of 1.5 €/l, comprehensive of the transportation cost, has been assumed.

The MG sizing options in Table 14.4 have been considered in the optimization.

The realization of hybrid systems containing both renewable and non-renewable generators can be considered an economically viable alternative to the

Table 14.4 Case study 2: MG sizing options

MG component	Capital cost (€/kW) *(€/kWh)	O&M cost (€/kW·year), *(€/year) **(€/h)	Size options (kW) *(Ah)
Photovoltaic	1,000	30	40–60–80–100
Wind 20 kW	6,000	180	1–2–3 turbines
Wind 30 kW	4,170	125	1–2 turbines
Storage	187*	100*	5,160–6,000–7,800–9,900–11,700*
AC/DC converter	270	100*	54
Diesel generator	500	0.013**	60

installation of autonomous diesel supply systems to satisfy the energy demand of rural communities, located in remote and sparsely populated areas.

Table 14.5 shows the first ten hybrid system architectures extrapolated from the results of the simulations, assuming a 5 m/s yearly average wind speed; the configurations are sorted by their NPC.

PV arrays are always included in the optimal system architectures: the gradually decreasing costs of PV technology permits to optimally exploit the good solar potential with low capital and maintenance costs.

The results of this study also demonstrate that the wind profile has a great impact on the configuration identified as the optimal one; for different average wind speed values architectures with different number and sizes of WT, PV arrays and batteries are chosen.

The economic parameters NPC and LCOE are considerably lower than those of the RC for all the configurations; the environmental impact, represented by the emissions in atmosphere, is also reduced. Due to the high renewables' production, usually large-sized batteries are necessary.

14.11 Case study 3: MG as an alternative to grid extension

14.11.1 Case study description

MG can be considered as an alternative to grid extension in remote areas. The MG of the previous case study has been considered as RC. In order to evaluate the break-even grid extension distance, three parameters must be provided to the planner: the initial capital cost for the extension of the grid (€/km), the annual operating and maintenance cost for the grid extension (€/year/km) and the price for purchasing the electricity from the grid (€/kWh).

The cost for the extension of the grid may vary from a minimum of about 5,000 €/km for single-wire-earth-return (SWER) solutions to a maximum of 25,000 €/km for more complicated three-phase systems. To take into consideration this wide range of variability, a sensitivity analysis is performed considering five different capital cost values: 5,000, 8,000, 15,000, 20,000 and 25,000 €/km.

Table 14.5 Case study 2: optimization results (wind speed AVG 5 m/s)

Case#	MG configuration				NPC	LCOE	Renewable fraction	Genset operation	Excess electricity	CO_2 emissions
	PV (kW)	Wind (kW)	Genset (kW)	Storage (Ah)	€€	€/kWh	–	h	%	kg/year
1	100	2 × 20	60	11,700	838,211	0.299	0.96	1,025	28.1	21,600
2	100	2 × 30	60	11,700	854,370	0.305	0.96	1,047	34.3	22,168
3	80	2 × 20	60	11,700	857,785	0.306	0.94	1,335	21.5	28,319
4	100	2 × 20	60	9,900	859,455	0.307	0.95	1,193	28.6	24,909
5	80	2 × 30	60	11,700	872,426	0.311	0.95	1,343	28.8	28,690
6	80	2 × 20	60	9,900	873,359	0.312	0.94	1,459	22	30,964
7	100	2 × 30	60	9,900	876,797	0.313	0.96	1,230	34.7	25,756
8	100	2 × 20	60	7,800	889,658	0.318	0.94	1,504	29.5	31,152
9	80	2 × 30	60	9,900	891,478	0.318	0.94	1,496	29.2	31,807
10	80	2 × 20	60	7,800	899,733	0.321	0.93	1,742	23	36,598

The yearly O&M cost is considered as 2% of the corresponding capital cost. A sensitivity analysis is also performed on the grid power price, thus three different fees are considered: 0.06, 0.08 and 0.1 €/kWh.

The possibility to use an off-grid traditional generation system, constituted by a diesel generator running continuously, is assumed as the RC for this study. This supply option corresponds to a NPC of 1,999,215 €, a LCOE of 0.713 €/kWh and a level of CO_2 emissions of 258,745 kg/year. During the 20 years considered, three replacements of the generator are needed because of its intensive utilization.

During the optimization process, the possibility to install PV systems of different sizes and one or more WT with rated power values of 20 and 30 kW is evaluated.

14.11.2 Break-even grid distance

The break-even distance is calculated by the comparison between the NPC of stand-alone solutions and the NPC related to the connection to the grid; a sensitivity analysis is performed on the capital cost of extension and on the grid power price.

Considering a constant capital cost, it is possible to evaluate the effect of the energy price variation. Figure 14.12 shows an example, obtained considering 5,000 €/km capital cost for building new infrastructures.

Similarly, Figure 14.13 graphically shows the effect of the capital cost variation on the break-even distance by considering a constant grid power price (0.06 €/kWh).

Grid extension could be considered as an alternative to stand-alone systems; the assessment of the optimal design depends on the comparison between the economic benefits achievable with a hybrid system and the costs for the extension of the grid and the subsequent purchase of energy from it. Then, the distance of the village from the nearest grid connection point plays an important role in the definition of the best supply solution.

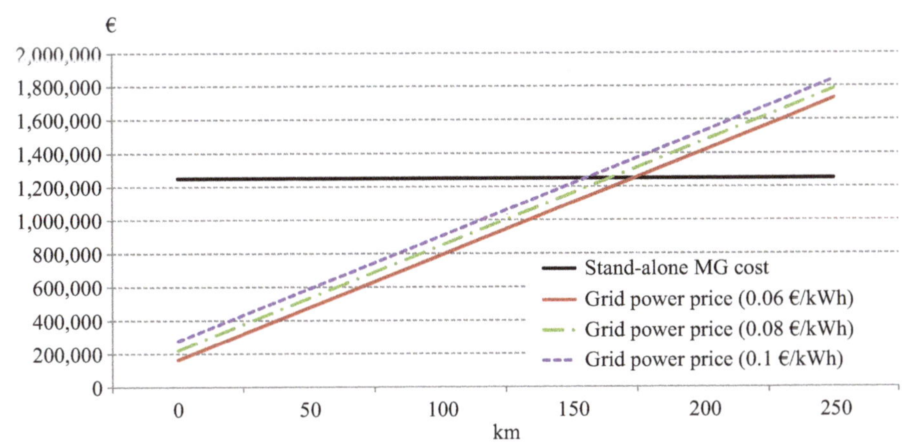

Figure 14.12 Break-even distances for different grid power prices

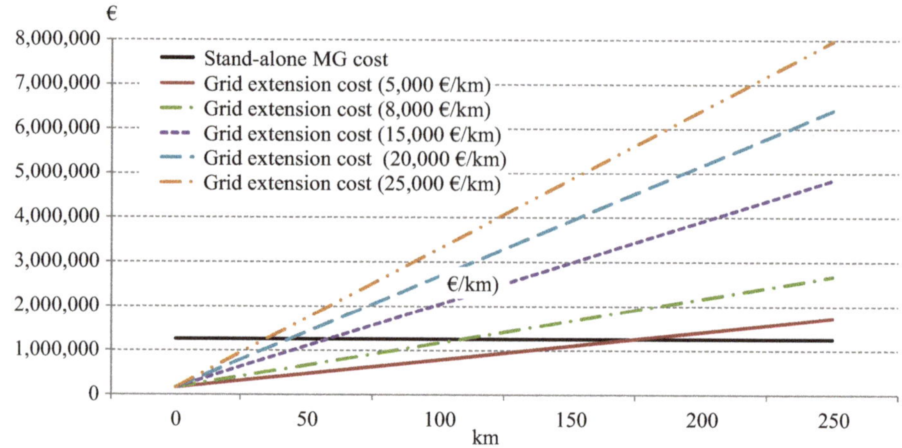

Figure 14.13 Break-even distances for different grid extension capital costs

14.12 Case study 4: industrial district with MG arrangement

14.12.1 Case study description

The scope of this case study is the definition of the optimal MG solution that could be used in an industrial district.

In particular, the tests have been performed by simulating the behaviour of industrial and commercial loads sited in the area destined to industrial/commercial activities. The test distribution network supplies 11 nodes and it is operated as an open loop. The layout of the network is represented in Figure 14.14.

The peak of the electrical demand is equal to 5.22 MW. To reduce the cost of services paid by customers in the area, it has been hypothesized to install DG in the district.

Table 14.6 shows the data assumed for the generators that are of different typologies: WT, internal combustion engine (ICE), combined heat and power (CHP) and micro-turbine (MT). All renewable generators produce electrical energy burning natural gas (NG).

Table 14.6 also reports for each unit the capital expenditures (CAPEX), the fixed O&M costs (O&MF), the variable ones (O&MV) and the electrical efficiency (η). The last two terms (O&MV and η) are differentiated by the I or II subscript in order to take into account that the relationship between power and cost of production is piecewise linear. Subscript I refers to the range between 0% and 50% of the nominal power (P_n), while subscript II refers to the remaining range (50% and 100% of P_n).

It is worth to note that the high values of η for the CHP unit represent the total (and not only the electrical) efficiency that includes both power and heat production.

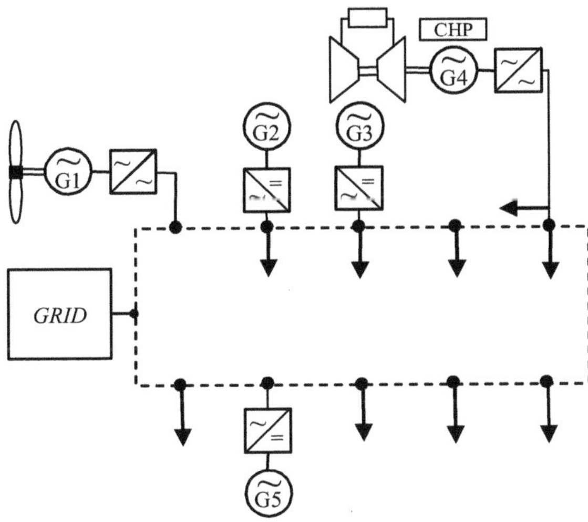

Figure 14.14 Industrial district MG

Table 14.6 Case study 3: generators data

N°	Type	CAPEX M€/MW	$O\&M_F$ €ct/h	$O\&M_{IV}$ €ct/kWh	$O\&M_{IIV}$ €ct/kWh	η_I	η_{II}	Primary source
G_1	WT	1.40	46.80	–	–	20	20	Wind
G_2	MT	0.60	85.06	10.00	7.88	26	33	NG
G_3	ICE	0.50	85.06	9.63	6.50	27	40	NG
G_4	CHP	0.50	85.06	9.63	6.50	71	74	NG
G_5	MT	0.60	85.06	10.00	7.88	26	33	NG

The cost of DG power generation for the generic *i*th generator, C_{DGi}, excluding the wind power one, depends on the despatched power P_i according to (14.6).

$$C_{DG,i}(P_i) = (\mathrm{O\&M}_{v_i} + C_{h_i} \cdot \eta_i)P + \mathrm{O\&M}_{Fi} \tag{14.6}$$

The hourly power production (*E*) of the CHP unit (G4 in Figure 14.14) is related to the hourly heat demand (*H*) by means of (14.7).

$$H = \alpha \cdot E \tag{14.7}$$

where α is the inversed electrical index (net electricity divided by the useful heat, assumed equal to 0.92 in this application).

The electrical (Figure 14.15) and thermal (Figure 14.16) loads are modelled according to their typical daily curves.

In the proposed application, weekday and holiday curves have been used for both commercial and industrial loads. The thermal demand is treated in the same manner considering weekday and holiday behaviours.

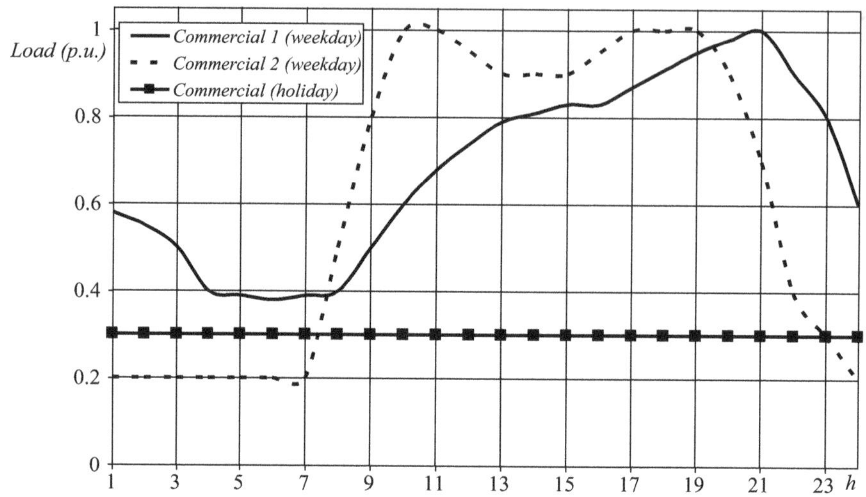

Figure 14.15 Typical daily curves of electrical load: commercial case

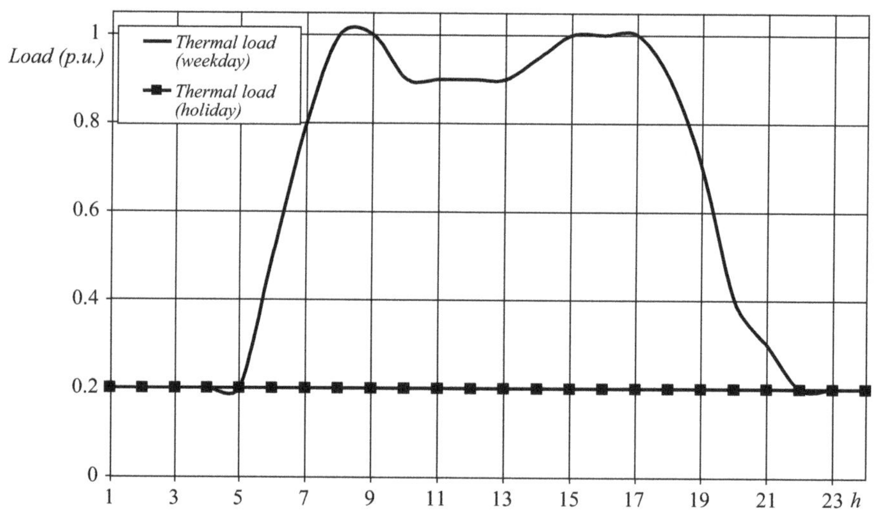

Figure 14.16. Thermal load in the MG

14.13 Economic analysis

This section is devoted to compare several scenarios represented by different kinds and penetration levels (PLs) of DG sources from an economic point of view, by hypothesizing that they differ in capital and operating costs. For the sake of simplicity, the economic analysis is based on equity investments, disregarding taxes.

Table 14.7 Case study 3: PLs and nominal DG powers

	Penetration level (%)	Nominal power (MW)				
		G_1	G_2	G_3	G_4	G_5
Case 1 (no MG)	0.0	–	–	–	–	–
Case 2	74.7	1.50	0.30	1.00	1.00	0.10
Case 3	89.1	1.50	0.55	1.25	1.00	0.35
Case 4	103.4	1.50	0.80	1.50	1.00	0.60
Case 5	74.7	–	0.80	1.50	1.00	0.60

Table 14.8 Case study 3: economic results (Case 1)

Annual cost for electrical loads	1,529.8 k€
Annual cost for thermal loads	546.4 k€
Total annual cost	2,076.2 k€

Such simplification is reasonable whether it is observed that in the area of the MG, which is already destined to industrial facilities, probably there are no more municipal taxes to pay for land use. Depreciation is included in the fixed O&M costs.

The economic evaluation criteria adopted to assess the profitability of the investment are the NPC, the IRR and the DPT, as described in Section 14.1.

Several simulation cases have been performed to evaluate the economic feasibility of the MG under test. The cases differ for the DG PL, in terms of generators capacity (i.e. nominal active power), as reported in Table 14.7.

In the first case, no MG operation is implemented and loads act in the market to purchase energy; all the other cases are managed by the EMS proposed in [13]. The DG PLs of the third and fourth cases are obtained by an increase in the size of G_2, G_3 and G_5 with a step of 250 kW for each. The WT generator (G_1) and the CHP generator (G_4) are not modified because the nominal power of these DG units depends on the windiness of the site and the thermal demand, respectively. Furthermore, it should be observed that G_4 cannot be fully despatched by the EMS because it has the primary duty to supply heat demand. Finally, Case 5 of Table 14.7 evaluates the MG feasibility without RES.

In the following subsections, all cases of Table 14.7 will be commented.

14.13.1 Case 1: no DG is installed in the area

The loads are not coordinated and each facility purchases electrical energy at market prices. The thermal loads are satisfied by conventional boilers. In this case, the global annual energy bill is equal to 2,076.2 k€ and can be calculated considering only the cost of the energy purchased in the energy market and the cost of oil to be burnt for heat demand (Table 14.8).

14.13.2 Cases 2–4: different DG configuration

In Cases 2–4, the DG PLs and the nominal power of each DG unit are shown in Table 14.8. In order to evaluate a condition of power production surplus with respect to load demands, Case 4 with such a high value of PL (103.4%) has been considered.

In all the cases, the global energy bill is obtained by summing the production cost for each DG unit (despatched by the EMS) and the cost of the energy purchased from the bulk power system in those hours where the internal production cannot satisfy the load demand due to technical or economic reasons. Obviously, the global energy bill can be reduced due to the revenues achieved by selling energy in some profitable hours.

Table 14.9 shows the economic results for the three cases. The values of the indicators adopted for the economic evaluation of MG are reported for each case. The CAPEX (IO), corresponding to the DG configuration of Table 14.9 (column 3), are calculated with the data of Table 14.6.

It is assumed a cumulative saving during 15 years, based on the estimated life of the system, and an increase of the cost saving of 5% a year (conservative value). The profits in Table 14.8 are calculated as the difference between the annual energy bills with respect to Case 1. In other words, profits represent the possible savings that can be obtained by changing from the traditional to the MG operation.

14.13.3 Case 5: Case 2 without wind generation

In this case the DG PL is the same as in Case 2, but the WT is not installed.

As it is apparent from Table 14.10 (similar to Table 14.9), despite the reduction of the initial investment, the absence of renewable generation consistently reduces the profit (27.2%) and the NPC, and increases the time required to recover the initial investment (DPT).

Table 14.9 Case study 3: economic results (Cases 2–4)

	PL (%)	**IO (M€)**	**Profit (%)**	**DPT (years)**	**NPV (M€)**	**IRR (%)**	***NPV/IO***
Case 2	74.7	3.34	72.3	2.66	13.5	49.50	4.0
Case 3	89.1	3.77	74.5	2.56	13.8	44.46	3.7
Case 4	103.4	4.19	76.7	2.24	14.0	41.22	3.3

Table 14.10 Case Study 3: economic results (Case 5)

	PL (%)	**IO (M€)**	**Profit (%)**	**DPT (years)**	**NPC (M€)**	**IRR (%)**	***NPV/IO***
Case 5	74.7	2.09	27.2	3.89	6.0	31.05	2.9

14.13.4 Discussion

First, it is important to remark that the implementation of an MG equipped with an optimal control allows, in all the studied cases, to achieve significant savings through the active participation in the energy market by selling and buying energy from the bulk power system and reliably satisfying the load demand. The results confirmed that all the economic indicators adopted to evaluate the investment are always favourable.

Obviously, the proposed cases have initial investments that increase with the penetration of DG (CAPEX are proportional to DG size, Table 14.6). The profit also increases with the DG PL because it depends on the difference between the installed capacity and the electrical loads: the bigger this difference, the greater the energy availability for the sale in the peak hours and the higher the achievable profit. It is important to observe that due to the particular choices of the DG units (Table 14.7), the profit does not rise linearly with the PL.

The DPT decreases with the increase of the DG PL, but the reduction is very small. In particular, between Cases 2 and 3 the difference is roughly 1 month (0.1 years), following an increase in CAPEX of 400 k€. Thus, although all the investments proposed are profitable from the DPT point of view, a difference of 1 or 2 months in the payback period does not compensate such a CAPEX growth.

The comparison among all the cases demonstrates that all the investments proposed are profitable thanks also to the high and similar values of NPC and IRR. However, whether there are no specific constraints to implement one case rather than the others, the second case is the more profitable. The last column in Table 14.10 which shows a measure of the cost–benefit ratio makes apparent that Case 2 is the one that represents the best compromise.

The proposed feasibility analysis has been applied to a case study derived from the real world. Despite the high investments required to install the DG and to implement an efficient communication system in the MG, the economic benefits are so significant that CAPEX may be fully paid back within a reasonable time. The best condition, 2 years DPT, depends on a high production from the WT. In Case 5, the annual profit is calculated without the wind generation and the DPT is roughly 4 years. Obviously, such figures for DPT are strongly dependent on the market conditions and on the distributed energy resources available. However, according to literature and practical experience, it may be affirmed that the DPT does not exceed 5 years but is normally comprised between 3 and 5 years.

In conclusion, MGs in the industrial districts have great technical and economic potential. They can promote the integration of DERs in the distribution systems by limiting their negative impact on the system.

14.14 Conclusion

This chapter has shown several case studies of MGs, the value of each MG system in providing energy efficiency and electricity bill reduction as a good alternative to feeding by public networks in rural and isolated communities. In particular, the possibility offered by such systems to satisfy the energy load requirements of several classes of electric users, ranging from buildings and business facilities, urban/industrial districts and rural communities has been presented and discussed.

References

[1] N. Hatziargyriou (Editor), *Microgrids: Architectures and Control*, Wiley-IEEE Press, Hoboken, NJ, February 2014.

[2] E. Ghiani, F. Pilo, and S. Cossu, "Evaluation of photovoltaic installations performances in Sardinia", *Energy Conversion and Management* 76, 1134–1142, 2013.

[3] M. Šúri, T.A. Huld, E.D. Dunlop, and H.A. Ossenbrink. 2007. "Potential of solar electricity generation in the European Union member states and candidate countries", *Solar Energy*, 81, 1295–1305.

[4] European Commission. Photovoltaic Geographical Information System. JRC Solar database. Available at http://re.jrc.ec.europa.eu/pvgis/cmaps/eur.htm

[5] T. Lambert, P. Gilman, and P. Lilienthal. "Micropower system modeling with HOMER", in *Integration of Alternative Sources of Energy*, John Wiley & Sons, Inc., Hoboken, NJ, 2005, pp. 379–418.

[6] A.J. Litchy, C. Young, S.A Pourmousavi, and M.H. Nehrir, "Technology selection and unit sizing for a combined heat and power microgrid: comparison of WebOpt and HOMER application programs," in *2012 North American Power Symposium (NAPS)*, Champaign, IL, pp. 1–6, 2012.

[7] IEA, *World Energy Outlook*, 2014.

[8] D. Schnitzer, D.S. Lounsbury, J.P. Carvallo, R. Deshmukh, J. Apt, and D.M. Kammen. "Microgrids for Rural Electrification", United Nations Foundation, February 2014.

[9] "Programme d'Electrification Rurale Globale", PERG, Office Nationald'Electricité (ONE). Available at www.one.org.ma

[10] E. Ghiani, C. Vertuccio, and F. Pilo, "Optimal sizing of multi-generation set for off-grid rural electrification", in *Proc. 2016 IEEE PES General Meeting*, Boston, MA, 17–21 July 2016.

[11] http://www.weather-and-climate.com

[12] E. Ghiani, C. Vertuccio, and F. Pilo, "Optimal sizing and management of a smart microgrid for prevailing self-consumption", *PowerTech 2015 Conference*, Eindhoven, The Netherlands, 29 June–2 July 2015.

[13] F. Pilo, G. Pisano, and G.G. Soma, "Neural implementation of microgrid central controllers", in *Proc. INDIN 2007 – 5th International Conference on Industrial Informatics*, Wien, Austria, 23–27 July 2007.

Chapter 15

Smart grid road map and challenges for Turkey

Musa Yilmaz[1] *and Heybet Kilic*[2]

The power of a conventional plant and that of a photovoltaic (PV) system are the same. In typical applications, a limited grid network, heating, ventilation and air-conditioning systems with remote water pumping for remote islands include the supply of electricity to villages. Turkey's energy dependence on imports, mainly on oil and natural gas (NG), has been increasing because of the growth in energy demand. Turkey has achieved the highest growth rate of energy demand in OECD countries over the last 12 years. Currently, Turkey is able to meet only about 26% of the total energy demand from its own domestic resources. On the one hand, the institutions responsible for the transmission and distribution of electricity and NG are privatized in Turkey, but on the other hand, works for the optimization of transmission and distribution networks are in progress. Although applications are made for the exploitation of transmission and distribution networks at an optimal level, the competent public authorities enforce new regulations in this context, with the aim of ensuring bilateral information flows between consumers and suppliers. Turkey has two interconnection points with the East European Transmission Grid. The test period for a synchronous parallel operation between the Turkish and European power systems had started on 1 June 2011 and ended in September 2012. At this moment, the trade is limited to 400 MW from Bulgaria and Greece to Turkey and 300 MW from Turkey to Europe via these countries. In order to provide a stable, low-cost, reliable, efficient, robust, sustainable and environment-friendly electrical energy system to consumers, a fully operational smart grid (SG) system needs to be established in Turkey. If classical grids in Turkey were transformed into SGs, not only would the above-mentioned benefits be achieved, but also Turkey would be able to attract a huge amount of investment to boost its economy. The Turkish grid system would then become a powerful player in the energy market in Europe.

[1]Electrical and Electronics Engineering Department, Engineering and Architect Faculty, Batman University, Batman, Turkey
[2]Electrical and Energy Department, Diyarbakır Vocational School of Technical Science, Dicle University, Diyarbakır, Turkey

Nomenclature

BIST	Borsa İstanbul
DEG	decentralized power generation
DER	distributed energy resources
DSM	demand-side management
EMRA	Energy Market Regulatory Authority
EMS	energy management systems
EUAS	Production of Electricity Co.
EV	electric vehicle
EXIST	Energy Exchange Istanbul
GHG	greenhouse gases
GIS	geographic information system
LCC	landscape conservation cooperative
MENR	The Ministry of Energy and Natural Resources
MG	micro-grid
NG	natural gas
NIST	National Institute of Standards and Technology
PV	photovoltaic
RE	renewable energy
RES	renewable energy source
SEG	smart energy grid
SG	smart grid
TEA	Turkish Electricity and Transmission Co. (Türkiye Elektrik Anonim Şirketi)
TEDAS	Turkish Production Electricity Distribution Co. (Türkiye Elektrik Dağıtım Anonim Şirket)
TEIAS	Turkish Electricity Transmission Co. (Türkiye Elektrik İletim Anonim Şirketi)
TEK	Turkish monopoly electricity (Türkiye Elektrik Kurumu)

15.1 Introduction

The consensus among climate scientists is that greenhouse gases (GHG) are causing dangerous climate change. Hence, limited production or without production of GHG or more efficient use of energy should include some ways to generate electricity.

Developing a method to produce electric power and providing a better and more efficient force that helps grow more green are global concerns about the economic feasibility of the world. Investment in the energy sector, commodity prices in the market due to the flightiness of energy prices and a high degree of uncertainties have become evident from renewable energy (RE) sources (RESs), with the emergence of SGs, the development potential of fossil fuels. RES, the utility industry, is one of the most interesting options for the future [1].

The new generation of power generation and distribution systems is the first step in the development of RESs. In the long run, it helps to mitigate the effects of climate change. However, the image of stability, reliability and cost must be taken into consideration. SG technology can help one to integrate RESs into future power systems. A European Technology Platform, SG, is a source of electrical energy through an integrated and defined system.

In the recent past, the world has witnessed an increase in demand for electricity and energy. From the studies of environment and climate change, it is clear that the impact of RES on the environment is limited. Supported by economic and environmental factors (solar, PV, wave, tidal, fuel cells, biogas and hydrogen). RE increasing proportion of an increasing reliance on the depletion of resources new, renewable sources of energy alternatives to traditional fossil harmless to the environment and friendly colleagues with a fundamental shift RE, clean fossil fuels and energy which leads to fuel switching. PV and distributed energy resource (DER) production programmes, and micro-grid (MG) are regarded as the most suitable and affordable options for the production of electric energy. The power of a conventional plant and that of a PV system are the same. In typical applications, the limited grid network, heating, ventilation and air-conditioning systems with remote water pumping for remote islands include the supply of electricity to the villages [2–5].

As part of the future of the realization of the smart energy grid (SEG), it is significant to demonstrate regional SEGs and micro-energy grids with RE as a decentralized generation. This decision includes the development of a dynamic and evolving SEG. To achieve this goal, we propose modelling and model simulation for target design and evaluation SEG. MG-distributed production structures are assigned to a geographic information system (GIS) that provides a bottleneck in the advancement and realization of real-world SEGs. In addition, there are no technologies to support the evaluation of different web applications at different scales inside the grid, such as a hybrid electric vehicle (EV). In addition, most of the design and operation decisions are based on the safety and reliability of physical structures. The current modelling and simulation technologies are not supported on the integrity of assets, models that are associated with energy/energy-modelling and the limits for the evaluation of various designs and have created scenarios in terms of security, reliability and availability. SEG physical structure dynamic model's simulation capabilities lead to limitations in the precise optimization of network performance.

Robust SEGs planning can be realized through automated and intelligent monitoring, planning and distributed modelling and simulation to support SEG designs and operation. The solution is real-time modelling and simulation of the SEG superstructure at two levels: the first level of the model grid and the low component level. It is a distributed data management module designed to provide the following elements – SEG-built models and vote: trellis of the physical-asset superstructure and models, energy–energy models, the asset integrity chart and model reliability, security and protection of models and operational and control models [6].

The proposed technology supports the functions of enterprises: intelligent multi-sensors measuring design assistance, distributed control and operations support, decentralized power generation (DEG) planning and asset optimization, degradation, and the upgraded synthesis and verification scenario to implement a security component in the component levels.

The technological characteristics of the MGs model, and the modelling of MGs with operational capacity, and alternative scenarios can be defined on the basis of performance indicators and protection strategies. The objective of this contribution is to ensure a framework for the plan and exploitation of scenarios for scenario designs and exploitation. The user interface occurs via accessible mobile phones to allow mobile access from remote locations and to identify the optimal opportunities for energy supply and energy consumption within SEGs. Deterministic and predictive SEG models are developed, using the static and dynamic parameters of connected SEG-superstructure physics and semantic energy networks, which are used for the synthesis of best practices and scenarios for the MG of the generation, protection and integration of applications. The modelling and simulation project is assigned to GIS for the geographical and ecological analysis of real-time data and risk-index-chains-calculus with power scenarios – monitored and analysed [1,2].

There are several scenarios that can be adopted for the analysis of energy efficiency in connection with the founder of MGs with different thermal, gas-to-electricity conversion and use strategies to meet regional energy needs. These strategies include (a) MG designs and control, (b) thermal storage systems, (c) NG-applications for transport and facilities, (d) gas-to-electricity conversion technologies and (e) energy saving within MGs. Researchers, technology providers and industries investigated in each of these track-effective technologies. Investors have also planned several technologies for hydrogen production and conversion [7–10].

The analysis of the energy efficiency in the integration of MGs with the existing energy infrastructure has the most impact on the environment, in terms of control, protection, power quality and sustainability. Figure 15.1 shows an integrated view of SEGs, in which the DESs are integrated in every step in the processes of sources, transport, treatment, conversion, production, storage and use. MGs and their integrations with grid topologies are considered as the island and network connections (LCCs [landscape conservation cooperatives]). This chapter presents the results of the LCCs. There are several scenarios to implement a fully integrated thermal gas flow networks, and to demonstrate local distributed energy systems, within the MGs, the proper analysis of the various energy supply and production (i.e. production) scenarios on the basis of the life cycle analysis, and energy policy and provisions for thermo-electricity networks and their operation. MGs, their optimal performance in the clear identification of potential and effective protection and control systems in relation to existing network infrastructures, energy resources, and the latest forecast demand in the area are analysed [11].

The proposed support will be provided for SEG infrastructures, the thermal gas-electricity networks and the efficient supply of energy to meet local or regional requirements. In addition, an energy data centre will support the evaluation of design, control and operating scenarios included with this built-in thermal gas – flow

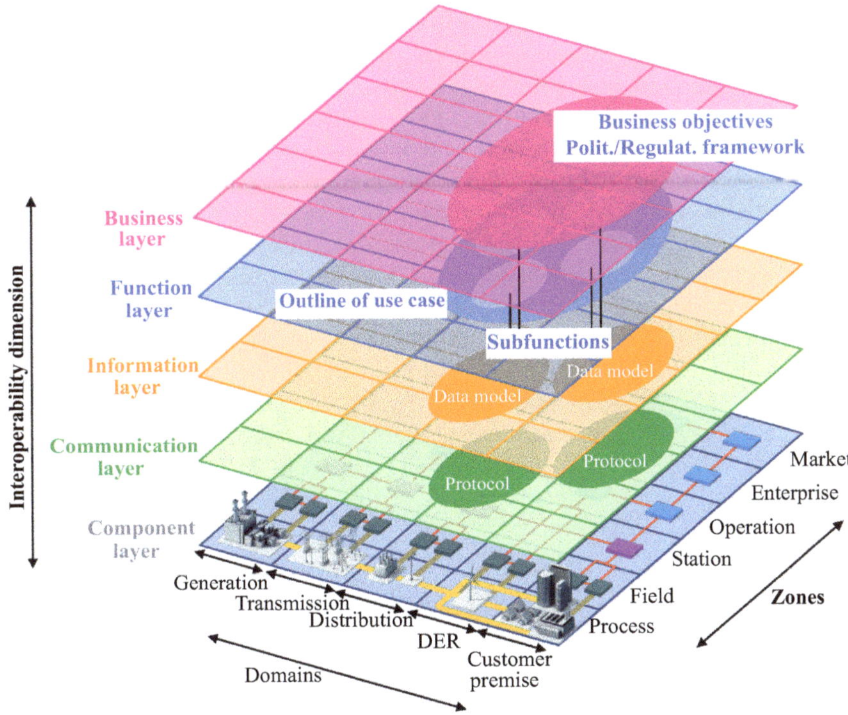

Figure 15.1 Smart grid architecture [12]

networks. It will also support the optimization of SGs and MGs process parameters, to achieve the economic needs of the most advanced and potential solution for the regional energy.

Energy efficiency encompasses SGs and MGs, such as (a) an MG design on the basis of the available energy resources, the existing network topology and the load and energy demand forecast; (b) dynamic protection with sufficient risk-based error propagation analysis in terms of the dynamic nature of MGs in island and grid connections; (c) cost–benefit analysis for various energy technologies, with a realistic profit planning; and (d) consideration of international standards and national schemes for energy efficiency limits and targets with multi-view analysis on the basis of detailed business models. In order to provide a logical justification for the development of interconnected MGs, modelling and simulation tools are required, as explained in the following section.

15.2 Comparison of conventional and smart grid

Established power supply systems, developed in the past 70 years, which have high-voltage interconnected network transformers are known as the transmission network. Each unit generator, whether hydroelectric powered, nuclear powered or fossil powered, has a capacity of 1,000 MW wide. The transmission network is used with the transport of electrical energy, sometimes over considerable distances, and the current is extracted, then for the final circuit for delivery to the end customer, a range of distribution transformers is passed through [13].

The consensus among climate researchers is clear that the emission of GHG will cause dangerous climate change. Therefore, the means used to use energy more efficiently and to produce electricity without CO_2 must be produced. An effective control of the load and reduction of energy losses and waste should be made. Smart meters are a crucial part of the SG because they can provide information about loads, and therefore power across the network. As soon as all the parts of the network power to monitor come to an observable state, there are many control facilities [14].

Although the 'SG' is defined likewise all over the world, it is obvious that there is a big segregation between it and the conventional energy system. In general, efficient, reliable, inter-SG is active and compatible. In Table 15.1, differences in efficiency, reliability, compatibility, interaction and openness are compared.

Table 15.1 Comparison of smart grid and traditional grid [15]

Feature	Conventional grid	Smart grid
Efficiency	The asset management and accomplishment of process data are poor	Efficient process: enlarge the data collection and attention to accident interception in order to decrease the negative effect on the customer and improve process efficiency Asset optimization: introduction of advanced data and control method to optimize the use of facilities and hardware, lower operating and maintenance expenses
Reliability	The ability of self-improvement is bad and it cannot handle the natural catastrophes	Self-improvement: sustained assessment can detect, analyse, react and even re-power the manufacture of the equipment or anomalous process of local area grids Security: effectively the damage of exterior assault of real systems or virtual systems
Compatibility	It is especially contrived for concentrated energy production and DERs	RESs can easily access the network
Openness	Force access to equipment	They authorize free access to the network and the coaction between equipment and the network
Interactivity	It is one sided and the customer must accept the activities	You notice your clients and embolden users in the electricity market

15.3 Grid modernization and current trends

The locution 'SG' refers to a modernization of the central power system and end-user transmission and distribution grids, MGs with RES (energy storage). Because of bi-directional current flow and data in the power grid and communication networks, the system must be standardized to implement bi-directional information flow. For more information, please contact us. In the United States, the National Institute of Standards and Technology (NIST) defines the SG as a transition process between the current power grid in the next; information and communication technology focuses on the objective. For the future of transition from the electricity distribution network to the SG, some standards as to be a consensus on the core technology can be considered as standards. There are different approaches to identify basic sets for the future. These approaches focus on dissimilar core fields, depending on the conditions. The information is presented in Table 15.2.

15.4 Smart grid integration and its definitions

The traditional electricity grid is not flowing in the natural unidirectional: the electricity from power generation plants to end users. This system has been providing good services for the last 100 years. In recent times, it was, however, subject to a number of technical, economic and environmental issues. Modern society demands that this system is reliable, scalable and manageable, as well as it is cost-effective, secure and interoperable. The next-generation power grid called the 'smart grid' is a promising solution for the long-term development of the industry. The SG will revolutionize the electrical power generation, transmission and distribution. Moreover, the current power grid is system supplemented by RESs, including wind, solar and biomass and is more environment friendly compared to fossil fuels. In many masses electrical power generation is used in the equipment. In this work, we describe the power of the energy system as a function of the strength of the current network [20–25].

In general, an SG is a combination of a traditional sales grid, a two-way communication network for sensing, monitoring and distribution of information on energy consumption. A model of communication structure in SG is shown in Figure 15.2. As shown in the figure, all typical SG-based energy generation units and energy-consuming entities are connected via a network. The generators feed the electricity into the grid, and consumers draw energy from the grid [23–26].

A key characteristic of an SG, the use of information and communication technologies is to collect automated information and to act in order to increase the efficiency, reliability, cost-effectiveness and sustainability of the power generation, transmission and distribution of electrical energy.

A hand component of the SG leaves the possibility of customer participation in the total network energy management. This participation occurs over the term of the demand or of the demand-side management (DSM), in which (a) energy suppliers provide motivation for clients to shift their load in time, and (b) energy from/

Table 15.2 The future smart grid [16–19]

Standards		Approach								
Specification	**Description**	**DKE Normungs-roadmap Smart Grid 1.0**	**NIST IOP Framework Roadmap 1.0**	**SMB SG 3 IEC**	**BMWi E-Energy Programme Recom-mendations OFFIS**	**BDI – Internet der Energie**	**Microsoft SERA**	**CIGRE D2.24**	**CEN/ CEN ELEC M/441**	**Japan's Roadmap to International Standardization for Smart Grid**
IEC 61970/ 61968	Common information model (CIM) for energy management system interfaces	Y	Y	Y	Y	Y	Y	Y	_	Y
IEC 61850	Substation automation systems and distributed energy resources (DERs)	Y	Y	Y	Y	Y	Y	Y	_	Y
IEC 62351	Smooth integration of smart grid security	Y	Y	Y	Y	Y	Y	Y	_	Y
IEC 62357	Seamless integration of smart grid security architecture	Y	Y	Y	Y	_	_	Y	_	_
IEC 60870	Communication and transport protocols	Y	_	_	Y	_	_	Y	Y	_
IEC	61400-25	Wind power communication EMS, DSM, DER	Y	_	Y	Y	Y	_	Y	_
IEC 61334	Distribution line message specification	_	_	_	Y	Y	_	Y	Y	Y
IEC 62056	Companion specification for energy metering	_	_	Y	Y	Y	_	_	Y	Y
IEC 62325	Market communications using CIM	Y	_	_	Y	_	_	Y	_	_
IEEE C37.11	Phasor measurement unit communications	_	Y	_	_	_	Y	_	_	Y

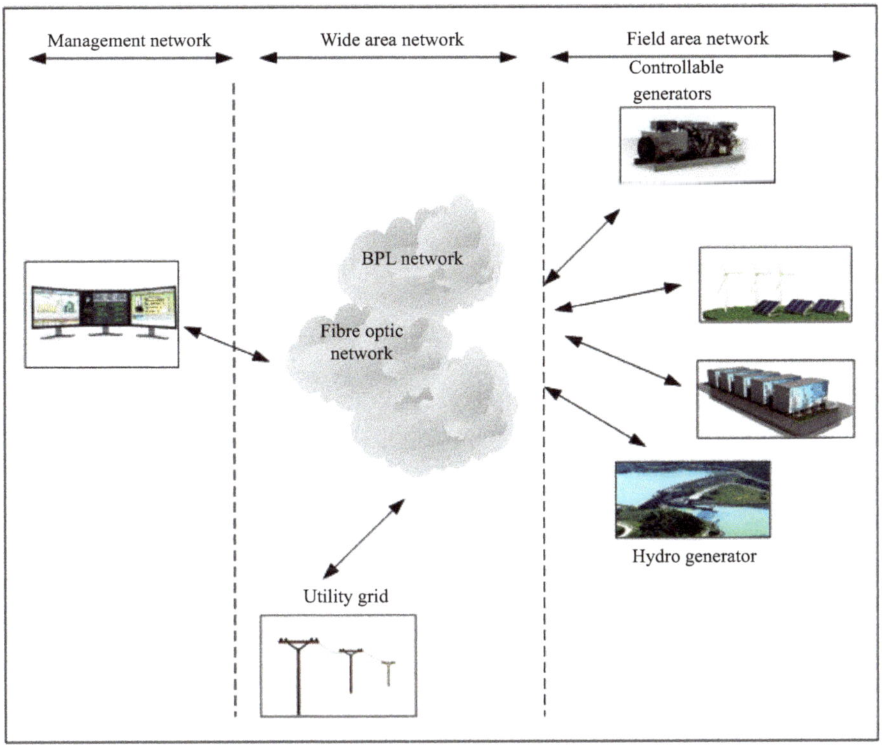

Figure 15.2 Smart grid communication framework

to the network. So in every SG mechanism it is imperative to take into account the demand–response models and their associated challenge.

The most recent intelligent MG-distribution network is a network of electric energy distribution that includes a group of loads, distributed generators (e.g. RESs such as solar cells and wind turbines), transmission, distribution and energy storage systems. An MG can dynamically respond to the changes in the energy supply by adapting the demand and the production itself [26].

15.4.1 Integration of distributed energy resources

The characterization of SG can be achieved by a wide penetration of distributed and RESs. High penetrations of RESs increase the uncertainty of generating resource availability. In addition, most of these resources are installed on the distribution side. The power system control model was developed to control a small number of generating stations of the central control systems. The current tax model for an RESs application that needs to be transferred to a wide geographical area, taking into account, and the amount of data to the monitoring stations has technical and economic challenges. The results of this study are summarized in the following table. The computing power needed to process this information in real time, and to take the necessary measures in real time is another challenge. The implementation of such control and communication networks will be costly [2].

In addition, the centralized control model suffers from the reliability and the only point of the error problems. The failure of the central server or of the communication channel can lead to a serious system problem. To solve this problem, the control model of centralized control should be changed to a decentralized control model. Instead of central headquarters, it is possible to have a local intelligent control. The local controller processes the data locally and performs the necessary control in your area [26–29]. Neighbouring areas can share information to coordinate the operation in the area and to support the overall system stability. Transferred-level data and control commands between intelligent controllers and control centres are highly required. Through the local processing of data and the implementation of local control actions for data transmission and communication, the bandwidth is reduced. The required computing power for control centres is low, and the system reliability is improved and a single point of failure is attempted to be overcome. However, the optimal level of decentralized control must be identified. A decentralized control model may vary from a division of the system into sub-areas with local controls, as shown in Figure 15.3(a), to distribute a multi-agent controller, complete, as shown in Figure 15.3(b). In the first topology, the system is

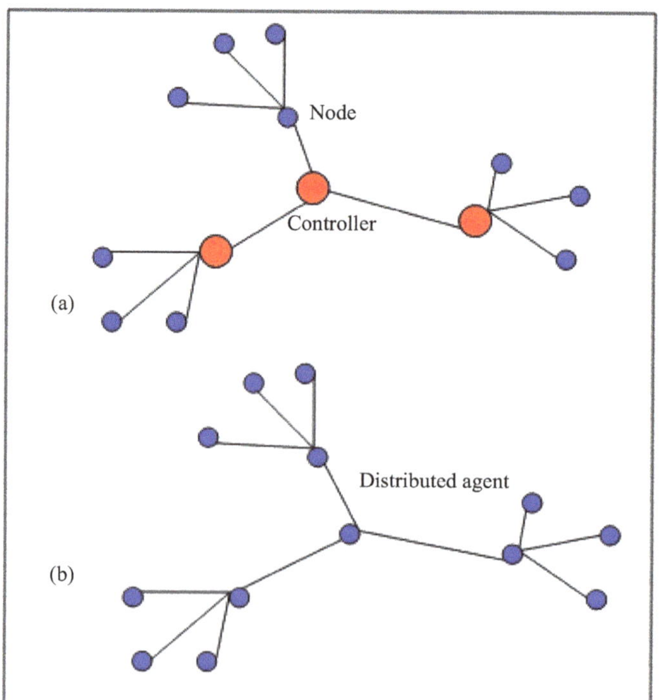

Figure 15.3 (a) Decentralized and (b) distributed control

divided into several control areas, with a central control for each area. The local controllers need to coordinate with the central controller. If there are any problems in one field, you can work in the other [28–30].

15.4.2 Interoperability challenges

The integration of a variety of systems is governed by different rules and assets; the ownership of the grid has an interoperability challenge. Several protocols and standards for the process automation and supervisory control and data acquisition (SCADA) systems, IEC61850 for substation automation, C37.118 for phasor measurement and OPC UA for the machine-to-machine communication as a DNP3 for the SG are successfully operated. In many cases, it is necessary to assign data from one protocol to another. A common data bus or an interface that can be used by application developers to develop SG applications such as energy management systems (EMSs) or DSM systems is missing. In addition, most of the installed components have been used for decades. It is not easy to replace these components due to economic factors. The interoperability with legacy devices and protocols should be taken into account for the improvement of new communication protocols and standards in order to meet the new requirements for SGs [31–34].

15.4.3 Challenges in communication

The communication network for SG applications must take into account the special requirements of the real-time control. The intelligent network control is based on various kinds of signals, and each type of signal has a specific requirement in terms of bandwidth, delay and availability. Data should be made available at the right time. The communication infrastructure design should be coordinated with the control requirement. Some of the data are sensitive to delays, such as protection and feedback signals in low inertia systems, such as MGs with RE. The other signals are more tolerant to delay, such as power consumption and smart meter data [33–35].

15.5 Communication technologies and protocols in smart grid

Information for many applications for SGs is very important, and for this reason, information technologies play a crucial role in the SG approach. Numerous of these technologies, such as wireless communication and communication networks, are already used in other areas. These technologies have now reached a degree of matureness so they can be used in the electricity grid [36–38].

The future of SG applications requires more measurements from the network and, therefore, be used more. The latest developments in wireless communication technology and microelectromechanical systems have allowed the development of recent types of wireless sensors technology, which combine with the detection, signal processing and communication of parts, and over short distances can communicate with each other. In addition, their low cost and ease of wireless placement facilitate the nature of these devices, detecting the fine steps required by Intelligent

Network applications. Wireless Detection Technology is also very crucial for one's home to store energy, information management applications such as internal temperature and movement of people [2].

In addition to the speed of communication and detection capabilities, the implementation of Intelligent Networks requires a powerful management infrastructure that is able to deliver dynamic smart control fixations. This control substructure must perform a full analysis of all events to be able to quickly set any adverse situation to respond. Nowadays, electricity grids are controlled by centralized management systems, which were designed decades ago, and realized, and in central systems, there is a lack of scalability and flexibility. Inherently, they are slow to respond to emergencies. The SG will be a significant evolution of this centralized distributed approach control and local frameworks and control point coordinates. In this model, the control of service centre processes to be thrust to smart control units, which are placed at key points inside the grid. The management software is implemented using technology software [36–41].

The new technologies and activities of the SG will be aided by the control system and the power grid. The introduction of DEG allows for more efficient power supply and to be more efficient and proactive control of the network with customers. The aim of this contribution is to ensure a structure for the integration of advanced communication technologies into the current network. Control processes will count in the future on the real-time information of the grid [15].

However, the ability to communicate is required by this new management system in the power grid current, in order to achieve the level of penetration. The present invention relates to a method for controlling the power of the electricity grid. The other problem that is based on the integration of complex technologies and implementations to achieve an intelligent structure is the huge number of data generated by different applications, which are used for additional analysis, control and real-time price [42–49].

15.6 Case study: smart grid applications, market and policy in Turkey

Turkey's energy dependence on imports, mainly on oil and NG, is increasing because of the increase in energy demand. Turkey has the highest growth rate of energy demand in OECD countries over the last 12 years (Figure 15.4). Currently, Turkey has been able to meet only about 26% of total energy demand from its own domestic resources [50–52].

On the one hand, the institutions responsible for the transmission and distribution of electricity and NG are privatized in Turkey, but on the other hand, works for the optimization of transmission and distribution networks are in progress. Although applications are made for the exploitation of transmission and distribution networks at an optimal level, the competent public authorities are expected to enforce new regulations in this context, with the aim of ensuring bilateral information flows between consumers and suppliers [46].

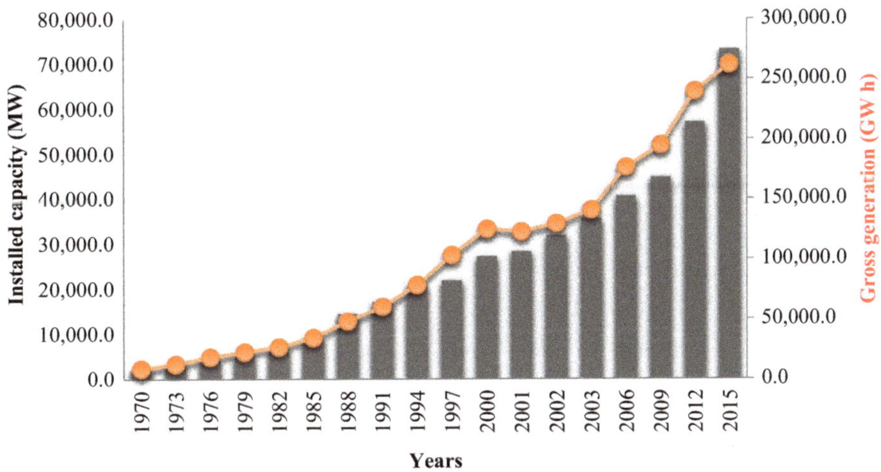

Figure 15.4 Turkey's energy generation and installed capacity

With regard to the implementation of the SG system, some regional applications were underway in Turkey, with an attempt to monitor networks and distribution companies that carried out a cost–benefit analysis for SG systems. In addition, in an attempt to develop a metrological measurement infrastructure to support the successful implementation and operation of SG in Europe, the National Institute of Metrology (TÜBITAK-UME) took part in the International European Metrology Research Program [47,48]. At this stage, investments in SGs are being made in Turkey through ready automatic meters and various SCADA systems. Investment activities for grid monitoring and distribution network management conducted independently for real-time monitoring and grid management are also included in the above scope. In addition to organized industrial areas and electricity and NG, municipal water distribution network administrations also began to switch to the automatic meter reading system. In this context, the 70,000 subscribers were included in the automatic meter reading of the system by the electricity distribution companies, 32,000 by the water administrations and 4,000 by the NG distribution of companies. The automatic meter reading system is mainly installed in the organized industrial zones and industrial facilities with a high level of electricity, NG and drinking water in our country, whereas the work concerning its implementation in residences is still going on [49–55].

It is felt that a country with big goals, like Turkey, should be among the countries that regulate changes, not a country exposed to the effects of technological changes. The same country is planned to be not only the one which uses SGs but also develops projects and produces such systems. In this context, Turkey is intended to guide the transition process of other countries in its region, the intelligent network of systems and to encourage domestic and foreign companies to invest in its area. In the medium- and long term, all meters are intended to be replaced by smart meters under the 2023 strategic vision [56–60].

In Turkey, there prevail a lack of common minimum criteria for meters and communication equipment, one of the most important components of intelligent network system, as well as a lack of criteria for data security in the intelligent and undetermined network standards for a functioning compatibility of the entire system to constitute a practical obstacle to an intelligent network [61–67].

15.6.1 History of the Turkish electricity market

Men started using electricity in their daily lives in 1878. The first power plant was commissioned in London in 1882. Hepbasli [66] reports that the first Turkish electric generator is a 2-kW dynamo connected to a water mill in Tarsus, Turkey, in 1902. Hepbasli [68] also reports that the first large power station was installed in Silahtaraga, Istanbul, in 1913. Since then, the following have been the main developments in the Turkish market of electricity [66,67]. The electricity industry was heavily dependent on foreign investment, which flowed in the country, as it sought to develop a liberal economy between 1923 and 1930. Mainly German, Belgian, Italian and Hungarian companies were provided electricity by Turkey. The first Turkish electricity company, 'Kayseri ve Civari Elektrik Turk Anonim Sirketi', was established in 1926 [68]. The 1930s demonstrated the beneficial effects of public ownership as the world's electricity industry was restructured.

A 5-year industrial plan was implemented by the government in 1933, which was an important active role in the electricity sector. It began with researching hydroelectricity and thermal energy for electricity generation. This initiation was followed by the publication of some of the legislation to enable municipalities to build and operate power plants. In 1935, the Electrical Power Resources Administration Survey was founded; later, the Bank of Provinces and State Hydraulic Works (DSI) were established. The government's development plan has been more of a national than global policy. The Ministry of Energy and Natural Resources of Turkey (MENR), which was responsible for Turkey's energy policy, was created in December 1963. It was followed by the adoption of a law building the Turkish monopoly electricity, Turkish Electricity Commission Administration (TEK).

All generations of power plants, transmission lines and equipment distribution were transferred to a single economic state of the company, the Turkish Electricity Authority (TEK), which benefited from a legal monopoly for the generation, transmission, distribution and retail sale of electricity.

The institutional structure of the Turkish electricity market, characterized by a high level of centralization and state control (as was the case in Europe at the time), was supported by global trends in restructuring of electricity markets and by technological improvements that have allowed small-scale electricity plants to be built, leading to more and more efficient and less expensive energy production. But after only 12 years, TEK legal monopoly till 1982 and TEK itself were divided into several generations, transmission and distribution companies [69]. As seen in Figure 15.5, TEK was able to maintain its vertically integrated structure, till 1993, when it was unbundled by

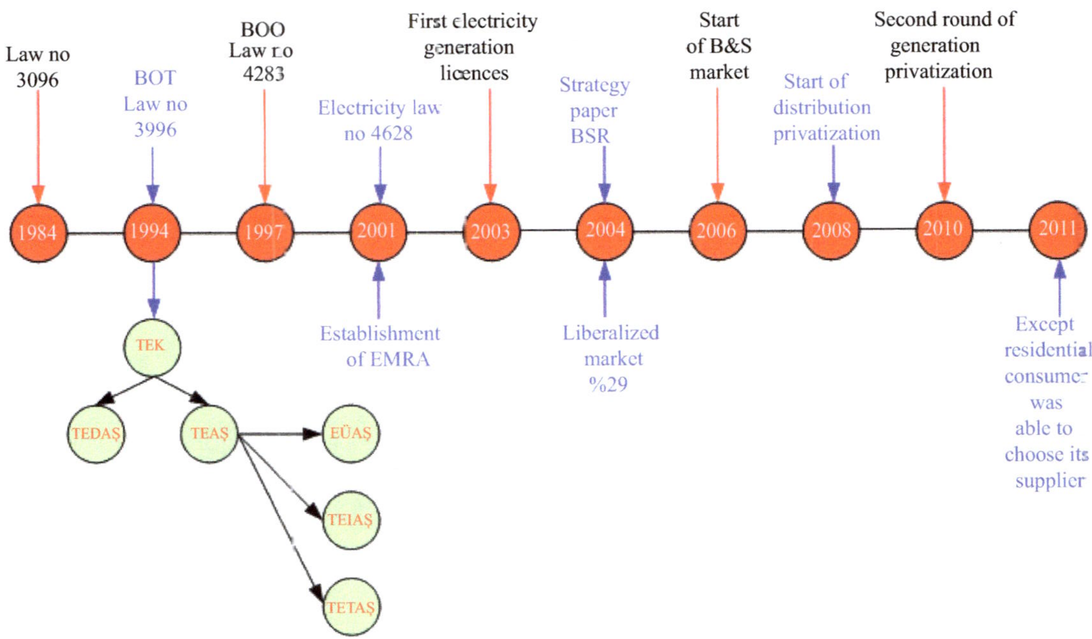

Figure 15.5 Milestones in Turkish electricity market reform

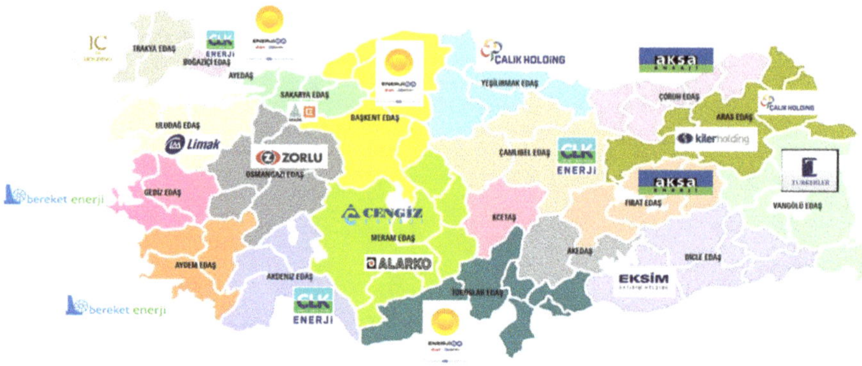

Figure 15.6 Twenty-one electricity distribution regions in Turkey

government decree into two state economic enterprises: Turkish Electricity and Transmission Co. (TEAŞ) and Turkish Production Electricity Distribution Co. (TEDAS). In 2000, the production and transmission of the markets were separated, and TEAŞ was divided into three state economic enterprises: the production of Electricity Co. (EUAS), the Turkish Electricity Transmission Co. (TEIAS) and Turkish Electricity Trading and Trading Co. (TETAS). TEDAS has also been reorganized, and 21 affiliated companies (Figure 15.6) have been established in accordance with Turkey's privatization programme for the privatization of administration [69].

It has been deemed necessary to establish a sound, stable and transparent financial market plan for electricity, competition and specific laws to sufficiently ensure high quality, continuity, low cost and respect of the market. The environment should be made conducive for the supply of electricity to consumers and to maintain an independent regulator and supervisory framework. The new law must cover the production, transmission and distribution and wholesale and retail sale of electricity and related services. It should also cover electricity, imports and exports and the rights and responsibilities of all entities in relation to these services. The new policy has created an Electricity Market Regulatory Body and defines the procedures and directors that underpin this body. The law also built procedures for the privatization of electricity generation and the distribution of assets [66,67].

Milestones in the reformation of the Turkish electricity market are shown in Figure 15.5 [67].

Three years ago, with an attempt to privatize AYEDAŞ Elektrik Istanbul on the European side of the grid, Boğaziçi Elektrik fell flat when MMEKA, a joint venture between Mehmet Emin Karamehmet and Mehmet Kazanci pledged 4.8 billion dollars to privatization tender two girls, failed to pay the agreed amount. This time, the winning bid for AYEDAŞ Elektrik was 1.81 billion.

Enerjisa also placed more than 1.7 billion bid for the privatization of grid operations in the Adana region of southern Turkey. Toroslar is the second largest grid among those privatized ones, which can provide the same day services to more than

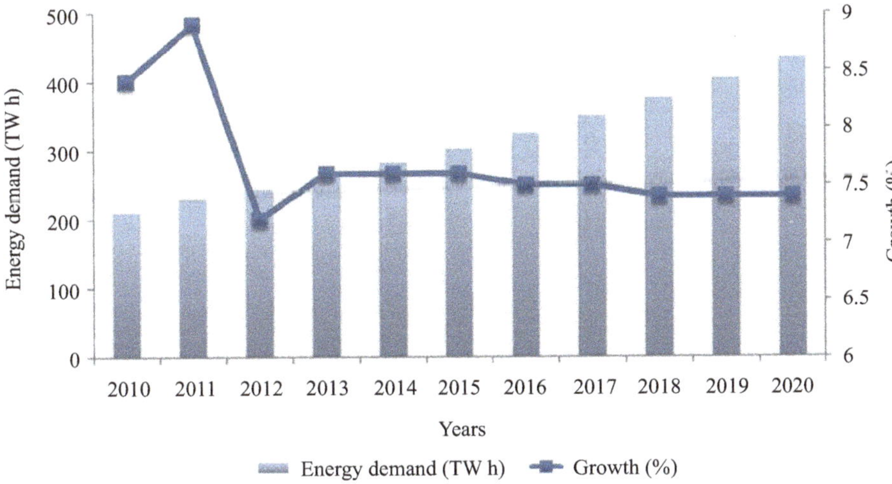

Figure 15.7 Electricity demand with growth forecast for Turkey, 2010–20

2.8 billion people living in Adana, Gaziantep, Hatay, Mersin, Osmaniye and Kilis. The loss and leakage rate in the region is on par with the Turkish general at 11%.

In Turkey, ever-increasing electricity demand trend will be preserved with strong GDP growth, industrialization and urbanization supported by structural reforms. Electricity demand has come across impressive growth in-line with economic developments, by industrialization and urbanization. Economic growth along with population growth expectations shows great potential in addition to growing electricity demand (Figure 15.7).

Economic development and increased electricity demand fuelled electricity generation investments. Installed capacity in Turkey has increased more than sevenfold since 1984. In July 2016, Turkey's installed capacity reached 77 GW (Figure 15.8 and Table 15.3).

EUAS State Corporation has still been an important player in the market, but the diversity of ownership exists. As of 2017, the EUAS is still the largest electricity generation company in Turkey. However, this will change once its assets are acquired by private investors during the continuation of privatization.

The second group is independent of the production companies, which will allow us to see their actions grow in the future as companies proceed with the commissioning of new factories and the acquisition of power plants within the privatization of wallet.

15.6.2 Coal market

From the 12.4-GW capacity produced by Turkey, an amount of 8.5 GW is used by domestic coal reserves (Figure 15.9). Turkey has 11.8 billion tons of reserves: 1.3 billion tons of coal and 10.8 billion tons of lignite.

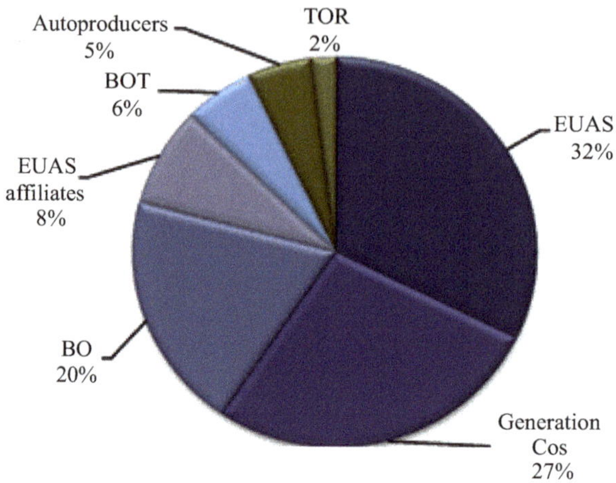

Figure 15.8 Market shares and capacity of leading electricity generation producers (%), 2016. EUAS, state-owned power generation company; BOO, build–operate–own; BOT, build–operate–transfer; IPP, independent power producers; TOR, transfer of operation rights

Table 15.3 Turkey's electricity generation producers

Producers	**Generation (GW h)**
EUAS	79,509
Generation Cos	68,077
BO	47,938
EUAS affiliates	19,823
BOT	14,810
Autoproducers	13,708
TOR	5,566
Total	249,431

The coal-fired power plants built between 1980 and 1990 are currently being privatized. Another 7 GW of capacity is doubled until the private sector is entrusted.

Despite government positions on the use of domestic coal, a large amount of investment is expected based on coal imports, determined from the licenses currently being assessed and reviewed. Project developers prefer to use more calorific value of coal, in order to ensure higher yields and a shorter ROI. A recent study provides 5 GW of additional coal capacity to be online by 2023 (Figure 15.10).

15.6.3 Gas market

On the supply side, Russia is the main source of gas, but Turkey also has connections to the Caucasus and Iran. Although Turkey is heavily dependent on imports, it is also an important transit country to Europe. Belonging to the state Botas dominates the

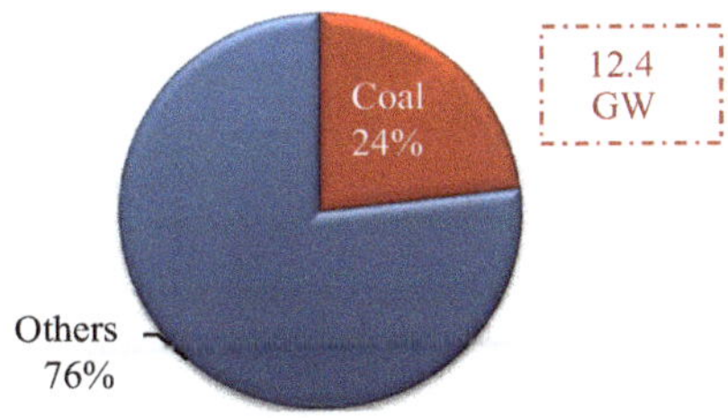

Figure 15.9 Installed capacity by a share of coal, 2011

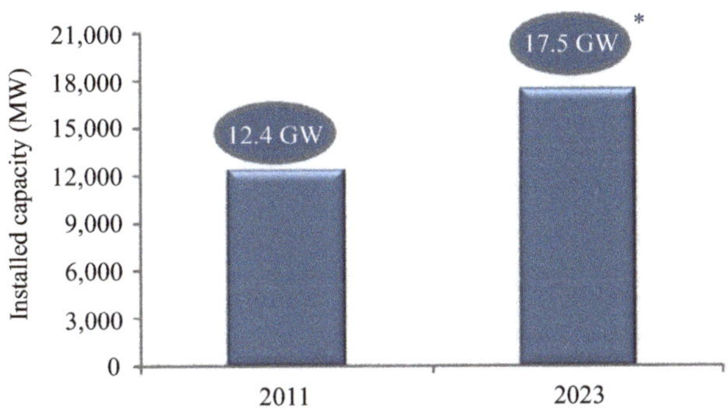

*Figure 15.10 Expansion of coal, 2011 and 2023. *Installed capacity in 2023 is calculated conservatively taking into account only the licenses under construction and approved. The licenses that are under consideration and in applied category have not been included*

local supply market, but there are plans to liberalize the market in the future, to create new opportunities. Although gas represents only 31% of capacity, 48% of electricity was produced by gas-fired plants in 2011 (Figures 15.11 and 15.12). A total of 8,120 MW of licenses have been approved, and a lot of this is under construction.

15.6.4 Hydropower market

In 2011, the installed capacity was 17.1 GW, whereas 15.2 GW of hydropower plants are under construction (Figures 15.13 and 15.14). On supply-side equipment, Turkey has a battle with Chinese manufacturers who have increased their presence in market, particularly in the supply of hydraulic turbines of below 15 MW, whereas for beyond 15 MW, investors tend to work with OEM countries such as Voith Hydro, Alstom and Andritz Hydro.

Challenges facing the market include

- poor coordination among the responsible corporations;
- public protest based on the environmental impacts;
- a lack of control mechanism of the hydropower plants during feasibility, construction and post-commissioning period.

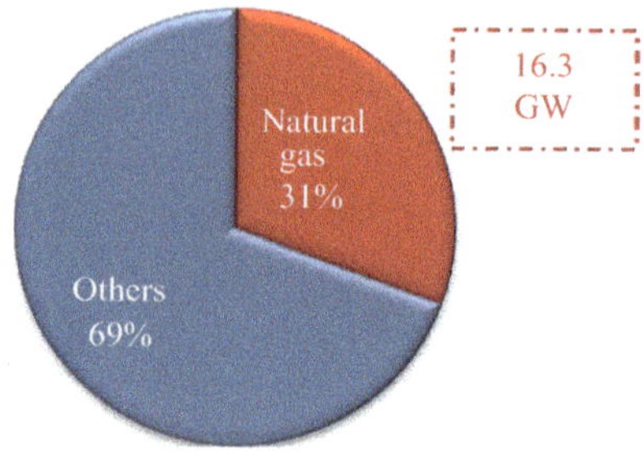

Figure 15.11 Installed capacity by a share of gas, 2011

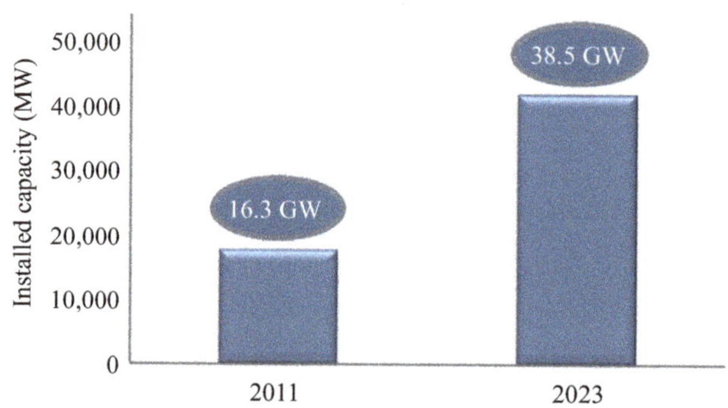

Figure 15.12 Turkey natural gas storage capacity and expansion, 2011 and 2023

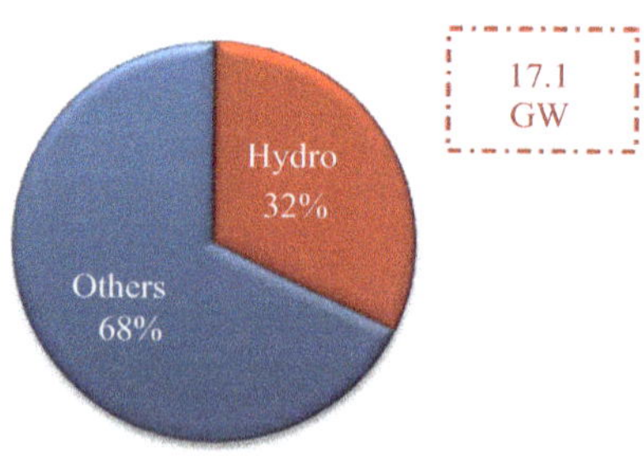

Figure 15.13 Installed capacity by a share of hydropower, 2011

In addition, Turkey has the potential to provide an overall figure of 430 TW h, natural, hydropower, which is about 1.1% of the world and 13.75% of the European potential. However, only about 30% (130 TW h) of this potential is known to be economically feasible. Recent surveys of small hydropower plants assume an additional economic energy potential of 38 TW h/year. About 97% of this economic potential is located in the 14 basins of the 26 water rivers in Turkey. The main rivers of Turkey are shown in Figure 15.15 [57]. Most of them are located in

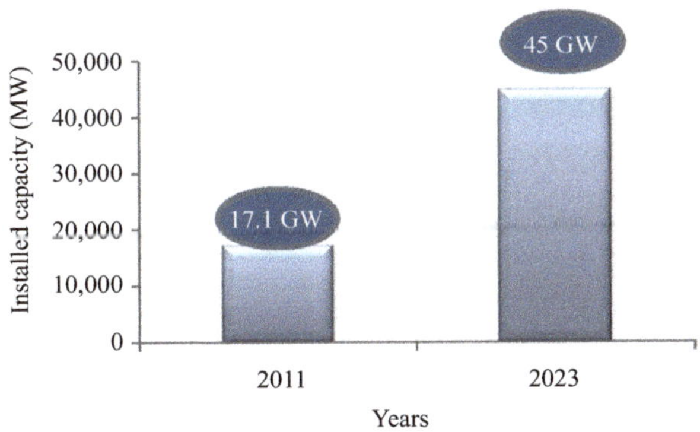

Figure 15.14 Turkey hydropower capacity and expansion, 2011 and 2023

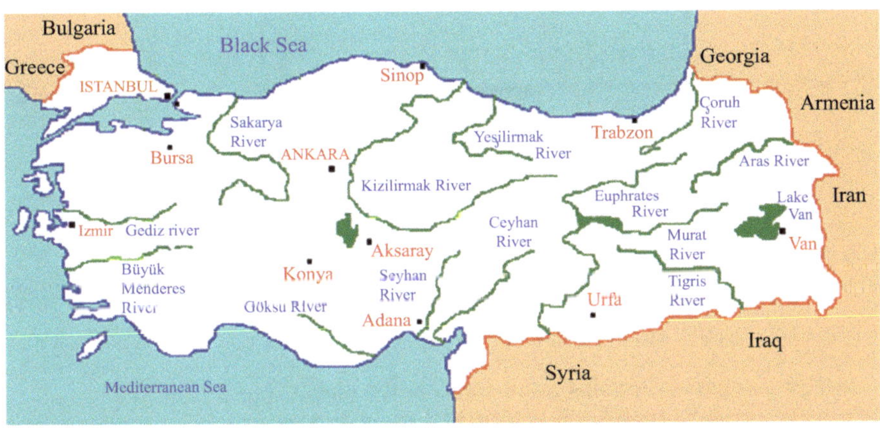

Figure 15.15 Main rivers in Turkey

the eastern regions of Turkey. Basically, the Euphrates and the Tigris River with their various watersheds and greater elevation contribute to the country's abundant potential, allowing large power stations to be built. In addition, small power plants on low-lying rivers and with small drainage areas, mainly located in the western regions, are suitable for generating electricity.

15.6.5 Wind market

As of the end of 2016, EPDK has granted 209 licenses corresponding to 7,479 MW capacity, including the operational and under-construction plants. The Turkish government has set a target of 20-GW wind capacity by 2023 (Figures 15.16 and 15.17) [69].

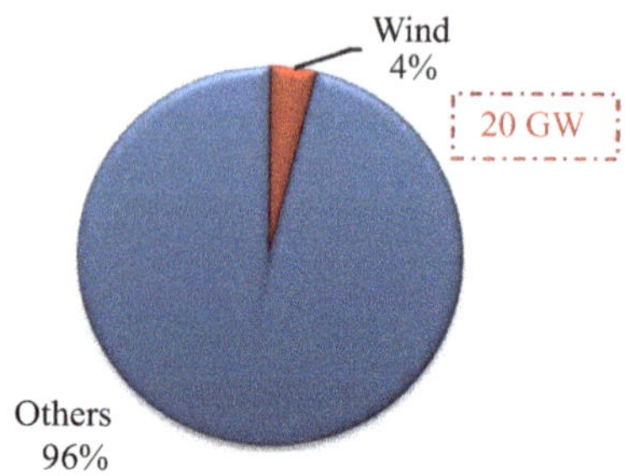

Figure 15.16 Installed capacity by a share of wind, 2011

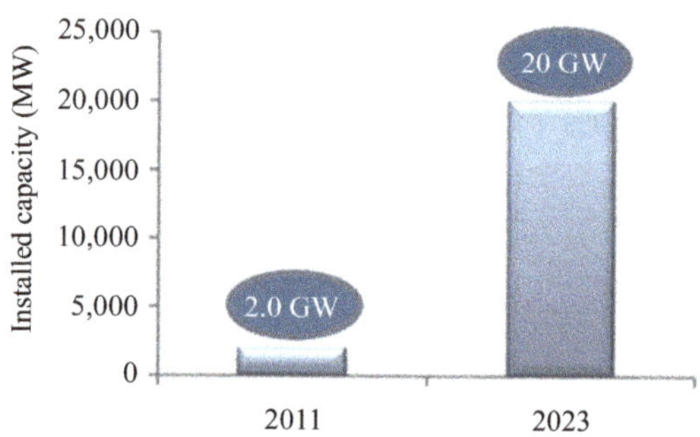

Figure 15.17 Turkey wind power capacity and expansion, 2011 and 2023

However, major challenges faced in the market include

- a high number of permits at the construction stage;
- priority given to exhaustible resources (gas and coal);
- obligation to provide an estimate of electricity output 13 h in advance, which is hard to provide with accuracy considering the intermittency of wind power;
- bureaucratic difficulties during the license period.

15.7 Policy state and progress relevant to smart grid in Turkey

Main objective and principal business activity is to 'Plan, establish, develop and manage energy market within the market operation license in an effective, transparent, reliable manner that fulfils the requirements of energy market and to be an energy market management that procures reliable reference price without discriminating equivalent parties and maximizes the liquidity with increasing number of market participants, product range and trading volume as well as allowing to merchandise by means of market merger'.

Energy Exchange Istanbul (EXIST) is an energy exchange company legally incorporated under the Turkish Electricity Market Law and enforced by the Energy Markets Operation License granted by the Energy Market Regulatory Authority (EMRA) of Turkey. EPİAŞ is responsible for managing and operating energy markets, including power and gas commodities. EPİAŞ ensures transparent, reliable and trustable market conditions as well as equal access for all market participants by providing a counterparty guarantee of the transactions.

Following the settlement calculations and in-line with the participants' intraday balancing practices as well as for every day of a billing period, EPİAŞ – in accordance with Takasbank – will notify participants regarding the daily down payments and the following day's day-ahead market's down-payment statements, which involve amounts either payable to EPİAŞ by participants or vice versa.

Market settlement is performed by the market operator EXIST in a fast, reliable and transparent manner. Shareholder structures are TEIAŞ (TSO) 30%, Borsa İstanbul (BIST) 30%, market participants 40%.

One important strategic document about the SGs policy in Turkey is the Ministry of Energy and Natural Resources Strategic Plan (2010–14). One of the formulated aims within this strategy, e.g., is 'Increasing the share of the RE resources within the energy supply'. But it is not stated in detail, which role SGs has or could have to reach this goal [70].

R&D support programmes related to SGs are the Research/Education Communities Programs and the Technology and Innovation Funding Programs. TUBITAK has a National Research Infrastructure Information System with detailed project information. But no public access to the database is foreseen. The institution mainly dealing with SGs issues is the Turkish Electricity Transmission System Operator (TEIAS) [70].

15.8 Smart grid opportunities in Turkey

Turkey has 21 electricity distribution regions, run by 21 electricity distribution companies, including 13 private companies as illustrated in Figure 15.18. The privately owned distribution companies serve 52%–53% of Turkish customers, corresponding to 55% of electricity consumed, according to the figures from the end of 2016. The Turkish government earned approximately USD 6 billion in revenues from the privatization of 13 regional companies.

On average, 16% of electrical energy in Turkey is lost or stolen (the 'loss and theft ratio'). Uludag DISCO has the lowest loss and theft ratio of approximately 7%. Dicle has the highest ratio of approximately 61%. Turkey needs to introduce smart metering to reduce losses and thefts in distribution, to increase the quality of supply and efficiency and to solve the problems encountered in operating day-ahead and balancing markets. By 2017, over $ 5 billion is due to be invested in SG systems [69].

All distribution companies have targets for dealing with technical and non-technical losses. These targets are set to create incentives to reduce the level of

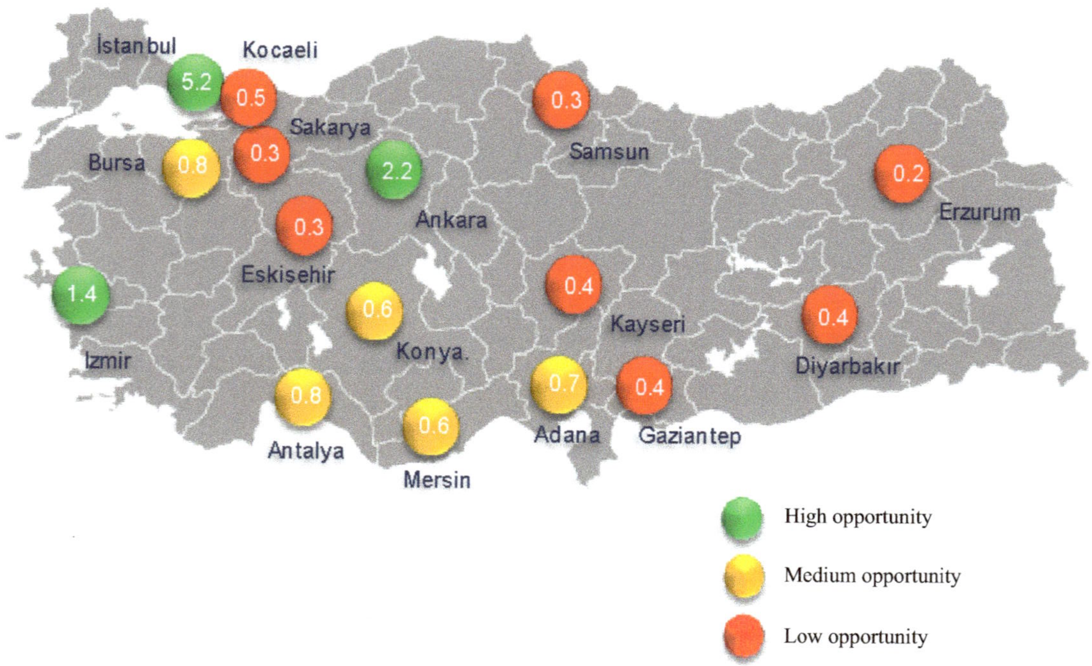

Figure 15.18 Market opportunity overview

losses. If they can reduce the level of losses below the targets, distribution companies can earn higher profits than the regulated profit.

In 2011, overall losses stood at 16%, of which technical losses were represented by 7%–8% and non-technical losses were represented by 8%–9%. The aim is to reduce overall losses to 10% by 2015 [67].

15.9 Energy from waste and biomass market opportunity and partner selection study in Turkey

The European utility wanted to investigate the market opportunity for energy from waste incineration plants and biomass plants in Turkey. It also wanted to identify and profile potential partners that could reinforce its market entry efforts.

The analysis was structured primarily from discussions with key market players, local municipalities, government ministries and private investors. A detailed assessment of the Turkish energy sector was carried out to assess market opportunities and challenges for waste incineration and biomass and position its development in the fuel hierarchy (Figure 15.18).

15.10 Strategic analysis of the wind power services market in Turkey

A leading global supplier of wind turbines has recently entered the market in Turkey. Given its late entrance to the market, the company wanted to gain detailed insights into the market trends and dynamics to ascertain the realistic market opportunity. In addition, their client wanted to understand the operational activities of competitors in order to ramp up its business in an optimal manner through best practices and lessons learned in the market.

Project objectives

- To assess the market opportunity of the wind power plant services market in revenues.
- To forecast the market size to 2017 (with 2012 as the base year).
- To identify key trends, revenues drivers/enablers and restraints and how this will impact the market.
- To identify key competitors and provide detailed profiles on their operational architecture.

Outcome and business impact

The client will use the data to plan its next operational investments in Turkey to build its wind power service networks most effectively incurring the least investments that will have the maximum coverage for its customers. The intelligence will feed into the structured rollout of team build up and the ramp up of its capabilities.

15.11 Strategy of Turkish electricity transmission system operator (TEIAS)

Turkey has two interconnection points with the East European Transmission Grid. The test period for synchronous parallel operation of the Turkish and European power systems started on 1 June 2011 and ended in September 2012. At this moment, the trade is limited to 400 MW from Bulgaria and Greece to Turkey, and 300 MW from Turkey to Europe via these countries.

TEIAS currently has the following actions regarding SG applications:

- one national control centre (Ankara);
- one emergency national control centre (Ankara);
- nine regional control centres;
- more than 200 remote terminal units and
- approximately 12 remotely controlled substations.

The following systems offer the best prospects for an SG in Turkey:

- automated meter reading systems;
- SCADA systems;
- geographical information systems (GIS);
- a wind power management system capable of managing up to 20,000 MW, as required by TEIAS;
- primary and secondary reserve control systems, as required by TEIAS;
- data communications systems and network;
- communication control centres;
- wind energy grid management systems;
- wind generation forecast, wind generation monitoring, static analysis and dynamic analysis systems;
- primary and secondary frequency control, instantaneous demand control (load-shedding: how much, how, automatically) and reactive power control systems;
- a protection control centre;
- solar energy grid management systems.

Opportunities

- R&D programmes and investments undertaken by the government to support SG
- Growth in the RE is expected to modernize the grids
- Research and development trends in SG technologies in Turkey
- Construction of new power plants (thermal and hydro, renewable, small power plants)
- European Union accession process and harmonization studies.

15.12 Key trends in the Turkish market

- The market reform process is progressing with private companies keen to expand their presence in the market.
- Demand for power will continue to rise at approximately 7% per year, necessitating new capacity additions.
- Turkey's fuel mix will diversify, with greater investment in renewables. Coal and gas will remain a key element of the fuel mix.
- Once a functioning bureaucracy is in place, the solar power market is expected to boom fuelled particularly by the 'below 500 kW regulation'. This regulation will unlock a significant amount of investment.
- Power generation is dominated by local, diversified industrial groups. Although several international utilities are partnering with Turkish companies, most projects are run solely by local companies, involved in the construction industry.
- Although new competitors are continuing to emerge, international investors are also keen to expand their presence.
- With most equipment imported from abroad, Turkey is an attractive market for equipment OEMs. The presence of Chinese players is increasing.
- There remains a bottleneck around financing, but this can be overcome as global lending conditions improve.

In order to provide a stable, low-cost, reliable, efficient, robust, sustainable and environment-friendly electrical energy system to consumers, a fully operational SG system needs to be established in Turkey.

An SG system would provide the following advantages and accelerate the full integration of the Turkish grid system with that of Europe:

- increase the quality and efficiency of supply;
- solve problems encountered in running day-ahead and balancing markets;
- enable frequency control by responding to spontaneous energy demand and holding energy in reserve;
- reduce technical losses in the network and thefts;
- enable more responsive load control of the energy transmission line;
- allow power to be cut off and turned on remotely;
- enable grid energy transmission capacity to be increased by reducing losses and controlling the active and reactive energy transmitted;
- help policymakers, transmission system operators and end users to prepare day-ahead plans;
- improve the ease with which the grid can be overseen and controlled;
- increase capacity to host DER;
- increase the ratio of using RE in the grid;
- provide flexibility in demand and supply;
- integrate EV into SGs ensuring that the charging and communication infrastructure works properly rather than testing vehicle-to-grid services;

- enhance the use of storage systems as an additional source of grid and
- improve collaboration between companies, universities and consumers.

If classical grids in Turkey were transformed into SGs, not only would the above-mentioned benefits be achieved, but Turkey would be able to attract a huge amount of investment and to boost its economy. The Turkish grid system would then be a powerful player in the energy market in Europe.

References

[1] K. Bakirci. Models for the estimation of diffuse solar radiation for typical cities in Turkey. Energy 2015;82:827–38.

[2] O. Kisi. Modeling solar radiation of Mediterranean region in Turkey by using fuzzy genetic approach. Energy 2014;64:429–36.

[3] Y. S. Güçlü, M. Ö. Yeleğen, İ. Dabanlı, and E. Şişman. Solar irradiation estimations and comparisons by ANFIS, Angström–Prescott and dependency models. Sol Energy 2014;109:118–24.

[4] Y. S. Güçlü, İ. Dabanlı, E. Şişman, and Z. Şen. HARmonic–LINear (HarLin) model for solar irradiation estimation. Renew Energy 2015;81:209–18.

[5] S. Maharjan, Q. Zhu, Y. Zhang, S. Gjessing, and T. Başar. Demand response management in the smart grid in a large population regime. IEEE Trans Smart Grid 2016;7(1):189–99.

[6] W.-T. Li, C. Yuen, N. Ul. Hassan, *et al.* Demand response management for residential smart grid: from theory to practice. Spec Sect Smart Grids: Hub Interdiscip Res 2015;3:2431–40.

[7] M. Erol-Kantarci, and H. T. Mouftah. Energy-efficient information and communication infrastructures in the smart grid: a survey on interactions and open issues. IEEE Commun Surv Tutorials 2015;17(1):179–96.

[8] E. Spanò, L. Niccolini, S. Di Pascoli, and G. Iannaccone. Last-meter smart grid embedded in an internet-of-things platform. IEEE Trans Smart Grid 2015;6(1):468–75.

[9] A. K. Marnerides, P. Smith, A. Schaeffer-Filho, and M. Andreas. Power consumption profiling using energy time-frequency distributions in smart grids. IEEE Commun Lett 2015;19(1):46–9.

[10] G. Rietveld, J. P. Braun, R. Martin, *et al.* Measurement infrastructure to support the reliable operation of smart electrical grids. IEEE Trans Instrum Meas 2015;64(6):1355–63.

[11] D. Li, and S. K. Jayaweera. Distributed smart-home decision-making in a hierarchical interactive smart grid architecture. IEEE Trans Parallel Distrib Syst 2015;26(1):75–84.

[12] M. Gottschalk, M. Uslar, and C. Delfs. The Use Case and Smart Grid Architecture Model Approach: The IEC 62559-2 Use Case Template and the SGAM Applied in Various Domains. Cham: Springer; 2017.

[13] L. Mariam, M. Basu, and M. F. Conlon. Microgrid: architecture, policy and future trends. Renew Sustain Energy Rev 2016;64:477–89.

[14] A. R. Khan, A. Mahmood, A. Safdar, Z. A. Khan, and N. A. Khan. Load forecasting, dynamic pricing and DSM in smart grid: a review. Renew Sustain Energy Rev 2016;54:1311–22.

[15] C. Wan, J. Zhao, Y. Song, Z. Xu, J. Lin, and Z. Hu. Photovoltaic and solar power forecasting for smart grid energy management. CSEE J Power Energy Syst 2015;1:38–46.

[16] The Climate Group. Current Situation and Prospect of the Development of Smart Grid in China, Beijing, China: The Climate Group; 2016.

[17] ETP. Smart Grids – Strategic Deployment Document for Europe's Electricity Networks of the Future; 2016.

[18] U. Rohjans, M. Uslar, R. Bleiker, *et al.* Survey of Smart Grid Standardization Studies and Recommendations. In: Proceeding of IEEE conference on smart grid communications. Germany: IEEE; 2010.

[19] M. G. Kanabar, I. Voloh, and D. McGinn. Reviewing smart grid standards for protection, control, and monitoring applications. In: 2012 IEEE PES Innovative Smart Grid Technologies (ISGT). Washington, DC: IEEE; 2012; 1–8.

[20] SMB Smart Grid Strategic Group (SG3). IEC Smart Grid Standardization Roadmap; 2016.

[21] G. X. Lin. An Overview on the Development of Smart Grid and Distributed Energy, vol. 1. Beijing, China: China High-Tech Enterprises; 2014; 141–2.

[22] G. Brando, A. Dannier, A. Del Pizzo, L. P. Di Noia, and C. Pisani. Grid connection of wave energy converter in heaving mode operation by supercapacitor storage technology. IET Renew Power Gen 2016;10:88–97.

[23] T. T. Nguyen, H. J. Yoo, and H. M. Kim. Applying model predictive control to SMES system in microgrids for eddy current losses reduction. IEEE Trans Appl Supercond 2016;26(4):1–5.

[24] A. R. Khan, A. Mahmood, A. Safdar, Z. A. Khan, and N. A. Khan. Load forecasting, dynamic pricing and DSM in smart grid: a review. Renew Sustain Energy Rev 2016;54:1311–22.

[25] L. X. Meng, E. R. Sanseverino, A. Luna, T. Dragicevic, J. C. Vasquez, and J. M. Guerrero. Microgrid supervisory controllers and energy management systems: a literature review. Renew Sustain Energy Rev 2016;60:1263–73.

[26] A. Kaur, J. Kaushal, and P. Basak. A review on microgrid central controller. Renew Sustain Energy Rev 2016;55:338–45.

[27] K. M. Tan, V. K. Ramachandaramurthy, and J. Y. Yong. Integration of electric vehicles in smart grid: a review on vehicle to grid technologies and optimization techniques. Renew Sustain Energy Rev 2016;53:720–32.

[28] M. Stadler, G. Cardoso, S. Mashayekh, *et al.* Value streams in microgrids: a literature review. Appl Energy 2016;162:980–9.

[29] Z. K. Shuai, Y. Y. Sun, Z. J. Shen, *et al.* Microgrid stability: classification and a review. Renew Sustain Energy Rev 2016;58:167–79.

[30] J.-F. Martínez, J. Rodríguez-Molina, P. Castillejo, and R. De Diego. Middleware architectures for the smart grid: survey and challenges in the foreseeable future. Energies 2013;6:3593–621.

[31] M. H. Cintuglu, A. T. Elsayed, and O. A. Mohammed. Microgrid Automation Assisted by Synchrophasors. In: Innovative smart grid technologies conference (ISGT). New York: IEEE Power and Energy Society; 2015.

[32] D. M. Laverty, D. J. Morrow, R. Best, and P. A. Crossley. Telecommunications for Smart Grid: Backhaul Solutions for The Distribution Network. In: IEEE PES General Meeting. Providence, RI: IEEE; 2010; 1–6.

[33] P. Owen, Powering the Nation: Household Electricity Using Habits Revealed, London: Energy Saving Trust; 2012.

[34] P. P. Parikh, M. G. Kanabar, and T. S. Sidhu. Opportunities and challenges of wireless communication technologies for smart grid applications. In: IEEE PES General Meeting. Providence, RI: IEEE; 2010; 1–7.

[35] A. Zaballos, A. Vallejo, and J. M. Selga. Heterogeneous communication architecture for the smart grid. IEEE Network 2011;25(5):30–7.

[36] P. Bertoldi, B. Hirl, and N. Labanca. Energy Efficiency Status Report on Electricity Consumption and Efficiency Trends in the European Union. Status Report 2012. Luxembourg: European Commission, Joint Research Centre, Institute for Energy; 2012. Available from: http://iet.jrc.ec.europa.eu/energyefficiency/sites/energyefficiency/files/energy-efficiencystatus-report-2012.pdf.

[37] P. Owen, and R. Foreman. Powering the Nation: Household electricity using habits revealed. London: Energy Saving Trust/DECC/DEFRA; 2012. https://energysavingtrust.org.uk/file/17511/download?token=QRfEXv2r.

[38] J. Froehlich, E. Larson, S. Gupta, G. Cohn, M. Reynolds, and S. Patel. Disaggregated end-use energy sensing for the smart grid. IEEE Pervasive Comput 2011;10(1):28–39.

[39] A. Hahn, A. Ashok, S. Sridhar, and M. Govindarasu. Cyber-physical security testbeds: architecture, application, and evaluation for smart grid. IEEE Trans Smart Grid 2013;4(2):847–55.

[40] N. Komninos, E. Philippou, and A. Pitsillides. Survey in smart grid and smart home security: issues, challenges and countermeasures. IEEE Commun Surv Tutorials 2014;16(4):1933–54.

[41] NIST. NIST Framework and Roadmap for Smart Grid Inter-Operability Standards, Release 1.0, Special Publication 1108R2, National Institute of Standards and Technology Std.; 2012.

[42] NIST. Preliminary Discussion Draft – NIST Framework and Roadmap for Smart Grid Inter-Operability Standards, Release 3.0, Special Publication 1108R3, National Institute of Standards and Technology Std.; 2014.

[43] Y. Gobena, A. Durai, M. Birkner, V. Pothamsetty, and V. Varakantam. Practical architecture considerations for Smart Grid WAN network. In: 2011 IEEE/PES Power Systems Conference and Exposition. Phoenix, AZ: IEEE; 2011; 1–6.

[44] IEEE. IEEE Standard Communication Delivery Time Performance Requirements for Electric Power Substation Automation, IEEE 1646-2004 Std.; 2011.

[45] J. G. Deshpande, E. Kim, and M. Thottan. Differentiated services QoS in smart grid communication networks. Bell Labs Tech J, 2011;16(3):61–81.

[46] Q. D. Ho, Y. Gao, and T. Le-Ngoc. Challenges and research opportunities in wireless communications networks for smart grid. IEEE Wireless Commun 2013;20(3):89–95.

[47] Q. D. Ho, and T. Le-Ngoc. Smart grid communications networks: wireless technologies, protocols, issues and standards, Chapter 5. In: Mohammad S. O., Alagan A., Isaac W. (eds.) Handbook on Green Information and Communication Systems, Amsterdam, The Netherlands: Elsevier; Summer 2012.

[48] M. Hildebrandt. Legal Protection by Design in the Smart Grid. Privacy, Data Protection and Profile Transparency, Arnhem: Smart Energy Collective; 2013. http://pilab.nl/wp-content/ uploads/2013/05/KEM-64P707-BRO-LPbD-in-SmartGrid_A4_FC_v4.pdf.

[49] C. Cuijpers, and B.-J. Koops. Smart metering and privacy in Europe: lessons from the Dutch case. In: Gutwirth S., Leenes R., De Hert P., Poullet Y. (eds.) European Data Protection: Coming of Age; 2013; 269–293. doi: 10.1007/ 978-94-007-5170-5_12.

[50] I. Brown. Britain's smart meter programme: a case study in privacy by design. Int Rev Law Comput Technol 2013;28(2):172–84. doi: 10.1080/ 13600869.2013.801580.

[51] A. Ashok, M. Govindarasu, and V. Ajjarapu. Online detection of stealthy false data injection attacks in power system state estimation. IEEE Trans Smart Grid 2016;9(3):1636–46.

[52] Y. Chakhchoukh, and H. Ishii. Enhancing robustness to cyber-attacks in power systems through multiple least trimmed squares state estimations. IEEE Trans Power Syst 2016;31(6):4395–405.

[53] S. Li, Y. Yılmaz, and X. Wang. Quickest detection of false data injection attack in wide-area smart grids. IEEE Trans Smart Grid 2015;6(6):2725–35.

[54] S. Li, Y. Yilmaz, and X. Wang. Sequential Cyber-Attack Detection in the Large-Scale Smart Grid System. In: 2015 IEEE international conference on smart grid communications; 2015; 127–32.

[55] N. G. Bretas, and A. S. Bretas. A two steps procedure in state estimation gross error detection, identification, and correction. Int J Electr Power Energy Syst 2015;73:484–90.

[56] Y. Zhang, J. Wang, F. Hu, and Y. Wang. Comparison of evaluation standards for green building in China, Britain, United States. Renew Sustain Energy Rev 2017;68:262–71. doi:10.1016/j.rser.2016.09.139. http://linkinghub.elsevier.com/retrieve/pii/S1364032116306499.

[57] V. Gazis. A survey of standards for machine-to-machine and the internet of things, IEEE Commun Surv Tutorials 2017;19(1):482–511. doi:10.1109/ COMST.2016.2592948. http://ieeexplore.ieee.org/document/7516570/

[58] W. Han, and Y. Xiao, Non-Technical Loss Fraud in Advanced Metering Infrastructure in Smart Grid, Cham: Springer; 2016; 163–72. doi:10.1007/ 978-3-319-48674-1_15.

[59] I. Colak, R. Bayindir, G. Fulli, I. Tekin, K. Demirtas, and C. F. Covrig. Smart grid opportunities and applications in Turkey. Renew Sustain Energy Rev 2014;33:344–52.

[60] I. Colak, G. Fulli, S. Sagiroglu, M. Yesilbudak, and C. F. Covrig. Smart grid projects in Europe: current status, maturity and future scenarios. Applied Energy 2015;152:58–70.

[61] H. Nazaripouya, Y. Wang, P. Chu, H. R. Pota, and R. Gadh. Optimal sizing and placement of battery energy storage in distribution system based on solar size for voltage regulation. In: 2015 IEEE Power and Energy Society General Meeting. Denver, CO: IEEE; 2015; 1–5.

[62] H. Nazaripouya, and S. Mehraeen. Modeling and nonlinear optimal control of weak/islanded grids using FACTS device in a game theoretic approach. IEEE Trans Control Syst Technol 2016;24(1):158–71.

[63] B. Khaki, B. Ranjbar-Sahraei, N. Noroozi, and A. Seifi. Interval type-2 fuzzy finite-time control approach for chaotic oscillation damping of power systems. Int J Innovative Comput Inf Control 2011;7(12), 6827–35.

[64] A. Ozdemir, J. Y. Lim, and C. Singh. Post-outage reactive power flow calculations by genetic algorithms: constrained optimization approach. IEEE Trans Power Syst 2005;20(3):1266–72.

[65] M. Yilmaz, and F. Kentli. Increasing of electrical energy with solar tracking system at the region which has Turkey's most solar energy potential. J Clean Energy Technol 2015;3:287–91.

[66] M. Yilmaz. The Prediction of Electrical Vehicles' Growth Rate and Management of Electrical Energy Demand in Turkey. In: Green technologies conference (GreenTech), 2017 ninth annual IEEE. IEEE; 2017; 118–23.

[67] S. Serter, B. Gümüş, M. Ünlü, *et al.* Prevalence of testicular microlithiasis in an asymptomatic population. Scand J Urol Nephrol 2006;40(3):212–14.

[68] A. Hepbasli. Development and restructuring of Turkey's electricity sector: a review. Renew Sustain Energy Rev 2005;9:311–43.

[69] Turkish Electricity Transmission Co. (TEIAS). Short History of Electrical Energy Development in Turkey and Some Statistical Figures. Ankara, Turkey; 2019. Available from: https://www.eigm.gov.tr/tr-TR/Istatistik-Raporlari.

[70] Republic of Turkey Ministry of Energy and Natural Resources. Ankara, Turkey; 2019. https://www.enerji.gov.tr/en-US/Pages/Electricity.

Chapter 16

Nanogrids: good practices and challenges in the projects in Colombia

Maria V. Vives[1], Harold R. Chamorro[2], Diego Ortiz-Villalba[3], Fernando Jiménez[1], Francisco M. Gonzalez-Longatt[4], Guillermo Jimenez-Estevez[5], Josep Guerrero[6], Angela Cadena[1] and Vijay K. Sood[7]

The Republic of Colombia is a country situated in the northwest of South America, with territories in Central America. In 2005, the interconnected electricity system served 87% of the population, a percentage that is below the 95% average for Latin America and the Caribbean. The Colombian government is making massive efforts to increase the electrification, especially at very remote and rural places. However, there are still several rural communities isolated from the main grid in Colombia. Also, although some communities have already been connected to the distribution system, the security and reliability are still an issue.

One key element in the effort of electrification is the use of very small microgrids projects called "nanogrids."

A nanogrid is a small power system that uses a combination of renewable and conventional energy sources to supply power to small local loads. The total load in a nanogrid is typically less than 20 kW, as an industrial site, small rural village. or a household. The generators are primarily based on clean energy such as solar arrays, wind turbines, and fuel cells. In Colombia, the Caribbean coastline, Andes Mountains, and strong agriculture provide the country with abundant distributed renewable energy resources that might largely surpass the fast-growing electricity demand. In this framework, the nanogrids emerge as a solution to supply energy for

[1]Universidad de los Andes, Bogotá, Colombia
[2]KTH Royal Institute of Technology, Stockholm, Sweden
[3]Universidad de Chile—Universidad de las Fuerzas Armadas ESPE, Santiago, Chile
[4]Centre for Renewable Energy Systems Technology (CREST), Loughborough University, Loughborough, UK
[5]Centro de Energía Universidad de Chile, Santiago, Chile
[6]Aalborg University, Aalborg, Denmark
[7]University of Ontario, Ontario, Canada

some communities, located far away from the network, improving the inhabitant′s quality of life. Moreover, the nanogrids might improve the security and reliability levels of the distribution system.

This chapter discusses good practices and proposed some solutions to overcome challenges detected in nanogrid projects developed in Colombia. In essence, this chapter is a case study of nanogrid systems; it focuses on a full detailed explanation of 23 nanogrid projects developed in Colombia considering location, installed power, purpose, etc. Then the nanogrid projects are evaluated, and it gives the opportunity to identify some characteristics that hinder or benefit the operation of these systems. Some of them are the growth of the demand, the appropriate and inadequate use of energy and drinking water, the changes of habits in the users in the presence of energy or the increase in the reliability of the energy supply and how it affects the maintenance or lack this, the operation of the systems.

16.1 Introduction

Rural electrification programs have attempted to solve the problem of providing access to electricity to communities that, due to their geographical characteristics, have remained excluded from electrification programs. However, there are still several rural communities isolated from the main grid in Colombia. In addition, although some communities have already been connected to the distribution system the security and reliability are still an issue.

On the other hand, a nanogrid is a small power system that uses a combination of renewable and conventional energy sources to supply power to small local loads. The total load in a nanogrid is typically less than 20 kW, as industrial site, small rural village, or a household. The generators are primarily based on clean energy such as solar arrays, wind turbines (WTs), and fuel cells. In Colombia, the Caribbean coastline, Andes Mountains, and strong agriculture provide the country with abundant distributed renewable energy resources that might largely surpass the fast-growing electricity demand. In this framework, the nanogrids emerge as a solution to energy supply for some communities, located far away from the network, improving the inhabitant's quality of life. Moreover, the nanogrids might improve the security and reliability levels of the distribution system.

This chapter discusses good practices and proposed some solutions to overcome challenges detected in some nanogrid projects developed in Colombia.

16.2 Literature review

Nanogrids are a sustainable option for the distributed generation interconnection in rural areas where the load power consumption may not be reachable by the distribution system network [1]. Approximately 1.4 billion people in the world currently lack access to electricity. Furthermore, 85% of the people without access to electricity live in rural areas [2].

Current governmental programs are attacking the power supply by increasing the use of photovoltaics (PV). However, local communities are socially impacted. Some important efforts are presented below, remarking that many countries around the world have adopted this technology while trying to solve the quality access to electricity for people in remote areas.

Jha *et al.* [1] show a demonstration pilot for rural electrification in Nepal. There are different viable options of forming rural microgrids by the interconnection of distributed energy resources such as micro-hydropower plant (MHP), WT, and PV systems. It discusses the socioeconomic challenges for the sustainability of the rural microgrid.

The SPEED consortium is making a modeling effort to capture the renewable energy resources available in rural areas in India [2]. In Graber *et al.* [3], it is assessed consumer valuation of different attributes of electricity supply to elucidate the conflict between solar microgrids and the centralized utility grid, as well as to provide insight into supporting government policies and structures. The study contributes significantly to the understanding of the role of microgrids in complementing a centralized system and its value as a sustainable energy solution for development.

In Brazil, the problem is addressed by considering a set of farmers located in a remote location, so local wind sources are considered to generate electric power showing its feasibility [4].

Chile has been taking an approach for implementing microgrid projects at the institutional level by means of a specific entity that uses methods that engage the community in microgrid operation and maintenance (O&M), which ensures long-run benefits. The first step, related to macro-level barriers, is addressed by building a complete cadaster of isolated communities, while the second, at the micro level, focuses on business models for covering investment and O&M costs [5].

Regarding the question of the future of DC rural microgrid, specifically solar DC microgrids [6] versus the usual AC framework, the cost of energy can be significantly lower in the case of a DC grid when considering the socioeconomic condition of rural areas as in the case of Bangladesh [7].

On the other hand, the feasibility of microgrid operation as a potential solution to improve the power supply reliability and grid resiliency for remote rural areas is recommended in the USA [8].

A stand-alone farm in Spain is based on intensive use of information to maximize the penetration of renewable energy through two main strategies: anticipation and opportunity. The loads are classified into four different types according to priority and are controlled by intelligent sockets interconnected in a mesh network. Each socket measures the main electric parameters, communicates to the local computer bidirectionality by using ZigBee communications, and is able to switch on/off the loads. After 5 years of operation, the results have shown the importance of real-time management system to reduce the operational costs of the microgrid [9].

In Pearce [10], it is shown an innovative solar company that is using microgrids to deliver power to villages deep in the African bush in Kenya. There small-scale microgrids are increasingly seen as the most promising way to bring electricity to people who currently lack it [11]. As expected South Africa's rural microgrids using combinations of wind, solar PV, and biogas combustion are technologically feasible, but they will require subsidization from government or nongovernmental organization (NGO) sources to be economically viable [12,13].

In [14], it is presented the Pegasus project. In this project, 10 partners from Mediterranean countries are working together to study microgrids in more detail, focusing on rural and island areas.

The development of microgrid on the basis of renewable energy sources in the Central European part of Russia not connected to centralized electrical network could provide comfortable level of the population living even that in Russia the market of renewables is extremely small [15].

Colombia has been facing several electricity challenges and several social problems have raised. One of the biggest problems in the Colombian distribution systems is the partial absence of infrastructure in certain areas which causes an economical–social imbalance. Therefore, the inclusion of renewable energy sources can potentially help with the electricity supply and contribute to the local communities involved. Gaona *et al.* [16] present a review about microgrids around the world, particularly analyzing cases installed in rural areas as a solution to the energy access problem in isolated areas. It describes and analyzes two main problems affecting the electric system in Colombia: the coverage of electric infrastructure and the energy supply due to El Niño phenomenon. The chapter includes the solutions implemented by the national government to install microgrids. Finally, the subsidies offered in Colombia and other countries are shown, giving a comprehensive description of the productive potential of renewable energy, the current legislation and regulations in the country for FNCE promotion (Non-Conventional Energy Sources), and the incorporation to the SIN (National Interconnected System), proposing that there should be focus on training, research, and implementation of microgrids in order to allow their deployment in Colombia.

Private and academia sectors are increasingly contributing worldwide to the development of this technology [17–19].

To empower the applications in the real world, a methodology for planning and design rural nanogrids is then required [15,20]; advanced energy management systems should be strongly considered [21–24]; and innovative solutions should be supported [25–28].

The interconnection of multiple micro/nanogrids leads to several benefits at both a local and global level such as reliability and resiliency, independency of the utility grid, and reduction of the global costs. In order to achieve these objectives, control strategies for the energy trading among microgrids should continue to be developed. At a high level, the networked microgrids problem is understood as the coupling of two problems: optimal management of energy resources among microgrids and trading promotion in the network. The former problem is solved through a resource allocation strategy that seeks to optimize local and global objectives such as operation costs, while the latter is solved through the design of incentive mechanisms that are able to balance the economic interests of the micro/nanogrids in the network [29].

Finally, economic development has been strongly correlated with increasing energy use and growth of greenhouse gas emissions. Increasing the use of renewable energy through nanogrids technology can help decouple that correlation. It offers the opportunity to improve access to modern energy services for the poorest members of society, which is crucial for the achievement of sustainable development goals. Nevertheless, knowledge regarding the nanogrids technologies is still very limited. Finding answers to the question of how to achieve effective, economically efficient, and socially acceptable transformations of the energy system will require a much closer integration of insights from social, natural, and economic sciences (e.g., through risk analysis approaches) in order to reflect the different dimensions of sustainability [30].

16.3 Nanogrids projects in Colombia

Colombia is making an approach to energy microgrids projects through small projects that we call nanogrids. These nanogrids will show the behavior of users in the presence of energy for the solution of daily needs, the challenges in design and operation, and the social changes that they have. The analysis of nanogrids might help predict behavior in larger networks in sectors with similar characteristics. The projects that will be analyzed are located on the north coast of Colombia as shown in the map of Figure 16.1. This zone presents energy problems due to lack of coverage, low reliability, and high energy prices.

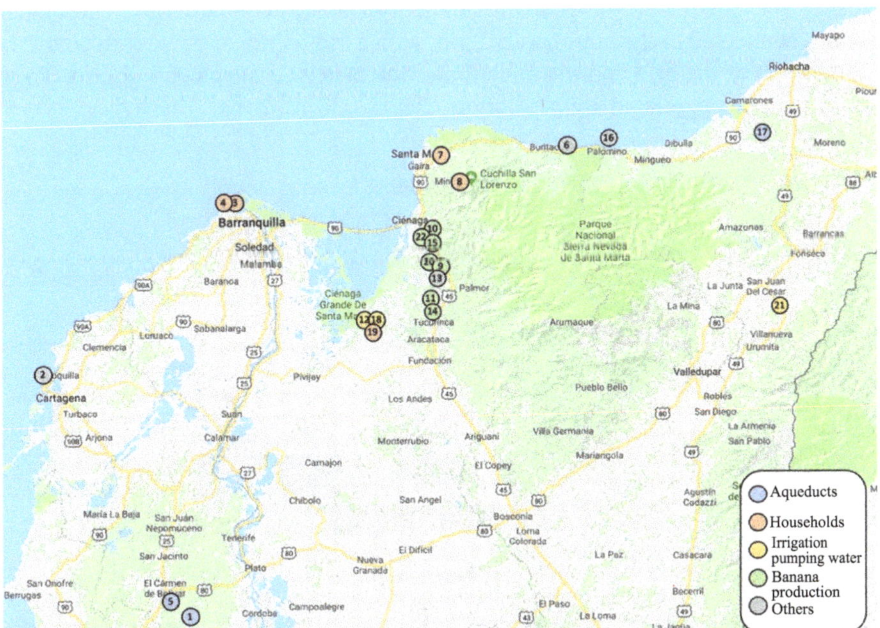

Figure 16.1 Nanogrids map

The nanogrid had been designed in four ways depending on the necessity: on-grid system, off-grid system, hybrid system, and solar water pumping.

16.3.1 On-grid system

The systems supply electric power to households, commercial establishments, farms, among others. For this case, they are connected to the commercial electrical grid. The PV cells (solar panels) generate energy that is injected to the loads directly and the surpluses are injected into the electrical grid. These systems do not have battery banks because they are designed to save a percentage of daily consumption. The interconnected systems in Colombia are regulated by resolution CREG 030 of 2018 and it is mandatory to have a bidirectional counter to count the grid consumption and the injection of the panels to the electricity grid (Figure 16.2).

16.3.2 Off-grid system

Off-grid systems are designed with a battery bank. These systems generate energy in the light hours that will be consumed throughout the day. The system operates autonomously during day time with solar energy and at night works with the energy stored in the battery bank. These systems are commonly used for households, country houses, commercial establishments, farms, among others (Figure 16.3).

16.3.3 Hybrid system

The systems supply electric power through a system of self-generation of energy that uses PV cells (solar panels) and an energy backup with batteries. The system operates connected to the commercial grid, with a bidirectional counter to count the consumption of the grid and the injection of the panels to the electrical grid. These connections are regulated by resolution CREG 030 of 2018. The hybrid system

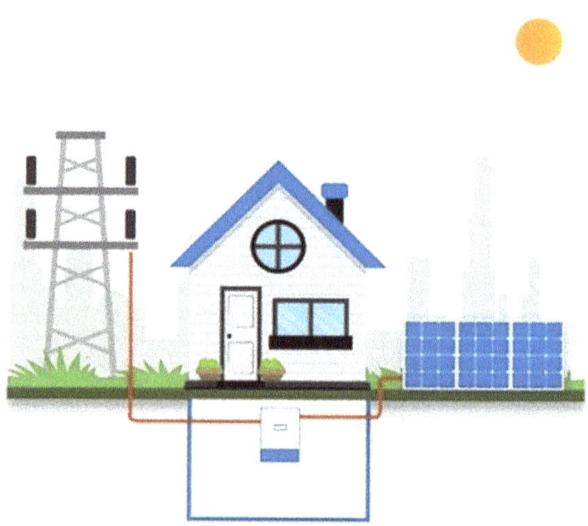

Figure 16.2 On-grid system

Figure 16.3 Off-grid system

Figure 16.4 Hybrid system

injects the energy from the solar panels to the loads and the surplus saves it in batteries or injects it directly into the grid. In the night hours it uses power from the electric grid, and in case if it fails it uses the backup of the batteries. These systems are mostly used by places with low energy quality such as the Banana Zone (Figure 16.4).

16.3.4 Solar water pumping

Water pumping systems are mainly used for livestock irrigation, crop irrigation, households, and rural cabins located at sites far from water sources. The proposed systems are automatic and configurable to the needs of each client, and they can be installed with solar DC pumps or conventional AC pumps powered by solar energy (Figure 16.5).

Some of the installed projects are connected to the grid as a solution to power cuts and to increase the reliability of the systems. In these projects, on-grid systems or hybrid systems, a battery bank is required to ensure the continuous flow of energy in the loads. In these projects, the battery banks are in 12, 24, and 48 V depending on the project. The diagram of these projects is shown in Figure 16.6.

Other projects are completely disconnected to the network due to lack of coverage, high costs of connection to the national electricity grid, or distrust of the quality of the region's electricity. It is necessary to have a battery bank to ensure

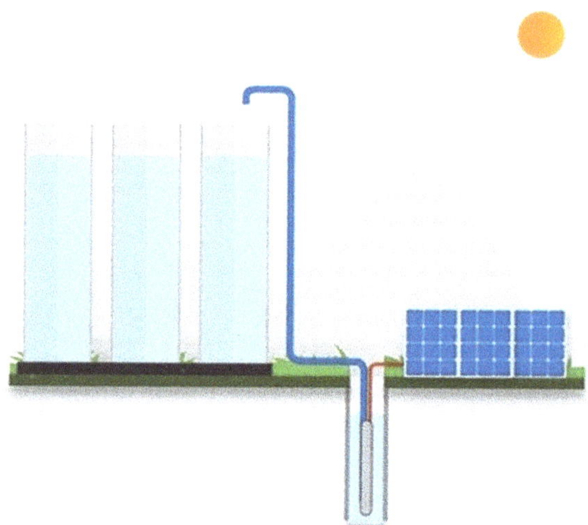

Figure 16.5 Solar water pumping system

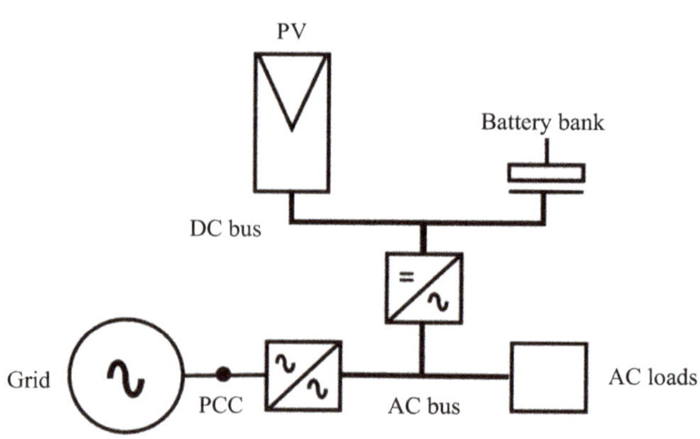

Figure 16.6 Nanogrid connected to the grid

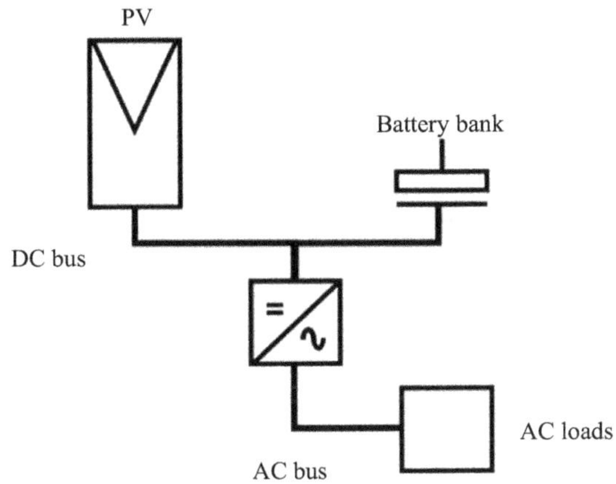

Figure 16.7 Nanogrid disconnected of the grid

constant flow. These are configured in 12, 24, and 48 V depending on the project. The diagram of these projects is shown in Figure 16.7.

The projects to be analyzed are mainly divided into aqueducts, homes, pumping water for irrigation, solutions for the banana sector, and others. Table 16.1 summarizes the details of nanogrid projects in Colombia to be analyzed.

16.3.5 Aqueducts

Projects 1 (Figure 16.8), 5, and 17 belong to aqueducts projects. Project 1, Pumping Aqueduct El Salado, is a project connected to the network that energizes the deep well pump and a chlorination system for the purification of water for the consumption. This aqueduct was powered by electricity, and due to the high energy costs and the intermittent availability of energy the community pumped the water every third day. With the solar energy system, the community can pump the necessary water daily. Project 5 has two pumps for connecting a rainwater collection system to an elevated tank, which by gravity provides water to the toilets of two schools. Project 17 is disconnected from the network, with a battery bank, and there is no longer any electricity network coverage in the pumping sector and the cost of extending the network is high. This project is for the pumping of wastewater from the population of Matita, Guajira in Colombia. In the case of aqueducts, the solar energy projects have a specific destination and direct connection to the pumping systems. In these cases, the user has no control over the use of energy, so these projects have been successful in their operation. However, in the case of project 1, not everyone in the community has been responsible for the use of water.

16.3.6 Household

Within the projects for households or occasional (holiday) homes, we have projects connected and disconnected from the electricity grid. Projects 3 (Figure 16.9) and 8

Table 16.1. Nanogrids in Colombia

No.	Project name	Description	Place	Installed capacity (kWp)	Commissioning date
1	Pumping Aqueduct El Salado	Pumping of 150 000 l of water per day for the aqueduct of El Salado, Bolívar from a well of 48 m deep. This project makes the pumping taking advantage of the hours of the day and has a battery backup to pump up to 2 h at night if necessary.	El Salado, Bolívar	15.0	2/11/16
2	Solar energy for Yacht	Solar energy and energy backup for yacht in Cartagena. This allows the yacht to have an independent generating energy with solar panels and backup for night hours.	Cartagena, Bolívar	0.50	23/06/16
3	House in Puerto Colombia	Hybrid solar project of 4.3 kWp. Energy is generated with solar panels and has energy backup (backup) for housing with air conditioning on the Colombian coast (Puerto Colombia, Atlántico).	Puerto Colombia, Atlántico	4.29	27/09/16
4	House outside Network	"The house that does not worry about the blackout." Power generation project with solar panels and energy backup for a house 100% disconnected from the national electricity grid. This project is a house located in Puerto Colombia, Atlántico.	Puerto Colombia, Atlántico	2.0	1/02/16
5	Pumping water for schools	Pumping of water for schools in Santa Clara and El Bálsamo, department of Bolívar. With the help of Fundación Semana, the community built rainwater storage systems to supply schools even in times of drought. The pumping of water from these schools is done by a solar pump without batteries.	Santa Clara y El Bálsamo, Bolívar	0.6	13/01/17

(Continues)

6	Ecotourism hotel	Solar energy for an ecotourism hotel disconnected 100% of the national electricity grid. The hotel is located on the banks of the Don Diego River just outside of Santa Marta, Magdalena. The system with solar panels supplies fans, lights, cell phone chargers, among others.	Don Diego, Santa Marta, Magdalena	1.0	16/10/17
7	Country house in Tigrera	Recreation house 100% disconnected from the network of Tigrera, Sierra Nevada de Santa Marta. The house has a solar energy system with PV solar panels that supply the energy of the refrigerator, ceiling fans, lights, and other equipment.	Tigrera, Sierra Nevada de Santa Marta	0.54	1/05/17
8	Country house in the Sierra Nevada	House for guests in the dives of the Sierra Nevada de Santa Marta. This is at 1 920 m above sea level. The solar project is for energy generation and support for the lighting of the holiday home.	Sierra Nevada de Santa Marta	0.54	15/09/17
9	Offices for agriculture	Hybrid solar project for power generation and backup for offices and packing plant of a banana farm. The system of panels and batteries provides energy equipment such as computers, electronic scales, digital recording cameras, and saving type lights.	Sevilla, Zona Bananera	1.08	11/07/17
10	Banana packing	Project of generation of solar energy and energy backup for equipment, luminaires, and security cameras of a banana packing plant in Zona Bananera, Magdalena. The generation is approximately 170 kWh monthly.	Orihueca, Zona Bananera	1.08	11/07/17

(Continues)

Table 16.1. (Continued)

No.	Project name	Description	Place	Installed capacity (kWp)	Commissioning date
11	Banana packer outside of Network, Tucurinca	Solar energy project, disconnected from the grid, for generation and energy backup for the processing and packing of bananas in a farm located in Tucurinca, Magdalena. The monthly electricity consumption of this farm is approximately 1,000 kWh per month.	Tucurinca, Magdalena	2.70	8/08/17
12	Water pumping Palma farm	Pumping 100% solar water where 279 m^3 of water is obtained per day for the irrigation of Finca de Palma in El Retén, Magdalena by installing 11 kWp in PV panels. Well 50 m deep.	El Retén, Magdalena	10.80	4/12/17
13	Generation and backup for CDI	Energy generation and backup for a centro de desarrolloinfantil (CDI) in the village of Sevilla, Magdalena. This solar energy system provides electrical systems such as fans, lights, and others.	Sevilla, Zona Bananera	3.24	22/11/17
14	Banana packing expansion	Extension of generation project for a farm disconnected from the network. The expansion doubles the capacity of the system panels, thus generating more energy. This expansion will generate additional energy for new equipment installed such as conveyor belts, air conditioners, pumps, among others.	Tucurinca, Magdalena	2.93	23/11/17
15	Security system for banana farm	Power generation for security camera system for banana farm. The system generates energy for 25 security cameras and complementary equipment. It is a hybrid system that generates energy for 24 h of operation and has an energy backup of at least 14 h.	Río Frio, Zona Bananera	0.8	3/04/18

(Continues)

16	Hostel in Palomino	Generation and energy backup for a hostel in Palomino, La Guajira. The system generates energy with solar panels for lights, fans, a refrigerator, and a motor pump. It has an energy backup of batteries for approximately 22 h of all charges.	Palomino, Guajira	0.5	15/03/18
17	Wastewater pumping system	Pumping of wastewater with solar energy for Matitas, La Guajira township. This system has emergency energy backup with batteries which allows working in a time of little or no solar radiation.	Matita, Guajira	2.4	3/04/18
18	Expansion of water pumping	Expansion project of a solar pumping system. This project initially pumped approximately 279 m^3 of water per day, and by expanding and changing the pumping system it is possible to pump 864 m^3/day. This is achieved with a maximum flow rate of 35 l/s. Palm irrigation project.	El Retén, Magdalena	0	23/04/18
19	Solar energy for housing	System of solar panels for a small dwelling house of the guard of a palm farm for lighting, cellular and television chargers near El Retén, Magdalena.	El Retén, Magdalena	0.54	22/05/18
20	Baler, Office and housing in Zona Bananera	Power generation system with solar panels for a banana farm. The solar hybrid system includes the generation of energy for the banana packing house, the farm office, and the family housing of the estate caretaker.	Orihueca, Zona Bananera	2.43	22/05/18
21	Water pumping for livestock farm	Water pumping system for livestock farm without electric power. The 100% solar pump with 15.8 kWp of solar panels pumps 840 m^3 of water per day from a deep well to irrigate 10 hectares for livestock grazing.	San Juan del Cesar, Guajira	15.84	30/06/18
22	Security cameras for video surveillance	The project consists of the energization with solar panels of the video surveillance systems of two banana plantations in Magdalena. These hybrid projects have energy support for night hours, low radiation, and failures of the national electricity grid.	Río Frio, Zona Bananera	1.12	11/07/18

Figure 16.8 Pumping Aqueduct El Salado

Figure 16.9 House in Puerto Colombia

are connected to the grid which uses solar energy to ensure electric power supply in the constant power cuts. Projects 4, 7, and 19 are 100% solar housing due to the high costs of connection to the national electricity grid. It is important to mention that in these nanogrids, awareness-raising work has been carried out on the efficient use of energy, thus changing the family's consumption behavior. They generate changes in the typical power consumption curve by moving loads at the hours of highest solar radiation.

16.3.7 Pumping water for irrigation

Projects 12, 18, and 21 are water pumping projects for irrigation. Project 12 is a water pumping project for a palm farm in northern Colombia. This project was designed to pump water at a flow rate of up to 10 l/s for use in sprinkler irrigation. Later, the user modified the design of the irrigation and with the same installed capacity in PV solar panels, changing the pump head, this project manages to pump up to 35 l/s (project 18) for irrigation by flood. Project 21 (Figure 16.10) is a 10-hectare irrigation project for livestock grazing. These projects, like the aqueducts, are solar projects directed to a single load (pumping system). This limits the manipulation of the use of energy, resulting in a successful project in its objective.

16.3.8 Banana sector

The banana sector located in Zona Bananera, Magdalena in Colombia is in a sector with a high number of power cuts that have long periods of time for restoration. This forces the banana farms to have an energy backup so as not to stop the productive process of banana selection and packing. Previously this support was through power plants with fossil fuels, however, some have changed to systems with PV solar panels and batteries. The productive process requires loads such as lighting, scales, security camera system, among others. Projects 9, 10, 15, 20, and 22 are projects in the banana sector connected to the electricity grid to generate energy during the day and to charge batteries. The batteries are used to supply power in hours of power cuts. Project 11 and its extension (project 14—Figure 16.11) is a banana farm that due to the high cost of connection to the grid and low reliability decided to generate all its energy through solar panels. It is a

Figure 16.10 Water pumping for livestock farm

Figure 16.11 Banana packing expansion

challenge that a banana farm completely disconnected from the network because you must do a social work to raise awareness among users in a strict use of loads in times of higher radiation.

16.3.9 Others

There are other projects in nanogrids for boats such as the case of project 2, nanogrids for hotels disconnected from the network (projects 6 and 16), and projects for educational organizations (project 13).

16.4 Problematic situations

The majority of the nanogrids are designed and constructed without a proper methodology, and some of them are built using methodologies applied to the microgrids (Figure 16.12). The cost of some stages consider into the design of the microgrid are high compared with the total cost of a nanogrid; therefore, several nanogrids lack a detailed transdisciplinary diagnosis due to owners and communities who do not have enough resources in order to consider co-construction teams into the design stage as proposed in Diego Jiménez *et al.* [20].

The operation of the nanogrids has given the opportunity to identify some characteristics that hinder or benefit the operation of these systems. Some of them are the growth of the demand, the appropriate and inadequate use of energy and drinking water, the changes of habits in the users in the presence of energy, or the increase in the reliability of the energy supply; moreover, the operation of the system might be affected by the lack or inadequate maintenance.

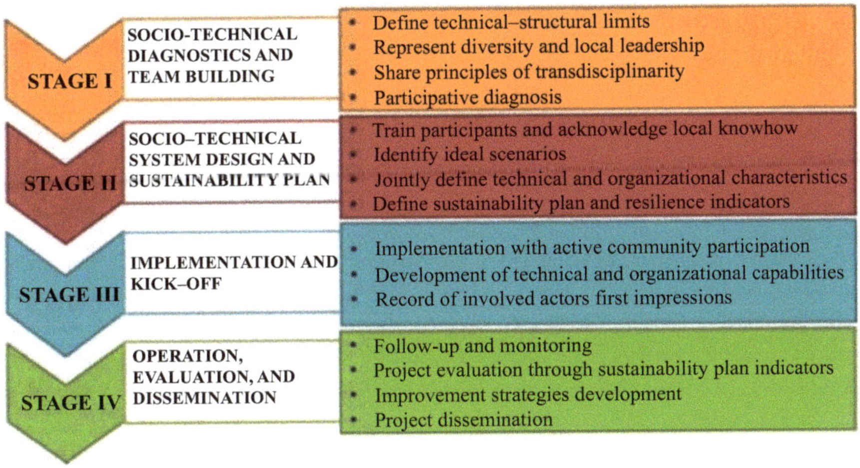

Figure 16.12 Co-construction stages [20]

16.4.1 Growth of the demand

This trend has been seen in most projects, but in some cases it has been controlled, training key people in the operation of the system. This happens both in electricity projects and in water pumping projects. The increase in the electricity demand occurs in users who did not have it or in those who had limited access. People in the presence of electric power tend to use more equipment than expected in the original system. For example, in the banana sector, projects have been designed for security cameras and other smaller loads, but at the time of electrical faults in the distribution network they usually connect additional loads that belong to the production process. In the case of ecological homes and hotels without electric power, the projects are sized for small loads; however in the presence of electric power, new needs arise such as fans, blenders, refrigerators, among others. In the case of water pumping projects using solar energy, ignorance and mistrust in technology mean that users sub-dimension the real needs they have, and when operating the project due to the proper functioning of the system they require expanding the size of the same, which translates into a higher frequency of pumping water and causes a redesign of the electrical system and pumping.

16.4.2 Rational use of energy/water

A nanogrid based on solar energy can supply demand only at certain times of the day. In project 1, it is observed that some users of the aqueduct, due to mistrust in the pumping system, unnecessarily used reserve tanks, some of these were filled, overflowing their capacity and causing waste of water. In the case of project 3, this is connected to the distribution network; a nanogrid based on solar energy was installed with the purpose of supplying part of its electricity demand. However, the lack of change in their consumption habits has not allowed efficient use of the electric power coming from the system during the day, which has caused that there is no significant saving at the canceling of the electric power account in the distribution company. Other similar projects where they already have electricity, the users have not changed their consumption behavior, which means that the electric power available at times of day with greater radiation is not used.

16.4.3 Changes in habits

One of the main challenges has been the dimensioning of projects in communities that do not have electricity or drinking water or have limited access to them. When people receive a supply of better quality they assume it unlimited, this causes changes in habits in their consumption. Electrical equipment that was not used is connected to the nanogrid, and new needs and expectations of the system are created. In the pumping for aqueducts, for example, in project 1, some people started using water to irrigate gardens they did not have it previously. In case of homes and hotels, the projects were sized for basic loads such as lights, cellular and radio chargers, or small televisions; however, with the nanogrids new needs appeared and there was an increase in demand with the use of appliances such as fans, blenders, refrigerators, among others.

16.4.4 Without current problems

In Figure 16.13, there are nine projects in which none of the problems mentioned earlier have been identified. Some of these projects, 14, 18, and 22, are consequential stages or extensions of previous projects. In these cases, the users have had an approach to the nanogrids and take more care of the energy resource. Other projects are specific and controlled uses such as projects 5 and 17; for this reason the operation of the project is correctly fulfilled, besides the number of users that intervene with the system is limited. Project 2 is an isolated vessel with a fixed load and use limited to the hours of navigation. This system has not presented problems so far. Finally, projects 11, 13, and 21 have not presented problems throughout the operation time. The loads were defined correctly, and the uses and schedules of these were respected and no new needs were generated.

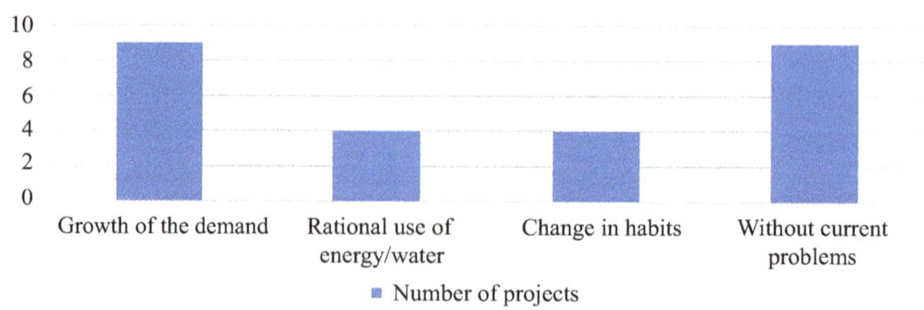

Figure 16.13 Problem by project

16.4.5 Number of users

The number of users is a characteristic that depends on the type of nanogrid. For example, project 17 benefits many users, but the number of people directly related to the system is low, so it is considered a limited number of users. On the other hand, the aqueduct of project 1 is considered a high user number because the use of water is directly determined by each person; this makes difficult the control of the use of the resource, as well as controlling changes in the consumption behavior of each user. The number of projects with each problem is seen in Figure 16.13.

16.5 General methodology to overcome current issues

The natural design of the nanogrids, due to its size, brings with it some problems that a microgrid does not have. The nanogrids are designed in conjunction with one or a few interested people who usually do not include the end user. As a result, end users are unaware of the capacity of the designed system, causing the actual and projected demand to differ significantly. In addition, the design is made with limited loads due to the initial investment so that the growth of the load is a problem for most of the cases. Each nanogrid is independent so it is necessary to perform an independent analysis for each case. For this, follow the steps as shown in Figure 16.14; identify the problem, propose a solution, implement the solution, and finally monitor and review this solution. The monitoring and review stage is used to verify if the proposed solution has had the expected results or not, and in case of failure, a new solution is proposed. The diagram of the methodology is shown in Figure 16.14.

16.5.1 Identifying the problem

The main problems identified are those listed in Sections 16.4.1, 16.4.2, 16.4.3, and 16.4.5. Identifying in detail the real problem helps to correctly propose a solution in the future.

16.5.2 Propose a solution

For the "growth of the demand" and "change in habits," we must consider growth of energy consumption at the time of designing. The Mining and Energy Planning Unit in Colombia (UPME) presents a study through surveys comparing the consumption of households of similar size and socioeconomic conditions in Colombia. We compare homes that are in on-grid, off-grid (with diesel plants), and dispersed on-grid (on-grid but not found in the municipal capitals). In Figure 16.15, the consumption for different altitudes is shown.

We can appreciate that on-grid households ("interconectados") have a 57% higher consumption than the consumption of an off-grid household. This may be

Figure 16.14 Methodology to correct nanogrid problems

Altitude (masl)	Rural and sustainable energization plans (PERS)	On-grid	Off-grid	Dispersed on-grid	Number of surveys
0–500	115.57	130.48	82.94	127.34	2 628
500–1 000	102.28	104.72	87.71	103.98	357
1 000-2 000	74.04	77.79	33.94	76.87	1 555
>2 000	57.19	59.58	36.03	56.48	1 516

Figure 16.15 Consumption for different altitudes

due to the fact that off-grid households can have interrupted energy (less than 24 h/day with a diesel plant). It can be inferred that the growth in electricity consumption between off-grid place and one that has constant and reliable energy (electric grid or microgrid) can be of this same magnitude. This could be considered at the time of designing the nanogrid, and it also serves as a correction factor for the nanogrids already installed at the time of a necessary extension.

Due to the fact that nanogrids are designed with a specific consumption profile that takes advantage of the sun hours, it is necessary to make a "rational use of energy/water." This problem can be possibly solved with explanatory talks about the correct usage of the system for the final, to maximize the solar radiation, avoiding the use of engines or other relevant loads at low radiation hours. The total daily consumption profile could be modified, using the loads at appropriate times to avoid battery wastage or to take advantage of the use of grid power when required (in the case of on-grid nanogrid). This practice becomes more important when there is a high "number of users" as it increases the uncertainty of the consumption profile.

16.5.3 Implementing the solution

The implementation of the solution must correct the problems in the design, so at this stage it is advisable to involve all the agents involved in the nanogrid as investors or the nanogrid owners, the end users, the operators of the system, among others.

16.5.4 Monitoring and review

After taking the corrective measures, a follow-up should be done approximately 1 month later to verify if the user profile changed and was maintained, decreased, or increased. Additionally, talk with end users to analyze if some problems are maintained or if they have been corrected. With the conclusions of the monitoring and review, you must determine if it is necessary to start the process and identify new problems to find different possible solutions.

16.6 Solutions for each type of issues

This section presents different solutions put in place and other proposed solutions for the problems of growth of the demand, rational use of energy, and change in habits.

16.6.1 Monitoring and review

One of the main problems is the growth of demand. In order to solve this problem already presented, it is necessary to identify the load that is producing the growth: What additional charges are being connected and why? Identify how important these new loads are and if it is necessary to maintain them or if it is possible to remove or limit them (Figure 16.16).

16.6.2 Limit the load

For project 15, it was necessary to limit the load. It was identified that the load that had been increased was a vacuum that is part of the banana packing process. This vacuum was intended to be used in the nanogrid in the days of failure in the grid. In spite of being an important load for the banana packing process, this is not part of the design for video surveillance system. For the limitation of the load, the designation of a responsible person in the farm was implemented to manage the equipment connected to the system and limit or avoid loads not initially sized. This is executed by the farm administrator who with a key limited the equipment connected to the system.

16.6.3 Load shifting

In the case of project 4, an off-grid household, the increase in the load was due to a higher volume of cloth washing with a washing machine and video games that were not considered in the design project stage. It was not economically viable to increase the generation or install backup system, so it was necessary to consider load shifting, for instance, using the washing machine at noon in order to use the solar radiation. Another example is to limit the schedule of video games for children at day time. On rainy days, washing machines and video games are not used to reserve the energy for important loads such as lights and fans.

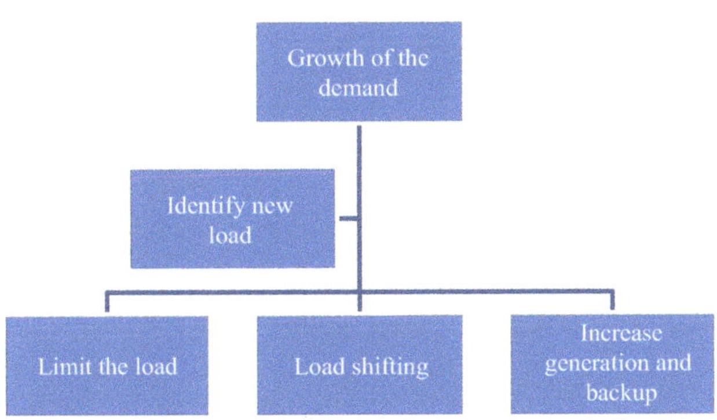

Figure 16.16 Decision-making diagram for growth on

16.6.4 Increase generation and backup

In many cases, the increase in the load is due to important needs identified later, so it is necessary to increase the system. After identifying the new load, its schedule and recurrence of use redesign the system and additional equipment is purchased for the new consumption profile. This solution is the most expensive of the above, but in some cases it is the necessary one. An increase was implemented in the system for projects 3, 4, 5, 10, 12, and 14.

It is important to identify these problems as soon as possible to avoid additional wear problems. In project 10, an early replacement and increase of the battery bank were necessary due to the fact that the increase of the load produced rapid and deep discharges to the batteries higher than those recommended by the manufacturer due to the use of high-power consumption equipment.

Although the control of the demand has only been executed with system administrators who identify when they are consuming more than what is estimated, we propose other systems of control of the demand with a device that shows to the user the renewable energy availability. The device shows a specific color, describing the energy availability for determined hour. It is also possible to do this by means of an automation in the use of loads through timers or load shedding schemes when there is little generation or low-energy backup is available. To implement these possible solutions for the control of demand, it is necessary to incur an investment cost that will depend on the level of automation they have as shown in Figure 16.17.

16.6.5 Rational use of energy/water

For the problem of the rational use of energy, the solution found was to talk with the users; in this way they become aware of the use of energy and water optimally. In the case of project 3, it is connected to the grid and a misuse of energy is reflected in a higher value in the bill, so it is easier to make users aware of the

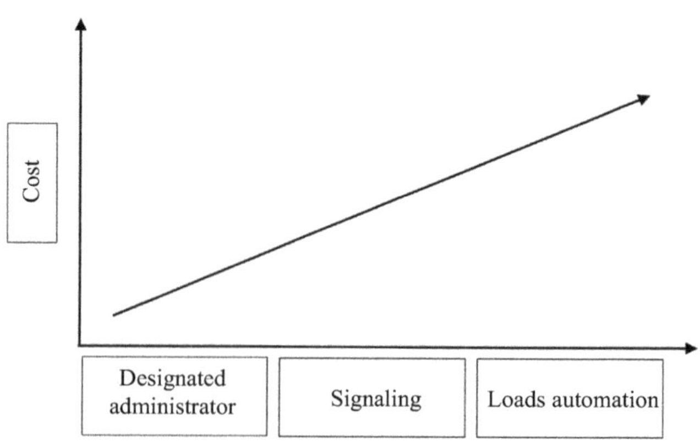

Figure 16.17 Decision-making diagram according to investment cost

consequences of the misuse of energy. In the case of aqueducts and the rational use of water, this problem does not generate additional costs for users, so it is less likely that people will easily identify their error. This problem was addressed with training in the efficient use of the system for the community and the environment.

16.6.6 Changes in habits

Changes in habits produce a growth in demand; therefore the same as what is posed in Figure 16.16 must be done; identify the new burdens that produce growth and define their level of importance to decide whether to eliminate them, redistribute burdens (e.g., schedules of use of each load), or increase the system.

16.7 Conclusions, lessons learned, and perspectives

The nanogrid projects have a positive effect on the communities, and their inhabitants can perform economic and recreational activities that require electricity. They can undertake new economic activities or enhance the existing ones, for instance the nanogrids contribute to the development of tourism and associated services, agriculture through irrigation technology, among others.

On the other hand, nanogrids also have challenges that need to be overcome, and there are some key issues that need to be included in the design and implementation stages of the nanogrids in order to guarantee their performance in the long term.

In order to identify some issues, the operation after commissioning was completed, and the nanogrid projects were divided into five groups: aqueducts, households, irrigation pumping water, banana production, and others.

As aforementioned there are some irrigation pumping water and household nanogrid projects in which the inhabitant's consumption behavior needs to be improved. In these cases, a lack of capacitation was identified before the projects were commissioned. The consumption behavior might be improving, optimizing the electricity and water consumption, using technological solutions as smart meters or demand side management programs [1]; in addition, the inhabitants might be trained as well.

On the other hand, there are some nanogrid projects where the inhabitant's consumption behavior has been modified, yielding a demand increase. This is due to the inhabitants who have increased their consumption using additional appliances in their household or hostels, exceeding the nanogrid installed capacity. In these cases, the design procedure might be improved considering a better demand estimation method [2,4] in order to achieve a better sizing of the nanogrid installed capacity. In addition, it is important to consider some social aspects; for instance social organizations that provide support on the inhabitants' development might be a link between the community and the nanogrid.

Through the inhabitants training, the communities can develop their own, nanogrid projects by means of the community renewable energy concept [5,7]. This concept comprises involvement or direct participation of the community beyond investments or shareholding. Community energy means that members are project owners and executes control actions on it, either through a cooperative or as landowners, or as residents and owners of the nanogrid.

References

[1] S. K. Jha, P. Stoa and K. Uhlen, "Socio-economic impact of a rural microgrid," *2016 4th International Conference on the Development in the in Renewable Energy Technology (ICDRET)*, Dhaka, 2016, pp. 1–4.

[2] J. Hurtt, D. Jhirad and J. Lewis, "Solar resource model for rural microgrids in India," *2014 IEEE PES General Meeting | Conference & Exposition*, National Harbor, MD, 2014, pp. 1–5.

[3] S. Graber, T. Narayanan and J. Alfaro, DebajitPalit, "Solar microgrids in rural India: Consumers' willingness to pay for attributes of electricity," *Energy for Sustainable Development* 42, 32–43, February 2018.

[4] M. F. Z. Souza, "On rural microgrids design—a case study in Brazil," *2015 IEEE PES Innovative Smart Grid Technologies Latin America (ISGT LATAM)*, Montevideo, 2015, pp. 160–164.

[5] K. Ubilla, G. A. Jiménez-Estévez, R. Hernádez, *et al.*, "Smart microgrids as a solution for rural electrification: ensuring long-term sustainability through cadastre and business models," *IEEE Transactions on Sustainable Energy*, 5 (4), 1310–1318, October 2014.

[6] P. Loomba, S. Asgotraa and R. Podmore, "DC solar microgrids—a successful technology for rural sustainable development," *2016 IEEE PES PowerAfrica*, Livingstone, 2016, pp. 204–208.

[7] M. R. Khan, "A stand alone DC microgrid for electricity supply in rural Bangladesh," *2nd International Conference on the Developments in Renewable Energy Technology (ICDRET 2012)*, Dhaka, 2012, pp. 1–4.

[8] H. Li, V. Singhvi, A. Maitra, *et al.*, "National grid microgrid feasibility evaluation: case study of a rural distribution feeder," *2014 IEEE PES General Meeting | Conference & Exposition*, National Harbor, MD, 2014, pp. 1–5.

[9] P. Salas, J. M. Guerrero and F. Sureda, "Mas Roig mini-grid: a renewable-energy-based rural islanded microgrid," *2014 IEEE International Energy Conference (ENERGYCON)*, Cavtat, 2014, pp. 975–982.

[10] F. Pearce, "African lights: solar microgrids bring power to Kenyan Villages," https://e360.yale.edu/features/african_lights_microgrids_are_bringing_power_to_rural_kenya, October 27, 2015.

[11] S. G. Fikari, S. G. Sigarchian and H. R. Chamorro, "Modeling and simulation of an autonomous hybrid power system," *2017 52nd International Universities Power Engineering Conference (UPEC)*, Heraklion, 2017, pp. 1–6.

[12] H. Patel and S. Chowdhury, "Review of technical and economic challenges for implementing rural microgrids in South Africa," *2015 IEEE Eindhoven PowerTech*, Eindhoven, 2015, pp. 1–6.

[13] *An Analysis of the Feasibility and Impacts of Implementing a Microgrid in South Africa, Using HOMER to Model and Predict Outcomes of Microgrid Implementation, Bass Connections Energy & the Environment Design and Innovation*, Final Report, April 30, 2017.

[14] *Developing and Promoting Microgrids in Rural and Island Areas*, Pegasus Project, https://www.fedarene.org/developing-promoting-microgrids-rural-island-areas-25425

[15] P. P. Bezrukikh, S. M. Karabanov and D. V. Suvorov, "Development of design principles of microgrid on the basis of renewable energy sources for rural settlements in Central European part of Russia," *2017 IEEE International Conference on Environment and Electrical Engineering and 2017 IEEE Industrial and Commercial Power Systems Europe (EEEIC/ I&CPS Europe)*, Milan, 2017, pp. 1–5.

[16] E. E.Gaona, C. L.Trujillo and J. A.Guacaneme, "Rural microgrids and its potential application in Colombia," *Renewable and Sustainable Energy Reviews*, 51, 125–137, November 2015.

[17] N. J. Williams, P. Jaramillo, J. Taneja, and T. S. Ustun, "Enabling private sector investment in microgrid-based rural electrification in developing countries: a review," *Renewable and Sustainable Energy Reviews* 52, 1268–1281, 2015.

[18] F. C. Robert, U. Ramanathan, Mukundan, P. Durga and R. Mohan, "When academia meets rural India: lessons learnt from a MicroGrid implementation," *2016 IEEE Global Humanitarian Technology Conference (GHTC)*, Seattle, WA, 2016, pp. 156–163.

[19] A. Hirsch, Y. Parag and J. Guerrero, "Microgrids: a review of technologies, key drivers, and outstanding issues," *Renewable and Sustainable Energy Reviews*, 90, 402–411, July 2018.

[20] J. Diego Jiménez, S. M. Vives, E. G. Jiménez and A. P. Mendoza, "Development of a methodology for planning and design of microgrids for rural electrification," *2017 CHILEAN Conference on Electrical, Electronics Engineering, Information and Communication Technologies (CHILECON)*, Pucon, 2017, pp. 1–6.

[21] H. R. Chamorro and J. F. Jimenez, "Use of petri nets for load sharing control in distributed generation applications," *2012 3rd IEEE International Symposium on Power Electronics for Distributed Generation Systems (PEDG)*, Aalborg, 2012, pp. 731–736.

[22] I. Tank and S. Mali, "Renewable based DC microgrid with energy management system," *2015 IEEE International Conference on Signal Processing, Informatics, Communication and Energy Systems (SPICES)*, Kozhikode, 2015, pp. 1–5.

[23] S. A. Arefifar, M. Ordonez and Y. Mohamed, "Energy management in multi-microgrid systems—development and assessment," *2017 IEEE Power & Energy Society General Meeting*, Chicago, IL, 2017, pp. 1–1.

[24] M. RafieeSandgani and S. Sirouspour, "Energy management in a network of grid-connected microgrids/nanogrids using compromise programming," *IEEE Transactions on Smart Grid* 9 (3), 2180–2191, May 2018.

[25] A. Kumar, A. R. Singh, Y. Deng, X. He, P. Kumar and R. C. Bansal, "A novel methodological framework for the design of sustainable rural microgrid for developing nations," *IEEE Access* 6, 24925–24951, 2018.
[26] S. Teleke, L. Oehlerking and M. Hong, "Nanogrids with energy storage for future electricity grids," *2014 IEEE PES T&D Conference and Exposition*, Chicago, IL, 2014, pp. 1–5.
[27] F. Gonzalez-Espin, V. Valdivia, D. Hogan, D. Diaz and R. F. Foley, "Operating modes of a commercial and industrial building microgrid with electrical generation and storage," *2014 IEEE 5th International Symposium on Power Electronics for Distributed Generation Systems (PEDG)*, Galway, 2014, pp. 1–5.
[28] R. H. Stanev, T. S. Asenov and G. I. Vacheva, "Micro and nanogrid active power management under stand alone and grid connected operation," *2017 XXVI International Scientific Conference Electronics (ET)*, Sozopol, 2017, pp. 1–4.
[29] J. L. Mirez, H. R. Chamorro, C. A. Ordonez and R. Moreno, "Energy management of distributed resources in microgrids," *2014 IEEE 5th Colombian Workshop on Circuits and Systems (CWCAS)*, Bogota, 2014, pp. 1–5.
[30] J. Sathaye, O. Lucon, A. Rahman, *et al.*, "Renewable energy in the context of sustainable energy," in *IPCC Special Report on Renewable Energy Sources and Climate Change Mitigation*, O. Edenhofer, R. Pichs-Madruga, Y. Sokona, *et al.* (eds). Cambridge University Press, Cambridge and New York, NY.

Chapter 17

Distributed generation deployment in the Libyan MV network: adverse impacts on practised protection scheme in the DN

Naser El-Naily[1], Saad M. Saad[1], Amina Elwerfaliy[2], Fatima Amara[3], and Tawfiq Hussein[2]

One of the most exceptional perspectives of the industrial growth taking place worldwide is the continued and abnormal increase in the energy demand. Concerns regarding the economic and environmental impacts of conventional energy sources have driven researchers and inventors to create more efficient, and less costly approaches for delivering electric energy. Utilizing renewable energy-based distributed generations (DGs) are likely to be employed and integrated into the distribution network (DN) increasingly as a prominent alternative to conventional energy sources used so far. The increased penetration of bulky DG units to the electrical network created fundamental shifts and changes in the philosophy of the operation and protection of DNs. The design of a particularly suitable protection scheme is one of the most significant technical problems associated with the integration of DG. In this study, The effects associated with the integration of DGs into The Libyan medium voltage-DN have been highlighted and discussed in detail. In particular, the insufficiency of the utilized protection practices thoroughly investigated and explained. Generally applicable approaches are recommended, which can enhance the applied protection schemes in the interconnected-DN.

17.1 Introduction

Currently, the energy sector in Libya is sustaining with the significant shifts and developments in line with global trends in energy engineering and covering the increasing growth in energy demand in the Libyan society. Dissimilar to the rest of the North African region, where less than 10 percent of population has access to

[1]Electrical Technology Department, College of Electrical and Electronics Technology-Benghazi, Benghazi, Libya
[2]Electrical and Electronics Engineering Department, University of Benghazi, Benghazi, Libya
[3]Sustainable Development Department, National Oil Corporation, Tripoli, Libya

electricity, Libya is a fully electrified country with current electricity consumption of 4,360 kWh per capita [1].

Many oil-producing countries in north-Africa, including Libya, have adopted a trend not to rely completely on fossil fuels as a source of energy, and to adopt cleaner, more sustainable and less overpriced energy sources to provide electricity to the Libyan electric grid. Unlike the other countries in North Africa, Libya provides electricity to all the population. This causes a challenge for Libya's electricity grid operators and expends huge amounts of fossil fuel resources to provide electricity to the huge areas of the Libyan state [2].

Reports and information on the geographical environment of Libya approve the possibility of exploiting the various clamps of Libya to provide a large amount of sustainable energy through wind and solar energy. Libya's prime location in the North African region, with a total area of 175,954 km^2, including large empty desert areas could be used to provide high solar energy. A 1,900 km^2 coastline also provides the possibility of extracting wind energy from the wind speed. Moreover, it is estimated that 1 km^2 of the Libyan desert can provide solar energy equivalent to 1.5 million barrels of oil [3]. As a result, and according to the reports gathered from the Renewable Energy Authority in Libya, there is a plan for penetrating large-share of renewable energy with traditional energy reaches 30% by the year 2030 which essentially depend on wind energy, concentrating solar power, photovoltaic (PV), and solar water heating. All the recommended renewable energy sources due to its connection to the distribution level could be considered as DG [4–7]. The renewable DG of electricity is a new important predisposition in energy systems, which should be considered as an alternative for traditional energy production. This concept is significant to prevent power failures and cost of grid maintenance.

In Libya, the utilization of renewable energy for electricity generation is still in its early days. Many projects were planned and contracted to be installed, 14 MW PV system in a Hun is conducted, 50 MW [8,9], and 50 MW grid-connected PV power plant in Kufra [10]. 25 MW DG planned to be connected in Benghazi [11–15]. Thus, Al-Fattaih wind farm project is planned at the end of 2010 for production power of 60 MW as the first stage of renewable energy development [16–18].

Thus, the problem of DG planning has recently received much attention by power system researchers to garner the maximum benefit from this upcoming power generation technology without violating the existing power system infrastructure. This study presents a state of the art of DG deployment techniques and their influence on the continuing research efforts in this field. To relieve the imminent researchers from the difficulties of availing appropriate guidance, an elfin attempt has put forward through this study by presenting critical information of research work done in this field. Figure 17.1 shows various impacts and constraints in countered DG installation.

17.1.1 Protection issues in DPGS

There are significant issues always rising when distributed power-generation systems integrated into the DN. This is because of the radial design of these DNs.

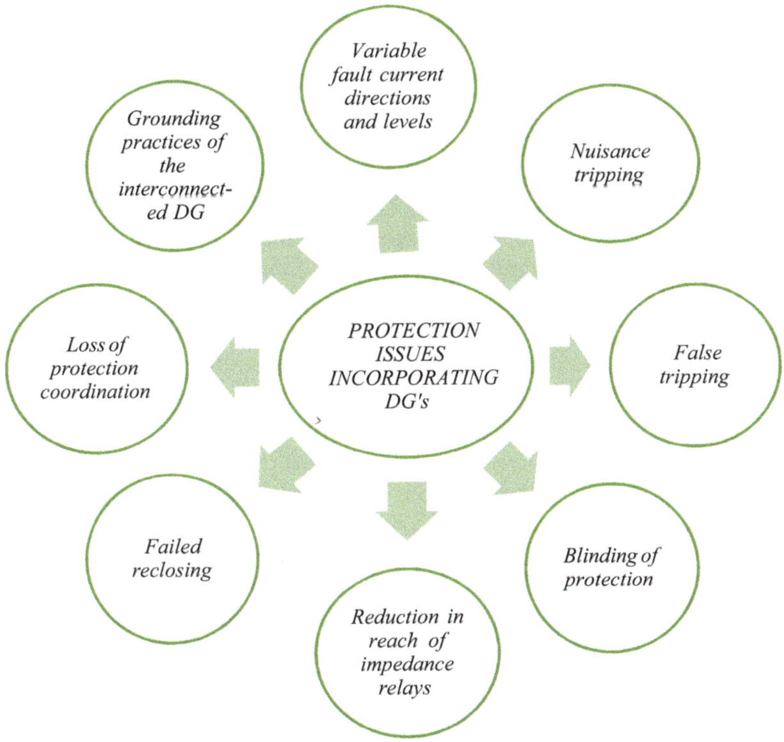

Figure 17.1 Common issues associated with DN with DG

In this system design, the electric power is delivered from substations to loads requirements, which is recognized by a unidirectional path [19]. In order to protect this kind of system from damage, various devices are required such as reclosers and fuses, as well as the coordinated operation of circuit breakers (CB) with overcurrent relays is required to prevent any damage. The damage can be permanent or temporary. These protection methods are carried out widely [19–22]. The radial design in distribution systems is no longer used, therefore, the conventional protective methods become inappropriate for DN interconnected to DGs at high penetration level. In the DN, there are some operating impacts of the DGs, which will be discussed in this study.

17.1.1.1 Variation in fault current directions and levels

There is a significant change in both value and direction of fault current because of the intelligence of the distributed energy resources (DER). The fault level dependents highly on different DGs technologies either converter interfaced or synchronous such as the capacity of the DG unit, or at which the point of common coupling (PCC) is positioned [23–26]. Some of the most important and fundamental problems will be discussed in the following subsequent sections:

- **Increase of fault current level**

The overall efficiency of the DN protection system due to integrating DG units will be decline. Since the size of either the CB or power equipment and protection relays' values have been changed and become inaccurate [27]. In particular, the breaking capacity of CBs might be less than the predictable maximum fault current. The latter can cause a considerable resistance in the short-circuit at buses and switchgears in the primary substations [28].

Furthermore, in the DN, it will be difficult to maintain the coordination of the primary-backup protective relay-pairs due to the rise in both minimum and maximum fault current. Nevertheless, updating the protection setting values and verifying existent equipment ought to be done in order to sustain the system efficiency, which must be done after any fluctuation in the grid. Both reliability and safety can be affected by the short circuit variation. The latter can be influenced by the fault current of DGs. The inverters' fault contribution relies on both the level of its operating current and the trip settings of DGs under-voltage protection. The inverter-DG output current continues to be at the fixed point of the load current, apart from the transient period of time [29].

At the single-phase fault, the current contribution is higher than the one that occurred at the three-phase fault, whereas IEEE-1547 demonstrates that this current must be less than 100% of the DGs three-phase fault current. For the zero sequence, the DG is either a fixed impedance or a source of voltage, while for the positive one it acts as an ideal current source. The significant induction machines' current remains for a few cycles, which is the result of the pre-fault voltage and transient reactance division [30–32].

In distribution overcurrent conditions, ignoring fault current contribution of either distributed or induction is the more practical way. The outcomes demonstrate the DG is substantially less than that of the motor loads. In this way, there is a plentiful point of reference for considering current-controlled inverter-based DG as an irrelevant short-circuit current condition [33]. In any case, the fault effect of DGs should be re-examined since the DG controls may be changed to achieve different utilities as a voltage source [34–36].

For synchronous generators, the fault contributions need exciters and field compelling circuits to give maintained fault current essentially over their full load current. Therefore, for a few establishments, it is important to depend on the distribution protection to overcome any dispersion circuit fault and consequently seclude the DG plant, which is then stumbled on over/under-voltage or loss-of-mains security [37,38].

- **Effects of fault currents reduction**

The prior section examined the DGs fault contribution to the fault location and the significant contribution in both synchronous and induction DGs. The inverse occurs when DG is associated through an electronic inverter interface [34,38].

In the case that an electronic inverter is used, a protection system relying on overcurrent scheme can find it difficult to recognize the fault current, particularly when it is very small. Both PVs and wound rotor induction generator are considered as this type of DGs. In addition, there are other common cases such as DFIG and fully rated converter wind turbine generators [34].

An issue, which could be experienced when DG is associated with the system, is the point at which the fault is downstream as illustrated in Figure 17.2. At the point when the fault happens downstream and its current from the network is not large enough to be detected by the CB1, which is located at the end of the line [34]. The higher the minimum fault current is, the less challenging will be to recognize it. As can be seen in Figure 17.2, DG can be a fast protection compared to the one that is at CB1. When DGs overcurrent protection clears the fault the generator will go into an uncontrolled island operation.

- **Bidirectional fault current flow: in DG-penetrated**

Equally as in Figure 17.3, in recent DNs with implanted DG, the direction of power flow alters when the consumption is less than the generation. In DN, because of the traditional radial nature, generating a unique forward path, which is eliminated, and then the fault current will move to both sides in the line [28]. In order to overcome the protective issue, using directional OC relays (OCRs) instead of other protective systems of medium and fuses. This can be an appropriate replacement [39].

17.1.1.2 Protection blinding

The main substation is considered as the only one outside source that makes a contribution in the fault occurrence in the conventional distribution grid. The overcurrent relays can recognize a relay reach, which is known as a down specific distance of the radial feeder. Increase in the overall short-circuit level in the system can be caused by the Integrated-DG units to the DN. The direction of both fault current from the grid and DG will be changed [39].

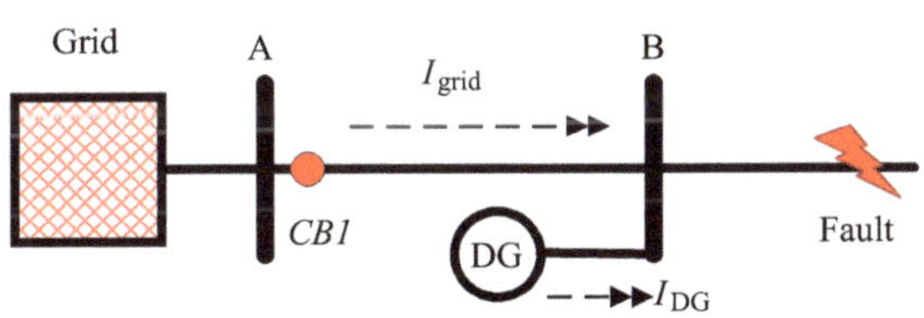

Figure 17.2 DG contribution during fault incidence

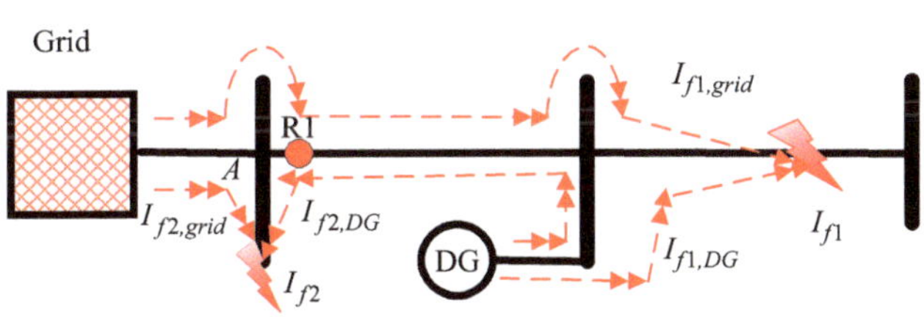

Figure 17.3 Bidirectional impact due to DG penetration

In addition, there will be a decrease in a fault current contribution, which generated by the upstream grid. The reason behind this drop is both the current division and feeder relay's sensitivity. Protection blinding is considered as unwanted impact due to the decrease in the fault current contribution from the main feeder, and it calls under-reach protection because of the reduction in feeder relay reach [40–45].

Also, the interfaced-DGs converter takes part in short circuit for a short period of time with a less value of current, which results in less serious impact.

Figure 17.4 shows the issue related to the protection blinding with taking into the account symmetrical fault that occurred at Bus A. $I_{f,sys}^{with\,DG}$, which is flown from the upstream grid, is less than $I_{f,sys}^{no\,DG}$. However, the overall fault current is risen owing to the added $I_{f,DG}$, as well as the feeder relay's primary reach is then decreased.

Several major factors can be investigated to quantify the protection blinding. These factors have an impact on its intensity, for instance, both DGs location and capacity; and fault position. Taking into account that the system is not loaded beyond the fault occurred, therefore, both short-circuit current and bus voltage can be found by the superposition theorem [45]. The overall fault current is described as follows before and after integrating the DG, respectively:

$$I_{f,tot}^{no\,DG} = I_{f,sys}^{no\,DG} = V_f/(Z_{sys} + Z_L) \tag{17.1}$$

$$I_{f,tot}^{with\,DG} = V_f/(Z_{sys} \times Z_{DG})/(Z_{sys} + Z_{DG}) + Z_L \tag{17.2}$$

Where:

All bus voltage are equal to V_f, which is known as the pre-fault magnitude. The equivalent impedance upstream is described as Z_{sys} at Bus A. The impedance between the Bus A and B is known as Z_L. The inserted impedance because of the DG1 is expressed as Z_{DG}.

It can be noticeable from both (17.1) and (17.2) that $I_{f,tot}^{with\,DG} > I_{f,tot}^{no\,DG}$. By using Kirchhoff laws for both current and voltage, therefore, the main system contribution and the DG unit one can be expressed as follows:

$$I_{f,tot}^{with\,DG} = V_f/\left(Z_{sys} \times \left(\frac{Z_L}{Z_{DG}} + 1\right)Z_L\right) \tag{17.3}$$

$$I_{DG} = (V_f - Z_L \times I_{f,sys}^{with\,DG}/(Z_L + Z_{DG})) \tag{17.4}$$

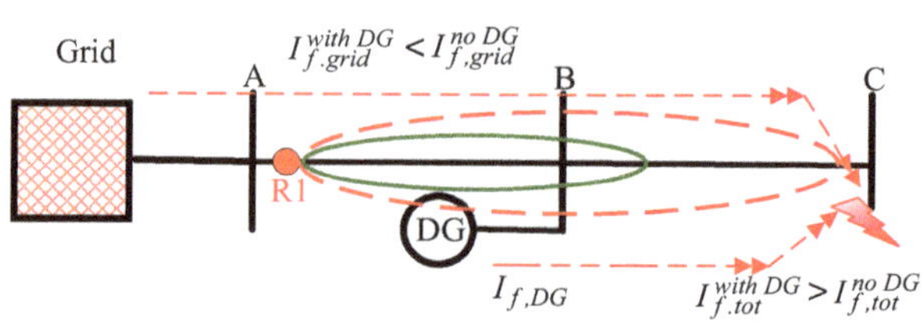

Figure 17.4 Blinding protection due to DG penetration

By making a comparison between (17.1) and (17.3), $I_{f,tot}^{no\,DG} > I_{f,tot}^{with\,DG}$. $Z_{DG} = m \times Z_{sys}$, $Z_L = n \times Z_{sys}$ express both the DG and distribution line impedances, respectively; they are proportional to the upstream grid impedance. Both (17.3) and (17.4) formulas can be used to derive the K ratio for main contribution with integrated DG or without it. However, the K ratio can be derived by using 17.3 and 17.4 as follows:

$$K(m,n) = \frac{I_{f,sys}^{with\,DG}}{I_{f,sys}^{no\,DG}} = \frac{(m + m * n)}{m * n + m * n} \tag{17.5}$$

$$I(m,n) = (I_{f,tot}^{with\,DG})/(I_{f,tot}^{no\,DG}) = \frac{(m+1) * n + m + 1}{(m+1) * n + m} \tag{17.6}$$

Both Figures 17.5 and 17.6 show 3-D plots of both ratios. When the fault is positioned at PCC, then $n = 0$ and $k = 1$, as the mains of contribution are not influenced by the DG1. If n is considered as a constant value, there will be a decrease in K ratio based on the decrease in m value. The mains of contribution are low when DG is large, and Z_{DG} is low. In addition, considering m is non-changeable value, there will be a decrease in the mains of contribution, as the fault position changes to the downside of the feeder. It can be noted that there will be a rapid increase in the fault current as both Z_{DG} and Z_L decreased. The total fault current will have a very small value since the integrated DG is small. Additionally, the short circuit current is not positioned close to the main substation [44].

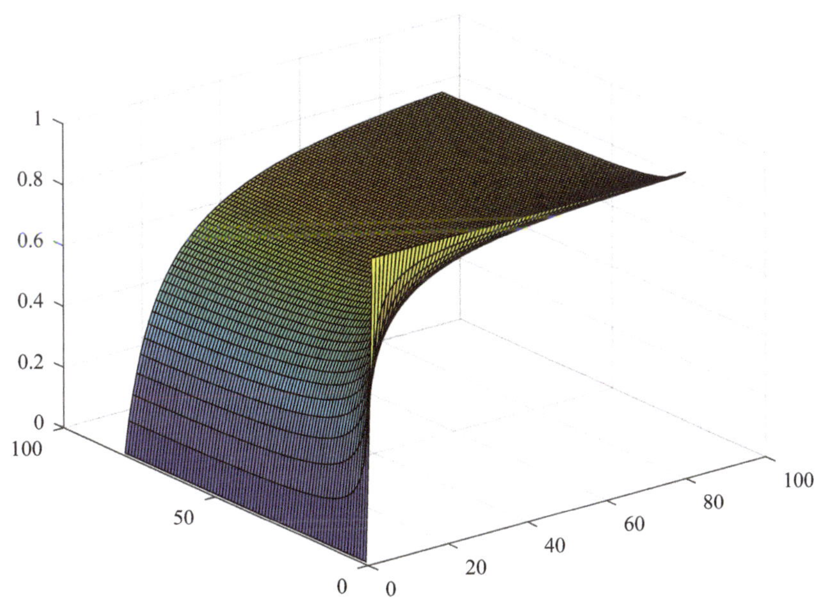

Figure 17.5 3D plot of K ratio

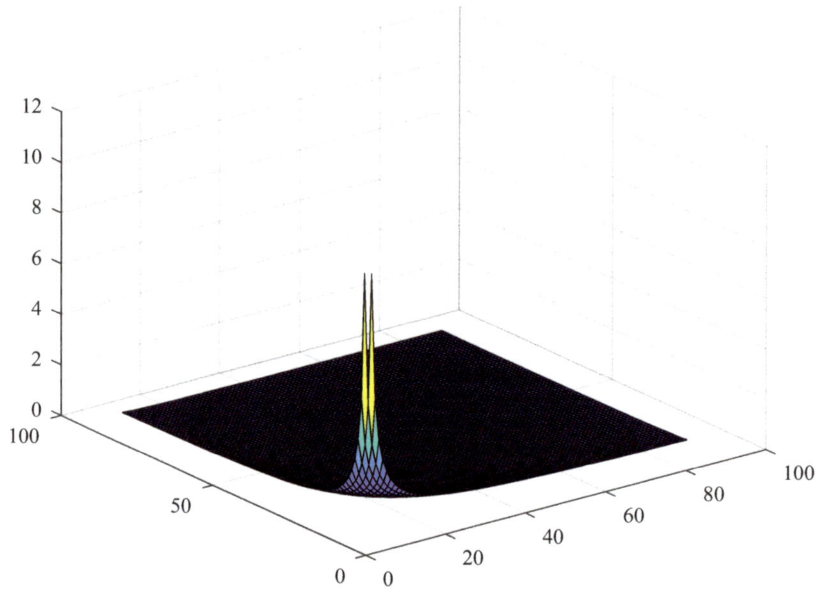

Figure 17.6 3D plot of r ratio

17.1.1.3 Sympathetic tripping

There is an undesirable operation known as sympathetic tripping that related to the feeder relay impact. This is caused by the back-feed of DG to a fault occurrence behind its protection region, this can lead to a power cut in the worked system [46,47]. This original wickedness is known as false tripping, which causes a wide range of tripping issues. Figure 17.4 shows a principle structure of the sympathetic tripping. The fault position is on the Feeder 2, as well as the short-circuit current is fed by the DG1 throughout the substation bus.

When the DG capacity is large, sometimes the CB R1 works before the R2. This issue increases during the use of the non-directional overcurrent relays. Thus, the direction of fault current flow cannot be maintained.

In distribution systems with traditional integrated-DG unit, the sympathetic tripping is likely to happen, as well as generating continuous short-circuit currents. This issue can occur at which feeder overcurrent relays are not the same as the inverse-time features, with either different pickup current or values of the time settings. The impact of both the DG capacity and fault position must be investigated in order to calculate the sympathetic tripping. Figure 17.7 illustrates a three-phase fault that positioned at bus 2, based on this, Z_{L2} is considered as a feeder 2 impedance. This is the primary operating case, and there is no integrated DG, thus the overall fault current is shown as:

$$I_{f,tot}^{no\,DG} = I_{f,sys}^{no\,DG} = V_f/(Z_{sys} + Z_{L2}) \tag{17.7}$$

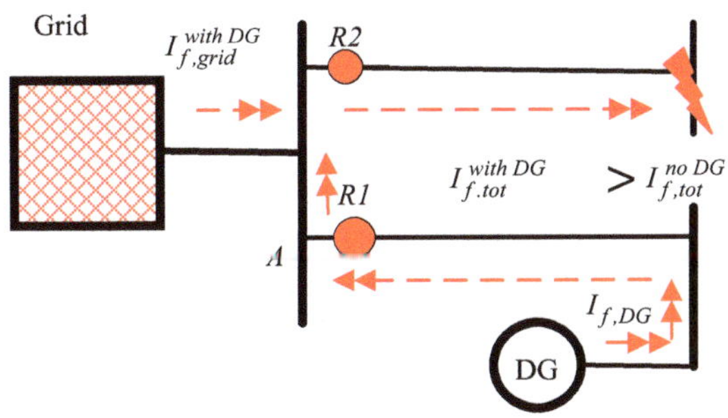

Figure 17.7 Sympathetic tripping due to DG penetration

When the DG is integrated close to the substation ($Z_{L1} \cong 0$), this leads to the following, which is known as the overall fault current:

$$I_{f,tot}^{with\,DG} = V_f/(Z_{L2} + (Z_{sys} * Z_{DG})/(Z_{sys} + Z_{DG}) \tag{17.8}$$

By using Kirchhoff laws, the following expressions can be obtained:

$$I_{f,sys}^{with\,DG} = V_f/\left(Z_{sys} + \left(1 + \frac{Z_{sys}}{Z_{DG}}\right) * Z_{L2}\right) \tag{17.9}$$

$$I_{f,DG} = V_f/\left(Z_{DG} + \left(1 + \frac{Z_{DG}}{Z_{sys}}\right) * Z_{L2}\right) \tag{17.10}$$

The fault current is high when the fault position is close to the substation. Using 17.9 and 17.10 in order to derive the r ratio of the total fault current. Taking into account the exact relation between both Z_{DG} and Z_{L2} with Z_{sys}, which is the case of protection blinding:

Figure 17.8 shows 3D-r(m,n) plot diagram. $r \to 1$ is resulted from $m \to 0$, which can occur when the DG impedance has a significant small value. Therefore, both R_1 and R_2 have the same tripping times during a fault occurrence at Bus 2. R_2 will trip slower than R_1 due to the selection of exact values for the two relays. There is an increase prediction that sympathetic tripping happens due to the large DG capacity. When a less time dial setting is selected [44].

17.1.1.4 Nuisance tripping of DG

It is the disturbance of either DG or any other units such as breakers, which can happen at the protective zones. This tripping can occur because of either the surges of the power or the external fault of DG [48]. These surges can occur because of large loss in load, for example, motors with existence of a DG. Losing the large load can cause a flow of undesirable power to the grid, which can lead relay to trip. If there is any fault that happens out of the protective zone, this can lead to nuisance tripping, and then result in an unexpected loss of supplies from DGs [49].

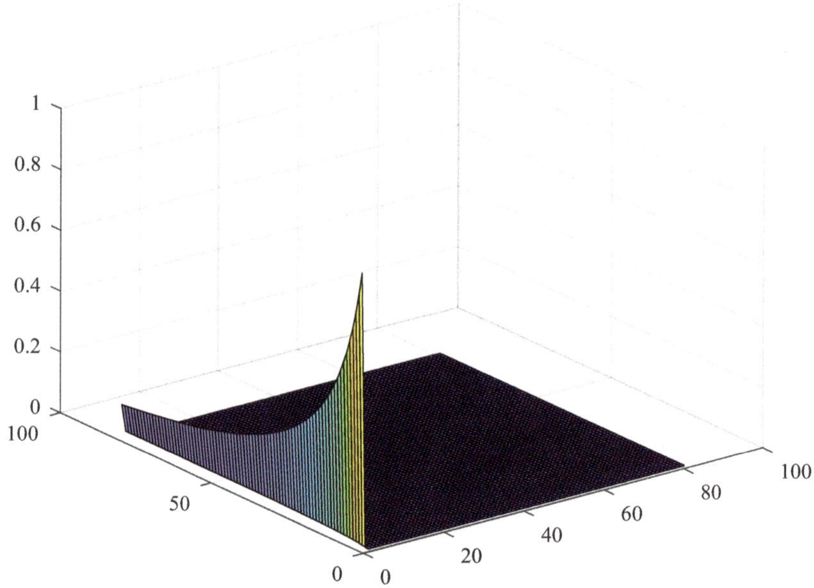

Figure 17.8 3D-r(m,n) plot diagram

17.1.1.5 Failed reclosing

As a result, the DG will be able to recognize and then isolate the faults, which is in earlier reclose interval. Otherwise, the fault will continue to flow, and there will be a delay in the arc and the fault will become permanent. So, there will be a power disconnection which is experienced by some customers, which should have been reported quickly. In addition, there will be frequent fluctuations due to the unbalanced active power, which occurred at the dead time of this reclosing, and the generators will be out of the synchronism [33].

17.1.1.6 Loss of fuse-saving scheme: furthermore, the presence

In distribution systems, the most challenging part is to coordinate between downstream lateral fuses and feeder reclosers. Fuses are considered the most common protective devices due to their cost and reliability, whereas the automatic circuit reclosers are classified as the ones that isolate temporary faults [50,51]. Majority of DN overcurrents have a good reputation as a fuse-saving technique. The reason behind this use is to meet their customers' needs.

In this scheme, during the fault occurrence and the recloser runs with a rapid curve as well as before the fuse begins to melt. After a certain time, delay, it will then reclose its poles. This can eradicate the transient fault and generate non-ionized and cold air on the fault position. During the permanent fault, at its slow curve the recloser will work, thus, the fault will be eliminated using load-side fuse. The recloser's current is considered to be the same as the downstream fuse's current. In distribution systems, because of the existence of DG, the overall current will rise. As illustrated in Figure 17.9, more current can be detected by the branch fuse before the upstream recloser. Therefore, before the recloser is operating the fuse may be burnt [52–55].

17.1.1.7 The effect of infeed on the distance protection

Based on the essential measurement of the impedance, distance relay is able to determine faults in any part of the network. Then this relay will make comparison between the fault current and the voltage, which is located on this relay. This comparison is important in order to calculate the impedance that is between the line and fault [56]. The infeed can contribute to the fault current, especially if the system has more than one generating source at the protection zone [57,58]. As it can be seen in Figure 17.10, the actual fault impedance is $Z_A + Z_B$, during the flow of IB, then the impedance can be expressed by $Z_A + Z_B + (I_B/I_A)Z_B$. The impedance has a great value, which is seen by the distance relay, at Bus B. This impact is known as 'infeed effect'.

The infeed effect can be generated by the existence of DG, the fault current that generated from integration the DG with the network will be added to the one

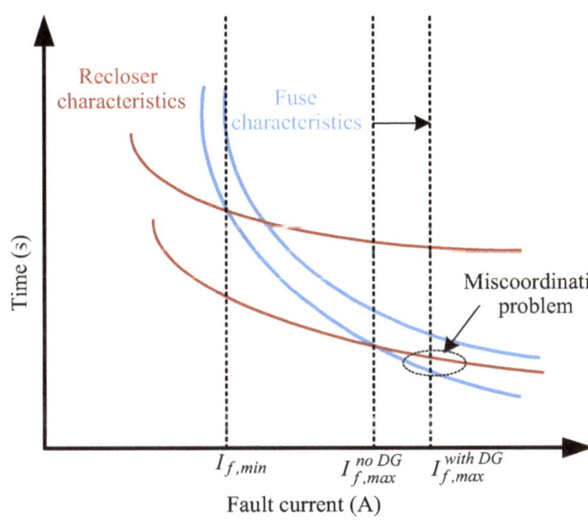

Figure 17.9 Fuse energy-saving scheme

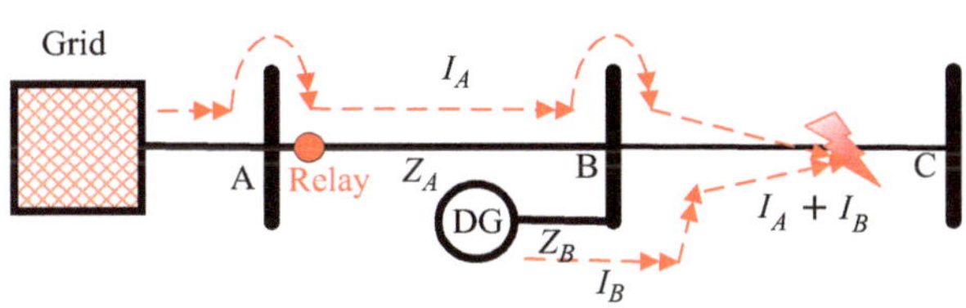

Figure 17.10 Impact of DG connection on distance protection

that provided from the utility. As seen in Figure 17.10, when the current moves from the connected-DG to bus C, the distance relay will not work based on the definite zone, because both the DG and infeed current have influence on the relay. Reaching the distance relay is reduced due to the DG. "When a fault occurs downstream of the bus where the DG is connected to the utility, the impedance measured by an upstream relay will be higher than the real fault impedance."

The following formula must be taken into account to find the relay setting for both Zone 2 and 3:

$$Z_{relay} = Z_A + (1 + K)Z_B \tag{17.11}$$

The infeed from the DG can be highly dependent on the protected zone. Several infeed constants, need to be calculated for the infeed for Zone 2 and then used for Zone 3 takes into account the infeed on the adjacent line. The calculation process in detail can be found below:

- **The effective cover of distance relays**

From Figure 17.11, when the supply infeeds are existent in integrated power system, then the value will not be in ohm for the effective reach of the distance relays. It can be calculated by the ratio between these two values by using the infeed constant, which are defined above [56]. Zone 2 and 3 distance values are calculated as follows:

$$Z_2 = Z_{L1} + (1 + K_1) * 2Z_{L2} \tag{17.12}$$

$$Z_3 = Z_{L1} + (1 + K_2) * 2Z_{L2} + (1 + K_3) * 3Z_{L3} \tag{17.13}$$

17.1.1.8 Loss of protection coordination

When two devices work on the basic backup mode at which the fault happens, this is defined as incoordination [59]. However, the protection devices' contribution may be lost because of both changeable fault level and its direction. This can be a result of the DG integration in distribution system. Figure 17.12 illustrates an example of this case [60].

As shown in Figure 17.12 if the fault occurs at either 1 or 2, the current that is flown from bus 1 will be the same as the one that flown from bus 2. When fault 1 occurs, its protection relay will be operating before the another one at bus 2. The latter will operate as a backup for this case. Vice versa will happen if the fault position is at bus 2. However, the coordination will be lost when one of the faults occur; in order to avoid this, only one time–current curve setting should be used.

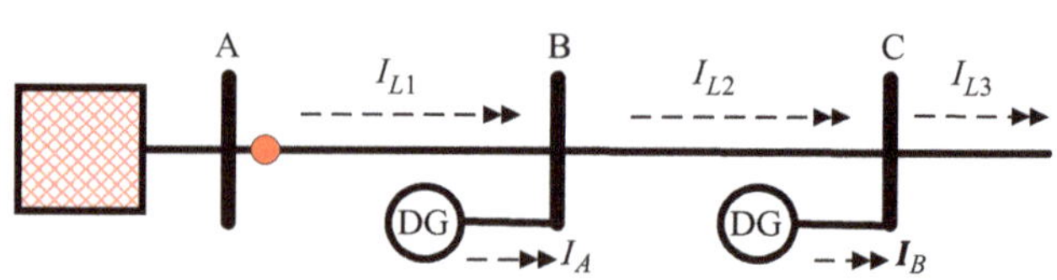

Figure 17.11 DG impact on effective cover of distance protection

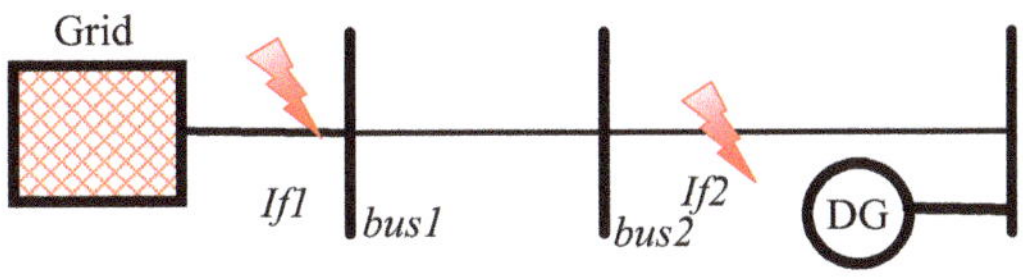

Figure 17.12 Loss of coordination due DG interconnection

- **Possible islanding operation**

According to IEEE 1547 standard, after 2 s the DGs must be separated as the island is noticed [61,62]. Local load can be supported by another way, which is a modern option due to the integration between operating island and DG. This can provide a reliable and flexible operation [63]. In distribution grids, however, this can cause other challenges in the protection system. The reason behind this is the low value of the short circuit capability, which is less than the one that can be generated by utility grid. The distribution systems are fed by DGs at islanding operation. Therefore, the grid-connected current is higher than both short circuit and fault current. In this case, the overcurrent protection's settings are high and then the protection devices will work instantly when the fault happens [64,65].

– **Resonance**

After completing a wide range of studies about the system resonance, DGs can be integrated into the network. In normal cases, resonance will not work in the system. Although in some cases with DG interconnection, resonance can happen in the distribution grid. This is due to the interaction, which can be a result of network impedance and DG's capacitor or any other capacitor in the system. As a result, overvoltage can occur, which can lead to damage to either the network or customer's equipment. There are two different resonances that can happen in DG existent.

– **Linear resonance**

Resonance can occur by interacting the generator reactance and network impedance, while the compensating capacitors are existing. When the asymmetrical fault occurs, and due to the different connection order of the sequence reactance, the overall reactance will change. Either series or parallel combination is a good example. This can result in resonance with compensation existent in the system. Switching either the DG or capacitors as well as changing the reactance of the system can overcome this effect [66].

– **Ferro-resonance**

Ferro-resonance happened due to the interaction between system capacitance and varying the value of inductance, which can be produced by the transformer. This can be provided by the core saturation of the ungrounded network. This saturation happens because of the DG offset, which produced by switching systems' equipment or energizing them with DG existence. In the system, overvoltage can be caused by Ferro-resonance. This can cause damage to protective devices such as DG transformers, surge arresters, and instrument transformers. Therefore, in order to overcome this issue detect overvoltage immediately that can occur from instrumentation transformer. This can be achieved by isolating the DG from the grid operating the relay. In this case, a damping resistor is utilized at the secondary part with an open delta structure [66].

- **Distribution system transient stability**

The transient stability is known as "the ability of the power system to maintain synchronism among synchronous generators when subjected to a severe transient disturbance such as a fault on the facilities" [67]. In order to detect the fault at a short period of time, the protection system is needed, this is known as the critical clearing time (CCT). The transient stability is a serious problem of the transmission network, but because of the passive property the DN does not have this issue. However, a direct integrated-DG to a synchronous generator can cause this issue in distribution system. This opposes using the CCT as a protection system. It can create more limitations on the protective devices' speed [68,69].

17.1.1.9 Impact of interconnection transformer

In DN, there is a considerable effect on the protection system due to the DG connected-winding of transformers, especially during a ground fault element of protection. There is no better way yet to find a perfect solution [33]. Therefore, an appropriate vector group is imposed by distribution network operators (DNOs). This is due to both their operation and interconnection needs. These kinds of groups have both benefits and drawbacks.

Zero-sequence current source cannot be provided because of the ungrounded high side of interconnection transformers, for example, Dd, Dyn, Yd. Thus, there is no effect on either sensitivity or ground fault protection components in the system. There are different issues that can happen due to the overvoltage in the situation of CBs feeder tripping as the islanded side is generated by the ungrounded supply. However, at ground faults, overvoltage issue is overcome by grounded high side YNd of the interconnection transformers. For ground overcurrent relays its both sensitivity and coordination are disturbed by establishing a zero-sequence current supply. 'YNyn' connection can be used in order to eliminate the overvoltage that occurred at ground fault. Thus, in distribution system the zero-sequence current supply will be added to this system. If the DG system is grounded. The main disadvantage of this transformer type is that it allows the ground overcurrent element of the feeder protection to pick up and respond to ground faults inside the DG station [70,71].

- **Ground faults**

There is a wide effect of the grounding on both the protective DG's systems or EPS when the ground faults occur. In addition, it depends on selecting vector groups of the DG's interconnection transformers, the ESP's transformers supply, the ways of either earthling or grounding elements. According to Figure 17.13, the transformer of EPS (T1) is generally Dyn (Delta (HV)/Wye-Gnd (LV)). However, T2 is considered as an interconnection transformer, which might be in Dd, Dyn, Yd, Ynd or Ynyn. Various vectors groups and grounding problems will be viewed according to earth loss.

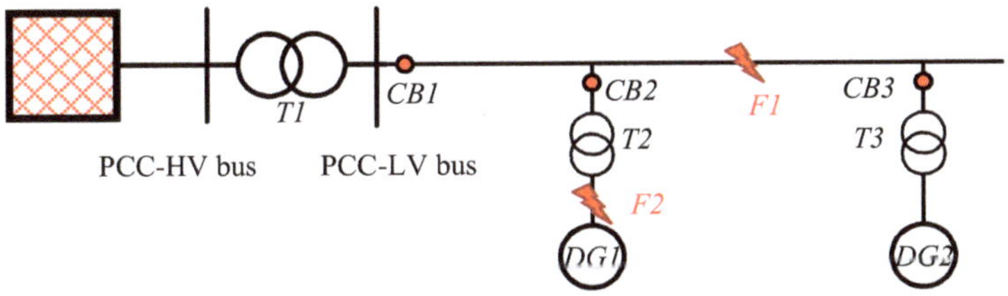

Figure 17.13 A typical DG interconnection with the corresponding protection schemes and protection issues due to fault events

• **Loss of Earth**

The loss of earth is an issue and is also known as loss of grounding reference, which is for a single point and grounded DN. This issue will be demonstrated below. When there is a loss in the utility earth connection, the result will be grounding the system. One of the potential safety hazards is a result of loss of the earth connection for a DG and that part of the network left connected to it [70].

This can occur when the EPS connection is lost. The issue can be increased if the single point of earth system in EPS is used. During DG's operation, it is preferred to utilize the earth connection that is positioned at either PCC-LV or grid connection point, which is a similar case for the DG's earth connection to stay open. Hence, quick response in ensuring the detection of the loss of DG's earth case is required, which is considered as a suitable switching response and in order to ensure that the system is safe. In order to prevent a possible safety hazard and to avoid compromising the EPS's earthing arrangements, this will entail earthing the DGs star point and isolating the DGs site from the EPS. Using a single point earth is highly considered as a standard method for MV networks. This has more than one benefit and it can give a reasonable way to maintain the earth fault current, which is flown throughout the network. There is a possibility to maintain the circulating of currents, at which their frequencies are fundamental or harmonic. Therefore, both the performance and sensitivity will have a proper design in detecting the earth faults.

• **Overvoltage**

As shown in Figure 17.13, when DG1 with interconnection transformer T2 is following the following vector group, Dd, Dyn, or Yd. During the ground fault at F2, therefore, overvoltage case will occur. In addition, the EPS will feed the fault current, this can be done throughout the CB3. The latter is linked with its productive relay, while the DG is linked. As a result, this system will not be grounded effectively. Line-to-neutral voltages on the unfaulted phases approach the normal line to line voltages. This can cause a severe overvoltage on the unfaulted phases.

There will be a serious harm to the equipment because the linked-equipment insolation is not chosen probably at these voltage levels. T2, therefore, is saturated and destroyed. The breaker bushing is failed as the flashover will occur for both insulator and lightning arresters. It is generally accepted that if a DG is rated at less than half of the minimum load on the circuit, it will be unable to sustain more than line to ground voltages. Therefore, the ungrounded T2 should only be considered if the DG is rated at less than half of the load on the circuit. If this type of vector group is used, voltage relays must trip the DG for an overvoltage condition.

• **Ground fault coordination**
According to Figure 17.13, when DG1 with interconnection transformer T2 is following the following vector group, Ynd Wye-Gnd (HV)/Delta (LV). Thus, there is a need for a ground fault coordination as the ground fault happened at F2. This coordination is needed between both ground relays CB3 and CB4. In the case that T3 is on the Ynd vector, therefore, more than one ground current path is required. Besides, between CB3, CB4, and CB7 a needed ground fault coordination will be connected. It can be noticed that because of the delta connection at the secondary of T2, it results in independence in the zero-sequence current source from the status of both DG1 and DG2 generators, and their breakers. Furthermore, there is a required ground coordination, which ground fault one. This should be connected at both CB3 and CB4 between their ground relays. This will be done when there is a fault at F4. T2 has a wye-gnd connection at its secondary part. It results in dependence in zero sequence current source from the status of both DG1 and DG2 generators, and their breakers [70,71].

17.1.1.10 Interconnection standards

According to the interconnection standards, there are technical details that are required in the electric power systems (EPS). The latter with which integrated with distribution generation (DGs). For example, fuel cells, PV, microturbines, reciprocating engines, wind generators, and synchronous generators are considered as a DG integration. In addition, plenty of international standards and recommendations have been carried out, although there are various rules regarding the number of national and utility-specific interconnection that are published [61,62,72–78].

These standardization efforts included in a series of standards:

- IEEE 1547 -2003 Standard for interconnecting distributed resources with EPS.
- IEEE-1547 (multiple guides; base, .1 to .8) A “universal” DG interconnection protection document set to be used as a minimum technical requirement base.
- IEEE Std 1547.1 provides conformance test procedures for equipment interconnecting DER with EPS.
- IEEE Std 1547.2 is an application guide for IEEE Std 1547.
- IEEE Std 1547.3 provides guidance for monitoring, information exchange, and control of DER interconnected with EPS.
- IEEE Std 1547.4 provides guidance for design, operation, and integration of distributed resource island systems with EPS.
- IEEE Standard 1547.5 has not been issued, yet. Its intended scope is to address issues when interconnecting electric power sources of more than 10 MVA to the power grid.
- IEEE Std 1547.6 is a recommended practice for interconnecting DER with electric distribution secondary networks.

- IEEE Std 1547.7 provides guidance for conducting distribution impact studies for DER interconnection.
- IEEE Std P1547.8 Draft recommended practice for establishing methods and procedures that provide supplemental support for implementation strategies for expanded use of IEEE Std 1547-2003.

17.2 Recommendations and suggestions

The aforementioned protection problems can be eliminated and the overall protection scheme can be considerably enhanced, performing the following actions: revision of short-circuit study considering all DG units in service in order to verify the adequacy of the existing primary and protective equipment. In addition, revision of relay setting values to prevent blinding incidents, where possible, paying particular attention to minimum faults. Feeder protection with directional overcurrent relays or distance relays to ban sympathetic incidents, even though the installation of voltage transformers augments the upgrade cost. Replacement of main lateral fuses with sectionalizers based on the results of cost–benefit analysis, where loss of fuse-saving scheme or fuse–fuse miscoordination is inevitable. Utilization of communication-based transfer trip schemes to provide more reliable anti-islanding protection and prohibit failed reclosing incidents.

17.3 Conclusion

DGs are electric power sources connected directly to the distribution systems or at the customer site. These DGs are introduced typically to complement central generation and provide benefits to the user. Since conventional distribution systems are not designed to accommodate DGs and interconnection issues pose specific challenges, protection issues should be identified and studied thoroughly at the planning stage The paper investigated the role and implementation of protection systems in DG interconnections. It presents a comprehensive review of protection issues, and protection system requirements, functions and implementation. The discussion includes consideration of the impact of DGs on the security of EPS. It proposes a comprehensive design methodology for a systematic approach to protection system design. This approach is based on the definition of relevant events and consequences and makes use of a knowledge base as an aid to protection relay design and setting.

References

[1] Mohamed AM, Al-Habaibeh A, and Abdo H. Future prospects of the renewable energy sector in libya. 2016.

[2] Faraj B. Alternative Energy Sources and its role in the production of electric power in Libya. Research Centre for Renewable Energy and Water Desalination-Tajoura. Tajoura, Libya; 2009.

[3] GPCFAAL. The General People's Committee of The Financial Audit Authority. Tripoli-Libya; 2007.
[4] Khalil E, Mohamed M, Belkasem E, *et al.* Demand Side Management in Libya-A Case Study of the General Electric Company of Libya: Transmission System Operation and Control Department. Tripoli-Libya; 2009.
[5] "Renewable energy in Libya"; 2012. Renewable Energy Authority of Libya, annual activity report.
[6] Asheibe A, and Khalil A. The renewable energy in Libya: Present difficulties and remedies. In: the Proceedings of the World Congress; 2013.
[7] Khalil A, and Asheibe A. The chances and challenges for renewable energy in Libya. In: the Proceedings of the Renewable Energy Conference; 2015. p. 1–6.
[8] Kagilik AS, and Tawel AM. Performance Analysis of 14 MW Grid-Connected Photovoltaic System. Solar Energy and Sustainable Development. 2015; 4(1):11–21.
[9] Kadem A, Rajab Z, Khalil A, *et al.* Economic feasibility, design, and simulation of centralized PV power plant. In: Renewable Energy Congress (IREC), 2018 9th International. IEEE; 2018. p. 1–6.
[10] Aldali Y, Henderson D, and Muneer T. A 50 MW very large-scale photovoltaic power plant for Al-Kufra, Libya: energetic, economic and environmental impact analysis. International Journal of Low-Carbon Technologies. 2011;6(4):277–293.
[11] El Naily N, Saad SM, Elhaffar A, *et al.* Mitigating the impact of Distributed Generator on medium Distribution Network by adaptive protection scheme. In: Renewable Energy Congress (IREC), 2017 8th International. IEEE; 2017. p. 1–6.
[12] Elhaffar A, El-Naily N, and El-Arroudi K. Management of Distribution System Protection with High Penetration of DGs. In: Energy Systems and Management. Cham: Springer; 2015. p. 279–291.
[13] Saad SM, El Naily N, Elhaffar A, *et al.* Applying adaptive protection scheme to mitigate the impact of distributed generator on existing distribution network. In: Renewable Energy Congress (IREC), 2017 8th International. IEEE; 2017. p. 1–6.
[14] El Naily N, Saad SM, Hussein T, *et al.* Minimizing the impact of distributed generation of a weak distribution network with an artificial intelligence technique. Applied Solar Energy. 2017;53(2):109–122.
[15] Saad SM, El Naily N, Wafi J, *et al.* An optimized proactive over-current protection scheme for modern distribution grids integrated with distributed generation units. In: Renewable Energy Congress (IREC), 2018 9th International. IEEE; 2018. p. 1–6.
[16] Rajab Z, Almaktar M, Al-Naily N, *et al.* Modeling approach to evaluate wind turbine performance: Case study for a single wind turbine of 1.65 MW in Dernah Libya. In: Renewable Energy Congress (IREC), 2017 8th International. IEEE; 2017. p. 1–5.
[17] Naily NE, Saad SM, Rajab Z, *et al.* An intelligent protection scheme to mitigate the impact of integrating large share wind energy resources in a weak distribution network. Wind Engineering. 2017;41(6):383–396.

[18] Saleh IM. Prospects of renewable energy in Libya. In: Solar Physics and Solar Eclipses (SPSE 2006); 2006. p. 153–161.

[19] Soshinskaya M, Crijns-Graus WH, Guerrero JM, *et al.* Microgrids: experiences, barriers and success factors. Renewable and Sustainable Energy Reviews. 2014;40:659–672.

[20] Choudhary NK, Mohanty SR, and Singh RK. A review on microgrid protection. In: Electrical Engineering Congress (iEECON), 2014 International. IEEE; 2014. p. 1–4.

[21] Al-Nasseri H, Redfern M, and Li F. A voltage based protection for microgrids containing power electronic converters. In: Power Engineering Society General Meeting, 2006. IEEE. IEEE; 2006. 7 pp.

[22] González J, Dysko A, Chanelière J, *et al.* The impact of renewable energy sources and distributed generation on substation protection and automation. Electra, N. 2012.

[23] Papaspiliotopoulos VA, Korres GN, and Hatziargyriou ND. Adverse impact of distributed generation on protection of the Hellenic MV network–recommendations for protection scheme upgrade. CIRED-Open Access Proceedings Journal. 2017;(1):934–938.

[24] Walling R, Saint R, Dugan RC, *et al.* Summary of distributed resources impact on power delivery systems. IEEE Transactions on Power Delivery. 2008;23(3):1636–1644.

[25] Blaabjerg F, Yang Y, Yang D, *et al.* Distributed power-generation systems and protection. Proc IEEE. 2017;105(7):1311–1331.

[26] Talebizadeh Sardari K. Enhancement of microgrid reliable operation using an adaptive protection strategy. California State University, Northridge; 2018.

[27] Wippenbeck T, Jäkel M, Schmidt T, *et al.* Development and cross-validation of short-circuit calculation methods for distribution grids with high penetration of inverter-interfaced distributed generation. In: 23rd Int. Conf. Electricity Distribution; 2015.

[28] Hussain B, Sharkh S, Hussain S, *et al.* Integration‘ of distributed generation into the grid: Protection challenges and solutions. 2010.

[29] Glinka F, Wippenbeck T, Schmidt T, *et al.* Challenges and solutions for MV and LV protection in grids with large amount of distributed generation–a final report from the German research project ProFuDis. CIRED-Open Access Proceedings Journal. 2017;(1):976–980.

[30] van der Walt HLR, Bansal RC, and Naidoo R. PV based distributed generation power system protection: a review. Renewable Energy Focus. 2018;24:33–40.

[31] Nikkhajoei H, and Lasseter RH. Microgrid protection. In: Power engineering society general meeting, 2007. IEEE; 2007. p. 1–6.

[32] Saint R, and Friedman N. The application guide for distributed generation interconnection-the NRECA guide to IEEE 1547. In: Rural Electric Power Conference, 2002. IEEE; 2002. p. D2–1.

[33] Barker PP, and De Mello RW. Determining the impact of distributed generation on power systems. I. Radial distribution systems. In: Power Engineering Society Summer Meeting, 2000. IEEE. vol. 3. IEEE; 2000. p. 1645–1656.

[34] Nassif A. An analytical assessment of feeder overcurrent protection with large penetration of distributed energy resources. IEEE Transactions on Industry Applications. 2018;54(5):5400–5407.

[35] Hung DQ, and Mithulananthan N. Multiple distributed generator placement in primary distribution networks for loss reduction. IEEE Transactions on industrial electronics. 2013;60(4):1700–1708.

[36] El Naily N, Saad SM, Hussein T, *et al.* On-line adaptive protection scheme to overcome operational variability of DG in smart grid via fuzzy logic and genetic algorithm. In: Renewable Energy Congress (IREC), 2018 9th International. IEEE; 2018. p. 1–6.

[37] Zay H, Mohamed A, Shareef H, *et al.* Impact of distributed generations on power system protection performance. International Journal of Physical Sciences. 2011;6(16):3999–4007.

[38] Kennedy J, Ciufo P, and Agalgaonkar A. A review of protection systems for distribution networks embedded with renewable generation. Renewable and Sustainable Energy Reviews. 2016;58:1308–1317.

[39] Coffele F, Booth C, Dyśko A, *et al.* Quantitative analysis of network protection blinding for systems incorporating distributed generation. IET Generation, Transmission & Distribution. 2012;6(12):1218–1224.

[40] Dugan RC, and Mcdermott TE. Distributed generation. IEEE Industry Applications Magazine. 2002;8(2):19–25.

[41] Zhou W, Zhang Y, Xia C-j, and Wang Q. Effect of distributed generation on relay protection of distributed system [J]. Power System Protection and Control. 2010;3.

[42] Papaspiliotopoulos V, Korres G, and Hatziargyriou N. An adaptive protection infrastructure for modern distribution grids with distributed generation. CIGRé. 2016.

[43] Papaspiliotopoulos V, Kleftakis V, Kotsampopoulos P, *et al.* Hardware-in-the-loop simulation for protection blinding and sympathetic tripping in distribution grids with high penetration of distributed generation. 2014.

[44] Papaspiliotopoulos VA, Korres GN, Kleftakis VA, *et al.* Hardware-in-the-loop design and optimal setting of adaptive protection schemes for distribution systems with distributed generation. IEEE Transactions on Power Delivery. 2017;32(1):393–400.

[45] Saadat H. Power System Analysis. vol. 232. WCB/McGraw-Hill Singapore; 1999.

[46] Jennett KI, Booth CD, Coffele F, *et al.* Investigation of the sympathetic tripping problem in power systems with large penetrations of distributed generation. IET Generation, Transmission & Distribution. 2014;9(4):379–385.

[47] Coster EJ, Myrzik JM, Kruimer B, *et al.* Integration issues of distributed generation in distribution grids. Proceedings of the IEEE. 2011;99(1):28–39.

[48] Seegers T, Birt K, Beazer R, *et al.* Impact of distributed resources on distribution relay protection. Rep to Line Protection Subcommitee, Power System Relay Committee, Power Engineering Society, IEEE. 2004.

[49] Brearley BJ, and Prabu RR. A review on issues and approaches for microgrid protection. Renewable and Sustainable Energy Reviews. 2017;67:988–997.

[50] Telukunta V, Pradhan J, Agrawal A, *et al.* Protection challenges under bulk penetration of renewable energy resources in power systems: a review. CSEE Journal of Power and Energy Systems. 2017;3(4):365–379.

[51] Haron AR, Mohamed A, Shareef H, *et al.* Analysis and solutions of over-current protection issues in a microgrid. In: Power and Energy (PECon), 2012 IEEE International Conference on. IEEE; 2012. p. 644–649.

[52] Naiem A, Hegazy Y, Abdelaziz A, *et al.* A classification technique for recloser-fuse coordination in distribution systems with distributed generation. IEEE Transactions on Power Delivery. 2012;27(1):176–185.

[53] Hussain B, Sharkh S, Hussain S, *et al.* An adaptive relaying scheme for fuse saving in distribution networks with distributed generation. IEEE Transactions on Power Delivery. 2013;28(2):669–677.

[54] Mascarella D, Arroudi K, Li S, *et al.* Real-Time Testing of Relay-Fuse Coordination Implemented with IEC 61850 GOOSE Messaging Using Industrial Relays. In: CIGRÉ Canada Conference; 2015.

[55] Wafi J, Saad SM, El Naily N, *et al.* Applicability analysis of proactive recloser-fuse scheme for wind distributed generation into modern distribution network. In: Renewable Energy Congress (IREC), 2018 9th International. IEEE; 2018. p. 1–6.

[56] Gers J, and Holmes E. Protection of Electricity Distribution Networks, (IEE Power and Energy Series). Institution of Electrical Engineers (IET); 2004.

[57] Mahadanaarachchi VP, and Ramakuma R. Impact of distributed generation on distance protection performance-A review. In: Power and Energy Society General Meeting-Conversion and Delivery of Electrical Energy in the 21st Century, 2008 IEEE. IEEE; 2008. p. 1–7.

[58] El Naily N, Saad SM, Elsayed RE, *et al.* Planning and application of distance relays coordination for IEC microgrid considering intermediate in-feed factor. In: Renewable Energy Congress (IREC), 2018 9th International. IEEE; 2018. p. 1–6.

[59] Wan H, Li K, and Wong K. An adaptive multiagent approach to protection relay coordination with distributed generators in industrial power distribution system. IEEE Transactions on Industry Applications. 2010;46(5):2118–2124.

[60] Girgis A, and Brahma S. Effect of distributed generation on protective device coordination in distribution system. In: Power Engineering, 2001. LESCOPE'01. 2001 Large Engineering Systems Conference on. IEEE; 2001. p. 115–119.

[61] IEEE Standards Board. IEEE 1547 standard for interconnecting distributed resources with electric power systems. IEEE Standards Association, Piscataway. IEEE; 2003.

[62] IEEE Application Guide for IEEE Std 1547(TM), IEEE Standard for Interconnecting Distributed Resources with Electric Power Systems. IEEE Std 15472-2008. 2009; p. 1–217.

[63] Mahat P, Chen Z, and Bak-Jensen B. Review on islanding operation of distribution system with distributed generation. In: Power and Energy Society General Meeting, 2011 IEEE. IEEE; 2011. p. 1–8.

[64] Justo JJ, Mwasilu F, Lee J, *et al.* AC-microgrids versus DC-microgrids with distributed energy resources: A review. Renewable and Sustainable Energy Reviews. 2013;24:387–405.

[65] Mahat P, Chen Z, Bak-Jensen B, *et al.* A simple adaptive overcurrent protection of distribution systems with distributed generation. IEEE Trans Smart Grid. 2011;2(3):428–437.

[66] Piasecki W, Florkowski M, Fulczyk M, *et al.* Mitigating ferroresonance in voltage transformers in ungrounded MV networks. IEEE Transactions on Power Delivery. 2007;22(4):2362–2369.

[67] Kundur P, Balu NJ, and Lauby MG. Power System Stability and Control. vol. 7. McGraw-Hill New York; 1994.

[68] Salman SK, and Rida IM. Investigating the impact of embedded generation on relay settings of utilities electrical feeders. IEEE Transactions on Power Delivery. 2001;16(2):246–251.

[69] Xyngi I, Ishchenko A, Popov M, *et al.* Transient stability analysis of a distribution network with distributed generators. IEEE Transactions on Power Systems. 2009;24(2):1102–1104.

[70] Mozina CJ. Update on the current status of DG interconnection protection: What IEEE P1547 doesn't tell you about DG interconnection protection. In: Beckwith Electric Co., Inc., Ninth Annual Protection Seminar Sept. IEEE; 2006. p. 24–28.

[71] Pettigrew B. Interconnection of a" Green Power" DG to the Distribution System, A Case Study. In: 2005/2006 IEEE/PES Transmission and Distribution Conference and Exhibition; 2006.

[72] IEEE Standard Conformance Test Procedures for Equipment Interconnecting Distributed Resources with Electric Power Systems - Amendment 1. IEEE Std 15471a-2015 (Amendment to IEEE Std 15471-2005). 2015;p. 1–27.

[73] IEEE Guide for Monitoring, Information Exchange, and Control of Distributed Resources Interconnected with Electric Power Systems. IEEE Std 15473-2007. 2007;p. 1–160.

[74] IEEE Guide for Design, Operation, and Integration of Distributed Resource Island Systems with Electric Power Systems. IEEE Std 15474-2011. 2011; p. 1–54.

[75] IEEE Recommended Practice for Interconnecting Distributed Resources with Electric Power Systems Distribution Secondary Networks. IEEE Std 15476-2011. 2011;p. 1–38.

[76] IEEE Guide for Conducting Distribution Impact Studies for Distributed Resource Interconnection. IEEE Std 15477-2013. 2014;p. 1–137.

[77] Photovoltaics DG, and Storage E. IEEE Standard for Interconnection and Interoperability of Distributed Energy Resources with Associated Electric Power Systems Interfaces.

[78] Rathore AK, and Rajagopal PU. Power Electronic Applications, Grid Codes, Power Quality Issues, and Electricity Markets for Distributed Generation. In: Handbook of Distributed Generation. Cham: Springer; 2017. p. 631–684.

Chapter 18

Conclusions and future scopes

Rajeev Kumar Chauhan[1], *Kalpana Chauhan*[2], *and Sri Niwas Singh*[3]

18.1 Conclusions

The book chapters are mostly centralized on case study-based research and/or solution for rural electrification. The content of the book also covered the energy storages along with their integration to DC microgrid (MG) and the related challenges. The low-voltage DC distribution system for various applications like charging electric vehicle has been discussed. The chapter-wise conclusion of the book is presented in the following sections.

18.1.1 Overview of microgrids for rural areas and low-voltage applications

It is concluded that the electrical distribution infrastructure is not very well extended in rural areas. But the technical and economical developments during the last decades have established an opportunity to create a new competitive distribution system based on modern power electronic technology. The renewable energy resources required to be provided with a certain reliability of supply and a certain level of stability. Many grid operators have to develop DC MGs, which treat photovoltaic power generation in a special manner.

18.1.2 Microgrid architectures

The general form of MG has been presented with considering the analysis of the factors that have given an impetus to the development of such systems. The basic characteristics and structures of MGs have been identified with an insightful literature review of the existing forms of these systems.

[1]Department of Electrical Engineering, Dayalbagh Educational Institute, Agra, India
[2]Department of Electrical Engineering, Central University of Haryana, Mahendragarh, India
[3]Department of Electrical Engineering, Indian Institute of Technology, Kanpur, Uttar Pradesh, India

18.1.3 The microgrid investment and planning in rural locations

Changes in typical MG planning formulation addressing the specific challenges of rural MGs were introduced in this chapter. The case studies have shown that these changes have a significant impact on the MG planning results and are able to better represent the reality of rural MGs.

18.1.4 Assessment of energy saving for integrated CVR control with large-scale electric vehicle connections in microgrids

The finding through the results shows that the implementation of coordinated controls of voltage reduction (CVRC) has effectively improved the energy reduction. However, the energy reduction capability gradually decreases with the increase in EV penetration. In comparison with different EV charging strategies, the impact of vehicle-to-grid (V2G) strategy on CVRC energy reduction capability is the least. In addition, the implementation of CVRC also helps to improve the EV penetration capability on both MGs.

18.1.5 Application of energy storage technologies on energy management of microgrids

The study of different energy storage (ESSs) technologies characteristics has been done. The advantages and disadvantages of ESSs types for energy management in the MG are also described. The implementation of different ESS technologies has been done for the MG application. In addition, the output power rate, storage capacity, efficiency, response time, ESS scale, capital cost, life cycle, and the operation mechanism of ESSs technologies have been explained.

18.1.6 Control technique for integration of smart DC microgrid to the utility grid

The proposed control scheme of three-phase voltage source inverter maintains the rated sinusoidal load–voltage with low total harmonic distortion, and ensures the required power flow between the DCMG and the utility grid has been discussed in this chapter. During various AC faults on the utility grid, the proposed control technique does not allow any adverse impacts on the generations, loads in the DCMG, and on the DCMG voltage, except the load in faulted area.

18.1.7 Energy management system of a microgrid using particle swarm optimization (PSO) and communication system

An insight into the role of the energy management system (EMS) for an MG has been discussed. The hierarchy of the various controllers utilized in the EMS is presented. The requirements for a communication system within the MG have been discussed. A test case with a PSO technique has also been provided for understanding.

18.1.8 Reconfiguration of distribution network using different optimization

An augmented grey wolf optimization (AGWO) algorithm that is used to determine the optimal switches and can be open to reconfigure the distribution system and to obtain the minimum system power loss has been proposed. It is observed that the proposed algorithm has achieved superior results when compared to other optimization techniques. The convergence characteristic of the proposed method proved that AGWO is able to solve difficult optimization problems in different fields.

18.1.9 Intelligent optimization scheduling of isolated microgrids considering energy storage integration, traditional generation, and renewable energy uncertainty costs

Mitigation of issues due to intermittency of renewable energy sources through energy storage systems is demonstrated using a non-dominated sorting genetic algorithm (NSGA II). The test system used corresponds to an MG in an island with similar features of a real MG in Dongfushan Island in China. The results of the evaluation are: optimization of the MG with plentiful renewable sources and a few resources, with the inclusion of uncertainty costs for wind generation; and finally, the most complete evaluation, where uncertainty is included with both wind and solar component. It is, therefore, possible to make decisions based on the information tool regarding the environmental, political, and economic characteristics of the region where the power system is located.

18.1.10 ADMM-based consensus droop control and distributed pinning droop control of isolated AC microgrids

Two recently developed distributed droop control mechanisms, the alternating direction multipliers method (ADMM)-based consensus control and a distributed pinning droop, have been studied for improving the slow convergence rate of the conventional consensus-based droop control method. Performance of the ADMM implementation of consensus algorithm has also been compared to the first-order consensus algorithm, which shows faster convergence rates for both variants of the ADMMs. Results of pinning droop control indicate that the multi-agent system framework can indeed improve the autonomous operation of the MG by suppressing the errors that the conventional droop control is not able to counteract. Thus, this pinning droop control architecture can serve as a real-time tool for EMS in modern MG control and operations.

18.1.11 Extendable multiple outputs hybrid converter for AC/DC microgrid

A comprehensive mathematical modeling of magnetically coupled extendable multiple outputs hybrid converter for AC/DC MG converter has been given in this chapter for the series and parallel mode of operations. The detailed steady-state and dynamic analysis is also carried out for analyzing the steady-state and transient behavior of the proposed converter. The proposed concept is validated with simulation and experimental results on a 500 W laboratory prototype for two simultaneous AC and one DC outputs.

18.1.12 Multi-input converters for distributed energy resources in microgrids

The multi-input DC/DC converters described in this chapter have been divided into three main groups of magnetic, electric, and electromagnetic types that are compared with one another. The general comparison among the configurations including battery life, soft-switching, source utilization, the input and output isolation, and some elements employed in each topology has been done and concluded that converters are a desirable option when more than one DG unit are connected to each other because of the inherent characteristics such as less semiconductor utilization and occupying less volume.

18.1.13 Modeling and control of DC–DC converters for DC microgrid application

In this chapter a complete analysis of DC–DC converter interfaced between photovoltaic distributed energy source and DC bus has been presented. The experimental results match with the simulation results in close proximity proving the accuracy of mathematical model and controller designed. It has analyzed the MG system for a PV-source fed DC–DC buck converter prototype of low voltage and low power.

18.1.14 Case studies of microgrids systems

Several case studies of MGs have been done providing energy efficiency and electricity bill reduction as a good alternative to feeding by public networks in rural and isolated communities. In particular, the possibility offered by such systems to satisfy the energy load requirements of several classes of electric users, ranging buildings and business facilities, urban/industrial districts, and rural communities, has been presented and discussed.

18.1.15 Smart grid road map and challenges for Turkey

Turkey has two interconnection points with the East European Transmission Grid. The test period for synchronous parallel operation of the Turkish and European power systems started on June 1, 2011, and ended in September 2012. For the moment, trade is being limited to 400 MW from Bulgaria and Greece to Turkey and 300 MW from Turkey to Europe via these countries. In order to provide a stable, low cost, reliable, efficient, robust, sustainable, and environmental-friendly electrical energy system to consumers, a fully operational smart grid system needs to be established in Turkey.

18.1.16 Nanogrids: good practices and challenges projects in Colombia

The nanogrid projects have a positive effect on the communities; their inhabitants can perform economic and recreational activities that require electricity. They can undertake new economic activities or enhance the existing ones, for instance, the nanogrids contribute to the development of tourism and associated services, agriculture through irrigation technology, among others. On the other hand, nanogrids also have challenges that need to be overcome; there are some key issues that need to be included into the design and implementation stages of the nanogrids in order to guarantee their performance in the long term. In order to identify some issues regarding the operation after commissioning was completed, the nanogrid projects were divided into five groups in this study: aqueducts, households, irrigation pumping water, banana production, and others.

18.1.17 Distributed generation deployment in the Libyan MV network: adverse impacts on practised protection scheme in the DN

The role and implementation of protection systems in DG interconnections has been investigated. A design methodology for a systematic approach to protection system design has been proposed. This approach is based on the definition of relevant events and consequences and makes use of a knowledge base as an aid to protect relay design and setting.

18.2 Future scopes

As there is requirement of utilization of renewable energy resources to generate the electricity in rural areas, there are researches going on to fulfill the energy requirements in efficient way. So the book will be helpful for the researchers in this area, as well as for the local audience to get awareness about the MG with renewable energy resources and utilization of the electric vehicles to reduce the carbon emission generated by conventional resources in electrifying the large population. The future scopes of the case studies done in this book are described chapter-wise.

18.2.1 Overview of microgrids for rural areas and low-voltage applications

The rapid increment of DC compatible appliances in the buildings, emerge the photovoltaic energy as the fastest growing source of renewable energy and expected to see continued strong growth in the immediate future. The photovoltaic power is, therefore, required to be provided with a certain reliability of supply and a certain level of stability. Motivated by the above issues, many grid operators have to develop DC MGs, which treat photovoltaic power generation in a special manner.

18.2.2 Microgrid architectures

The basic characteristics and structures of MGs identified in this chapter will be helpful in selecting and designing the most suitable MG architecture for commercial and professional buildings.

18.2.3 The microgrid investment and planning in rural locations

With the help of the description given in the chapter about the changes in typical MG planning formulation and challenges of rural MGs, it will be easy to represent the MG planning for better setup in the rural areas.

18.2.4 Assessment of energy saving for integrated CVR control with large scale electric vehicle connections in microgrids

The analysis in this paper could guide the utilities to decide how to reasonably apply the coordinated controls of voltage reduction CVRC in the future distributed MGs with the substantial growth of EVs. Further work will consider the demand growth and more LCTs (i.e. heat pump and photovoltaic) to investigate the potential CVRC benefits on the future distributed MGs.

18.2.5 Application of energy storage technologies on energy management of microgrids

The present study of different ESSs technologies characteristics will be helpful for the researchers and professionals working in the ESS. The EMS for MGs can be designed taking into account previously considered advantages and disadvantages.

18.2.6 Control technique for integration of smart DC microgrid to the utility grid

During the DC fault on the DCMG bus, the proposed control technique scheme allows to maintain the rated sinusoidal three-phase AC voltage, and ensures the required power flow between the DCMG and the utility grid. As the faults are cleared, the DCMG system regains pre-fault conditions with faster response, and operates under normal mode even during the fault and post-fault periods. So the control techniques can be used for the protection and stability of future grids.

18.2.7 Energy management system of a microgrid using particle swarm optimization (PSO) and communication system

The proposed EMS of an MG using PSO will be helpful to ensure that the MG will operate in an economical and reliable manner. The control system will have the capability of scheduling and controlling all distributed energy resources to warrant the stability, reliability, and economical operation of the MG.

18.2.8 Reconfiguration of distribution network using different optimization

The study of reconfiguration of distribution network will be helpful in improving the power system performance because the loads need highly safe and dependable power supply, distribution network becomes too important and an active part in power system that can be optimized and controlled. With the help of the present study the estimation of the amount of total power losses in distribution system will be taken as a main tool for evaluating the system.

18.2.9 Intelligent optimization scheduling of isolated microgrids considering energy storage integration, traditional generation, and renewable energy uncertainty costs

There is a possibility for future extension of the results of this work to different types of batteries taking into account the values provided by the manufacturers. Otherwise, further studies should be done on the hyper volume analysis, which should be considered for possible improvement of the results in the Paretos.

18.2.10 ADMM-based consensus droop control and distributed pinning droop control of isolated AC microgrids

Future scope included to investigate the feasibility of our proposed two new droop control with random time-varying communication typologies and delays in the communication network. In addition, the robustness of these two droop control under various network parameters will also be an aspect worth studying in the near future.

18.2.11 Extendable multiple outputs hybrid converter for AC/DC microgrid

With the help of the proposed hybrid converter it will be possible to use it for single DC and *n* number of AC outputs simultaneously. This type of converter topologies will be implemented for rural-based MG application, where requirements are less maintenance, high power density, and less cost.

18.2.12 Multi-input converters for distributed energy resources in microgrids

In the future, the use of different types of DG sources will increase and the interest in using them together in one area increases. This matter will result in new topologies being introduced that can connect to more than two DG sources like multi-input converters. The use of these topologies will decrease converters, size drastically. In the future, research will focus on management systems for optimal utilization of DG sources to maximize system efficiency. The researchers and suppliers will try to design and manufacture converters that can supply DC and AC loads simultaneously, without complex control system are used in the converter.

The features discussed in the study considered by manufacturers and consumers are the number of inputs, the number of outputs, soft switching, battery-charging control, bidirectional power flow, less noise, hybrid operation, optimum and straightforward control, isolation between input and output, smaller volume, maximum use of small and large sources simultaneously, etc.

18.2.13 Modeling and control of DC–DC converters for DC microgrid application

This chapter has analyzed the MG system for a PV-source fed DC–DC buck converter prototype of low voltage and low power. However, the same procedure can be followed for high power level and other DC–DC converter configurations in future.

18.2.14 Case studies of microgrids systems

In this chapter, various case studies have been done. They concluded that by supplying high efficiency power and implementing the techniques to reduce the electricity bills, the electrification of the rural areas is possible in near future.

18.2.15 Smart grid road map and challenges for Turkey

The study shows that if classical grids in Turkey were transformed into smart grids, not only would the above-mentioned benefits be achieved, but Turkey would be able to attract huge amount of investment and boost its economy. The Turkish grid system would then be a powerful player in the energy market in Europe.

18.2.16 Nanogrids: good practices and challenges projects in Colombia

The future scope of nanogrids involves their commercialization. They can undertake new economic activities or enhance the existing ones, for instance, the nanogrids contribute to the development of tourism and associated services, agriculture through irrigation technology among others.

18.2.17 Distributed generation deployment in the Libyan MV network: adverse impacts on practised protection scheme in the DN

The study shows that the protection scheme can be considerably enhanced, performing the following actions: revision of short-circuit study considering all DG units in service in order to verify the adequacy of the existing primary and protective equipment in future. In addition, revision of relay setting values to prevent blinding incidents, where possible, paying particular attention to minimum faults. Feeder protection with directional overcurrent relays or distance relays to ban sympathetic incidents, even though the installation of voltage transformers augments the upgrade cost. Utilization of communication-based transfer trip schemes to provide more reliable anti-islanding protection and prohibit failed reclosing incidents.

Index

www.ingramcontent.com/pod-product-compliance
Lightning Source LLC
LaVergne TN
LVHW051047210926
848486LV00001B/44
* 9 7 8 1 7 8 5 6 1 9 9 8 4 *